Lecture Notes in Statistics 112

Edited by P. Bickel, P. Diggle, S. Fienberg, K. Krickeberg,
I. Olkin, N. Wermuth, S. Zeger

Springer

New York
Berlin
Heidelberg
Barcelona
Budapest
Hong Kong
London
Milan
Paris
Santa Clara
Singapore
Tokyo

Doug Fisher
Hans-J. Lenz
(Editors)

Learning from Data

Artificial Intelligence and Statistics V

Springer

Doug Fisher
Vanderbilt University
Department of Computer Science
Box 1679, Station B
Nashville, Tennessee 37235

Hans-J. Lenz
Free University of Berlin
Department of Economics
Institute of Statistics and
Econometrics
14185 Berlin, Garystre 21
Germany

CIP data available.
Printed on acid-free paper.

Camera ready copy provided by the author.
Printed and bound by Braun-Brumfield, Ann Arbor, MI.
Printed in the United States of America.

9 8 7 6 5 4 3 2 1

ISBN 0-387-94736-1 Springer-Verlag New York Berlin Heidelberg   SPIN 10534441

Preface

Ten years ago Bill Gale of AT&T Bell Laboratories was primary organizer of the first *Workshop on Artificial Intelligence and Statistics*. In the early days of the Workshop series it seemed clear that researchers in AI and statistics had common interests, though with different emphases, goals, and vocabularies. In learning and model selection, for example, a historical goal of AI to build autonomous agents probably contributed to a focus on parameter-free learning systems, which relied little on an external analyst's assumptions about the data. This seemed at odds with statistical strategy, which stemmed from a view that model selection methods were tools to augment, not replace, the abilities of a human analyst. Thus, statisticians have traditionally spent considerably more time exploiting prior information of the environment to model data and exploratory data analysis methods tailored to their assumptions. In statistics, special emphasis is placed on model checking, making extensive use of residual analysis, because all models are 'wrong', but some are better than others. It is increasingly recognized that AI researchers and/or AI programs can exploit the same kind of statistical strategies to good effect.

Often AI researchers and statisticians emphasized different aspects of what in retrospect we might now regard as the same overriding tasks. For example, those in AI exploited computer processing to a great extent and explored a wide variety of search control strategies, while many in statistics focused on principled measures for evaluating the merits of varying models and states, often in computationally-limited circumstances. There were numerous other differences in the traditional research biases of AI and statistics, such as interests in deductive versus inductive inference. These differences in goals, methodology, and vocabulary fostered a healthy tension that fueled debate, and contributed to excitement and exchange of ideas among participants of the early AI and Statistics Workshops.

This volume contains a subset of papers presented at the *Fifth International Workshop on Artificial Intelligence and Statistics*, which was held at Ft. Lauderdale, Florida in January 1995. The primary theme of the Fifth Workshop was 'Learning from Data', but a variety of areas at the interface of AI and statistics were represented at the Workshop and in this volume. The Fifth workshop saw some of the same kind of tension and excitement that characterized earlier workshops, with one participant confiding that this was the first time that he/she had felt 'intimidated' in presenting a paper. However, there has also been noticeable growth, with researchers from AI and statistics becoming aware of and exploiting research in the other field. In some cases, such as Causality and Graphical Models, the papers of this volume suggest that a dichotomy between AI and statistics is artificial. In cases such as Learning, researchers might still be reasonably described as falling into the statistics or AI camps, but there is heightened awareness of commonalities across the disciplines.

As we have noted, a variety of research areas are represented in this volume. This variety contributed to a minor dilemma for the editors, since many contributions fell into more than one subject category. In some cases, such as Natural Language Processing (NLP), a categorization was fairly easy. While all of the NLP papers could be viewed as learning from data, the application of statistical methods in NLP is currently attracting the enthusiastic attention of researchers in both AI and statistics – a trend that we want to highlight. In other cases, we have compromised in our classification, but the process of classifying the contributions, to say nothing of the paper selection process, led to some insights about the difference in meaning and emphases that probably remain in AI and statistics. We suspect, for example, that those in AI probably ascribe a more general meaning to the term 'Exploratory Data Analysis'.

We were both pleased to have helped organize the Fifth Workshop on Artificial Intelligence and Statistics; the excitement that we felt from the exchange of ideas made it worth the effort. We also extend our thanks to a several groups. Foremost, we thank the authors – those represented in this volume and the Workshop proceedings of preliminary papers. The program committee, which reviewed initial submissions to the Workshop, contributed mightily to the quality and vibrancy of the Workshop. We gratefully acknowledge the sponsorship of the *Society for Artificial Intelligence and Statistics* for covering initial costs, and the *International Association for Statistical Computing* for promulgating information about the Workshop. Finally, on this, the ten-year anniversary of the AI and Statistics Workshop series, we thank Bill Gale and his colleagues for their foresight in organizing the first workshop.

Doug Fisher, Vanderbilt University, USA
Hans-J. Lenz, Free University of Berlin, Germany

Previous workshop volumes:

Gale, W. A., ed. (1986) *Artificial Intelligence and Statistics*, Reading, MA: Addison-Wesley.

Hand, D. J., ed. (1990) Artificial Intelligence and Statistics II. *Annals of Mathematics and Artificial Intelligence*, **2**.

Hand, D. J., ed. (1993) *Artificial Intelligence Frontiers in Statistics: AI and Statistics III*, London, UK: Chapman & Hall.

Cheeseman, P., & Oldford, R. W., eds. (1994) *Selecting Models from Data: Artificial Intelligence and Statistics IV*, New York, NY: Springer Verlag.

1995 Organizing and Program Committees:

General Chair:	D. Fisher	Vanderbilt U., USA
Program Chair:	H.-J. Lenz	Free U., Berlin, Germany

Program Committee Members:

	W. Buntine	NASA (Ames), USA
	J. Catlett	AT&T Bell Labs, USA
	P. Cheeseman	NASA (Ames), USA
	P. Cohen	U. of Mass., USA
	D. Draper	U. of Bath, UK
	Wm. Dumouchel	Columbia U., USA
	A. Gammerman	U. of London, UK
	D. J. Hand	Open U., UK
	P. Hietala	U. Tampere, Finland
	R. Kruse	TU Braunschweig, Germany
	S. Lauritzen	Aalborg U., Denmark
	W. Oldford	U. of Waterloo, Canada
	J. Pearl	UCLA, USA
	D. Pregibon	AT&T Bell Labs, USA
	E. Roedel	Humboldt U., Germany
	G. Shafer	Rutgers U., USA
	P. Smyth	JPL, USA

Tutorial Chair:	P. Shenoy	U. Kansas, USA
Local Arrangements:	D. Pregibon	AT&T Bell Labs, USA

Sponsors:	Society for Artificial Intelligence and Statistics
	International Association for Statistical Computing

Contents

III Search Control in Model Hunting

IV Classification

Part I

Causality

1
Two Algorithms for Inducing Structural Equation Models from Data

Paul R. Cohen, Dawn E. Gregory, Lisa Ballesteros, Robert St. Amant

Computer Science Department, LGRC
University of Massachusetts
Box 34610
Amherst, MA 01003-4610

ABSTRACT We present two algorithms for inducing structural equation models from data. Assuming no latent variables, these models have a causal interpretation and their parameters may be estimated by linear multiple regression. Our algorithms are comparable with PC [Spirtes93] and IC [Pearl91a, Pearl91b], which rely on conditional independence. We present the algorithms and empirical comparisons with PC and IC.

1.1 Structural Equation Models

Given a dependent variable x_0 and a set of predictor variables $P = \{x_1, x_2, \ldots, x_k\}$, multiple regression algorithms find subsets $p \subset P$ that account for "much" of the variance in x_0. These are search algorithms, and they are not guaranteed to find the "best" p—the one that makes R^2 as big as possible, especially when p is a proper subset of P [Mosteller77, Cohen75, Draper66]. The question arises, what should be done with the variables in $T = P - p$? T is the set of variables that are not selected as predictors of x_0. In many data analysis tasks, the variables in T are used to predict the variables in p. For instance, we might select x_1 and x_3 as predictors of x_0; and x_2, x_5, x_6 as predictors of x_1; and x_4 as a predictor of x_2 and x_3; and so on. We can write $structural\ equations$:

$$
\begin{aligned}
x_0 &= \beta_{0,1}x_1 + \beta_{0,3}x_3 + u \\
x_1 &= \beta_{1,2}x_2 + \beta_{1,5}x_5 + \beta_{1,6}x_6 + v \\
x_2 &= \beta_{2,4}x_4 + w \\
x_3 &= \beta_{3,4}x_4 + z
\end{aligned}
$$

The principal task for any modeling algorithm is to decide, for a given "predictee," which variables should be in p and which should be in T. Informally, we must decide where in a structural equation model a variable does most good. For example, parents' education (PE) and child's education (CE) could be used as predictors of a child's satisfaction when he or she takes a job (JS), but we might prefer a model in which PE predicts CE, and CE predicts JS (or a model in which PE predicts CE and JS). This paper presents two algorithms that build structural equation models.

There are clear parallels between building structural equation models and building

---

[1] $Learning\ from\ Data:\ AI\ and\ Statistics\ V.$ Edited by D. Fisher and H.-J. Lenz. ©1996 Springer-Verlag.

causal models. Indeed, *path analysis* refers to the business of interpreting structural equation models as causal models [Li75, Wright21]. Path analysis has been heavily criticized (e.g., [Shaffer92, Holland86]) in part because latent variables can produce large errors in estimated regression coefficients throughout a model [Spirtes93]. Recent causal induction algorithms rely not on regression coefficients but on conditional independence [Pearl91b, Spirtes93]. These algorithms use covariance information only to infer boolean conditional independence constraints; they do not estimate strengths of causal relationships, and, most importantly from our perspective, they do not use these strengths to guide the search for causal models.

Our algorithms, called FBD and FTC, use covariance information, in the form of estimated standardized regression coefficients, to direct the construction of structural equation models and to estimate the parameters of the models. Because latent variables can result in biased estimates, our algorithms might be misled when latent variables are at work. In practice, FBD and FTC are more robust than, say, stepwise multiple regression. They often discard predictors that are related to the predictee only through the presence of a latent variable [Ballesteros94]. We have not yet shown analytically why the algorithms have this advantage. Until we do, our only claim for FBD and FTC is this: when latent variables are at work, our algorithms build multilevel regression models of heuristic value to analysts, just as ordinary regression algorithms build useful (but suspect) single-level models. If we can assume *causal sufficiency* [Spirtes93]—essentially, no latent variables—these models may be interpreted as causal models [Cohen94b].[2]

1.2   FBD and FTC

Structural equation models can be represented by directed acyclic graphs (DAGs) in which a link $x_1 \rightarrow x_0$ is interpreted to mean x_1 is a predictor of x_0. FBD and FTC rely on statistical *filter conditions* to remove insignificant links from the model being constructed. The two algorithms use the same filter conditions, differing only in how they are applied. The most important filter (and the only novel one) is the ω statistic:

$$\omega_{ij} = \left| \frac{r_{ij} - \beta_{ij}}{r_{ij}} \right| = \left| 1 - \frac{\beta_{ij}}{r_{ij}} \right|$$

Given a predictee, say x_i, and a set of predictors P, we first regress x_i on all the predictors in P. This yields a standardized regression coefficient β_{ij} for each predictor x_j. Now, β_{ij} is partial—it represents the effect of x_j on x_i when the influences of all the other variables are fixed. Following [Li75, Sokal81] we call β_{ij} an estimate of the *direct* influence of x_j on x_i. The correlation r_{ij} represents the *total* influence of x_j on x_i, so $r_{ij} - \beta_{ij}$ estimates the influence of x_j that is due to its relationships with other predictors—its *indirect* influence. Thus, ω_{ij} estimates the fraction of x_j's total influence on x_i that is indirect. If x_j has little direct influence on x_i relative to its indirect influence through, say, x_k, then x_k, not x_j, is heuristically more apt to be a direct cause of x_i. Thus we filter x_j when ω_{ij} exceeds a threshold T_ω.

Small values of ω are necessary but not sufficient to consider one variable a cause of another. This is because ω_{ji} will be very small if both r_{ij} and β_{ij} are small, but in this case

---

[2]FBD and FTC run in CLIP/CLASP [CLIP/CLASP], a Common Lisp statistical package developed at UMass. For more information on CLIP/CLASP, please contact clasp-support@cs.umass.edu.

the predictor x_j has little influence on x_i, direct or indirect, and should be filtered. Even if the β coefficients are relatively high, the regression of x_i on x_j and other variables might account for very little variance in x_i, so we filter x_j if it doesn't contribute enough (perhaps with other variables) to R^2 for x_i. Currently, the only other filter used by FBD and FTC is a test for *simple conditional independence* using the partial correlation coefficient. If x_k renders x_i and x_j independent, the partial correlation $r_{ij \cdot k}$ will be approximately 0, and we remove the link between x_i and x_j.

1.2.1   The FBD Algorithm

The FBD algorithm is told x_0, the sink variable, and works backwards, finding predictors for x_0, then predictors for those predictors, and so on.

1. Enqueue x_0 into an empty queue, Q.

2. Create M, an empty model (with n nodes and no links).

3. While Q is not empty, do:
   (a) Dequeue a variable x_j from Q
   (b) Find a set of predictors $P = \{x_i \mid x_i \neq x_j$ and $x_i \to x_j$ passes all filter conditions and $x_i \to x_j$ will not cause a cycle$\}$
   (c) For each $x_i \in P$, add the link $x_i \to x_j$ into M
   (d) For each $x_i \in P$, enqueue x_i into Q

FBD performs well (see below) but it has two drawbacks: First, we must identify the sink variable x_0; most model induction algorithms do not require this information. Second, predictors of a single variable are enqueued on Q in an arbitrary order, yet the order in which variables are dequeued can affect the structure of the resulting model. For example, the link $x_i \to x_j$ will be reversed if x_i is dequeued before x_j. These issues led to the development of a second algorithm, called FTC.

1.2.2   The FTC Algorithm

FTC deals with these problems by inserting links into the model in order of precedence, rather than in order of selection. Precedence is determined by a sorting function $S(x_j \to x_i)$. The FTC algorithm is as follows:

1. Let $L = \{x_j \to x_i : i \neq j, 1 \leq i \leq n; 1 \leq j \leq n\}$; i.e. L is the set of all potential links in a model with n variables.

2. For each link $x_j \to x_i \in L$, test each filter condition for $x_j \to x_i$. If *any* condition fails, remove $x_j \to x_i$ from L.

3. Sort the links remaining in L by some precedence function $S(x_j \to x_i)$.

4. Create M, an empty model (with n nodes and no links) to build on.

5. While L is not empty, do:
   (a) Remove the link $x_j \to x_i$, of the highest precedence, from L.
   (b) If $x_j \to x_i$ does not cause a cycle in M, add $x_j \to x_i$ to M. Otherwise, discard $x_j \to x_i$.

Experiments with different sorting functions led us to the following simple procedure: for a link $x_j \to x_i$, its score for sorting is R^2 from the regression of x_i on all the other variables. Thus, a link's precedence is the R^2 of its dependent variable. We justify this policy with the following observation: variables with high values of R^2 are less likely to have latent influences, so they are preferred as dependent variables.

The complexity of these algorithms is $O(n^4)$, where n is the number of variables. Most of this is attributed to the linear regressions, which have complexity $O(n^3)$ in our implementation.

1.3   Empirical Results

We have tested FBD and FTC under many conditions: on artificial data (using our own data generator and the TETRAD generator), on published data [Spirtes93], and on data generated by running an AI planning system called PHOENIX [Cohen94a]. We tested how well the ω heuristic selects predictors, measured in terms of variance accounted for in the dependent variable [Cohen94c]. Also, we have compared FBD and FTC with other algorithms—PC for FBD [Spirtes93] and IC for FTC [Pearl91b].[3] Finally, to assess how latent variables affected its performance, we compared FBD with stepwise linear regression as implemented in MINITAB [Ballesteros94].

1.3.1   Artificial Models and Data

We worked with a set of 60 artificially generated data sets: 20 data sets for each of 6, 9, and 12 measured variables. These were generated from the structural equations of 60 randomly selected *target models*. The advantage of this approach is that the model constructed by each algorithm can be evaluated against a known target.

The target models were constructed by randomly selecting m links from the set of potential links $L = \{x_i \to x_j \mid i \neq j\}$. For each model of n variables, m is chosen from the range $1.0(n-1)\ldots3.0(n-1)$; thus the target models have an *average branching factor* between 1 and 3. As each link is selected, it is inserted into the target model. With probability 0.3, the link will be a *correlation* link, indicating the presence of a latent variable.

Once the structure of each target model has been determined, the *structural equations* are created for each dependent variable x_j. For directed links $x_i \to x_j$, a path coefficient is randomly selected from the range $-1.0\ldots1.0$. For correlation links, a latent variable l_{ij} is created (these variables are not included in the final data set), and a path coefficient is selected for the links $l_{ij} \to x_i$ and $l_{ij} \to x_j$.

Finally, data are generated from the structural equations. For each independent and latent variable, a set of 50 data points is sampled from a Gaussian distribution with mean of 0 and standard deviation of 1 (we have also run experiments with variates selected from uniform distributions). Sample values for the dependent variables are computed from the structural equations, and a Gaussian error term is added to each (with mean 0 and standard deviation 1).

---

[3]We are very grateful to Professors Spirtes, Glymour and Scheines, and Professor Pearl, for making the code for their algorithms available.

1.3.2 How Good Is ω at Selecting Predictors?

Suppose we have a dataset with one dependent variable x_0 and a set P of predictor variables, and we want to find the best subset $p\text{-}best(k)$ of k predictors. Here, "best" means no other subset p' of k predictors accounts for more variance in x_0 than $p\text{-}best(k)$ does. We can find $p\text{-}best(k)$ by exhaustive search of all subsets of k predictors. Or, we can regress x_0 on all the predictors in P, and select the k predictors with the lowest ω scores. How well does the latter method perform in comparison with the former, exhaustive method?

For each of the 60 target models described in the previous section, we found by exhaustive search the best subsets of $k = 3, 4, 5$ predictors of the "sink" variable, x_0. Next we selected subsets of k predictors using ω scores. The *batch discarding* method regresses x_0 on $x_1, x_2, \ldots$, calculates ω for each of these predictors, and selects the k predictors with the best ω scores. This set is denoted $p\text{-}batch(k)$. The *iterative discarding* method repeatedly regresses x_0 on a set of predictors, discarding the predictor with the worst ω score, until only k predictors remain, denoted $p\text{-}iter.(k)$. We chose to try both methods since one can expect β coefficients to change substantially even if one predictor is removed from a regression.

If ω scores select good predictors, then $p\text{-}best(k)$, $p\text{-}batch(k)$, and $p\text{-}iter.(k)$ should contain the same predictors; if they don't, we would like $p\text{-}batch(k)$ and $p\text{-}iter.(k)$ to account for nearly as much of the variance in x_0 as $p\text{-}best(k)$.

Table 1.1 shows how many predictors $p\text{-}batch(k)$ and $p\text{-}iter.(k)$ have in common with $p\text{-}best(k)$. For 12-variable models and $k = 5$, the mean number of predictors shared by $p\text{-}batch(k)$ and $p\text{-}best(k)$ is 3.15, and, for $p\text{-}iter.(k)$ and $p\text{-}best(k)$ this number is 3.375. Thus, when batch discarding selects five variables, roughly two of them (on average) are not the best variables to select.

k	$Vars.$	$p - batch(k) \cap p - best(k)$	$p - iter.(k) \cap p - best(k)$
5	12	3.15 (1.16)	3.375 (.86)
5	9	3.65 (.49)	3.7 (.52)
5	6	5.0 (0)	5.0 (0)
4	12	2.3 (.57)	2.3 (.74)
4	9	3.05 (.36)	3.08 (.53)
4	6	3.33 (.23)	3.33 (.23)
3	12	1.48 (.61)	1.68 (.58)
3	9	2.18 (.46)	2.18 (.51)
3	6	2.28 (.36)	2.33 (.33)

TABLE 1.1. Means and (Standard Deviations) of the size of the intersections

On the other hand, Table 1.2 shows that the variables in $p\text{-}batch(k)$ and $p\text{-}iter.(k)$ account for almost as much of the variance in x_0 as those in $p\text{-}best(k)$. Let $R^2_{p\text{-}best(k)}$ be the variance in x_0 that is "available" to predictors in $p\text{-}batch(k)$ and $p\text{-}iter.(k)$. For instance, if the best k predictors account for only 50% of the variance in x_0, then we want to express the predictive power of $p\text{-}batch(k)$ as a fraction of 50%, not 100%. Table 1.2 therefore contains the ratios $R^2_{p\text{-}batch(k)}/R^2_{p\text{-}best(k)}$ and $R^2_{p\text{-}iter.(k)}/R^2_{p\text{-}best(k)}$. For example, for 12-variable models and $k = 5$, the predictors selected by batch discarding account for 85% of the available variance in x_0, on average, and those selected by iterative discarding account for 86%.

k	$Vars.$	$R^2_{p-batch(k)}/R^2_{p-best(k)}$	$R^2_{p-iter.(k)}/R^2_{p-best(k)}$
5	12	.845 (.03)	.86 (.023)
5	9	.94 (.006)	.94 (.005)
5	6	1.0 (0)	1.0 (0)
4	12	.80 (.04)	.82 (.03)
4	9	.94 (.008)	.94 (.005)
4	6	.91 (.01)	.91 (.01)
3	12	.73 (.06)	.80 (.04)
3	9	.91 (.02)	.92 (.01)
3	6	.88 (.03)	.89 (.03)

TABLE 1.2. Means and (Standard Deviations) of the R^2 ratios

Batch and iterative discarding do not find exactly the same predictors as exhaustive search for the best predictors, but the ones they find account for much of the available variance in the dependent variable. Bear in mind that exhaustive search for the best k of N variables requires N-choose-k multiple regressions with k predictors, whereas iterative discarding requires $N - k$ multiple regressions with between N and $k + 1$ predictors, and batch discarding requires just one regression with N predictors. Thus, batch discarding finds predictors that are nearly as good as exhaustive search, with a fraction of the effort.

We wondered whether something simpler than ω scores, such as sorting by beta coefficients, would perform equally well. In most cases, beta coefficients selected the same predictors as ω scores, but sometimes they recommended different sets with bad R^2 scores. We can see why (and also why ω is preferable) by considering four cases:

High r_{0i}, high β_{0i}: In this case ω is small in absolute value and x_i will be accepted as a predictor of x_0. Similarly, β is high, so by this criterion x_i will also be accepted as a predictor. Since x_i should be accepted in this case, both ω and β do the right thing.

High r_{0i}, low β_{0i}: Here, ω is large in absolute value and β small, so both statistics will reject x_i, which is the right thing to do.

Low r_{0i}, high β_{0i}: Here, ω is large in absolute value so x_i will be rejected. However, x_i's β score suggests accepting it. The correct action is to reject x_i because the only way to get, say, $r_{0i} = 0$ and $\beta_{0i} = .8$ is for x_i's direct influence (.8) to be cancelled by its indirect influence (-.8) through other predictors. We want predictors that have large direct influence and small indirect influence, so we ought to discard x_i. In this case, β coefficients make the wrong recommendation.

Low r_{0i}, low β_{0i}: In this case, ω is small but FBD and FTC will discard x_i as a predictor because β is small.

1.3.3 Comparative Studies

FBD and FTC were compared to the PC [Spirtes93] and IC [Pearl91a, Pearl91b] algorithms, respectively. We chose these algorithms for comparison due to their availability and strong theoretical support. Both PC and IC build models from constraints imposed by *conditional independencies* in the data.

In order to provide a comprehensive evaluation of the algorithms, we used a variety of performance measures for each model. The measures we used are shown in table 1.3. See [Cohen94b, Cohen94c] for details of these and other dependent measures.

Measure	Meaning
Dependent R^2	The variance accounted for in x_0, the sink variable.
ΔR^2	The mean of the absolute differences in R^2 between all the predictees in the target model and the model being evaluated.
Correct %	The percentage of the directed links in the target model that were correctly identified.
Wrong/Correct	The ratio of wrong links found for every correct one.
Wrong Reversed	The number of links $x_i \rightarrow x_j$ that should have been reversed, $x_i \leftarrow x_j$.
Wrong Not Reversed	The number of links $x_i \rightarrow x_j$ that should not have been drawn in either direction.

TABLE 1.3. Dependent measures and their meanings.

We compared FBD with the PC algorithm [Spirtes93], because PC can be given exogenous knowledge about causal order, specifically, it can be told which is the sink variable x_0.[4] FTC was compared to the IC algorithm, since neither uses external knowledge about the data. Both PC and IC take a least-commitment approach to causal induction, conservatively assigning direction to very few links in order to avoid misinterpretation of potential latent influences. FBD and FTC, on the other hand, commit to a direction in all cases. Hence it was necessary to evaluate statistics like $Dependent R^2$ for PC and IC models in two ways—by interpreting undirected links as directed (choosing the most favorable direction, always), or simply ignoring undirected links. In Table 1.4 the last two rows, denoted $w/undirected$, give scores obtained by interpreting undirected links as directed in the most favorable way.

$Measure$	$Algorithm$	$6vars$	$9vars$	$12vars$
$Dependent R^2$	$FBD*$	0.734 (0.187)	0.787 (0.146)	0.728 (0.278)
	$PC*$	0.301 (0.396)	0.403 (0.307)	0.363 (0.320)
	FTC	0.734 (0.187)	0.787 (0.146)	0.728 (0.278)
	IC	0.326 (0.389)	0.451 (0.338)	0.303 (0.363)
$w/undirected$	$PC*$	0.607 (0.337)	0.685 (0.266)	0.684 (0.269)
	IC	0.603 (0.31)	0.811 (0.12)	0.739 (0.216)
ΔR^2	$FBD*$	0.118 (0.087)	0.134 (0.077)	0.179 (0.130)
	$PC*$	0.333 (0.151)	0.321 (0.102)	0.372 (0.100)
	FTC	0.118 (0.087)	0.107 (0.087)	0.147 (0.086)
	IC	0.347 (0.128)	0.345 (0.137)	0.348 (0.095)
$w/undirected$	$PC*$	0.185 (0.084)	0.166 (0.101)	0.193 (0.051)
	IC	0.199 (0.104)	0.124 (0.063)	0.172 (0.063)

TABLE 1.4. Means and (Standard Deviations) of R^2 scores for several causal induction algorithms (* = uses additional knowledge)

FBD and FTC attained significantly higher $Dependent R^2$ and ΔR^2 scores than PC and IC (much higher, when we ignore undirected links). This is to be expected: FBD and FTC are driven by covariance information and PC and IC use covariance only to find conditional independence relationships. (Significance was tested with paired-sample t tests, $p < .05$.) FBD and FTC have significantly higher $Correct\%$ scores than PC and IC. FBD has a higher $Wrong/Correct$ score than PC, although not significantly so, because it commits to more directed links than PC. As the models become larger, FTC becomes significantly better than IC on $Wrong/Correct$ scores. Although FBD and FTC include more links than the

---

[4]We thank Professor Glymour for suggesting this comparison.

other algorithms, FTC maintains a low rate of incorrect identifications. Roughly 72% of PC's wrong links are wrong because they are backwards (i.e., *Wrong Reversed*); conversely, only 28% of PC's links should not have been drawn in either direction. For IC, 44% of the wrong links are backwards, and for FBD and FTC, the numbers are 33% and 30%, respectively. Clearly, when FBD and FTC draw an incorrect link, it generally shouldn't be in the model pointing in either direction, whereas PC and to a lesser extent IC err by drawing links backwards. Keep in mind, though, that this is an analysis of wrong links, only, and that FTC has better *Wrong/Correct* performance than the other algorithms.

Measure	*Algorithm*	*6vars*	*9vars*	*12vars*
Correct%	*FBD*∗	0.653 (0.186)	0.651 (0.184)	0.528 (0.248)
	PC∗	0.284 (0.181)	0.273 (0.172)	0.167 (0.109)
	FTC	0.658 (0.203)	0.682 (0.223)	0.566 (0.211)
	IC	0.293 (0.18)	0.272 (0.19)	0.165 (0.11)
Wrong/Correct	*FBD*∗	0.685 (0.542)	0.932 (0.391)	1.396 (1.082)
	PC∗	0.458 (0.534)	0.617 (0.921)	1.276 (1.315)
	FTC	0.467 (0.627)	0.696 (0.496)	0.809 (0.288)
	IC	1.48 (0.944)	1.50 (1.11)	2.17 (2.10)
Wrong Reversed	*FBD*∗	1.200 (1.056)	2.200 (1.824)	4.100 (3.323)
	PC∗	0.650 (0.813)	0.950 (1.146)	1.950 (1.146)
	FTC	0.600 (0.681)	1.450 (1.849)	2.650 (1.981)
	IC	1.55 (1.05)	1.35 (1.27)	2.0 (1.30)
Wrong Not Rev.	*FBD*∗	1.900 (1.165)	4.900 (1.997)	9.100 (4.388)
	PC∗	0.250 (0.444)	0.300 (0.470)	0.950 (0.945)
	FTC	1.350 (0.988)	3.500 (1.821)	6.400 (2.798)
	IC	1.2 (1.06)	2.5 (2.09)	2.9 (1.8)

TABLE 1.5. Means and (Standard Deviations) of link scores for several causal induction algorithms (* = uses additional knowledge)

1.3.4   *Performance with Latent Variables*

Regression coefficients are unstable, especially when unmeasured or *latent* variables influence them. Selecting variables by their ω scores lessens this problem. In our lab, Ballesteros [Ballesteros94] has obtained good results with models Spirtes et al. [Spirtes93, page 240] show are difficult for ordinary regression. For these models, regression often chooses predictors whose relationships to the dependent variable are mediated by latent variables or common causes (we refer to these variables as proxy variables). Ballesteros generated 12 different sets of coefficients for the structural equations for each of the four Spirtes et al. models, and data sets for each having 100 to 1000 variates. FBD's performance on these data was measured by the number of times it chose the correct predictors and number of times it incorrectly chose a proxy variable. The data from these experiments are given in Table 1.6. FBD correctly chose true predictors 90% of the time, and correctly rejected proxy variables 80% of the time.

	chosen	rejected	Total
True Predictor	108	12	120
Proxy Variable	12	48	60
Total	120	60	180

TABLE 1.6. Contingency Table FBD Choices

As a comparison, Ballesteros used MINITAB to run stepwise regressions (significance

= .05) on each latent variable dataset to find out how often the algorithm chose a proxy variable as a predictor. The data are given in Table 1.7.

	chosen	rejected	Total
True Predictor	118	2	120
Proxy Variable	42	18	60
Total	160	20	180

TABLE 1.7. Contingency Table Stepwise Choices

MINITAB correctly selected 98% of the true predictors, but it correctly rejected proxies only 30% of the time. We are trying to understand why stepwise regression performs so poorly in the presence of latent variables and why FBD is less susceptible. Our best guess is that stepwise regression is a "step-up" procedure that starts with no predictors and adds them one at a time, whereas FBD starts with all predictors, computes regression coefficients, and then removes some predictors with filter conditions. We think that a regression coefficient for a predictor is less likely to be biased when it is one of several variables in a regression, than when it is one of very few predictors. If so, step-up procedures will be more susceptible to bias than, say, backward elimination and FBD. Ballesteros has some evidence to support this conjecture.

1.4   Conclusions

FBD and FTC are simple, polynomial-time algorithms that construct models without searching the entire space of models. Our empirical results show that ω does remarkably well as a heuristic for selecting predictors. In fact, it performs so well that FBD and FTC build models that are in some respects superior to those constructed by PC and IC. Admittedly, neither FBD nor FTC infers the presence of latent variables, which may be a significant drawback for some applications. However, we have shown that FBD will often avoid predictors that are connected to the variables they predict via a common but latent cause.

Acknowledgments

This work is supported by ARPA/Rome Laboratory under contract #'s F30602-91-C-0076 and F30602-93-C-0010, and by a NASA GSRP Training Grant, NGT 70358. The U.S. government is authorized to reproduce and distribute reprints for governmental purposes not withstanding any copyright notation hereon.

1.5   REFERENCES

[CLIP/CLASP] Anderson, S. D., & Carlson, A., & Westbrook, D. L., & Hart, D. M., & Cohen, P. R. (1993). CLIP/CLASP: Common Lisp Analytical Statistics Package/Common Lisp Instrumentation Package. Technical report 93-55, Department of Computer Science, University of Massachusetts, Amherst, MA. This document is available under http://eksl-www.cs.umass.edu/publications.html.

[Ballesteros94] Ballesteros, L. (1994). Regression–Based Causal Induction with latent variable models. In *Proceedings of the Twelfth National Conference on Artificial Intelligence*. AAAI Press / MIT Press

[Cohen75] Cohen, J., & Cohen, P. (1975). *Applied Multiple Regression/Correlation Analysis for the Behavioral Sciences* Hillsdale, NJ: LEA

[Cohen94a] Cohen, P. R., & Hart, D. M., & St.Amant, R., & Ballesteros, L., & Carlson, A. (1994). Path Analysis Models of an Autonomous Agent in a Complex Environment. In P. Cheeseman and R. W. Olford (Eds.), *Selecting Models from Data: AI and Statistics IV* Springer-Verlag, 243–251.

[Cohen94b] Cohen, P. R., & Ballesteros, L., & Gregory, D., & St.Amant, R. (1994). Regression Can Build Predictive Causal Models. Technical report 94-15, Department of Computer Science, University of Massachusetts, Amherst, MA. This document is available under http://eksl-www.cs.umass.edu/publications.html.

[Cohen94c] Cohen, P. R., & Ballesteros, L., & Gregory, D., & St.Amant, R. (1994). Experiments with a regression–based causal induction algorithm. EKSL memo number 94-33, Department of Computer Science, University of Massachusetts, Amherst, MA. This document is available under http://eksl-www.cs.umass.edu/publications.html.

[Draper66] Draper, N. R., & Smith, H. (1966). *Applied Regression Analysis.* New York, NY.

[Holland86] Holland, P. (1986). Statistics and Causal Inference. *Journal of the American Statistical Association*, 81:945–960.

[Li75] Li, C. C. (1975). *Path analysis–A primer* Boxwood Press.

[Mosteller77] Mosteller, F., & Tukey, J. W. (1977). *Data Analysis and Regression: A Second Course in Statistics* Reading, MA: Addison Wesley.

[Pearl91a] Pearl, J., & Verma, T. S. (1991) A Theory of Inferred Causation In J. Allen and R. Fikes and E. Sandewall (Eds.), *Principles of Knowledge Representation and Reasoning: Proceedings of the Second International Conference* (pp 441–452). Morgan Kaufmann.

[Pearl91b] Pearl, J., & Verma, T. S. (1991) A Statistical Semantics for Causation. In *Statistics and Computing*, 2: 91–95.

[Shaffer92] Shaffer, J. P. (Ed.) (1992). *The Role of Models in Nonexperimental Social Science: Two Debates* Washington DC: American Educational Research Association and American Statistical Association

[Sokal81] Sokal, R. R., & Rohlf, F. J. (1981). *Biometry: the Principles and Practice of Statistics in Biological Research.* New York: W. H. Freeman and Co. second edition.

[Spirtes93] Spirtes, P., & Glymour, C., & Scheines, R. (1993). *Causation, Prediction, and Search.* Springer-Verlag.

[Wright21] Wright, S. (1921) Correlation and causation. *Journal of Agricultural Research*, 20: 557–585

2
Using Causal Knowledge to Learn More Useful Decision Rules From Data

Louis Anthony Cox, Jr.

U S WEST Advanced Technologies
4001 Discovery Drive, Boulder, Colorado, 80303
(303)-541-6043 (phone) (303)-388-0609 (Fax) tony@uswest.com

2.1   Introduction

One of the most popular and enduring paradigms in the intersection of machine-learning and computational statistics is the use of recursive-partitioning or "tree-structured" methods to "learn" classification trees from data sets (Buntine, 1993; Quinlan, 1986). This approach applies to independent variables of all scale types (binary, categorical, ordered categorical, and continuous) and to noisy as well as to noiseless training sets. It produces classification trees that can readily be reexpressed as sets of expert systems rules (with each conjunction of literals corresponding to a set of values for variables along one branch through the tree). Each such rule produces a probability vector for the possible classes (or dependent variable values) that the object being classified may have, thus automatically presenting confidence and uncertainty information about its conclusions. Classification trees can be validated by methods such as cross-validation (Breiman et al., 1984), and they can easily be modified to handle missing data by constructing rules that exploit only the information contained in the observed variables.

Despite these powerful advantages, classification tree technology, as implemented in commercially available software systems, is often more useful for pattern recognition than for decision support. Practical business and engineering decisions require some new considerations to be incorporated into the recursive partitioning paradigm. The most important ones include

(i) *Costs of information acquisition* (Cox and Qiu, 1994). If a classification tree requires tests that are infeasible or that are too expensive to perform in practice, then it will be rejected by practitioners no matter how well it performs on training samples.

(ii) *Ability to make changes based on the tree.* If the best tree for predicting a response uses information (e.g., about time-varying covariates) that does not become available in practice until after key decisions must be made, then the inferences supported by the tree, while perhaps very valuable for scientific research purposes, will not be suitable for real-time decision support and guidance of actions. An example from the domain of cancer risk prediction is as follows. The best predictor of liver carcinomas in mice exposed to certain chemicals turns out to be the presence of liver adenomas. This is useful to scientists studying the relation between benign and malignant tumors, but it is useless for

---

[1] *Learning from Data: AI and Statistics V.* Edited by D. Fisher and H.-J. Lenz. ©1996 Springer-Verlag.

predicting whether a specific mouse will develop a liver carcinoma, since neither adenomas nor carcinomas can be observed until autopsy.

(iii) *Confidence that interventions prescribed by a tree will cause their intended consequences.* Statistical associations may not reflect causal relations, and the causal relations relating acts to their consequences may be uncertain. If a classification tree prescribes taking a certain action, how certain is it that the probability distribution over consequences will shift as the tree predicts?

(iv) *Pursuit of multiple objectives that may be differently affected by the actions taken.* Actions that affect the values of variables in a causal model may lead to (perhaps probabilistic) changes in several outcome measures. For example, in an analysis of employee survey data presented below, the multiple objectives include improving employee job satisfaction, increasing the productivity of work groups and the quality of work done, reducing job-related stress, and improving the perceived performance of management. Although these goals are largely consonant, there is some conflict among them, e.g., between reducing job stress and increasing productivity. The challenge in this setting – to find a small core set of actions that is undominated by other available actions in improving measurements of all these performance dimensions simultaneously – is typical of many real-world applications.

In summary, a theory is needed of how to learn *prescriptive*, rather than merely descriptive, classification trees and models from data. The goal is to identify and recommend changes (actions or decisions) that are predicted to have high impacts on improving the criteria. This contrasts with the usual goal of existing classification tree systems, which focus on recommending tests that will be most useful in accurately predicting the values of some quantities from observations of the values of other quantities.

This paper presents results of an applied research effort focused on how to modify conventional recursive partitioning programs (e.g., CART, Knowledge Seeker, ID3) so that they will learn useful decision rules (prescriptions for action) from data. The desired output is no longer a probabilistic prediction of the value of a dependent variable, based on the observed values of independent variables. Instead, it is a prescription for what potentially costly actions to take (i.e., what values to assign to different controllable input variables), based on (perhaps costly) measurements of multiple independent variables, so as to bring several dependent variables simultaneously into a desired "efficient" (undominated) target set of joint values. The resulting methodology has been applied to several real problems in the telecommunications industry, including selecting actions to improve customer service (Cox and Bell, 1995) and identifying a strategy for improving employee performance and morale, discussed in the last part of this paper.

2.2   Problem Formulation

Let Y be a vector of dependent variables, X a set of controllable independent variables, and Z a set of observable (empirically measurable) but not controllable variables. Values of these variables may differ across the individuals ("cases" or "instances") to which the decision rule is applied. A reduced causal model describing the (in general probabilistic) dependence of Y on X and Z is defined by a pair of functions $[p(y|x,z), f(z)]$, where $p(y|x,z)$ is the conditional probability density that $Y = y$, given that $X = X$ and $Z = z$. The function $f(z)$ is the marginal frequency distribution of Z values in a population of individuals or cases under study, i.e., $f(z) = Pr(Z = z)$. Such a "reduced" form can always be obtained from a directed graph representation of a corresponding "structural" causal model, such as a path diagram or an influence diagram. The idea of causality in this context is that X is not merely passively associated with Y – e.g., because X and Y have a joint frequency distribution in the population that is not the product of their marginals – but in fact that Y will change when X is changed (Spirites and Glymour, 1994). Learning how to choose X so as to obtain desired Y values is an exercise in valid prediction of how changes will propagate from controlled variables to outcome variables – a task more difficult than statistical inference alone.

Let $v_j(y, z)$ denote the "value", as measured on criterion scale j, of obtaining response y from an individual described by covariate vector z. Let $v(y, z)$ be the value vector whose jth component is $v_j(y, z)$, for $j = 1, 2, ..., K$, where K is the number of criteria. Then the problem addressed in this paper is to learn directly from a data set how to choose x so that the frequency distribution of $\mathbf{v}(y, z)$ induced by x in the population of individuals is undominated (in the sense of multivariate first-order stochastic dominance, FSD). The desired output is thus a prescribed choice of x from a set A of feasible alternatives such that no other choice of x in A yields a higher probability of simultaneously obtaining at least as much of every criterion (assuming that criteria are oriented so that more is better). According to the standard von Neumann- Morgenstern (NM) theory of "rational" decision making (and to many more recent normative theories with less stringent axioms), all rational decision makers who prefer more to less of each criterion will prefer x to x' if and only if x dominates x' by multivariate first-order stochastic dominance (Hirschleifer and Riley, 1992). We will abbreviate this relation as x FSD x'.

If the reduced model (p, f) were known with certainty, then the preceding problem could be formulated as a standard multicriteria statistical decision theory problem, e.g., as the vector optimization problem

$$\max_{x \in A} E_{p,f}[v(y, z)|x] \tag{1}$$

or as the scalar optimization problem

$$\max_{x \in A} E_{p,f}[u(v(y, z))|x] \tag{2}$$

where u is a multiattribute utility function mapping the vectors $\mathbf{v}(y, z)$ into "utilities" representing risk-adjusted individual preferences for outcomes (Hirschleifer and Riley, 1992). Costs of control (i.e., the cost needed to assign a specific value to a controllable variable) may be included in either formulation either as a penalty term or as another criterion. (Either leads to the same set of undominated actions.) If the detailed individual preferences and risk attitudes required to construct a multiattribute utility function are not available, then the vector optimization problem (1) can still be solved by identifying values of x that are undominated with respect to the FSD criterion. The resulting solution is typically a set of several x values, each of which may be preferred by some rational decision maker to any of the others.

The significance of the concept of causality in this setting is that the standard statistical decision theory formulation may generate incorrect recommendations, as we shall now show.

Consider a causal model represented by the diagram

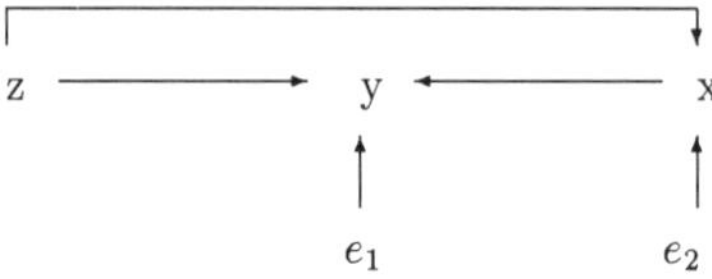

and described by the equations

$$y = z - x + e_1 \tag{3}$$

$$x = z/3 + e_2, \tag{4}$$

which together imply that

$$y = 2x + (e_1 - 3e_2). \tag{5}$$

Assume that $E(e_1) = E(e_2) = 0$ and that these error terms are uncorrelated. Then the regression relation between x and y is found from equation (5) to be

$$E(y|x) = 2x. \tag{6}$$

On the other hand, the expected change in y when x is increased from 0 to $x^* > 0$ directly (without changing z) is found from equation (3) to be

$$E(y|x = x^*) - E(y|x = 0) = [E(z) - x^*] - [E(z) - 0] = -x^*. \tag{7}$$

Thus, what might be termed the "causal regression line" relating y to x is

$$E(y||x) = -x \tag{8}.$$

(Here, "$||$" denotes a "causal conditional", i.e., $E(y||x)$ is the expected value of y given x as a value that has been set, rather than merely being observed.) The causal association between x and y has the opposite sign from their statistical association: $E(y||x) \neq E(y|x)$. This continuous version of Simpson's paradox, adapted from Lindley (1990), can readily be extended to probabilities by interpreting y as the logit for occurrence of a binary event (e.g., cancer among individuals in a population exposed occupationally to a dose, x, of an anticarcinogenic chemical whose ambient concentration is positively correlated with the ambient level of a carcinogenic chemical, z.) Treating conditional probabilities (or conditional expected values of y) given x as the values that would be obtained by setting values of x will in general lead to incorrect decisions because of misspecification errors in the model relating decisions to their probable consequences.

Even if the causal relation between controllable variables and outcomes is known, however, there is another pragmatic difficulty. In applications ranging from statistical quality control to design of new products, neither the population "mixing distribution" f nor the probabilistic response function (or probabilistic choice function, in marketing applications), p, is likely to be well known. Instead, an approximation to the model must be learned directly from the data. Thus, the full problem is to examine a sample of cases with different (x, y, z) values, some of which may be missing, and then to make a decision based on the observed sample values – and on any knowledge, beliefs, and information (e.g., on mental models of the causal relations among the x, y, and z variables) – that are expected to lead to a more favorable subjective causal distribution of outcomes. The decision to be made is typically a choice of either a specific *value* of x to apply to the population as a whole – for example, in decision contexts such as public health decision making, where x might represent a permissible exposure limit for a chemical that has been statistically associated with adverse health outcomes – or else choice of a decision rule to be applied to individual cases – for example, a rule for examining and treating patients arriving with certain symptoms. In either case, the novel challenge faced by practitioners is to select a *decision rule* or value based on an incompletely known probability model that has been partially revealed through sample values of the (x, y, z) variables.

Before turning to strategies for solving this problem, it is worth noting the following additional complexities that often arise in practice.

> The choice set, A, of possible values of x may be specified through a complex set of rules or constraints, e.g., stating that choosing some components of x restricts the permissible values of the remaining components.

> Some of the Z variables may be observable only if others have already been observed, or only if certain actions have already been taken.

More generally, the feasible sequences of activities (observations and actions) may be partially ordered by a "causally enables" relation between sets of completed activities and as-yet unattempted activities. Such constraints embody the types of temporal/causal restrictions that have been incorporated in most automatic planning systems since STRIPS, in which the "enables" relation is interpreted in terms of satisfaction of preconditions. If such constraints are important in an application, the task of discovering a minimum expected cost undominated strategy may be computationally intractable. Indeed, even without these restrictions, discovery of minimum expected cost decision rules is in general NP-hard (Cox and Qiu, 1994). Therefore, we turn next to a class of heuristics that have proved useful for solving the types of problems described here.

2.3   A Solution Heuristic

We next present an algorithm, somewhat similar in spirit to Box's EVOP methodology for combined experiment design and process improvement (Box and Draper, 1969), that has proved useful in learning robust, efficient decision rules (and reduced models supporting them) from data. The process requires searching through a space of classification trees to discover locally efficient (undominated) ones that are also robust to small changes in the observed data. Tree evaluation and selection criteria include

- *Actionability.* The tree should consist of nodes that represent controllable variables, and the actions that it prescribes should not require information or other preconditions that violate causal knowledge.

- *Causality and efficiency.* The actions in the prescribed tree should lead to an undominated improvement in the evaluation criteria of the multi-criteria design or decision problem.

- *Robustness.* The set of recommended actions in the prescribed tree should remain the same or nearly the same if another "neighboring" efficient tree is selected instead. This criterion may be formalized in terms of "core strategies", i.e., sets of actions that constitute kernels or quasi-kernels of underlying binary comparison digraphs.

The focus of the following discussion is on methods that have been used in practical applications at U S WEST over the past several years. Therefore, our solution approach is presented as a general "meta-heuristic", i.e., a generalized algorithm whose steps can be instantiated in various ways. An implementation that we have used to analyze several data sets consisting of responses to customer and employee surveys is then described in the context of an application to finding actions to improve employee survey results. Given the complexity of the class of problems addressed by this method, many interesting theoretical and empirical research questions remain open about the best techniques for carrying out the steps in the meta-heuristic. These are topics of ongoing research at U S WEST.

The following terms will be used. An *action* consists of a measurement that can be performed (its preconditions are satisfied) or an act that can be taken. (Thus, "actions" and "activities" are synonymous in this paper.) Actions may be partially ordered by precedence constraints reflecting causal enablement. (For example, mouse liver tumors cannot be observed – a measurement – until the mouse has been sacrificed – an act.) There may be a cost (dollars, time, or other resources) required in order to undertake an act. Effective heuristics for identifying low-cost inspection strategies when inference, rather than action, is the goal have been presented by Cox and Qiu (1994). The following discussion focuses on cases in which economic cost is only one of many criteria and the goal is to discover effective sequences of actions (i.e., "plans" that tend to produce desired outcomes).

Our proposed heuristic approach is as follows. (This version is for decision problems in which a set of actions is to be selected to apply to an entire population, e.g., an employee body or a set of customers. A similar heuristic applies when decision trees incorporating diagnostic tests or other possibly expensive observations are allowed and can be applied to individuals one at a time.)

2.4   Heuristic Search for Undominated Action Plans

0. Start with an empty set of actions. Search over action sequences of length 1 (i.e., individual actions) to identify actions that lead to an undominated (via FSD) improvement in the causal distribution of consequences. (As suggested previously, a "causal distribution" refers to the probability distribution that is expected to hold once the action has been taken. This distribution typically cannot be derived from observed data alone, but requires use of a hypothesized causal model, e.g., represented as an influence diagram or Bayesian belief net, relating actions to their probable consequences.) An action that leads to an undominated improvement in the performance criteria is called an "undominated action". It may also be called an *undominated partial plan of length 1.*

Phase 1: Generate undominated partial plans. For each undominated partial plan of length k, consider appending each action that is feasible once the actions in the undominated partial plan have been taken. If an action can be found that leads to a stochastically dominating (multivariate FSD) improvement in the performance criteria (with respect to the estimated causal distribution of consequences induced by the plan), then append it to the partial plan to create a new partial plan of length $k + 1$. Repeat until no further FSD improvements can be found. Delete all partial plans that have been improved (in the sense of FSD) by one or more extensions. The result of this phase is a set of undominated partial plans of various lengths.

Phase 2: Identify a "core set" of actions for initial implementation. Each undominated partial plan identified in Phase 1 consists of an ordered set of actions. To identify which specific actions to take first (continuing with the case where one or more actions must be selected for application to an entire population of cases), each initially feasible action is assigned a numerical score equal to the number of "votes" that it receives from the set of undominated partial plans (i.e., the number of such plans that it appears in). The initially feasible actions with the highest scores are recommended for implementation.

In practice, both phases are often carried out using only the benefits-related criteria, and then the top-scoring actions are evaluated for cost, practical feasibility, and confidence that the predicted consequences will occur – i.e., that conditional frequency distributions obtained from data are causally predictive. The output of Phase 2 is a set of top-ranked actions recommended for implementation. By construction, they meet the criteria of actionability (they are actions), efficiency (it is not easy to find plans that dominate the best plans starting with the recommended actions), causality (judgments about causal distributions have been incorporated into the search for FSD-improving extensions of partial plans), and robustness (since FSD is used as the criterion in generating plans for consideration).

Many refinements may be made in the implementation of this basic framework. In applications to large data bases (involving thousands of survey questionnaires – the "cases" in the training sample – and typically several dozen to over 100 variables, some actionable and some merely observable), we use a "best-first" rule to select partial plans for possible extensions. Then, we search among actions that have already been automatically rank-ordered in terms of their ability to improve prediction of one of the criteria, using the KNOWLEDGE SEEKER[TM] recursive partitioning algorithm (Biggs et al., 1991). Undominated distributions are screened for by looking for distributions that maximize the proportion of cases giving extremely high performance on the selected criterion. The process is repeated for one criterion at a time, since the KNOWLEDGE SEEKER[TM] program is constructed to deal with only one dependent variable at a time, and the intersection of the resulting recommended action sets is used to make a final recommendation. This multi-pass implementation is admittedly only an approximation to the ideal two-phase approach based on multivariate FSD, but it has proved successful in identifying small sets of recommended actions that have won immediate credence and strong support from decision makers knowledgeable about the selected application domains.

2.5   An Example: Using Employee Survey Data to Plan Improvements

We conclude with insights gleaned from a recent application to employee survey data. The study had multiple goals: to understand drivers of employee job satisfaction, morale, and performance and to recommend management actions that would improve all three. The data were obtained from questionnaires filled out by several thousand U S WEST employees covering many aspects of work life, including perceptions of one's own job performance, rating of one's immediate supervisor along various dimensions, descriptions of one's work group and the quality of its work, ratings of top management on different dimensions, and attitudes and beliefs about the company as a whole, both in absolute terms and compared to other companies. Each questionnaire contained approximately 110 detailed questions (including about a dozen demographic and career history questions). Most questions involved ratings on a conventional five-point scale (ranging from strongly disagree to strongly agree). Responses were treated as ordinal categorical variables. The sample was designed to be exhaustive and the response rates were high enough so that selection artifacts could not significantly affect the conclusions.

The large volume of data generated by this study was analyzed using the approach outlined in the previous section. Some instructive lessons from the analysis are as follows:

1. The automated KNOWLEDGE SEEKER$^{(TM)}$ procedure produced trees that clearly revealed the need to introduce causal knowledge into the analysis in order to achieve useful results. For example, a classification tree for predicting the rating of the GROUP'S WORK QUALITY variable (reflecting agreement with "My work group produces high-quality work") showed a mix of potentially actionable variables (such as "WE COOPERATE TO GET THE JOB DONE", which may be affected by strong team management) and factors that cannot be directly manipulated (such as "I LIKE THE KIND OF WORK I DO" or "MY WORK GIVES ME A FEELING OF PERSONAL ACCOMPLISHMENT").

2. We introduced *basic causal knowledge* in the simplest way: by separating variables into potentially actionable ones [represented by input nodes, i.e., nodes with only outward-directed arrows in a directed acyclic graph (dag) model of the causal relations among variables]; consequence or outcome measures (corresponding to output or "sink" nodes in the dag model), and intervening variables (represented by nodes with both inward and outward directed arrows.)

3. *A dag model representing causal relations* among variables was developed by an informal but useful process familiar to many researchers and practitioners in this field. First, a table of Spearman rank correlations was used to identify the pairs of variables that are most strongly associated. Then, "common sense" (i.e., knowledge of the meanings of the variables and how they might causally affect each other) was used to make tentative assignments of causal directions between strongly linked pairs. Next, other variables that might plausibly "explain away" the association were tested to see whether the binary association in fact disappeared when it was conditioned on the value of the new variable(s) (i.e., whether d-separability could be established). Conversely, third variables were interposed between strongly linked pairs (as in $x \rightarrow z \rightarrow y$, where z has been interposed between x and y) or introduced as common factors (as in $x \leftarrow z \rightarrow y$) where doing so would successfully "explain away" the association between the two variables. Human judgment was used to select variables to try interposing and to suggest other aspects of model structure, such as directions for arcs. This process was used to grow larger and

larger digraphs such that any pair of variables joined by a directed arc had a positive association not explained by any third variable. The largest dag generated is a model for the data. It is consistent both with the binary associations and the conditional independence structures implicit in the data, as well as with the causal knowledge used to suggest the graph structure (by selecting triples of variables to test for conditional independence). A more formal approach in a similar spirit is the TETRAD program (Spirites and Glymour, 1994 and references therein).

4. The causal knowledge represented in the dag model allowed the number of distinct variables in the analysis to be reduced from over 100 to 24. The reduction process (described by the jargon name "homomorphic aggregation") is conceptually simple. If any two variables are related to the rest of the variables in the dag model by identical directed arcs, then the two variables are aggregated into one combined variable and interpreted as two measures of the same underlying construct. The final set of 24 variables included 13 input variables, 3 output variables (intention to stay with the company, quality of work produced, and rating given to one's supervisor), and 8 intervening variables (including job satisfaction, job stress, and possession of skills needed to do a good job). The abbreviated titles of the input variables include "job security", "teamwork", "can speak one's mind around here", "urgency of job", "employees are well trained", "can get needed skills", "my boss asks my opinion", and "my boss rewards improved work quality".) Some of these inputs affected (either directly or along a chain) many output variables. For example, "my boss asks my opinion" directly affects "supervisor rating" and indirectly affects "quality of work" through the intervening variable "I can take action". It also indirectly increases (in the sense of FSD) "job satisfaction" and decreases "job stress" through paths that involve the intervening variable "my job makes good use of my skills and abilities".

5. Following these pre-processing steps, we applied the two-phase method of the previous section (using KNOWLEDGE SEEKER[TM] as a tool to search for FSD-undominated partial plans for one output variable at a time). The dag model was used to identify actionable input variables and to relate them not only to the three output variables, but also to intervening variables of interest such as job stress and job satisfaction.

The main results were that the method successfully identified a small set of core actions and implementable policies that simultaneously led to predicted improvements in all of the main outcome measures (job satisfaction, employee retention, perceived quality of work, and appraisal of management). Few undominated partial plans exceeded four actions in length. (Precedence constraints were not active in this problem, so plans were just unordered sets of activities.) A plan that remained undominated even when various subsets of variables were eliminated from the model (indicating a form of "robustness") emphasized the following three clusters of actionable principles (synthesized from items in the questionnaire):

Challenge employees to find new and better ways of doing things that affect external customers positively. (Listen to and actively seek their ideas, then follow up with actions.)

Enable employees to improve their skills so that they can do their jobs well. (Coach/develop them and keep them informed and involved in decisions that affect their work.)

Reward employees for demonstrating continuous quality improvement.

These recommendations, extracted from the large volume of data and many competing items on the questionnaire, are remarkably consistent with recent popular books on effective management of research and technology organizations. The dag causal model behind them identifies *sense of personal accomplishment* as the key intervening variable mediating between these types of inputs and the outcome criteria of job satisfaction and employee retention, perceived work quality, and evaluation of U S WEST management.

Specific plans based on the preceding analysis were predicted to dramatically increase the proportion of employees expressing top levels of job satisfaction and work quality. (This prediction involved a causal judgment: that the selected factors actually affect scores on outcomes. An alternative hypothesis that was consistent with the data was that an unmeasured latent variable – "toughness in grading" – could explain away some or all of the association between these inputs and the various outcome measures.) In 1994, management changes based on the preceding analysis were trialed in a small group of approximately 50 employees. Results collected in the third quarter showed that the trial group outperformed the rest of its department and the rest of the company in almost all (25 out of 27) measured categories. Although causality has not been proved (e.g., because no pre-measures were performed), use of the heuristic approach outlined in this paper so far appears to be associated with dramatic improvements in performance metrics.

2.6 REFERENCES

[Biggs91] Biggs, D., B. de Ville, and E. Suen, "A method of choosing multiway partitions for classification and decision trees", *Journal of Applied Statistics*, **18**, 1, 49-62, 1991.

[Box69] Box, G.E.P., and N.R. Draper. *Evolutionary Operations*. Wiley, 1969.

[Buntine93] Buntine, W., 1993. Learning classification trees. In D.J. Hand (Ed), *Artificial Intelligence Frontiers in Statistics: Artificial Intelligence and Statistics IV*. Chapman and Hall, 182-201.

[Cox94] Cox, L.A., Jr., and Y. Qiu, Minimizing the expected cost of classifying patterns by sequential costly inspections. In P. Cheeseman and R.W. Oldford (eds), *Selecting Models from Data: Artificial Intelligence and Statistics IV*. Springer-Verlag, 1994.

[Cox95] Cox, L.A., Jr., and G.E. Bell, "A machine-learning approach to process improvement decision-making," *Annals of Operations Research* (D. O'Leary, ed., forthcoming, 1995).

[Hirschleifer92] Hirschleifer, J., and J.G. Riley, *The Analytics of Uncertainty and Information*. Cambridge University Press, New York, 1992.

[Lindley90] Lindley, D.V., "Regression and correlation analysis" in J. Eatwell et al. (eds), *The New Palgrave: Time Series and Statistics*. W.W. Norton, New York, 1990.

[Quinlan86] Quinlan, J., 1986. Induction of decision trees. *Machine Learning*, 1, 81-106.

[Spirites94] Spirites, P., and C. Glymour. Inference, intervention, and prediction. In P. Cheeseman and R.W. Oldford (eds), *Selecting Models from Data: Artificial Intelligence and Statistics IV*. Springer-Verlag, 1994.

3
A Causal Calculus for Statistical Research

Judea Pearl

Cognitive Systems Laboratory
Computer Science Department
University of California, Los Angeles
Los Angeles, CA 90095-1596
USA

ABSTRACT
A calculus is proposed that admits two conditioning operators: ordinary Bayes conditioning, $P(y|X = x)$, and causal conditioning, $P(y|set(X = x))$, that is, conditioning $P(y)$ on holding X constant (at x) by external intervention. This distinction, which will be supported by three rules of inference, will permit us to derive probability expressions for the combined effect of observations and interventions. The resulting calculus yields simple solutions to a number of interesting problems in causal inference and should allow rank-and-file researchers to tackle practical problems that are generally considered too hard, or impossible. Examples are:

1. Deciding whether the information available in a given observational study is sufficient for obtaining consistent estimates of causal effects.

2. Deriving algebraic expressions for causal effect estimands.

3. Selecting measurements that would render randomized experiments unnecessary.

4. Selecting a set of indirect (randomized) experiments to replace direct experiments that are either infeasible or too expensive.

5. Predicting (or bounding) the efficacy of treatments from randomized trials with imperfect compliance.

Starting with nonparametric specification of structural equations, the paper establishes the semantics necessary for a theory of interventions, presents the three rules of inference, and proposes an operational definition of structural equations.

3.1 Introduction

The central aim of many empirical studies in the physical, behavioral, social, and biological sciences is the elucidation of cause-effect relationships among variables. It is through cause-effect relationships that we obtain a sense of a "deep understanding" of a given phenomenon, and it is through such relationships that we obtain a sense of being "in control," namely, that we are able to shape the course of events by deliberate actions or policies. It is for these two reasons, understanding and control, that causal thinking is so pervasive, popping up in everything from everyday activities to high-level decision-making: For example, a cigarette smoker would like to know, given his/her specific characteristics, to what degree his/her health would be affected by refraining from further smoking; a

---

[1] *Learning from Data: AI and Statistics V.* Edited by D. Fisher and H.-J. Lenz. ©1996 Springer-Verlag.

policy maker would like to know to what degree anti-smoking advertising would reduce costs of health care; and so on. Although a plethora of data has been collected on smoking and health, the appropriate methodology for extracting answers to such questions from the data has been fiercely debated, partly because some fundamental questions of causality have not been given fully satisfactory answers.

The two fundamental questions of causality are:

1. What empirical evidence is required for legitimate inference of cause-effect relationships?

2. Given that we are willing to accept causal information about a certain phenomenon, what inferences can we draw from such information, and how?

The primary difficulty is that we do not have a clear empirical semantics for causality; statistics teaches us that causation cannot be defined in terms of statistical associations, while any philosophical analysis of causation in terms of deliberate control quickly reaches metaphysical deadends over the meaning of free will. Indeed, Bertrand Russell noted that causation plays no role in physics proper and offered to purge the word from the language of science.

Philosophical difficulties notwithstanding, scientific disciplines that must depend on causal thinking have developed paradigms and methodologies that successfully bypass the unsettled questions of causation and that provide acceptable answers to pressing problems of experimentation and inference. Social scientists, for example, have adopted path analysis and structural equations, and programs such as LISREL have become common tools in social science research. Econometricians, likewise, have settled for stochastic simultaneous equations models as carriers of causal information and have focused most of their efforts on developing statistical techniques for estimating the parameters of these models. Statisticians, in contrast, have adopted Fisher's randomized experiment as the ruling paradigm for causal inference, with occasional excursions into its precursor – the Neyman-Robin model of potential response.

None of these paradigms and methodologies can serve as an adequate substitute for a comprehensive theory of causation, suitable for experimental research. The structural equations model is based largely on informal modeling assumptions and has hardly been applied beyond the boundaries of linear equations with Gaussian noise. The statisticians' paradigm, on the other hand, is too restrictive, as it does not allow for the integration of large body of substantive knowledge with statistical data.

Building on the paradigm that actions are a form of surgeries performed on autonomous mechanisms, this paper treats the emergence of the causal relation as an abbreviation for the surgery process, and demonstrates how causal graphs can resolve many of the confusions and contradictions that have prevented a workable theory of actions from evolving. Specifically, in Section 3.2 it will be shown that if qualitative causal information is encoded in the form of a graph, then it is simple to assess, from non-experimental data, both the strength with which causal influences operate among variables and how probabilities will change as a result of external interventions. Using graph encoding, it is also possible to specify conditions under which manipulative experiments are not necessary and where passive observations suffice. The graphs can also be queried to produce mathematical expressions for causal effects, or to suggest additional observations or auxiliary experiments from which the desired inferences can be obtained.

Section 3.3 will present a symbolic machinery that admits both probabilistic and causal information about a given domain, and produces probabilistic statements about the effect of actions and the impact of observations. The calculus admits two types of conditioning operators: ordinary Bayes conditioning, $P(y|X = x)$, which represents the observation $X = x$, and causal conditioning, $P(y|set(X = x))$, read: the probability of $Y = y$ conditioned on holding X constant (at x) by deliberate action. Given a mixture of such observational and causal sentences, together with the topology of the causal graph, the calculus derives new conditional probabilities of both types, thus enabling one to quantify the effects of actions and observations.

Finally, Section 3.4 will establish an operational definition of structural equations using the *set* operator.

3.2 Causal Theories, Actions, Causal Effect, and Identifiability

Definition 1 *A causal theory is a 4-tuple*

$$T =< V, U, P(\mathbf{u}), \{f_i\} >$$

where

(i) $V = \{X_1, \ldots, X_n\}$ *is a set of observed variables*

(ii) $U = \{U_1, \ldots, U_m\}$ *is a set of exogeneous unobserved variables which represent disturbances, abnormalities or assumptions,*

(iii) $P(\mathbf{u})$ *is a distribution function over* $U_1, \ldots, U_m$, *and*

(iv) $\{f_i\}$ *is a set of n deterministic functions, each of the form*

$$X_i = f_i(X_1, \ldots, X_n, U_1, \ldots, U_m) \quad i = 1, \ldots, n \tag{3.1}$$

each f_i representing an autonomous mechanism which can be perturbed (by the action) separately of other mechanisms. We will assume that the set of equations in (iv) has a unique solution for $X_i, \ldots, X_n$, given any value of the disturbances $U_1, \ldots, U_m$. Therefore the distribution $P(\mathbf{u})$ induces a unique distribution on the observables, which we denote by $P_T(\mathbf{v})$.

We will consider concurrent actions of the form $set(X = x)$, where $X \subseteq V$ is a set of variables and x is a set of values from the domain of X. In other words, $set(X = x)$ represents a combination of actions that forces the variables in X to attain the values x.

Definition 2 *(Effect of actions) The effect of the action $set(X = x)$ on a causal theory T is given by a subtheory T_x of T, where T_x obtains by deleting from T all equations corresponding to variables in X and substituting the equations $X = x$ instead.*

The "local surgery" described by Definition 2 reflects the understanding that interventions overrule mechanisms that tie the manipulated variables to their pre-intervention causes and replace them with new mechanisms representing the net effect of the action.

Definition 3 *(causal effect) Given two disjoint subsets of variables, $X \subseteq V$ and $Y \subseteq V$, the causal effect of X on Y, denoted $P_T(y|\hat{x})$, is a function from the domain of X to the space of probability distributions on Y, such that*

$$P_T(y|\hat{x}) = P_{T_x}(y) \tag{3.2}$$

for each realization x of X. In other words for each $x \in dom(X)$, the causal effect $P_T(y|\hat{x})$ gives the distribution of Y induced by the action set$(X = x)$.

Note that causal effects are defined relative to a given causal theory T, though the subscript T is often suppressed for brevity.

Definition 4 (identifiability) *The causal effect of X on Y is said to be* identifiable *if the quantity $P(y|\hat{x})$ can be computed uniquely from any positive distribution of the observed variables, that is, if for every pair of theories T_1 an T_2 such that $P_{T_1}(\mathbf{v}) = P_{T_2}(\mathbf{v}) > 0$, we have $P_{T_1}(y|\hat{x}) = P_{T_2}(y|\hat{x})$*

Identifiability means that $P(y|\hat{x})$ can be estimated consistently from an arbitrarily large sample randomly drawn from the distribution of the observed variables.

Definition 5 (back-door path) *A path from X to Y in a graph G is said to be a* back-door *path if it starts with an arrow pointing into X.*

Figure 1 illustrates a simple causal theory in the form of a diagram. It describes the causal relationships among the season of the year (X_1), whether rain falls (X_2), whether the sprinkler is on (X_3), whether the pavement would get wet (X_4), and whether the pavement would be slippery (X_5). All variables in this figure are binary, taking a value of either true or false, except the root variable X_1 which can take one of four values: Spring, Summer, Fall, or Winter. Here, the absence of a direct link between X_1 and X_5, for example, captures our understanding that the influence of seasonal variations on the slipperiness of the pavement is mediated by other conditions (e.g., the wetness of the pavement).

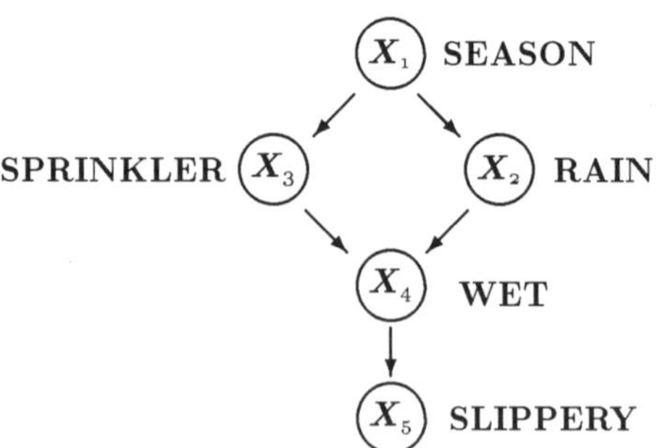

FIGURE 1. A diagram representing a causal theory on five variables.

The theory corresponding to Figure 1 consists of five functions, each representing an autonomous mechanism:

$$\begin{aligned}
X_1 &= U_1 & X_3 &= f_3(X_1, U_3) & X_5 &= f_5(X_4, U_5) \\
X_2 &= f_2(X_1, U_2) & X_4 &= f_4(X_3, X_2, U_4)
\end{aligned} \tag{3.3}$$

To represent the action "turning the sprinkler ON", $set(X_3 = \text{ON})$, we delete the equation $X_3 = f_3(X_1, U_3)$ from the theory of Eq. (3.3), and replace it with $X_3 = \text{ON}$. The resulting subtheory, $T_{X_3=\text{ON}}$, contains all the information needed for computing the effect of the actions on other variables. For example, it is easy to see from this subtheory that the only variables affected by the action are X_4 and X_5, that is, the descendant of the manipulated variable X_3.

The probabilistic analysis of causal theories becomes particularly simple when two conditions are satisfied:

1. The theory is recursive, i.e., there exists an ordering of the variables $V = \{X_1, \ldots, X_n\}$ such that each X_i is a function of a subset $\mathbf{pa}_i$ of its predecessors

$$X_i = f_i(\mathbf{pa}_i, U_i), \qquad \mathbf{pa}_i \subseteq \{X_1, \ldots, X_{i-1}\} \tag{3.4}$$

2. The disturbances $U_1, \ldots, U_n$ are mutually independent, $U_i \perp\!\!\!\perp U_j$, which also implies (from the exogeneity of the U_i's)

$$U_i \perp\!\!\!\perp \{X_1, \ldots, X_{i-1}\} \tag{3.5}$$

These two conditions, also called Markovian, are the basis of Bayesian networks [Pearl, 1988] and they enable us to compute causal effects directly from the conditional probabilities $P(x_i|\mathbf{pa}_i)$, without specifying the functional form of the functions f_i, or the distributions $P(u_i)$ of the disturbances. This is seen immediately from the following observations:

The distribution induced by any Markovian theory T is given by the product

$$P_T(x_1, \ldots, x_n) = \prod_i P(x_i|\mathbf{pa}_i) \tag{3.6}$$

where $\mathbf{pa}_i$ are the direct predecessors (called *parents*) of X_i in the diagram. On the other hand the distribution induced by the subtheory $T_{x'_j}$, representing the action $set(X_j = x'_j)$ is given by product

$$P_{T_{x'_j}}(x_1, \ldots, x_n) = \begin{cases} \prod_{i \neq j} P(x_i|\mathbf{pa}_i) = \frac{P(x_1, \ldots, x_n)}{P(x_j|\mathbf{pa}_j)} & \text{if } x_j = x'_j \\ 0 & \text{if } x_j \neq x'_j \end{cases} \tag{3.7}$$

because $T_{x'_j}$ is also Markovian, and the partial product reflects the surgical removal of the $X_j = f_j(\mathbf{pa}_j, U_j)$ from the theory of equation (3.4).

In the example of Figure 1, the pre-action distribution is given by the product

$$P_T(x_1, x_2, x_3, x_4, x_5) = P(x_1)P(x_2|x_1)P(x_3|x_1)P(x_4|x_2, x_3)P(x_5|x_4) \tag{3.8}$$

while the surgery corresponding to the action $set(X_3 = \text{ON})$ amounts to deleting the link $X_1 \rightarrow X_3$ from the graph and fixing the value of X_3 to ON, yielding the post-action distribution:

$$P_T(x_1, x_2, x_4, x_5|set(X_3 = \text{ON})) = P(x_1)\ P(x_2|x_1)\ P(x_4|x_2, X_3 = \text{ON})\ P(x_5|x_4) \tag{3.9}$$

Note the difference between the action $set(X_3 = \text{ON})$ and the observation $X_3 = \text{ON}$. The latter is encoded by ordinary Bayesian conditioning, while the former by conditioning a mutilated graph, with the link $X_1 \rightarrow X_3$ removed. This mirrors indeed the difference between seeing and doing: after observing that the sprinkler is ON, we wish to infer that the season is dry, that it probably did not rain, and so on; no such inferences should be drawn in evaluating the effects of the deliberate action "turning the sprinkler ON". The amputation of $X_3 = f_3(X_1, U_3)$ from (3.3) ensures the suppression of any abductive inferences from any of the action's consequences.

Note also that Equations (3.6) through (3.9) are independent of T, in other words, the pre-actions and post-action distributions depend only on observed conditional probabilities but are independent of the particular functional form of $\{f_i\}$ or the distribution $P(\mathbf{u})$ which generate those probabilities. This is the essence of identifiability as given in Definition 4, which stems from the Markovian assumptions (3.4) and (3.5). Section 3.3 will demonstrate that certain causal effects, though not all, are identifiable even when the Markovian property is destroyed by introducing dependencies among the disturbance terms.

Generalization to multiple actions and conditional actions are straightforward. Multiple actions $set(X = x)$, where X is a compound variable result in a distribution similar to (3.7), except that all factors corresponding to the variables in X are removed from the product in (3.6). Stochastic conditional strategies of the form

$$set(X_j = x_j) \text{ with probability } P^*(x_j | \mathbf{pa}_j^*)$$

where $\mathbf{pa}_j^*$ is the support of the decision strategy, also result in a product decomposition similar to (3.7), except that each factor $P(x_j | \mathbf{pa}_j)$ is replaced with $P^*(x_j | \mathbf{pa}_j^*)$.

The surgical procedure described above is not limited to probabilistic analysis. The causal knowledge represented in Figure 1 can be captured by logical theories as well, for example,

$$
\begin{aligned}
x_2 &\iff [(X_1 = \text{Winter}) \vee (X_1 = \text{Fall}) \vee ab_2] \wedge \neg ab_2' \\
x_3 &\iff [(X_1 = \text{Summer}) \vee (X_1 = \text{Spring}) \vee ab_3] \wedge \neg ab_3' \\
x_4 &\iff (x_2 \vee x_3 \vee ab_4) \wedge \neg ab_4' \\
x_5 &\iff (x_4 \vee ab_5) \wedge \neg ab_5'
\end{aligned}
$$

where x_i stands for $X_i = true$, and ab_i and ab_i' stand, respectively, for triggering and inhibiting abnormalities. The double arrows represent the assumption that the events on the r.h.s. of each equation are the *only* causes for the l.h.s.

It should be emphasized though that the models of a causal theory are not made up merely of truth value assignments which satisfy the equations in the theory. Since each equation represents an autonomous process, the scope of each individual equation must be specified in any model of the theory, and this can be encoded using either the graph (as in Figure 1) or the generic description of the theory, as in (3.3). Alternatively, we can view a model of a causal theory to consist of a mutually consistent set of submodels, with each submodel being a standard model of a single equation in the theory.

3.3   Causal Calculus

The identifiability of causal effects demonstrated in Section 3.2 relies critically on the Markovian assumptions (3.4) and (3.5). If a variable that has two descendants in the graph is unobserved, the disturbances in the two equations are no longer independent, the Markovian property (3.4) is violated and identifiability may be destroyed. This can be seen easily from Eq. (3.7); if any parent of the manipulated variable X_j is unobserved, one cannot estimate the conditional probability $P(x_j|\mathbf{pa}_j)$, and the effect of the action $set(X_j = x_j)$ may not be predictable from the observed distribution $P(x_1,\ldots,x_n)$. Fortunately, certain causal effects are identifiable even in situations where members of $\mathbf{pa}_j$ are unobservable and, moreover, polynomial tests are now available for deciding when $P(x_i|\hat{x}_j)$ is identifiable, and for deriving closed-form expressions for $P(x_i|\hat{x}_j)$ in terms of observed quantities [Pearl, 1995, Galles & Pearl, 1995].

These tests and derivations are based on a symbolic calculus to be described in the sequel, in which interventions, side by side with observations, are given explicit notation, and are permitted to transform probability expressions. The transformation rules of this calculus reflect the understanding that interventions perform "local surgeries" as described in Definition 2, i.e., they overrule equations that tie the manipulated variables to their pre-intervention causes.

Let X, Y, and Z be arbitrary disjoint sets of nodes in a DAG G. We say that X and Y are independent given Z in G, denoted $(X \parallel Y|Z)_G$, if the set Z d-separates X from Y in G. We denote by $G_{\overline{X}}$ the graph obtained by deleting from G all arrows pointing to nodes in X. Likewise, we denote by $G_{\underline{X}}$ the graph obtained by deleting from G all arrows emerging from nodes in X. To represent the deletion of both incoming and outgoing arrows, we use the notation $G_{\overline{X}\underline{Z}}$. Finally, the expression $P(y|\hat{x},z) \triangleq P(y,z|\hat{x})/P(z|\hat{x})$ stands for the probability of $Y = y$ given that $Z = z$ is observed and X is held constant at x.

Theorem 1 Let G be the directed acyclic graph associated with a Markovian causal theory, and let $P(\cdot)$ stand for the probability distribution induced by that theory. For any disjoint subsets of variables X, Y, Z, and W we have:

Rule 1 Insertion/deletion of observations

$$P(y|\hat{x},z,w) = P(y|\hat{x},w) \ \text{ if } \ (Y \parallel Z|X,W)_{G_{\overline{X}}} \tag{3.10}$$

Rule 2 Action/observation exchange

$$P(y|\hat{x},\hat{z},w) = P(y|\hat{x},z,w) \ \text{ if } \ (Y \parallel Z|X,W)_{G_{\overline{X}\underline{Z}}} \tag{3.11}$$

Rule 3 Insertion/deletion of actions

$$P(y|\hat{x},\hat{z},w) = P(y|\hat{x},w) \ \text{ if } \ (Y \parallel Z|X,\ W)_{G_{\overline{X},\ \overline{Z(W)}}} \tag{3.12}$$

where $Z(W)$ is the set of Z-nodes that are not ancestors of any W-node in $G_{\overline{X}}$.

Each of the inference rules above follows from the basic interpretation of the "$\hat{x}$" operator as a replacement of the causal mechanism that connects X to its pre-action parents by a new mechanism $X = x$ introduced by the intervening force. The result is a submodel characterized by the subgraph $G_{\overline{X}}$ (named "manipulated graph" in Spirtes et al. (1993)) which supports all three rules.

Rule 1 reaffirms d-separation as a valid test for conditional independence in the distribution resulting from the intervention $set(X = x)$, hence the graph $G_{\overline{X}}$. This rule follows from the fact that deleting equations from the system does not introduce any dependencies among the remaining disturbance terms.

Rule 2 provides a condition for an external intervention $set(Z = z)$ to have the same effect on Y as the passive observation $Z = z$. The condition amounts to $\{X \cup W\}$ blocking all back-door paths from Z to Y (in $G_{\overline{X}}$), since $G_{\overline{X}\underline{Z}}$ retains all (and only) such paths.

Rule 3 provides conditions for introducing (or deleting) an external intervention $set(Z = z)$ without affecting the probability of $Y = y$. The validity of this rule stems, again, from simulating the intervention $set(Z = z)$ by the deletion of all equations corresponding to the variables in Z (hence the graph $G_{\overline{XZ}}$).

Corollary 1 A causal effect $q: P(y_1, ..., y_k | \hat{x}_1, ..., \hat{x}_m)$ is identifiable in a model characterized by a graph G if there exists a finite sequence of transformations, each conforming to one of the inference rules in Theorem 1, which reduces q into a standard (i.e., hat-free) probability expression involving observed quantities. $\square$

Although Theorem 1 and Corollary 1 require the Markovian property, they do not require all variables to be observable and, hence, they can be applied to non-Markovian, recursive theories as well. To demonstrate, assume that variable X_1 in Figure 1 is unobserved, rendering the disturbances U_3 and U_2 dependent since these terms now include the common influence of X_1. Theorem 1 tells us that the causal effect $P(x_4|x_{\hat{3}})$ is identifiable, because:

$$P(x_4|\hat{x}_3) = \sum_{x_2} P(x_4|\hat{x}_3, x_2) P(x_2|\hat{x}_3)$$

Rule 3 permits the deletion

$$P(x_2|\hat{x}_3) = P(x_2), \text{ because } (X_2 \perp\!\!\!\perp X_3)_{G_{\overline{X}_3}},$$

while Rule 2 permits the exchange

$$P(x_4|\hat{x}_3, x_2) = P(x_4|x_3, x_2), \text{ because } (X_4 \perp\!\!\!\perp X_3|X_2)_{G_{\underline{X}_3}}.$$

This gives

$$P(x_4|\hat{x}_3) = \sum_{x_2} P(x_4|x_3, x_2) P(x_2)$$

which is a "hat-free" expression, involving only observed quantities.

In general, it can be shown [Pearl, 1995] that:

1. The effect of interventions can often be identified (from nonexperimental data) without resorting to parametric models,

2. The conditions under which such nonparametric identification is possible can be determined by simple graphical criteria, and,

3. When the effect of interventions is not identifiable, the causal graph may suggest non-trivial experiments which, if performed, would render the effect identifiable.

The ability to assess the effect of interventions from nonexperimental data has immediate applications in the medical and social sciences, since subjects who undergo certain treatments often are not representative of the population as a whole. Such assessments are also important in AI applications where an agent needs to predict the effect of the next action on the basis of past performance records, and where that action has never been enacted out of free will, but in response to environmental needs or to other agent's requests.

Historical background. An explicit translation of interventions to "wiping out" equations from linear econometric models was first proposed by Strotz & Wold (1960) and later used in Fisher (1970) and Sobel (1990) . Extensions to action representation in non-monotonic systems were reported in [Goldszmidt & Pearl, 1992, Pearl, 1993a]. Graphical ramifications of this translation were explicated first in Spirtes et al. (1993) and later in Pearl (1993b). A related formulation of causal effects, based on event trees and counterfactual analysis was developed by Robins (1986, pp. 1422-25). Calculi for actions and counterfactuals based on this interpretation are developed in [Pearl, 1994b] and [Balke & Pearl, 1994a], respectively.

3.4   The Operational Meaning of Structural Equations

Traditionally, statisticians have approved of only one method of combining subject-matter considerations with statistical data: the Bayesian method of assigning subjective priors to distributional parameters. To incorporate causal information within the Bayesian framework, plain causal statements such as "Y is affected by X" must be converted into sentences capable of receiving probability values, e.g., counterfactuals. Indeed, this is how Rubin's model has achieved statistical legitimacy: causal judgments are expressed as constraints on probability functions involving counterfactual variables.

Causal diagrams offer an alternative language for combining data with causal information. This language simplifies the Bayesian route by accepting plain causal statements as its basic primitives. These statements, which merely identify whether a causal connection between two variables of interest exists, are commonly used in natural discourse and provide a natural way for scientists to communicate experience and organize knowledge. It can be anticipated, therefore, that the language of causal graphs will find applications in problems requiring substantial use of subject-matter considerations.

The language is not new. The use of diagrams and structural equations models to convey causal information has been quite popular in the social sciences and econometrics. Statisticians, however, have generally found these models suspect, perhaps because social scientists and econometricians have failed to provide an unambiguous definition of the empirical content of their models, that is, to specify the experimental conditions, however hypothetical, whose outcomes would be constrained by a given structural equation. As a result, even such basic notions as "structural coefficients" or "missing links" become the object of serious controversy [Freedman, 1987] and conflicting interpretations (Wermuth, 1992; Whittaker, 1990, p. 302; Cox & Wermuth, 1993).

To a large extent, this history of controversy and miscommunication stems from the absence of an adequate mathematical notation for defining basic notions of causal modeling. For example, standard probabilistic notation cannot express the empirical content of the coefficient b in the structural equation $Y = bX + \epsilon_Y$ even if one is prepared to assume that ϵ_Y (an unobserved quantity) is uncorrelated with X. Nor can any probabilistic meaning be attached to the analyst's excluding from the equation certain variables that are highly correlated with X or Y.

The notation developed in this paper gives these notions a clear empirical interpretation, because it permits one to specify precisely what is being held constant and what is merely measured in a controlled experiment. (The need for this distinction was recognized by many researchers, most notably Pratt & Schlaifer (1988) and Cox (1992)). The meaning of b is simply $\frac{\partial}{\partial x}E(Y|\hat{x})$, namely, the rate of change (in x) of the expectation of Y in an experiment where X is held at x by external control. This interpretation holds regardless of whether ϵ_Y and X are correlated (e.g., via another equation: $X = aY + \epsilon_X$) and, moreover, the notion of randomization need not be invoked. Likewise, the analyst's decision as to which variables should be included in a given equation can be based on a hypothetical controlled experiment: A variable Z is excluded from the equation for Y if it has no influence on Y when all other variables, S_{YZ}, are held constant, that is, $pr(y|\hat{z}, \hat{s}_{YZ}) = pr(y|\hat{s}_{YZ})$. In other words, variables that are excluded from the equation $Y = bX + \epsilon_Y$ are not conditionally independent of Y given measurements of X, but rather conditionally independent of Y given settings of X. The operational meaning of the so called "disturbance term", ϵ_Y, is likewise demystified: ϵ_Y is defined as the difference $Y - E(Y|\hat{s}_Y)$; two disturbance terms, ϵ_X and ϵ_Y, are correlated if $pr(y|\hat{x}, \hat{s}_{XY}) \neq pr(y|x, \hat{s}_{XY})$, and so on.

The distinctions provided by the "hat" notation clarify the empirical basis of structural equations and should make causal models more acceptable to empirical researchers. Moreover, since most scientific knowledge is organized around the operation of "holding X fixed," rather than "conditioning on X," the notation and calculus developed in this paper should provide an effective means for scientists to communicate subject-matter information, and to infer its logical consequences.

3.5 REFERENCES

[Balke & Pearl, 1994a] Balke, A. and Pearl, J., "Probabilistic evaluation of counterfactual queries," in *Proceedings of the Twelfth National Conference on Artificial Intelligence (AAAI-94)*, Seattle, WA, Volume I, 230-237, July 31 - August 4, 1994.

[Cox, 1992] Cox, D.R.,"Some statistical aspects," *Journal of the Royal Statistical Society*, Series A, 155, 291–301, 1992.

[Cox & Wermuth, 1993] Cox, D.R. and Wermuth, N., "Linear dependencies represented by chain graphs," *Statistical Society*, 8, 204–218, 1993

[Fisher, 1970] Fisher, F.M., "A correspondence principle for simultaneous equation models," *Econometrica*, 38, 73–92, 1970.

[Freedman, 1987] Freedman, D., "As others see us: A case study in path analysis" (with discussion), *Journal of Educational Statistics*, 12, 101–223, 1987.

[Galles & Pearl, 1995] Galles, D. and Pearl, J., "Testing Identifiability of Causal Effects," in P. Besnard and S. Hanks (Eds.), *Uncertainty in Artificial Intelligence 11*, Morgan Kaufmann, San Francisco, CA, 185–195, 1995.

[Goldszmidt & Pearl, 1992] Goldszmidt, M. and Pearl, J., "Rank-based systems: A simple approach to belief revision, belief update, and reasoning about evidence and actions," in B. Nebel, C. Rich, and W. Swartout (Eds.), *Proceedings of the Third International Conference on Knowledge Representation and Reasoning*, Morgan Kaufmann, San Mateo, CA, 661-672, October 1992.

[Pearl, 1988] Pearl, J., *Probabilistic Reasoning in Intelligence Systems*, Morgan Kaufmann, San Mateo, CA, 1988.

[Pearl, 1993a] Pearl, J., "From Conditional Oughts to Qualitative Decision Theory" in D. Heckerman and A. Mamdani (Eds.), *Proceedings of the Ninth Conference on Uncertainty in Artificial Intelligence*, Washington, D.C., Morgan Kaufmann, San Mateo, CA, 12–20, July 1993.

[Pearl, 1993b] Pearl, J., "Graphical models, causality, and intervention," *Statistical Science*, 8 (3), 266–273, 1993.

[Pearl, 1994b] Pearl, J., "A probabilistic calculus of actions," in R. Lopez de Mantaras and D. Poole (Eds.), *Proceedings of the Tenth Conference on Uncertainty in Artificial Intelligence (UAI-94)*, Morgan Kaufmann, San Mateo, CA, 454-462, 1994.

[Pearl, 1995] Pearl, J., "Causal diagrams for experimental research, (with discussion)," *Biometrika*, 82(4), 669–709, December 1995.

[Pratt & Schlaifer, 1988] Pratt, J.W. and Schlaifer, R., "On the interpretation and observation of laws," *Journal of Econometrics*, 39, 23–52, 1988.

[Robins, 1986] Robins, J., "A new approach to causal inference in mortality studies with a sustained exposure period – applications to control of the healthy workers survivor effect," *Mathematical Modelling*, 7, 1393–1512, 1986.

[Sobel 1990] Sobel, M.E., "Effect analysis and causation in linear structural equation models," *Psychometrika*, 55(3), 495–515, 1990.

[Spirtes et al., 1993] Spirtes, P., Glymour, C., and Schienes, R., *Causation, Prediction, and Search*, Springer-Verlag, New York, 1993.

[Strotz & Wold, 1960] Strotz, R.H. and Wold, H.O.A., "Recursive versus nonrecursive systems: An attempt at synthesis," *Econometrica* 28, 417–427, 1960.

[Wermuth, 1992] Wermuth, N., "On block-recursive regression equations" (with discussion), *Brazilian Journal of Probability and Statistics*, 6, 1–56, 1992.

[Whittaker, 1990] Whittaker, J., *Graphical Models in Applied Multivariate Statistics*, John Wiley and Sons, Chinchester, England, 1990.

4
Likelihood-based Causal Inference

Qing Yao[†] and David Tritchler[††]

Department of Biostatistics[†]
Harvard School of Public Health
Ontario Cancer Institute and University of Toronto[††]

ABSTRACT A method is given which uses subject matter assumptions to discriminate recursive models and thus point toward possible causal explanations. The assumptions alone do not specify any order among the variables — rather just a theoretical absence of direct association. We show how these assumptions, while not specifying any ordering, can when combined with the data through the likelihood function yield information about an underlying recursive order. We derive details of the method for multi-normal random variables.

4.1 INTRODUCTION

Starting from Sewall Wright (1934), directed graphs have been used to represent structures in which variables 'cause' or 'influence' other variables. Nodes of the graph arc used to represent variables and an arrow from one variable to another indicates that the first has a direct causal influence on the second, an influence not blocked by holding constant others considered. If the graphs are restricted to directed acyclic graphs (DAGs) by prohibiting directed cycles, then there exists an ordering of the vertices in the DAG consistent with the direction of the edges, in that all variables are ordered after their causes in a causal or temporal sense. Conversely, an ordering of the variables can specify a recursive model for which statistical analysis is routine (see section 4.3), and which will define a DAG.

Thus there is a correspondence between causal descriptions, DAGs, and recursive statistical models. A search for likely causal explanations can be implemented as a search for good-fitting recursive models. The interpretation of such a model is controversial. Cox and Wermuth (1993) prefer to restrict the term causal to "situations when there is some understanding of an underlying process" and indicate that recursive models "could be consistent with a causal explanation". Pearl and Verma (1991) and Spirtes et al (1993) regard directed association in a good-fitting DAG as a definition of causality. Our view is similar to Cox and Wermuth's, but the causal interpretation of recursive models is not central to this paper; our focus is the identification of recursive models. Throughout, we use the term causal to indicate a directed edge in a DAG, as this corresponds with the intuition for such models; we make no claims about an underlying mechanism. We use the

[1] *Learning from Data: AI and Statistics V.* Edited by D. Fisher and H.-J. Lenz. ©1996 Springer-Verlag.

[2] The authors thank David Andrews, Paul Corey, Arthur Dempster, Michael Escobar, Wayne Oldford, Judea Pearl and Robert Tibshirani for helpful comments and discussion.

[3] Qing Yao was supported by a fellowship from the Natural Sciences and Engineering Research Council of Canada. David Tritchler was supported by a research grant from the Natural Sciences and Engineering Research Council of Canada.

term 'causal ordering' synonymously with 'recursive ordering', the sequence of variables in a recursive model, again in accordance with intuition.

It is standard practice to rely on purely subject matter considerations to specify a causal ordering for the variables in modelling observational data. An important question is: can we obtain any information about the ordering from the data? We argue in section 4.3.2 that, in the absence of external assumptions, we cannot. In this paper we give a method which uses subject matter assumptions to discriminate variously ordered recursive models. The assumptions themselves do not specify any order among the variables — rather just a theoretical absence of direct association. We show how they can lead to a partial order which can suggest causal explanations and guide subsequent model building.

Some researchers (Pearl and Verma, 1991; Spirtes et al., 1993) have emphasized stability assumptions (which a priori rule out certain probability models), under which causal inferences may be made. The relation of our method to causation is examined in the last section of the paper, where it is shown that our inferences about the order of variables have validity similar to inference about association from observational data. That is, they are still vulnerable to the confounding biases that randomization guards against. Thus the method should not be viewed as "causal inference" that has the strength of an experiment for predicting the effect of an intervention, but rather as a technique for investigating the directionality of relationships suggested by the data. The conclusions are no less, and no more, reliable than the associations of variables routinely determined from observational data.

Another approach to recursive causal model selection is given by Madigan and Raftery (1994). Their model selection criteria emphasize parsimony rather than the satisfaction of external constraints, and they focus less on making probability statements about precedence relations.

The next section introduces DAGs, linear orders, and recursive models. Section 3 describes the linear recursive analysis of normally distributed data and gives a method for extracting and interpreting information about causal order. The method is applied to an example in section 4. The last section discusses the connection of our method with causal inference.

4.2   DAGS AND RECURSIVE MODELS

A directed graph is a graph representing random variables and the statistical dependencies among them. Random variables are represented by nodes of the graph, and the direct influence of a variable U upon V is depicted by an arrow (synonymously, directed edge) from node U to node V. If there is a directed edge $U \rightarrow V$, we call U a parent and V a child of U, and U is considered a cause of V. A directed path is a sequence $Z_{i_1} \rightarrow Z_{i_2} \rightarrow \cdots \rightarrow Z_{i_n}$ of distinct nodes; here we say that Z_{i_n} is a descendant of Z_{i_1} and Z_{i_1} is an ancestor of Z_{i_n}. We denote the set of ancestors of V as $An(V)$, and define $An(U, V)$ to be $An(U) \cup An(V)$. A cycle is a directed path from a node to itself. A directed acyclic graph (DAG) is a directed graph with no cycles.

4.2.1   Markov properties of DAGs

Kiiveri and Speed (1982) related causal influences represented by a directed graph without cycles to conditional independence constraints. The connection is the (local) Markov condition, which specifies that every vertex is statistically independent of its non-descendants given its parents. This is equivalent to the recursive factorization of the joint distribution of the vertices

$$f(\mathbf{V}) = \prod_{v \in \mathbf{V}} f(v|parents(v)), \tag{4.1}$$

where $f(v) \neq 0$, which uniquely determines the joint distribution. The Markov condition is intuitively desirable in a model for causal influence, in that immediate causes should shield v from every variable that is not a consequence of v.

4.2.2   Orderings and recursive models

Suppose that we label the nodes by numbering them so that $i < j \Rightarrow i \in nd(j)$ in any feasible DAG model, where $nd(j)$ is the set of non-descendants of node j. That is, arrows always point from low numbers to higher numbers in accordance with causal or temporal order. Thus if $X(i)$ is a potential cause of $X(j)$, $i < j$. This labeling identifies a variable $X(i)$ with its temporal or causal order i. We write $X(i) \prec X(j)$ whenever $i < j$. For sets $A = \{a_k\}$ and $B = \{b_l\}$, we write $A \prec B$ whenever $a_k \prec b_l$ for all such pairs.

The numbering of variables defines a complete, irreflexive, asymmetric, transitive relation, which is formally referred to as a linear order or an ordering. Viewed as a relation, it is the set of all ordered pairs $(X(i), X(j)), i < j$. We may also write such an ordered pair as $X(i) \prec X(j)$ for emphasis. An ordering will be written as $X(1)X(2) \cdots X(n)$ or, alternatively, as $X(1) \prec X(2) \cdots \prec X(n)$. Another interpretation of an ordering is as a complete directed graph with edges $X(i) \to X(j)$ for $i < j$. The transitivity of the relation implies that the graph is acyclic, so the ordering is equivalent to a complete DAG. We can obtain DAG models for the data by deleting edges from the complete DAG specified by an ordering. Pearl (1988) and Lauritzen et al. (1990) show that the following property, the local well numbering property, is equivalent to the Markov property.

$$X(i) \perp \{X(j); j < i\} \mid parents(X(i)) \tag{4.2}$$

The set $\{X(j); j < i\}$, will be referred to as $pred(X(i))$, the predecessors of $X(i)$. Given an ordering we can determine the parents of every variable by a recursive series of analyses of the ordered sets $\{X(2), X(1)\}, \{X(3), X(2), X(1)\}, \cdots, \{X(n), X(n-1), \cdots, X(1)\}$, where each variable is regressed on its predecessors. The resulting graph will possess the Markov property due to the equivalence of the Markov property and (4.2).

4.3   INFERENCE ABOUT ORDER

4.3.1   Linear recursive regression

Recursive equations have been related to path analysis and the statistical theory of covariance selection (Wermuth, 1980), and later to graphical models (Wermuth and Lauritzen, 1983). We will only consider the case of Gaussian linear recursive regression models, although the general method to be proposed is appropriate for any recursive model for which the parameters are separable and express pairwise independence.

We assume a linear relationship between the variables in a p-dimensional random vector $\mathbf{X} = (X_1, X_2, ..., X_p)^T$. For simplicity, we also assume that X_i follows an origin-centered, non-degenerate continuous distribution with covariance matrix $\boldsymbol{\Sigma}$. For a linear order ϕ, if the variables are indexed as $X_i = X(i)$, the corresponding set of linear recursive equations is

$$
\begin{aligned}
X_1 &= \epsilon_1 \\
X_2 &= \beta_{21}X_1 + \epsilon_2 \\
X_3 &= \beta_{31\cdot2}X_1 + \beta_{32\cdot1}X_2 + \epsilon_3 \\
&\ \ \vdots \\
X_p &= \beta_{p1\cdot23\cdots p-1}X_1 + \beta_{p2\cdot13\cdots p-1}X_2 + \cdots + \beta_{p,p-1\cdot12\cdots p-2}X_{p-1} + \epsilon_p,
\end{aligned}
\tag{4.3}
$$

where ϵ_i's are errors assumed to be independently distributed with mean zero and variance d_{ii}. We can write the above equations as

$$
\mathbf{BX} = \epsilon,
\tag{4.4}
$$

where $\mathbf{B}$ is a lower triangle matrix with diagonal elements 1, lower off-diagonal elements the negative coefficients of the above equations, and $\epsilon \sim N(0, \mathbf{D})$ with $\mathbf{D}$ a full rank diagonal matrix (d_{ii}). In the equation corresponding to X_i, only predecessors of X_i appear on the right hand side, so the non-zero coefficients correspond to parents of X_i. Taking variances of both sides of equation (4.4) gives

$$
\boldsymbol{\Sigma} = \mathbf{B}^{-1}\mathbf{D}\mathbf{B}^{-T}.
\tag{4.5}
$$

For a particular order, $(\mathbf{B}, \mathbf{D})$ is a one-to-one transformation of $\boldsymbol{\Sigma}$. Hence the linear recursive equations completely capture the covariance structure of the data. Since ϵ (and thus $\mathbf{X}$) is normally distributed,

$$
\beta_{ij\cdot1\cdots j-1,j+1\cdots i-1} = 0 \quad \text{iff} \quad X_i \perp X_j | (X_1 \cdots X_{j-1}, X_{j+1} \cdots X_{i-1}).
\tag{4.6}
$$

We may express the above conditional independence as $X_i \perp X_j \mid pred(X_i, X_j)$.

Because of the separability of parameters (Whittaker, 1994) implied by (4.3), the MLEs of the recursive system can be estimated by each regression separately; the MLEs for each equation are the LSEs of each regression.

4.3.2 Discriminating orderings

The order of variables is often unknown, and it would be useful to be able to obtain some information about the causal ordering of the variables from the data. Unfortunately, without restrictions, all orderings give the same maximum likelihood when linear recursive equations are fit. This is because for any ordering, with no restrictions, the decomposition (4.5) holds and $(\mathbf{B}, \mathbf{D})$ is a unique one to one transformation of $\boldsymbol{\Sigma}$. Thus by the composition of functions there exists a unique one to one transformation between the parameters $(\mathbf{B_1}, \mathbf{D_1})$ and $(\mathbf{B_2}, \mathbf{D_2})$ for any two orderings. Thus the likelihood of these parameterizations will agree, and the observed data by itself will not yield information on ordering.

In a relatively well-understood domain, we may be able to specify that some variables can not directly influence each other. That is, we may state that for a pair of variables, neither variable can be a parent of the other. By Verma and Pearl (1990), then they are independent given their ancestors; we write this as $X_i \perp X_j \mid An(X_i, X_j)$. Although this information does not tell us anything directly about ordering of variables, we will show that it will give us information about the underlying structure when combined with the data through the likelihood function.

To assume X_i and X_j are not directly related means X_i and X_j are not adjacent in the underlying DAG, which is modeled in the recursive system as $\beta_{ij \cdot pred(X_i, X_j)} = 0$ according to (4.6). By applying such assumptions to the recursive system, i.e. restricting some regression coefficients to zero, for different orderings the likelihood will vary. The assumptions have injected some ordering information into the system by implying different parameterizations and restrictions for different orderings, in that the set $pred(X_i, X_j)$ varies for different orderings. Therefore given such assumptions or restrictions, the likelihood can tell us which orderings are consistent with the data and assumptions. Note that these assumptions are symmetric with respect to X_i and X_j; that is, they do not directly express any information about order.

For a particular ordering ϕ, define $l(\phi)$ to be the log-likelihood for the system (4.4) when the vector $\mathbf{X}$ is ordered as ϕ. The log-likelihood $l(\phi)$ is subject to restrictions on regression coefficients based on subject matter knowledge and maximized over the remaining parameters, for ordering ϕ. For large samples, goodness of fit is tested using the statistic

$$R(\phi) = -2[Max\{l(\phi)\} - Max\{l_0\}] \sim \chi^2_{d.f.}, \tag{4.7}$$

where ϕ is the linear order determining the recursive model, $Max\{l(\phi)\}$ is the maximum log-likelihood for the restricted model, $Max\{l_0\}$ is for the unrestricted model and does not depend on the ordering, and $d.f. =$ Number of restrictions (Whittaker, 1990). For different orderings the restrictions are on different parameters, i.e. $\beta_{ij \cdot pred(X_i, X_j)} = 0$ changes meaning since the set $pred(X_i, X_j)$ depends on the ordering. If the assumptions are true, in the limit the true ordering gives the same fit and likelihood as the unrestricted case (Yao, 1994). Theory does not, however, guarantee that an incorrect order will not fit; external information or subject matter knowledge are needed to choose among the orderings which yield a high maximized likelihood. If no ordering gives good fit, the assumptions should be re-investigated, or there may be important variables omitted from the analysis.

4.3.3 Equivalent sets

Some orderings are equivalent with respect to the restrictions, in that their maximum likelihoods are identical so they cannot be discriminated for any sample size. Equivalence is determined by the following theorems.

Theorem 1 *If we have assumptions* $X_{i_1} \perp X_{j_1} | pred(X_{i_1}, X_{j_1})$, $\cdots$, $X_{i_m} \perp X_{j_m} | pred(X_{i_m}, X_{j_m})$, *then for an order* $\mathbf{A_0} \prec X_{j_1} \prec \mathbf{A_1} \prec \cdots \prec X_{j_m} \prec \mathbf{A_m}$, *where* $\mathbf{A_k}$'s *are the sets between* $X_{j_{k-1}}$ *and* X_{j_k} *and all* $X_{i_k} \prec X_{j_k}$, *permutation within any* $\mathbf{A_k}$, *$k = 0, 1, \cdots, m$, does not affect the maximum likelihood.*

Theorem 2 *Under the conditions of Theorem 1, if X_{i_k} and X_{j_k} are consecutive in the ordering, they may be permuted without affecting the maximum likelihood.*

Proofs are given in Yao (1994). The theorems indicate the resolution of the method. When no assumptions are made, there is a single equivalent set composed of all orderings. As more assumptions are introduced, the ability to discriminate between orderings improves, in that the equivalent sets become smaller.

The following corollary gives some more insight into the influence of the assumptions.

Corollary 1 *If the only assumption is $X_i \perp X_j | pred(X_i) \cup pred(X_j)$, then for an order $\mathbf{A} \prec X_i \prec \mathbf{B} \prec X_j \prec \mathbf{C}$, switching X_i and X_j in the ordering does not affect the maximum likelihood*

Thus an assumption about X_i and X_j does not yield information about the relative order of X_i and X_j.

We call the set of orderings with the same maximized likelihood by Theorem 1 and 2, the equivalent set. The first step in interpreting the outcome of the above analysis is to eliminate orderings which conflict with subject-matter knowledge about order. We will assume that all infeasible orderings have been eliminated from the equivalent sets.

4.3.4   Partial orderings

An equivalent set can be viewed as a partial ordering of the variables. Any set Θ of orderings, and in particular an equivalent set, defines a partial order Γ as follows: for an ordered pair (x, y), $(x, y) \in \Gamma \Leftrightarrow (x, y) \in \phi$ for every ordering $\phi \in \Theta$ (Dushnik and Miller, 1941). We can thus interpret the equivalent set with the lowest value of $R(\cdot)$ given by (4.7) as defining the partial order of the variables which is the most consistent with the data and assumptions.

There may be multiple equivalent sets which are accepted by the test (4.7). These may be entertained as competing explanations of the data and judged individually on their scientific merit. Alternatively, we may pool a number of equivalent sets strongly supported by the data. This expanded set of orderings will determine a partial order in the same fashion as a single equivalent set, by intersecting ordered pairs according to Dushnik and Miller's definition. To guide the pooling of equivalent sets, we may use a hypothesis test based on equation (4.7) to form a test-based confidence set which is a union of equivalent sets. That is, for each order ϕ, use the null distribution (4.7) to accept or reject ϕ at level α. Then the set composed of all accepted equivalent sets will have confidence level $1 - \alpha$, since P(true order in the confidence set) = P(true order is not rejected by (4.7)) = 1 - P(order rejected | order is true) = 1 - α. When the true ordering is in the confidence set, if every member of that set shares an ordered pair (x, y), then the true order must possess that property. As a result, every pair in the resulting partial order agrees with the underlying true ordering at confidence level $1 - \alpha$.

As always, the confidence level can be misleading if the assumptions of the model are false. Suppose that (4.4) does not hold. Then there will be no orders that fit well and the confidence set will be relatively empty, which tends to result in relatively many ordered pairs being inferred. The extreme case is when a single ill-fitting order is accepted, in which case the complete order would be inferred - a very strong statement. A universal lack of fit results in anti-conservatism.

We note that a set of orderings can also imply non-binary relationships. For example, consider variables A, B, and C, along with the condition $A \perp B \mid C$. An equivalent set for that condition is $\{CAB, CBA, BCA, ACB\}$. The partial order that is the union of these four orderings is null — there is no common ordered pair. However, selecting that equivalent set does rule out the orderings ABC and BAC. This tells us that we should not consider models for which C is preceded by *both* A and B, which is not expressible as a binary relation.

4.3.5 A Monte Carlo method

Our method involves permutation and hence exponential computational complexity. Thus, to evaluate posterior probabilities of arbitrary properties of the underlying linear order, we derive an alternative Bayesian importance sampling method. Let $f(\mathbf{x}; \beta|\phi_i)$ be the restricted likelihood function for a given linear order ϕ_i where restrictions express assumptions as described in section 4.3. We denote $\hat{f}(\mathbf{x}; \beta|\phi_i)$ as the maximized estimate of the function. If we sample linear orders $\phi_i, i = 1, 2, \cdots, M$ from a prior distribution $p(\phi)$, then for any function $g(\phi)$, we prove in Yao (1994) the following convergence property,

Theorem 3 *Under regularity conditions such that the MLE of β is strongly consistent, as the number of observations n and M go to infinity,*

$$[\sum_{i=1}^{M} \hat{f}(\mathbf{x}; \beta|\phi_i)]^{-1} \sum_{i=1}^{M} g(\phi_i)\hat{f}(\mathbf{x}; \beta|\phi_i) \xrightarrow{a.s.} E_{\phi|\mathbf{x}}[g(\phi)], \qquad (4.8)$$

where $E_{\phi|\mathbf{x}}[\cdot]$ is an expectation over the posterior distribution of the orderings.

Let $g(\phi)$ be an indicator function for any given pair of variables, eg. for the pair X and Y, we define

$$g_{xy}(\phi) = \begin{cases} 1, & if \quad X \prec Y \in \phi \\ 0, & o.w. \end{cases} \qquad (4.9)$$

Then $E_{\phi|\mathbf{x}}[g_{xy}(\phi)]$ measures the posterior probability of "$X \prec Y$ implied in the ordering ϕ" under the restrictions. Geweke (1989) gives the importance sampling error of the estimated posterior probability:

$$\sqrt{\sum_{i=1}^{m}\{g_{xy}(\phi_i) - E_{\phi|\mathbf{x}}[g(\phi)]\}^2 w_i^2 / \sum_{i=1}^{m} w_i^2}, \qquad (4.10)$$

where $w_i = \hat{f}(\mathbf{x}; \beta, \phi_i)$.

4.4 AN EXAMPLE

Spirtes et al. (1993) analyze a data set describing the causes of publishing productivity. The variables, listed in temporal order, are *Sex*, *Ability* (based on undergraduate experience), *Gpq* (quality of graduate program), *Fjq* (quality of first job), *Pubs* (number of publications), and *Cites* (number of citations). The sample consisted of 76 female and 86 male academic psychologists. We will uncover features of the underlying temporal order using some reasonable assumptions. We assume that *Sex* is exogenous and is not directly associated with *Ability*, *Gpq*, or *Cites*. We leave open the possibility that there is sex discrimination in employment or refereeing. For these three restrictions on the linear recursive coefficients, we list the equivalent sets not rejected at significance level .2 with their goodness-of-fit statistics (on 3 degrees of freedom), using an obvious shorthand.

1)	$SAFGPC$	R = 3.47
2)	$SAPGFC$	R = 3.66
3)	$SGFAPC$	R = 3.98
4)	$SGAFPC, SGAPFC$	R = 4.03
5)	$SAPCGF, SAPGCF$	R = 4.09
6)	$SFAGPC, SFGAPC$	R = 4.12
7)	$SAFPGC, SAPFGC, SAPFCG, SAFPCG$	R = 4.14
8)	$SAGPCF, SGAPCF$	R = 4.45
9)	$SAPCFG$	R = 4.49

Recursive linear models appear to provide a reasonable fit to the data. The 80 percent confidence set of orderings defines a partial order with $Pubs \prec Cites$, $Ability \prec Pubs$, and $Ability \prec Cites$, in agreement with the temporal order.

The Monte Carlo method of section 4.3.5 was applied, using the same assumptions as above and equal prior probability for all orders with Sex occurring first, zero otherwise. The posterior probabilities of the correct precedence relations were: $Ability \prec Gpq$, .63; $Ability \prec Fjq$, .63; $Ability \prec Pubs$, .84; $Ability \prec Cites$, .74; $Gpq \prec Fjq$, .50; $Gpq \prec Pubs$, .60; $Gpq \prec Cites$, .62; $Fjq \prec Pubs$, .59; $Fjq \prec Cites$, .66; $Pubs \prec Cites$, .61.

4.5   CAUSATION

A key assumption attached to the system of equations (4.4) is that

$$\epsilon \sim N(0, \mathbf{D}), \tag{4.11}$$

with $\mathbf{D}$ a full rank diagonal matrix (d_{ii}). We now show how this assumption relates to causation, and to the validity of our method.

Stone (1993) shows that if

$$\epsilon_i \perp X_j \mid \{X_1, \ldots, X_{i-1}\} \backslash X_j, 1 \le j \le i-1 \tag{4.12}$$

in a linear model, then $\beta_{ij \cdot 1 \cdots j-1, j+1 \cdots i-1} \ne 0$ implies that X_j causes X_i, assuming that the temporal or causal order is consistent with the indexing of the variables. For any positive distribution, (4.12) holds for all $j = 1, \ldots, i-1$ if and only if

$$\epsilon_i \perp X_1, \ldots, X_{i-1}. \tag{4.13}$$

This is the usual requirement for the consistency of the least square estimate. It is easy to show that in the recursive system (4.4), the assumption (4.11) is sufficient to establish (4.13) and thus (4.12) for $i = 1, \ldots, p$. Thus the usual assumption about error structure implies that causation can be tested by examining the regression coefficients when the linear model is correct and the variables are ordered correctly.

The validity of our proposed method rests on the assumption that if subject-matter knowledge specifies that $X_i \perp X_j \mid An(X_i, X_j)$, i.e. they are not directly related in the underlying DAG, then $X_i \perp X_j \mid pred(X_i, X_j)$ in the correctly ordered recursive model. Note that $pred(X_i, X_j)$ consists of variables actually observed and may be a proper subset of $An(X_i, X_j)$ since the underlying system could include unobserved ancestors. The following theorem establishes the validity of the method.

Theorem 4 *Given the diagonal covariance structure (4.11), $X_i \perp X_j \mid An(X_i, X_j)$ implies $X_i \perp X_j \mid pred(X_i, X_j)$ in the model (4.4).*

Proof: For an assumption $X_i \perp X_j | An(X_i, X_j)$, arbitrarily assume that X_j precedes X_i and that the variables are ordered correctly as indexed, so $X_j \in pred(X_i) = \{X_1, \ldots, X_{i-1}\} \subset An(X_i) = An(X_i, X_j)$. The linear model specifies that $X_i = \beta(Pa(X_i) \cap pred(X_i)) + \epsilon_i$, where ϵ_i is a function of the variables in $Pa(X_i) \backslash pred(X_i)$, and $Pa(X_i)$ is the set of parents of X_i, observed or unobserved. Assume that $X_i \perp X_j | An(X_i)$ holds, and thus also $X_i \perp X_j | Pa(X_i)$. Since $X_i \perp pred(X_i) \backslash \{Pa(X_i) \cup X_j\} | Pa(X_i)$ by the Markov assumption, $X_i \perp X_j | Pa(X_i)$ implies that $X_i \perp X_j | \{Pa(X_i) \cup pred(X_i) \backslash Pa(X_i)\} \backslash X_j$, which can be written as $X_i \perp X_j | pred(X_i) \backslash X_j \cup Pa(X_i) \backslash pred(X_i)$ or $X_i \perp X_j | pred(X_i) \backslash X_j, \epsilon_i$. Since the independence of errors (4.11) implies (4.12), $X_i \perp X_j | pred(X_i) \backslash X_j$.

We stress that the inferences provided by our method are about precedence, not causation. The statement $A \prec B$ is consistent with A preceding B in the order but not necessarily directly causing B. The role of our method is in contributing information about the underlying recursive ordering. Further modeling is needed to investigate causal relationships. Our method can also be used to judge causal models generated by the algorithms of Pearl and Verma, and Spirtes, et al., since it relies on different assumptions.

Thus the usual assumptions about the error structure of the linear recursive system validate our method, and also allow non-zero regression coefficients to indicate causation given the true ordering. This is not to say that causal inferences may be made routinely. Rather, it emphasizes what an extremely strong assumption (4.11) is. It amounts to assuming that all relevant variables have been observed, so there is no confounding. For observational data the domain being studied must be very well-understood before we can assume that all relevant factors are being modeled and can make the symmetric independence assumptions required for our method. Clearly, these methods are exploratory and do not constitute an alternative to randomized experiments.

REFERENCES

Dushnik, B. and Miller, E.W. (1941) Partially ordered sets. *American Journal of Mathematics*, **63**, 600-610.

Geweke, J. (1989) Bayesian inference in econometric models using Monte Carlo integration. *Econometrica*, **57**, 1317-1339.

Kiiveri, H. and Speed, T. (1982) Structural analysis of multivariate data: A review. *Social Methodology* (ed. Leinhardt, S.). Jossey-Bass, San Francisco.

Lauritzen, S. L., Dawid, A. P., Larsen, B. N. and Leimer, H. G. (1990) Independence properties of directed Markov fields. *Networks*, **20**, 491-505.

Madigan, D. and Raftery, A. E. (1994) Model selection and accounting for model uncertainty in graphical models using Occam's Window. *J. Am. Statist. Ass.*, **89**, 1535-1546.

Pearl, J. (1988) *Probabilistic Reasoning in Intelligent Systems*. Morgan and Kaufmann, San Mateo, CA.

Pearl, J. and Verma, T. (1991) A theory of inferred causation. Principles of Knowledge Representation and reasoning: *Proceedings of the Second International Conference*, Morgan Kaufmann, San Mateo, CA.

Spirtes, P , Glymour, C, and Scheines, R. (1993) Causation, prediction, and search. *Lecture notes in statistics*, **81**. Springer-Verlag, New York.

Stone, R. (1993) The assumptions on which causal inferences rest. *J. Roy. Statist. Soc.* B, **55**, 455-466.

Verma, T.S. and Pearl, J. (1990) Equivalence and synthesis of causal models. In *Proceedings of the Conference on Uncertainty in AI*, Cambridge, MA., July, 1990.

Wermuth, N. (1980) Linear recursive equations, covariance selection and path analysis. *J. Am. Statist. Ass.*, **75**, 963-972.

Wermuth, N. and Lauritzen, S. (1983) Graphical and recursive models for contingency tables. *Biometrika* **72**, 537-552.

Whittaker, J. (1990) *Graphical Models in Applied Multivariate Statistics*. Wiley, New York.

Yao, Q. (1994) Inference about causal ordering. Unpublished Ph.D. dissertation, University of Toronto, Dept. of Preventive Medicine and Biostatistics.

Part II

Inference and Decision Making

5
Ploxoma: Testbed for Uncertain Inference

H. Blau

Department of Computer Science
University of Rochester
Rochester, New York 14627-0226
blau@cs.rochester.edu

ABSTRACT This paper compares two formalisms for uncertain inference, Kyburg's Combinatorial Semantics and Dempster-Shafer belief function theory, on the basis of an example from the domain of medical diagnosis. I review Shafer's example about the imaginary disease *ploxoma* and show how it would be represented in Combinatorial Semantics. I conclude that belief function theory has a qualitative advantage because it offers greater flexibility of expression, and provides results about more specific classes of patients. Nevertheless, a quantitative comparison reveals that the inferences sanctioned by Combinatorial Semantics are more reliable than those of belief function theory.

5.1   Introduction

This paper compares Kyburg's Combinatorial Semantics (CS) and Dempster-Shafer belief function theory (BFT), two formalisms for representing uncertain inference from statistical data. I reconsider a medical diagnosis example introduced by Shafer [Shafer82] concerning the imaginary disease *ploxoma*. In Section 2, I cite Shafer's example and sketch the steps of its treatment using BFT. Section 3 presents a précis of Combinatorial Semantics[2], and Section 4 shows how the ploxoma example can be handled in the CS framework. Section 5 draws conclusions from this comparison. One of the operations applied in the BFT analysis cannot be translated into CS, and CS groups together two classes of patients that BFT can distinguish. I compute new belief functions after eliminating from the example those elements that are problematic for CS. While BFT appears on qualitative grounds to be more flexible and informative, a quantitative comparison shows that the conclusions sanctioned by CS are more reliable. BFT's results may not conform to frequencies observed in the real world; those of CS do.

Many authors, including [Hunter87, Lemmer86, Pearl90], have observed that belief functions lack a simple frequency interpretation. The contribution of this paper is a comparison between BFT and CS, discussed here for the first time. In the tradition of Smets [Smets91, Smets94], I center my investigation on a concrete instance chosen not as a counterexample, nor as a philosophical conundrum like the three prisoner problem [Diaconis78, DiaconisZabell86], but as a basis for contrasting alternative analyses. A good example is a valuable tool to isolate the points of difference among theories, and to identify their strengths and weaknesses. Shafer's ploxoma story suits this purpose well because it is

---

[1] *Learning from Data: AI and Statistics V.* Edited by D. Fisher and H.-J. Lenz. ©1996 Springer-Verlag.
[2] Combinatorial Semantics [Kyburg95] provides the semantic underpinnings for Evidential Probability [Kyburg83].

fairly realistic and has a rich structure that calls for a variety of belief function operations.

5.2   Shafer's Ploxoma Example

Shafer [Shafer82, p. 331] uses the example of an imaginary disease *ploxoma* to illustrate how belief functions could be used to model uncertainty in medical diagnosis.

> Imagine a disorder called "ploxoma" which comprises two distinct "diseases": θ_1 = "virulent ploxoma", which is invariably fatal, and θ_2 = "ordinary ploxoma", which varies in severity and can be treated. Virulent ploxoma can be identified unequivocally at the time of a victim's death, but the only way to distinguish between the two diseases in their early stages seems to be a blood test with three possible outcomes, labelled x_1, x_2 and x_3. The following evidence is available:
>
> (i)   Blood tests of a large number of patients dying of virulent ploxoma showed the outcomes x_1, x_2 and x_3 occurring 20, 20 and 60 per cent of the time, respectively.
>
> (ii)   A study of patients whose ploxoma had continued so long as to be almost certainly ordinary ploxoma showed outcome x_1 to occur 85 per cent of the time and outcomes x_2 and x_3 to occur 15 per cent of the time. (The study was made before methods for distinguishing between x_2 and x_3 were perfected.) There is some question whether the patients in the study represent a fair sample of the population of ordinary ploxoma victims, but experts feel fairly confident (say 75 per cent) that the criteria by which patients were selected for the study should not affect the distribution of test outcomes.
>
> (iii)   It seems that most people who seek medical help for ploxoma are suffering from ordinary ploxoma. There have been no careful statistical studies, but physicians are convinced that only 5–15 per cent of ploxoma patients suffer from virulent ploxoma.

The three items of evidence are represented by three belief functions (BEL_1, BEL_2, BEL_3) whose frame of discernment is the cross product of disease and blood test result: $\Lambda = \{\theta_1, \theta_2\} \times \{x_1, x_2, x_3\}$. The study of virulent ploxoma can be represented by a belief function BEL_α defined over $\Omega = \{(\theta_1, x_1), (\theta_1, x_2), (\theta_1, x_3)\}$. But we need a belief function defined over all of Λ, not just a subset of Λ. We *conditionally embed* BEL_α in Λ to obtain BEL_1. The study of ordinary ploxoma can be represented by a belief function BEL_β defined over $\Omega = \{(\theta_2, x_1), (\theta_2, x_2), (\theta_2, x_3)\}$. To capture the experts' misgivings about the sample selection in this study, we *discount* BEL_β at the rate of 0.25, yielding BEL_γ. We conditionally embed BEL_γ in Λ to obtain BEL_2. The third item of information can be represented by a belief function BEL_δ defined over $\Omega = \{\theta_1, \theta_2\}$. We *minimally extend* BEL_δ to obtain BEL_3 defined over Λ. The belief functions for the three pieces of evidence are then combined using Dempster's rule of combination, yielding BEL_{123}. The last step is to condition BEL_{123} on the results of the patient's blood test (Table 5.1, corresponds to Shafer's Table 3 [Shafer82, p. 332]). Complete details of this derivation may be found in [BlauKyburg94].

	$BEL_{123}(\{\theta_1\}\|x_i)$	$BEL_{123}(\{\theta_2\}\|x_i)$	$BEL_{123}(\{\theta_1,\theta_2\}\|x_i)$
x_1	0.014	0.965	0.021
x_2	0.062	0.918	0.020
x_3	0.165	0.782	0.053

TABLE 5.1. BEL_{123} conditioned on results of blood test

5.3   Précis of Combinatorial Semantics

When we have data that are not entirely conclusive, but that support a particular hypothesis to some degree, we may choose to accept the the hypothesis, provided the uncertainty does not seem unacceptably high. Kyburg's Combinatorial Semantics [Kyburg95] seeks to formalize inference from a set of sentences Γ to a sentence ϕ, where Γ does not *entail* ϕ, but to some extent *justifies* ϕ.

5.3.1   Syntax of the Language

Kyburg's logic is two-sorted. The *empirical* sort is used to describe the domain of discourse. The *mathematical* sort is intended to represent real numbers. Predicate symbols, function symbols, and variables for the empirical part of the language are as in standard first order logic. The mathematical sort includes mathematical variables, the constants 0 and 1, and the standard operations and relations for real numbers. Kyburg introduces a special variable binding operator, denoted by '%', that relates formulas and real numbers. '%(ψ,ϕ,p,q)' is a well-formed formula if

1. ψ and ϕ are open formulas having no free mathematical variables, and the free empirical variables of ϕ include all those of ψ;

2. ψ belongs to the class of target formulas and ϕ to the class of reference formulas (see [Kyburg95, Section 3] for details);

3. p and q are mathematical variables or rigid real number designators.[3]

This type of formula will be called *statistical*, and we will refer to '%' as the *statistical operator*. This sentence means: among those objects in the empirical domain that satisfy ϕ, the proportion of objects that also satisfy ψ lies between p and q, inclusive. For example, the formula '%$(\texttt{practices\_sports}(x),\texttt{resident\_of}(x,\texttt{Colorado}),0.70,0.80)$' says that somewhere between 70% and 80% of Colorado residents practice sports.

5.3.2   Semantics for the Language

A model $\mathcal{M}$ for this language is a tuple $\langle D_m, D_e, \xi_m, \xi_e \rangle$ where D_m and ξ_m (D_e and ξ_e, respectively) are the domain and interpretation function for the mathematical (empirical) part of the language. The empirical domain D_e is constrained to be *finite* so that the "proportion" involved in the semantics of the statistical operator is well-defined.

---

[3]A rigid designator is one which designates the same object under every interpretation of the language.

The definition of truth for statistical statements involves a new concept, the *satisfaction set* of a formula ϕ with respect to a model $\mathcal{M}$: $SS_{\mathcal{M}}(\phi)$. If ϕ is a closed formula, then $SS_{\mathcal{M}}(\phi) = D_e$ iff ϕ is true in $\mathcal{M}$, and $SS_{\mathcal{M}}(\phi) = \emptyset$ otherwise. If ϕ is an open formula having n free empirical variables and no free mathematical variables, the satisfaction set of ϕ is the subset of D_e^n for which $\mathcal{M}$ makes ϕ true. $SS_{\mathcal{M}}(\phi)$ is undefined when ϕ has free mathematical variables. If ϕ consists of an n-ary empirical predicate symbol 'pred' followed by n empirical variables, then $SS_{\mathcal{M}}(\phi)$ is simply $\xi_e(\text{'pred'})$. For example, $SS_{\mathcal{M}}(\text{'plays\_sport}(p, s)\text{'})$ is the set of pairs $\langle p, s \rangle$ of elements in the empirical domain of $\mathcal{M}$ such that person p plays the sport s. But not all the terms following the predicate symbol need be variables. $SS_{\mathcal{M}}(\text{'plays\_sport}(p, \texttt{tennis})\text{'})$ is the set of tennis players. Note that the free variables in ϕ may occur as arguments to function symbols. $SS_{\mathcal{M}}(\text{'plays\_sport}(\texttt{brother\_of}(p, \texttt{john}), \texttt{tennis})\text{'})$ is the set of tennis-playing brothers of John.

A statistical formula '$\%(\psi, \phi, p, q)$' is true with respect to a model $\mathcal{M}$ and a variable assignment v exactly when (i) $SS_{\mathcal{M}}(\phi)$ is non-empty, and (ii) $p \leq \frac{|SS_{\mathcal{M}}(\phi \wedge \psi)|}{|SS_{\mathcal{M}}(\phi)|} \leq q$. If either condition fails to hold, the statistical formula is false in $\mathcal{M}$. Since the empirical domain D_e is constrained to be finite, $SS_{\mathcal{M}}(\phi)$ and $SS_{\mathcal{M}}(\phi \wedge \psi)$ are likewise finite and the ratio in condition (ii) is well-defined (so long as $SS_{\mathcal{M}}(\phi)$ is non-empty).

5.3.3 Partial Entailment

The criterion of validity for *deductive* inference is: Γ entails ϕ if and only if every model that makes the sentences of Γ true also makes ϕ true. This criterion is too strong for *uncertain* inference. Kyburg proposes the criterion of *partial entailment*, which captures the idea that the inference from Γ to ϕ is justified (although not valid) if ϕ is true in some proportion (almost all) of the models in which Γ is true. We say that Γ *partially entails, to the degree* $[p, q]$, the formula ϕ (written $\Gamma \models_{[p,q]} \phi$) if, among the models that make Γ true, the proportion that also make ϕ true lies between p and q, inclusive.[4] The proportion is well-defined: there can only be a finite number of models for the empirical part of the language, because the language itself is finite and the empirical domain is finite.

What is the relation between partial entailment and uncertain inference? If $\Gamma \models_{[p,q]} \phi$ and p is sufficiently close to 1, then we regard the inference from Γ to ϕ as warranted. Γ justifies the acceptance of ϕ. How do we decide if p is "sufficiently close" to 1? That depends on the situation: we set the threshold of acceptance at a level appropriate to the context in which we are reasoning.

5.3.4 Statistical Formulas and Partial Entailment

Suppose we know that between 70% and 90% of all lap swimmers wear swim goggles, and Jane is a lap swimmer.

$$\Gamma_1 = \left\{ \begin{array}{l} \%(\texttt{wears\_goggles}(x), \texttt{lap\_swimmer}(x), 0.70, 0.90) \\ \texttt{lap\_swimmer}(\texttt{jane}) \end{array} \right\}$$

---

[4]The idea of partial entailment explained here is a simplification: the full definition of this concept is much more complex (see [Kyburg95, Section 5]). But this simplified version will be adequate for our purposes.

What can we conclude about whether Jane wears swim goggles? Theorem 1 tells us that Γ_1 partially entails, to the degree [0.70,0.90], the sentence 'wears_goggles(jane)'. We may or may not want to accept this sentence based on our knowledge of the two premises. It depends whether we think 0.70 is sufficiently close to 1 for us to consider the inference warranted.

Theorem 1 $\left\{ \begin{array}{l} \text{'}\%(\text{A}(x), \text{B}(x), p, q)\text{'} \\ \text{'B(a)'} \end{array} \right\} \models_{[p,q]} \text{'A(a)'}$

Returning to the example, consider a model $M = \langle D_m, D_e, \xi_m, \xi_e \rangle$ that makes Γ_1 true. Suppose there are 100 lap swimmers in D_e: that is, $|SS_M(\text{'lap\_swimmer}(x)\text{'})| = 100$. If the statistical premise is true in M, then there must at least 70 and up to 90 of the lap swimmers who are in $SS_M(\text{'wears\_goggles}(x)\text{'})$. The empirical interpretation function ξ_e maps the constant symbol 'jane' to one of the 100 individuals in $SS_M(\text{'lap\_swimmer}(x)\text{'})$, but it could be any one of them. So there are 100 possible choices for $\xi_e(\text{'jane'})$, of which at least 70 and at most 90 are in $SS_M(\text{'wears\_goggles}(x)\text{'})$. Among the models that make the two premises true, the proportion that also make the conclusion 'wears_goggles(jane)' true lies between 0.70 and 0.90. The same argument applies regardless of the size of $SS_M(\text{'lap\_swimmer}(x)\text{'})$. So we do not have to know the size of D_e, so long as we know that it is finite.

It turns out that lap swimmers who prefer the backstroke do not need swim goggles as much as other lap swimmers do. So among backstroke lap swimmers, only 50% to 60% wear swim goggles. Furthermore, we learn that Jane is a lap swimmer who prefers the backstroke. Our database of knowledge now contains:

$$\Gamma_2 = \left\{ \begin{array}{l} \%(\text{wears\_goggles}(x), \text{lap\_swimmer}(x), 0.70, 0.90) \\ \%(\text{wears\_goggles}(x), \text{lap\_swimmer}(x) \land \text{backstroke\_swimmer}(x), 0.50, 0.60) \\ \text{lap\_swimmer(jane)} \\ \text{backstroke\_swimmer(jane)} \end{array} \right\}$$

Of course, we know that all backstroke lap swimmers are lap swimmers.

$$\forall x((\text{lap\_swimmer}(x) \land \text{backstroke\_swimmer}(x)) \rightarrow \text{lap\_swimmer}(x))$$

What do we now conclude about whether Jane wears swim goggles? We have conflicting statistics for lap swimmers in general and for lap swimmers who prefer the backstroke. But since we know that Jane herself prefers the backstroke, we select the more specific reference class instead of the broader one. Theorem 2 tells us that Γ_2 partially entails, to the degree [0.50,0.60], the sentence 'wears_goggles(jane)'.

Theorem 2, the principle of specificity, says that when we have conflicting statistical information, we guide our judgment by the most specific reference class. What is meant by "conflicting" statistical information? We say that two intervals $[p, q]$ and $[r, s]$ *differ* when neither is included in the other (they may partially overlap, or they may be disjoint).

Theorem 2 If $\text{DIF}([p, q], [r, s])$, then $\left\{ \begin{array}{ll} \text{'}\%(\text{A}(x), \text{B}(x), p, q)\text{'} & \text{'B(a)'} \\ \text{'}\%(\text{A}(x), \text{C}(x), r, s)\text{'} & \text{'C(a)'} \\ \text{'}\forall x(\text{C}(x) \rightarrow \text{B}(x))\text{'} \end{array} \right\} \models_{[r,s]} \text{'A(a)'}$

The case of two statistical statements in which the interval of one is wholly included in the interval of the other falls under a different principle, that of strength [Kyburg95, Section 8][Kyburg83, pp. 380–381]. The principle of strength requires a more complex definition of partial entailment than the one given above. I have summarized here only the concepts of CS that are necessary for the analysis of the ploxoma example.

5.4 Combinatorial Semantics Analysis of Ploxoma Example

To compute interval estimates, we need to know how many patients participated in the epidemiological studies. Suppose we were told that the study of virulent ploxoma involved a total of 200 patients, of whom 40 had blood test result x_1, 40 had test result x_2, and 120 had test result x_3. The study of ordinary ploxoma involved a total of 1000 patients, of whom 850 had test result x_1, and 150 had test result x_2 or x_3. Since ordinary ploxoma is more common than virulent ploxoma, it is not surprising that the second study had a much larger sample size than the first. From these data we calculate intervals at the 99% confidence level, which we then incorporate into the statistical formulas (1)–(5).

(1) $\%(\mathtt{blood\_x1}(y), \mathtt{virulent}(y), 0.137, 0.282)$
(2) $\%(\mathtt{blood\_x2}(y), \mathtt{virulent}(y), 0.137, 0.282)$
(3) $\%(\mathtt{blood\_x3}(y), \mathtt{virulent}(y), 0.509, 0.685)$
(4) $\%(\mathtt{blood\_x1}(y), \mathtt{ordinary}(y), 0.819, 0.877)$
(5) $\%((\mathtt{blood\_x2}(y) \lor \mathtt{blood\_x3}(y)), \mathtt{ordinary}(y), 0.123, 0.181)$

Formulas (6) and (7) capture the opinion of doctors concerning the prevalence of virulent ploxoma.

(6) $\%(\mathtt{virulent}(y), \mathtt{ploxoma}(y), 0.05, 0.15)$
(7) $\%(\mathtt{ordinary}(y), \mathtt{ploxoma}(y), 0.85, 0.95)$

What about the complication that scientists are only (as Shafer puts it) 75% sure that the results of the second ploxoma study are reliable? There is no good way to represent this information in the CS framework. One could consider two ways to take this aspect into account: (i) broaden the intervals in formulas (4) and (5), to indicate greater ignorance about the true proportion of blood test result x_1 (x_2, x_3) among ordinary ploxoma patients, or (ii) compute intervals at a lower confidence level. The problem with (i) is: by how much do we broaden the interval? What principle would sanction the choice of this number? The problem with (ii) is: the confidence level associated with an interval estimate does not represent a judgment of how well the experiment was designed. The 75% confidence interval would in fact be narrower than the 99% interval. All in all, it seems wise to refrain from adjusting our confidence intervals in ways that have no foundation in statistical theory. We will therefore proceed as if we had no doubts about the quality of the data from the second ploxoma study.

John walks into the doctor's office, suffering from ploxoma.[5] His blood test result is x_1. We cannot infer anything about John from the statistical statements (1)–(5), because we do not know whether John belongs to the reference class of virulent or of ordinary ploxoma patients: indeed, this is precisely what we want to find out. Obviously we need to turn these statistics around to get an estimate of the incidence of virulent ploxoma

---

[5]Shafer assumes that we already know the patient is suffering from ploxoma; the task is to estimate the risk of virulent ploxoma based on the result of the blood test. Apparently there is no danger of confusing ploxoma with some other disease.

among ploxoma patients with test result x_1. Bayes' Rule allows us to do this.

$$Pr(\theta_1|x_1) \;=\; \frac{Pr(\theta_1)Pr(x_1|\theta_1)}{Pr(\theta_1)Pr(x_1|\theta_1) + Pr(\theta_2)Pr(x_1|\theta_2)}$$

In this equation we consider only ploxoma patients, so θ_1 (virulent) and θ_2 (ordinary) together partition the sample space. We are interested in an interval, not a point estimate, for $Pr(\theta_1|x_1)$. To calculate the lower bound on this interval, we minimize the numerator of the fraction by taking the lower bound for $Pr(\theta_1)$ and $Pr(x_1|\theta_1)$, which forces us to take the upper bound for $Pr(\theta_2)$ and $Pr(x_1|\theta_2)$. To calculate the upper bound on this interval, we maximize the numerator of the fraction by taking the upper bound for $Pr(\theta_1)$ and $Pr(x_1|\theta_1)$, which forces us to take the lower bound for $Pr(\theta_2)$ and $Pr(x_1|\theta_2)$. Substituting the appropriate values from formulas (1), (4), (6), and (7), we obtain the desired statistical statement.

(8) $\%(\texttt{virulent}(y), (\texttt{ploxoma}(y) \wedge \texttt{blood\_x1}(y)), 0.008, 0.057)$

When we try to do the same with formula (2), we run into difficulty. To calculate $Pr(\theta_1|x_2)$, we would have to know $Pr(x_2|\theta_2)$. But we have only $Pr((x_2 \vee x_3)|\theta_2)$, because the x_2 patients and the x_3 patients were grouped together in the study of ordinary ploxoma. Instead of $Pr(\theta_1|x_2)$, we must content ourselves with $Pr(\theta_1|(x_2 \vee x_3))$. We have to re-examine the data from the first ploxoma study to compute a confidence interval for $Pr((x_2 \vee x_3)|\theta_1)$. Of the 200 virulent ploxoma patients who participated in the study, the combined total for x_2 and x_3 test results is 160. This yields the confidence interval in (9). We can now calculate the upper and lower bounds for $Pr(\theta_1|(x_2 \vee x_3))$, as well as $Pr(\theta_2|x_1)$ and $Pr(\theta_2|(x_2 \vee x_3))$.

(9) $\%((\texttt{blood\_x2}(y) \vee \texttt{blood\_x3}(y)), \texttt{virulent}(y), 0.718, 0.863)$
(10) $\%(\texttt{virulent}(y), (\texttt{ploxoma}(y) \wedge (\texttt{blood\_x2}(y) \vee \texttt{blood\_x3}(y))), 0.173, 0.553)$
(11) $\%(\texttt{ordinary}(y), (\texttt{ploxoma}(y) \wedge \texttt{blood\_x1}(y)), 0.943, 0.992)$
(12) $\%(\texttt{ordinary}(y), (\texttt{ploxoma}(y) \wedge (\texttt{blood\_x2}(y) \vee \texttt{blood\_x3}(y))), 0.447, 0.827)$

Now what can we say about John the ploxoma patient with blood test result x_1? There are two pairs of statistical statements that apply to John: (6) and (7), (8) and (11). By the principle of specificity (Theorem 2), (8) takes precedence over (6), and (11) takes precedence over (7). Our knowledge base partially entails, to the degree [0.008,0.057], the sentence '$\texttt{virulent(john)}$'; it partially entails, to the degree [0.943,0.992], the sentence '$\texttt{ordinary(john)}$'. Depending on our threshold for acceptance, we may want to accept the latter sentence. What about Peter, a ploxoma patient whose blood test result is x_2? Again, there are two pairs of statistical statements that apply to Peter: (6) and (7), (10) and (12). By the principle of specificity, (10) takes precedence over (6), and (12) takes precedence over (7). Our knowledge base partially entails, to the degree [0.173,0.553], the sentence '$\texttt{virulent(peter)}$'; it partially entails, to the degree [0.447,0.827], the sentence '$\texttt{ordinary(peter)}$'. We cannot accept either of these, because in each case the lower bound of the confidence interval does not exceed 0.5. If Peter's test result had been x_3, we would have drawn the same conclusion: we do not have enough information to distinguish between the x_2's and the x_3's when it comes to judging their risk of virulent ploxoma. In sum, a blood test result of x_1 strongly suggests that the patient has ordinary ploxoma, but a result of x_2 or x_3 does not give a clear indication either way.

	$BEL'_{123}(\{\theta_1\}|x_i)$	$BEL'_{123}(\{\theta_2\}|x_i)$	$BEL'_{123}(\{\theta_1,\theta_2\}|x_i)$
$\{x_1\}$	0.016	0.963	0.021
$\{x_2, x_3\}$	0.431	0.521	0.048

TABLE 5.2. BEL'_{123} conditioned on results of blood test

	Belief Functions	Combin. Semant.	Naive Probability
$\theta_1|x_1$	0.016 0.037	0.008 0.057	0.012 0.040
$\theta_1|x_2 \vee x_3$	0.431 0.479	0.173 0.553	0.219 0.485
$\theta_2|x_1$	0.963 0.984	0.943 0.992	0.960 0.988
$\theta_2|x_2 \vee x_3$	0.521 0.569	0.447 0.827	0.515 0.781

TABLE 5.3. Comparison of intervals

5.5   Comparison of the Two Approaches

Having represented the ploxoma example using belief functions and Combinatorial Semantics, we can compare these approaches both qualitatively and quantitatively. Qualitatively, BFT seems to be more flexible. The discounting operation (applied in the derivation of BEL_2) allows us to express the doubts we may have concerning the reliability of the evidence at our disposal. As we have seen, there is no counterpart to this operation in CS. With BFT, we can condition BEL_{123} on x_2 or on x_3 individually, even though the second ploxoma study does not distinguish between them. With CS, we have to merge these two classes of patients. We seem to be losing information that the first study gives us about the difference between x_2 and x_3: result x_3 is more common than x_2 among virulent ploxoma patients.

To make a meaningful quantitative comparison of the two techniques, we must be sure the numbers we compare are based on the same problem description. First, we must recompute BEL_2 without discounting the second item of evidence. This in turn forces us to recompute BEL_{123}, giving us BEL'_{123}. Second, we must condition BEL'_{123} on the set $\{x_2, x_3\}$ instead of the singleton sets $\{x_2\}$ and $\{x_3\}$. The results are shown in Table 5.2. (See [BlauKyburg94, Appendix] for the details of this recalculation.) We now consider the interval [belief,plausibility] for virulent and ordinary ploxoma according to the blood test result. Table 5.3 sets side by side the belief-plausibility intervals and the degrees of partial entailment from CS. In each case, the belief function intervals are properly included in the intervals derived by statistical methods. How shall we interpret this relationship?

Classical probability theory provides a common framework in which to evaluate BFT and CS. Kyburg shows in [Kyburg87] that to any belief function BEL, defined over a frame of discernment Ω, there corresponds a closed, convex set $\mathcal{P}$ of classical probability functions, where

$$\mathcal{P} = \{\, Pr : Pr \text{ is a probability function and } \forall A \subseteq \Omega \ (BEL(A) \leq Pr(A) \leq PL(A)) \,\}.$$

If we understand the belief-plausibility intervals in this way (although Shafer would disapprove), it might seem that they give us more information than CS, because they place tighter bounds on the frequencies that interest us. Unfortunately, in this case the "information" is simply misleading. Consider $Pr(\theta_1|x_2 \vee x_3)$: BEL'_{123} says that this probability is no lower than 0.431 and no higher than 0.479, but the 99% confidence interval extends as low as 0.173 and as high as 0.553. Given the evidence at hand, there is a 1% risk that the true value lies outside [0.173,0.553]. But the risk is greater that the true value lies outside [0.431,0.479]. BEL'_{123} is wrong, and CS right, about the possible worlds in which $Pr(\theta_1|x_2 \vee x_3)$ falls in [0.173,0.431) or in (0.479,0.553]. If we use BFT to guide us in making decisions—for instance, when choosing among various treatment protocols for ploxoma—we will make mistakes more often than if we rely on CS.

Suppose we now ignore the issue of confidence intervals entirely, and accept the observed frequencies of the three blood test results as point estimates for the corresponding probabilities. That is, since 20% of the virulent ploxoma patients tested x_1, we will take $Pr(x_1|\theta_1) = 0.20$, and similarly for the other percentages mentioned in the first and second items of evidence. Doubt remains about the incidence of the virulent form of the disease, which according to doctors strikes 5–15% of all ploxoma patients (the third item of evidence). Again we use Bayes' Rule to calculate probability intervals for $Pr(\theta_1|x_1)$, *etc.*, substituting point estimates for all the conditional probabilities, and the upper and lower limits for $Pr(\theta_1)$ and $Pr(\theta_2)$. The resulting intervals appear in the third column of Table 5.3. As one would expect, the "naive probability" intervals are narrower than CS's confidence intervals, because we have removed the uncertainty arising from the limited sample sizes in the epidemiological studies. However, the naive probability intervals are wide enough to properly include the belief-plausibility intervals.[6] Even if we had huge sample sizes and could get highly accurate estimates of the proportion of {virulent, ordinary} ploxoma patients with test results $\{x_1,x_2,x_3\}$, we still could not attain the level of certainty that BFT purports to give us.

5.6 Where's the Beef?

The preceding discussion supports the judgment that Combinatorial Semantics is superior to belief function theory as a formalism for uncertain inference (if we interpret belief functions in terms of convex sets of classical probability functions). CS may be less flexible in its representation of knowledge, may yield less "informative" conclusions, but at least the inferences it does sanction are not misleading. However, the practical usefulness of CS has yet to be demonstrated. Dempster and Kong [DempsterKong87, p. 33] emphasize that we cannot evaluate a theory of uncertain inference without reference to its practical applications.

> Belief function methodology does introduce more complexity into the class of available representations of uncertainty ... The important question is whether the added flexibility is necessary in practice to permit satisfactory representation of an analyst's state of uncertainty about the real world. We believe that it is literally impossible to answer the question outside the context of real examples based on attempts to construct formal representations of uncertainty

---

[6]This comes as no surprise in light of Theorem A.3 in [Kyburg87, pp. 286–287], which states that the lower and upper bounds on the conditional probability derived using Bayes' Rule are always as wide (and often wider) than the belief-plausibility intervals resulting from Dempster conditioning. This point is mentioned as early as [Dempster67].

reflecting actual uncertain knowledge of the real world.

I agree with Dempster and Kong that the true test of a formalism for uncertain reasoning is its application to real problems in the real world. Loui's experiment in predicting computer network usage [Loui90] is the only such application of Combinatorial Semantics reported to date.

5.7 Acknowledgements

I thank my advisor, Professor H. E. Kyburg, for his guidance in this research, and in particular for his patient explanations of Combinatorial Semantics. I am indebted to I. Macarie and C. M. Teng for their helpful comments, and to B. Murtezaoğlu for the use of his program to calculate the normal approximation of the binomial confidence interval. I am also grateful to Professors L. Schubert and N. Jochnowitz for their unfailing confidence in my work.

5.8 REFERENCES

[BlauKyburg94] H. Blau and H. E. Kyburg, Jr., "Ploxoma: Testbed for Uncertain Inference," Technical Report 537, Department of Computer Science, University of Rochester, December 1994.

[Dempster67] A. P. Dempster, "Upper and Lower Probabilities Induced by a Multivalued Mapping," *Annals of Mathematical Statistics*, 38(2):325–339, April 1967.

[DempsterKong87] A. P. Dempster and A. Kong, "Commentary on Papers by Shafer, Lindley, and Spiegelhalter," *Statistical Science*, 2(1):32–36, 1987.

[Diaconis78] P. Diaconis, "Review of "A Mathematical Theory of Evidence"," *Journal of the American Statistical Association*, 73(363):677–678, September 1978.

[DiaconisZabell86] P. Diaconis and S. Zabell, "Some Alternatives to Bayes's Rule," In B. Grofman and G. Owen, editors, *Information Pooling and Group Decision Making: Proceedings of the Second University of California, Irvine, Conference on Political Economy*, pages 25–38. JAI Press, 1986.

[Hunter87] D. Hunter, "Dempster-Shafer vs. Probabilistic Logic," In *Proceedings Third AAAI Workshop on Uncertainty in Artificial Intelligence*, pages 22–29, University of Washington, Seattle, July 1987.

[Kyburg83] H. E. Kyburg, Jr., "The Reference Class," *Philosophy of Science*, 50:374–397, 1983.

[Kyburg87] H. E. Kyburg, Jr., "Bayesian and Non-Bayesian Evidential Updating," *Artificial Intelligence*, 31:271–293, March 1987.

[Kyburg95] H. E. Kyburg, Jr., "Combinatorial Semantics," Technical Report 563, Department of Computer Science, University of Rochester, January 1995.

[Lemmer86] J. F. Lemmer, "Confidence Factors, Empiricism and the Dempster-Shafer Theory of Evidence," In L. N. Kanal and J. F. Lemmer, editors, *Uncertainty in Artificial Intelligence*, pages 117–125. North-Holland, 1986.

[Loui90] R. P. Loui, "Evidential Reasoning Compared in a Network Usage Prediciton Testbed: Preliminary Report," In R. D. Shachter, T. S. Levitt, L. N. Kanal, and J. F. Lemmer, editors, *Uncertainty in Artificial Intelligence 4*, pages 253–269. North-Holland, 1990.

[Pearl90] J. Pearl, "Reasoning with Belief Functions: An Analysis of Compatibility," *International Journal of Approximate Reasoning*, 4(5/6):363–389, September/November 1990.

[Shafer82] G. Shafer, "Belief Functions and Parametric Models," *Journal of the Royal Statistical Society, Series B (Methodological)*, 44(3):322–352, 1982.

[Smets91] P. Smets, "About Updating," In B. D. D'Ambrosio, P. Smets, and P. P. Bonissone, editors, *Proceedings of the 7th Conference on Uncertainty in Artificial Intelligence*, pages 378–385. Morgan Kaufmann, 1991.

[Smets94] P. Smets, "What is Dempster-Shafer's Model?," In R. R. Yager, J. Kacprzyk, and M. Fedrizzi, editors, *Advances in the Dempster-Shafer Theory of Evidence*, chapter 1, pages 5–34. John Wiley & Sons, 1994.

6

Solving Influence Diagrams Using Gibbs Sampling

Ali Jenzarli

College of Business
Box 15F
The University of Tampa
Tampa, FL 33606-1490, USA
Tel: (813) 253-3333, ext. 3306 Fax: (813) 258-7408
Internet: ali.jenzarli@resnet.fmhi.usf.edu

ABSTRACT   We describe a Monte Carlo method for solving influence diagrams. This method is a combination of stochastic dynamic programming and Gibbs sampling, an iterative Markov chain Monte Carlo method. Our method is especially useful when exact methods for solving influence diagrams fail.

6.1 Introduction

In this paper we describe a Monte Carlo method for solving influence diagrams. This method is a combination of stochastic dynamic programming and Gibbs sampling, an iterative Markov chain Monte Carlo method. Our method is especially useful when exact methods for solving influence diagrams fail. Examples include decision problems that are highly asymmetric (Smith, Holtzman and Matheson, 1993), and problems where we have continuous as well as discrete variables.

We organize this paper as follows. In Section 2 we review influence diagrams (IDs) and their properties. In Section 3 we describe solution algorithms for IDs. And, in Section 4 we show how to use Gibbs sampling to implement stochastic dynamic programming for an influence diagram.

6.2 Influence Diagrams

In this section we review influence diagrams (IDs) and their properties. We describe the assumptions on which the ID decision model is based. And, we illustrate our ideas with an example.

Let us begin by recalling the standard definition of an influence diagram.

Definition 1. An *influence diagram (ID)* is a DAG with variables as nodes, together with a specification for only *some* of the variables of conditionals given their parents. We call those variables with conditionals *chance* variables, and those without *decision* variables. For each chance variable, we specify a set of *possible values*. For each decision variable, we specify a set of admissible values that we call *decision alternatives*. An ID has a special chance node that is a sink and that is a deterministic function of its parents. We call this node the *value* node. Arrows into chance nodes are called *relevance arrows* while arrows into decision nodes are called *informational arrows*.

Figure 1 shows an ID, where chance nodes are circles, decision nodes are rectangles, the value node is a diamond, relevance arrows are solid, and informational arrows are dashed.

Figure 1. An example of an influence diagram

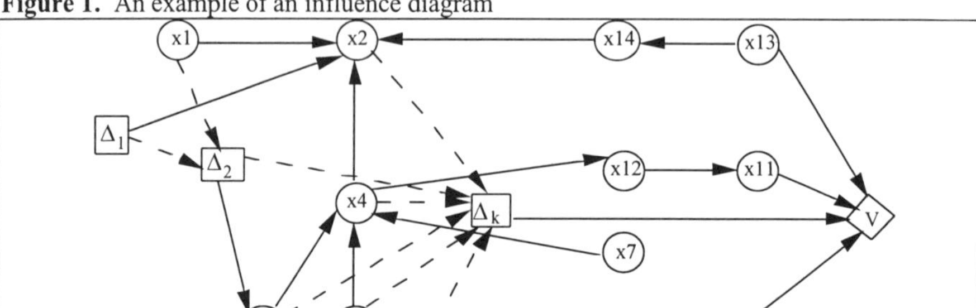

Following Clemen (1991), we distinguish between informational arrows and relevance arrows by representing the former with dashed arrows while still using solid arrows to represent the latter. Notice that decision variables are treated just like chance variables when they are parents. The difference between an informational and a relevance arrow depends on the kind of variable the arrow points to, not on the kinds of variables it comes from.

We follow Howard and Matheson (1981) and most of the ID literature in making two additional assumptions:

1. The decisions are all made by a single decision maker who remembers his or her previous decisions. This assumption is represented by the existence of a path in the DAG consisting only of all the decision variables. In other words, the decision variables are ordered, say $\Delta_1, \Delta_2, ..., \Delta_k$, so that there is an arrow from Δ_i to Δ_{i+1}, for i = 1, 2, ..., k–1. We summarize this by saying that the decision variables are *completely ordered* by the DAG.

2. The decision maker does not forget any previous information as he or she progresses through the decisions. This assumption is represented by the existence of an arrow from X to Δ_j whenever there is an arrow from X to Δ_i and i<j. In other words, each decision node inherits the parents of preceding decision nodes. This is called the *no-forgetting* assumption. Notice that a decision variable need not inherit the parents of chance variables that precede it.

A sink node other than the value node, V, is called a *barren node* (Shachter, 1986). According to Shachter, deleting barren nodes from an ID does not affect the optimal expectation of V, and results in another ID that is still a complete representation of the original decision problem.

Moreover, we observe that deleting all barren nodes from an ID makes V the only sink node in the DAG, and thus every remaining variable now has a directed path to V. Because of the two assumptions above, we can leave out the informational arrows from the DAG of an ID without barren nodes. We can do so only if we draw the DAG of an ID without barren nodes according to a layout that proceeds from left to right, or top to bottom, as follows. We draw chance variables observed before the first decision to the left, or on top, of the first decision node, chance variables observed after one decision and before another decision between the two decision nodes, and chance variables observed after the last decision or never after the last decision node and before the value node V. This layout, as we will see in Section 4, will make the explanation of our Gibbs sampling algorithm much easier. The reader is referred to Figure 2 in Section 4 for the ID of Figure 1 without informational arrows.

6.3 Solving Influence Diagrams

In this section we describe solution algorithms for influence diagrams (IDs) as elaborations of various forms of the principle of optimality in stochastic dynamic programming, which allows us to find the decision functions in problems of this type sequentially (Bellman and Dreyfus, 1962). In particular, we describe the standard version of the principle of optimality in stochastic dynamic programming. However, since this version is not quite adequate for the case of influence diagrams, we introduce a modification that allows us to determine an optimal decision function for each decision variable sequentially.

In its standard form, the principle of optimality in stochastic dynamic programming applies when we want to maximize or minimize the expectation of a real-valued variable V whose joint distribution with k+1 other variables $\Gamma_0, \Gamma_1, ..., \Gamma_k$ (which may each actually be vectors of variables) depends in a stagewise manner on k parameters (which also may be single numbers, vectors or functions) $\delta_1, ..., \delta_k$. More precisely, we assume that we can factor the joint probability for $\Gamma_0, \Gamma_1, ..., \Gamma_k$ and V in the form

$$P_{\delta_1,...,\delta_k}(\Gamma_0,...,\Gamma_k,V) = h_0(\Gamma_0)h_{\delta_1}(\Gamma_1|\Gamma_0)...$$

$$...h_{\delta_{k-1}}(\Gamma_{k-1}|\Gamma_0,...,\Gamma_{k-2})h_{\delta_k}(\Gamma_k,V|\Gamma_0,...,\Gamma_{k-1}), \quad (1)$$

where the factors are conditional probabilities. We must also assume that it is computationally feasible to compute

$$E_{\delta_k}\left(V|\Gamma_0,...,\Gamma_{k-1}\right)$$

from

$$h_{\delta_k}\left(\Gamma_k,V|\Gamma_0,...,\Gamma_{k-1}\right)$$

for each value of δ_k and each configuration of values of $\Gamma_0 \cup ... \cup \Gamma_{k-1}$, or at least to find for each configuration $\left(\gamma_0,...,\gamma_{k-1}\right)$ of $\Gamma_0 \cup ... \cup \Gamma_{k-1}$ the value of δ_k that optimizes

$$E_{\delta_k}\left(V|\Gamma_0 = \gamma_0,...,\Gamma_{k-1} = \gamma_{k-1}\right). \quad (2)$$

Finally, we must assume (this is crucial) that we can find a single value of δ_k that optimizes (2) for <u>all</u> $\left(\gamma_0,...,\gamma_{k-1}\right)$. Since the distribution of $\Gamma_0 \cup ... \cup \Gamma_{k-1}$ does not depend on δ_k, we have

$$E_{\delta_1,...,\delta_k}(V) = E_{\delta_1,...,\delta_{k-1}}\left(E_{\delta_k}\left(V|\Gamma_0,...,\Gamma_{k-1}\right)\right).$$

Therefore, this optimizing value of δ_k will also optimize the unconditional expectation $E_{\delta_1,...,\delta_k}(V)$ for any choice of $\left(\delta_1,...,\delta_{k-1}\right)$. And therefore, it can be extended to a choice of $\left(\delta_1,...,\delta_k\right)$ to optimize this unconditional expectation.

Suppose we fix this optimal value of δ_k eliminating it from our notation, and reducing (1) to

$$h_{\delta_1,...,\delta_{k-1}}(\Gamma_0,...,\Gamma_k,V) = h_0(\Gamma_0)h_{\delta_1}(\Gamma_1|\Gamma_0)...$$

$$...h_{\delta_{k-1}}(\Gamma_{k-1}|\Gamma_0,...,\Gamma_{k-2})h(\Gamma_k,V|\Gamma_0,...,\Gamma_{k-1}). \quad (3)$$

From this point, we proceed in either of two ways. We can sum or integrate Γ_k out of the expectation. Or we can incorporate Γ_k as part of Γ_{k-1}.

The first option, summing or integrating Γ_k out, means reducing (3) to

$$h_{\delta_1,\ldots,\delta_{k-1}}(\Gamma_0,\ldots,\Gamma_{k-1},V) = h_0(\Gamma_0)h_{\delta_1}(\Gamma_1|\Gamma_0)\ldots h'_{\delta_{k-1}}(\Gamma_{k-1},V|\Gamma_0,\ldots,\Gamma_{k-2}),$$

where

$$h'_{\delta_{k-1}}(\Gamma_{k-1},V|\Gamma_0,\ldots,\Gamma_{k-2}) = h_{\delta_{k-1}}(\Gamma_{k-1}|\Gamma_0,\ldots,\Gamma_{k-2})\int h(\gamma_k,V|\Gamma_0,\ldots,\Gamma_{k-1})d\gamma_k.$$

Once again, we assume that we can choose δ_{k-1} so as to optimize simultaneously

$$E_{\delta_{k-1}}(V|\Gamma_0 = \gamma_0, \Gamma_1 = \gamma_1,\ldots,\Gamma_{k-2} = \gamma_{k-2}) \tag{4}$$

for all $(\gamma_0,\ldots,\gamma_{k-2})$. Then, as before, the choice of δ_{k-1} can be extended to a choice of $(\delta_1,\ldots,\delta_{k-1})$ to optimize the unconditional expectation

$$E_{\delta_1,\ldots,\delta_{k-1}}(V).$$

So we may also fix this optimal value of δ_{k-1}, and reduce the problem further. We can continue in this way, choosing the δ_i sequentially, provided that the successive simultaneous optimizations like those in (2), (4), etc. are possible.

The second option means setting $\Gamma'_{k-1} = \Gamma_{k-1} \cup \Gamma_k$, and reducing (3) to

$$h_{\delta_1,\ldots,\delta_{k-1}}(\Gamma_0,\ldots,\Gamma_{k-1},V) = h_0(\Gamma_0)h_{\delta_1}(\Gamma_1|\Gamma_0)\ldots h'_{\delta_{k-1}}(\Gamma'_{k-1},V|\Gamma_0,\ldots,\Gamma_{k-2}),$$

where

$$h'_{\delta_{k-1}}(\Gamma'_{k-1},V|\Gamma_0,\ldots,\Gamma_{k-2}) = h_{\delta_{k-1}}(\Gamma_{k-1}|\Gamma_0,\ldots,\Gamma_{k-2})h(\Gamma_k,V|\Gamma_0,\ldots,\Gamma_{k-1}).$$

Again, if we can choose δ_{k-1} to optimize simultaneously (4) for all $(\gamma_0,\ldots,\gamma_{k-2})$, and so on, we can proceed to choose the δ_i sequentially.

This standard version of stochastic dynamic programming is not quite adequate for the case of influence diagrams. The reason is that although these diagrams involve factorizations that can be written in the form (1), the factors are not necessarily conditional probabilities.

The standard version of stochastic dynamic programming can be modified to fit influence diagrams, but there has been a considerable variety of opinion about how to do this. The oldest sequential solution algorithm for influence diagrams, the Olmsted-Shachter reduction algorithm (Olmsted, 1983; Shachter, 1986) goes considerably beyond stochastic dynamic programming, in order to maintain a representation of the influence diagram form as the algorithm proceeds. More recent algorithms, including the valuation network algorithm of Shenoy (1992 and 1993) and the potential influence diagram algorithm of Ndilikilikesha (1992), stay closer to stochastic dynamic programming.

The simulation algorithm we describe in Section 4 does not fit exactly into either Shenoy's or Ndilikilikesha's framework, primarily because their algorithms integrate Γ_k out, while our algorithm follows the second option described above, that of absorbing Γ_k into $h_{\delta_{k-1}}$. We could elaborate one of their frameworks in order to make our algorithm fit, but it will be simpler for us to deal directly with the necessary modification in the standard form of stochastic dynamic programming that we have just described.

Here is the modification that we require. Let us assume that the joint probability for $\Gamma_0,\Gamma_1,\ldots,\Gamma_k$ and V is proportional to a factorization of the following form

$$P_{\delta_1,\ldots,\delta_k}(\Gamma_0,\ldots,\Gamma_k,V) \propto h_0(\Gamma_0)h_{\delta_1(\Gamma_0)}(\Gamma_1|\Gamma_0)\ldots$$

$$\ldots h_{\delta_{k-1}(\Gamma_0,\ldots,\Gamma_{k-2})}(\Gamma_{k-1}|\Gamma_0,\ldots,\Gamma_{k-2})h_{\delta_k(\Gamma_0,\ldots,\Gamma_{k-1})}(\Gamma_k,V|\Gamma_0,\ldots,\Gamma_{k-1}) \tag{5}$$

Here we do not assume that the factors are conditional probabilities. But we do assume that the δ_i are functions; and we assume, as the notation indicates, that for fixed values of

$\Gamma_0, \Gamma_1, ..., \Gamma_{k-1}$, the factor $h_{\delta_k(\Gamma_0,...,\Gamma_{k-1})}$, regarded as a function of Γ_k and V, depends on δ_k only through the value δ_k assigns to those values of $\Gamma_0, \Gamma_1, ..., \Gamma_{k-1}$. This assumption, as we will see implies that the simultaneous optimizations at each step are possible.

Notice first that the factorization (5) implies that

$$h_{\delta_k(\Gamma_0,...,\Gamma_{k-1})}\left(\Gamma_k, V | \Gamma_0, ..., \Gamma_{k-1}\right),$$

for fixed values of $\Gamma_0, \Gamma_1, ..., \Gamma_{k-1}$, is at least proportional to the conditional probability distribution for Γ_k and V given these values of $\Gamma_0, \Gamma_1, ..., \Gamma_{k-1}$. To see this, recall that a conditional probability distribution is always proportional to the corresponding joint probability distribution. Thus

$$P_{\delta_1,...,\delta_k}\left(\Gamma_k, V | \Gamma_0, ..., \Gamma_{k-1}\right) = \lambda P_{\delta_1,...,\delta_k}\left(\Gamma_0, ... \Gamma_k, V\right), \tag{6}$$

where λ is constant with respect to Γ_k and V. (The other variables are thought of as fixed.) We usually write (6) with a symbol of proportionality:

$$P_{\delta_1,...,\delta_k}\left(\Gamma_k, V | \Gamma_0, ..., \Gamma_{k-1}\right) \propto P_{\delta_1,...,\delta_k}\left(\Gamma_0, ... \Gamma_k, V\right).$$

Since only the last factor of (5) involves Γ_k or V, (6) implies that

$$P_{\delta_1,...,\delta_k}\left(\Gamma_k, V | \Gamma_0, ..., \Gamma_{k-1}\right) \propto h_{\delta_k(\Gamma_0,...,\Gamma_{k-1})}\left(\Gamma_k, V | \Gamma_0, ..., \Gamma_{k-1}\right).$$

Again, this proportionality is to be interpreted by taking both sides as functions of Γ_k and V only, with the other variables fixed; we are able to omit the other factors only because they, as functions of the other variables, are also fixed and hence can be absorbed into the constant of proportionality.

Whenever a function is proportional to a probability distribution (or probability conditional), it contains all the information needed to find that conditional because the constant of proportionality is simply what is needed to make the function sum (or integrate) to one. Thus

$$h_{\delta_k(\gamma_0,...,\gamma_{k-1})}\left(\Gamma_k, V | \Gamma_0 = \gamma_0, ..., \Gamma_{k-1} = \gamma_{k-1}\right)$$

has, in particular, all the information needed to determine the conditional expectation of V given $\left(\gamma_0, ..., \gamma_{k-1}\right)$,

$$E_{\delta_k(\gamma_0,...,\gamma_{k-1})}\left(\Gamma_k, V | \Gamma_0 = \gamma_0, ..., \Gamma_{k-1} = \gamma_{k-1}\right). \tag{7}$$

We can choose the value of $\delta_k\left(\gamma_0, ..., \gamma_{k-1}\right)$ to optimize this expectation, and by doing this for each set of values $\left(\gamma_0, ..., \gamma_{k-1}\right)$, we will have chosen a function δ_k that simultaneously optimizes (7) for all $\left(\gamma_0, ..., \gamma_{k-1}\right)$.

Once this choice of δ_k has been carried out, we can proceed, as before, absorbing $h_{\delta_k(\Gamma_0,...,\Gamma_{k-1})}$ into $h_{\delta_{k-1}(\Gamma_0,...,\Gamma_{k-2})}$, first integrating γ_k out if we wish to do so. This means replacing the factor in the second line of (5) with a factor

$$h'_{\delta_{k-1}(\Gamma_0,...,\Gamma_{k-2})}\left(\Gamma'_{k-1}, V | \Gamma_0, ..., \Gamma_{k-2}\right),$$

where

(1) $\Gamma'_{k-1} = \Gamma_{k-1} \cup \Gamma_k$,

(2) Δ_k, which is in Γ_k, is now interpreted as a chance node, and

(3) $h'_{\delta_{k-1}(\Gamma_0,...,\Gamma_{k-2})}\left(\Gamma'_k, V | \Gamma_0, ..., \Gamma_{k-2}\right) = h_{\delta_{k-1}(\Gamma_0,...,\Gamma_{k-2})}\left(\Gamma_{k-1} | \Gamma_0, ..., \Gamma_{k-2}\right)$

$$\cdot h^*_{\Delta_k}\left(\Delta_k | \Gamma_0, ..., \Gamma_{k-1}\right) h_{\delta_k^*}\left(\Gamma_k, V | \Gamma_0, ..., \Gamma_{k-1}\right),$$

where δ_k^* is the optimal decision function and $h_{\Delta_k}^*$ is the zero-one conditional representing δ_k^*.

In order to fit influence diagrams into this version of stochastic dynamic programming, we write Γ_i for the set of variables consisting of Δ_i together with the chance variables observed by the decision maker between Δ_i and Δ_{i+1}, for $i = 1, ..., k-1$, we write Γ_0 for the chance variables observed before Δ_1 and Γ_k for the set of variables consisting of Δ_k together with the chance variables (other than V) observed after Δ_k (or never), and we write δ_i for the decision function for Δ_i. Then we set $h_0(\Gamma_0)$ equal to the product of conditionals for the chance variables in Γ_0. For $i = 1, ..., k-1$, we set $h_{\delta_i}(\Gamma_i | \Gamma_0, ..., \Gamma_{i-1})$ equal to the product of conditionals for the chance variables in Γ_i, times the conditional corresponding to the decision function δ_i (this conditional gives only probabilities of zero and one). And we similarly set $h_{\delta_k}(\Gamma_k, V | \Gamma_0, ..., \Gamma_{k-1})$ equal to the product of the conditionals for all variables in $\Gamma_k \cup \{V\}$. Since $h_{\delta_k}(\Gamma_k, V | \Gamma_0, ..., \Gamma_{k-1})$ depends on δ_k only through its value $\delta_k(\Gamma_0, ..., \Gamma_{k-1})$, this puts us in the framework just described.

4. Derivation of a General Solution Algorithm

In this section we show how to use Gibbs sampling (Geman and Geman, 1984; and Gelfand and Smith, 1990), an iterative Markov chain Monte Carlo algorithm (Hastings, 1970) to implement stochastic dynamic programming for an influence diagram. Since the stochastic dynamic program is iterative, it suffices to explain how to implement it using Gibbs sampling for a single step. We will explain how to implement it for the first step.

Our task, then, is to find the decision function δ_k. This means finding, for each configuration $(\gamma_0, \gamma_1, ..., \gamma_{k-1})$ of $\Gamma_0 \cup ... \cup \Gamma_{k-1}$, the value d_k of the decision Δ_k that optimizes

$$E_{d_k}(V | \Gamma_0 = \gamma_0, \Gamma_1 = \gamma_1, ..., \Gamma_{k-1} = \gamma_{k-1}). \tag{8}$$

(Notice that we write d_k in the place of δ_k as a subscript on the expectation operator; this is because the expectation for the configuration $(\gamma_0, \gamma_1, ..., \gamma_{k-1})$ of the predecessors depends only on the value d_k that δ_k assigns to this configuration.) To this end, we simply compute (8) for all d_k and choose the d_k that gives the optimal (largest or smallest depending on whether we are maximizing or minimizing) result.

To compute (8) for a particular d_k, we recall that the conditional joint distribution of $\Gamma_k \cup \{V\}$ is proportional to

$$h_{d_k}(\Gamma_k, V | \Gamma_0 = \gamma_0, ..., \Gamma_{k-1} = \gamma_{k-1}),$$

which is simply the product of the conditionals for $\Gamma_k \cup \{V\}$.

Leaving aside the variables Δ_k and V, which are deterministic in this conditional joint distribution (Δ_k is equal to the constant d_k, and V is a function of the other variables), we can say that the conditional distribution of the other variables (all chance variables) is the product of their original conditionals. We are not interested, however, in all these variables; we are really interested only in V. Hence we can discard the conditionals for any variables that are independent of V in the conditional joint distribution. Figure 2 gives the ID of Figure 1, where relevance arrows are shown and informational arrows are omitted.

Figure 2. The ID example of Figure 1 without informational arrows

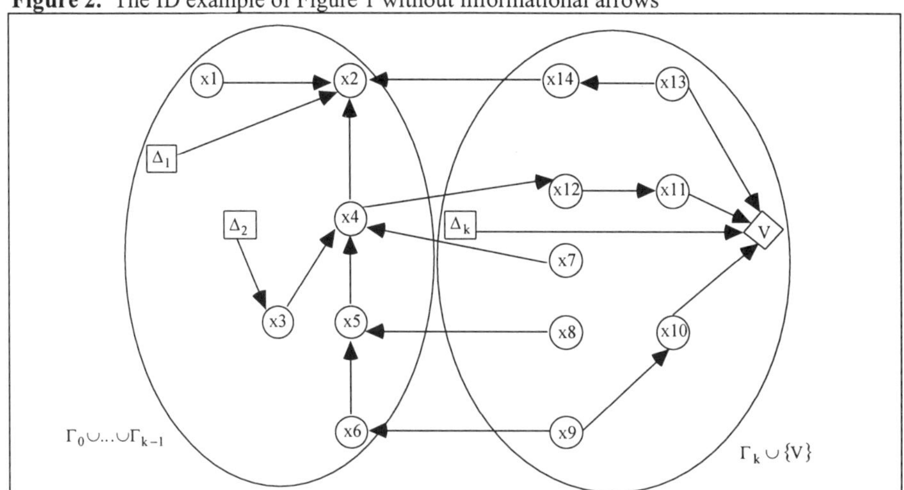

Here x_7 and x_8 are independent of V in the conditional distribution of $\Gamma_k \cup \{V\}$ because the distribution factors into parts involving only x_7 and x_8 and parts involving only the other variables. So we may omit their conditionals, effectively eliminating them from the problem, and reducing Γ_k to a smaller set Γ'_k.

The general rule for this reduction of our problem can be formulated graphically as follows. Consider the directed graph of the ID without informational arrows (as in Figure 2). Form the moral graph (Jensen et al., 1990). And omit any variables from Γ_k that are not connected with V in the subgraph of this moral graph determined by $\Gamma_k \cup \{V\}$. A variable X in Γ_k is not connected with V in this subgraph if there is no path in $\Gamma_k \cup \{V\}$ that connects X to V. Figures 3 and 4 show how this procedure applies to the example of Figure 2.

Figure 3. Moral graph with x_7 and x_8

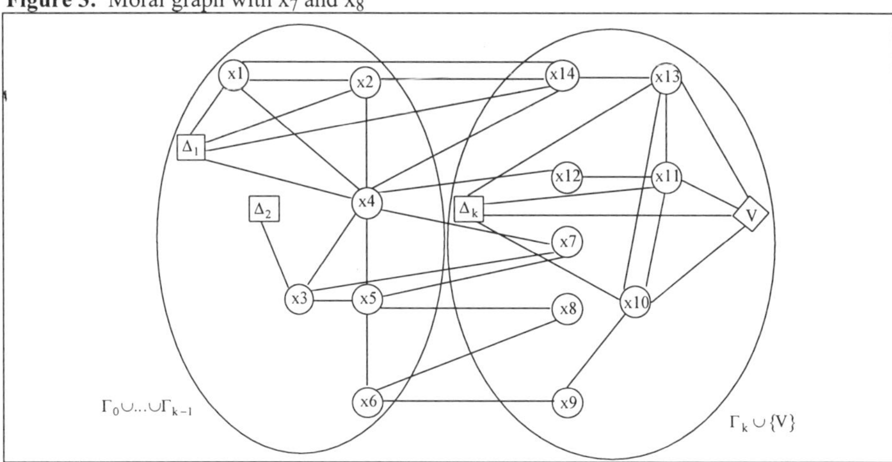

Figure 4. Moral graph without x_7 and x_8

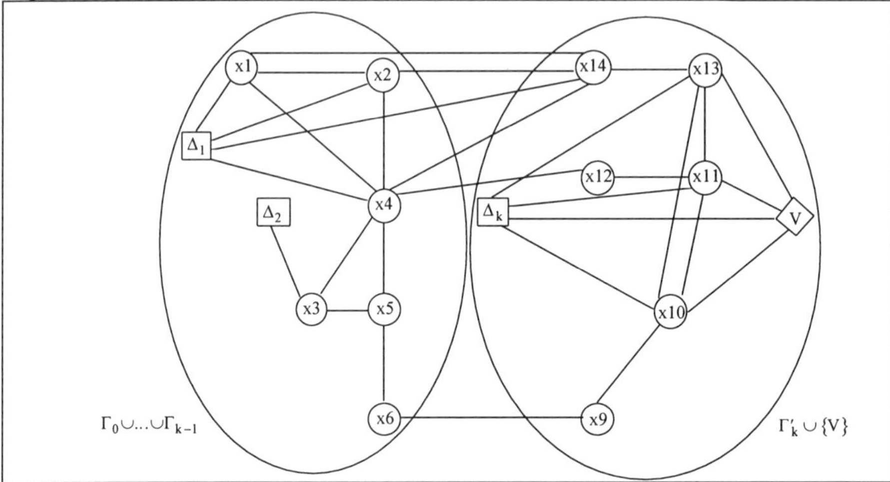

Next, notice that only some of the variables in $\Gamma_0 \cup \ldots \cup \Gamma_{k-1}$ will affect the expectation (8). Indeed, the only ones to affect it will be those involved in the factors that remain in the product. In our example, these factors are:

1. $f_1(x_{14}) = [x_2 = \bullet | x_1 = \bullet, x_4 = \bullet, x_{14}, \Delta_1 = \bullet]$
2. $f_2(x_{13}, x_{14}) = [x_{14} | x_{13}]$
3. $f_3(x_{12}) = [x_{12} | x_4 = \bullet]$
4. $f_4(x_{11}, x_{12}) = [x_{11} | x_{12}]$
5. $f_5(x_9, x_{10}) = [x_{10} | x_9]$
6. $f_6(x_9) = [x_6 = \bullet | x_9]$
7. $f_7(x_{13}) = [x_{13}]$
8. $f_8(x_9) = [x_9]$
9. $f_9(\Delta_k) = [\Delta_k = d_k | \Gamma_1 = \gamma_1, \Gamma_2 = \gamma_2]$
10. $f_{10}(V, \Delta_k) = [V | \Delta_k = d_k, x_{10}, x_{11}, x_{13}]$,

where $\bullet$ means that the variable in question is fixed at a certain value. The variables involved in these factors are x_1, x_2, x_4, x_6 and Δ_1. The other variables, Δ_2, x_3 and x_5, are not involved. Since the expectation (8) is not affected by the variables in $\Gamma_0 \cup \ldots \cup \Gamma_{k-1}$ that are not involved in these factors, we need not make our choice of d_k depend on them. In other words, we can make Δ_k a function only of the variables involved in the factors—x_1, x_2, x_4, x_6 and Δ_1 in our example.

The general rule for finding the variables on which Δ_k will depend can be described in terms of the moral graph we obtained previously: they are the neighbors of Γ_k in this graph. In our example (Figure 4), we see that x_1, x_2, x_4, x_6 and Δ_1 are the neighbors of Γ_k.

The factors that remain can be envisioned in terms of the directed subgraph determined by the variables that remain, as in Figure 5. They are the factors in which the variables remaining in the circle are parents or children.

Figure 5. Directed subgraph determined by the variables that remain

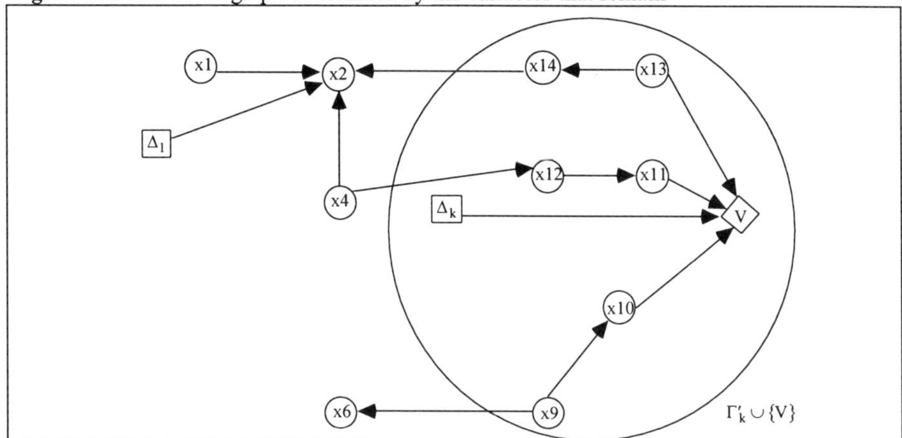

Now that the factors are identified , we do Gibbs sampling with these factors to simulate the joint distribution of $\Gamma'_k \setminus \{\Delta_k, V\}$. For the configuration of $\Gamma'_k \setminus \{\Delta_k, V\}$ obtained at each step of the Gibbs sampling, we compute V. This gives a sequence of values for V simulating a random sample from its conditional distribution, from which we may compute its conditional expectation.

When we move on to the next step of the stochastic dynamic program, we use the second of the two options discussed in Section 1. In other words, we absorb Γ_k into Γ_{k-1}, and we include the conditionals from Γ_k in the new factorization of $h_{\delta_{k-1}}$. In order to avoid zero probabilities that would interfere with the Gibbs sampling, we do not include the conditional for Δ_k corresponding to the decision function we have just found for Δ_k. Instead, we substitute this decision function in all the conditionals in which Δ_k appeared as a parent, thus eliminating Δ_k from the graph and producing arrows from the variables on which Δ_k depends to the variables for which it was a parent. Figure 6 illustrates the result for our example.

Figure 6. DAG with Δ_k absorbed into its direct successors

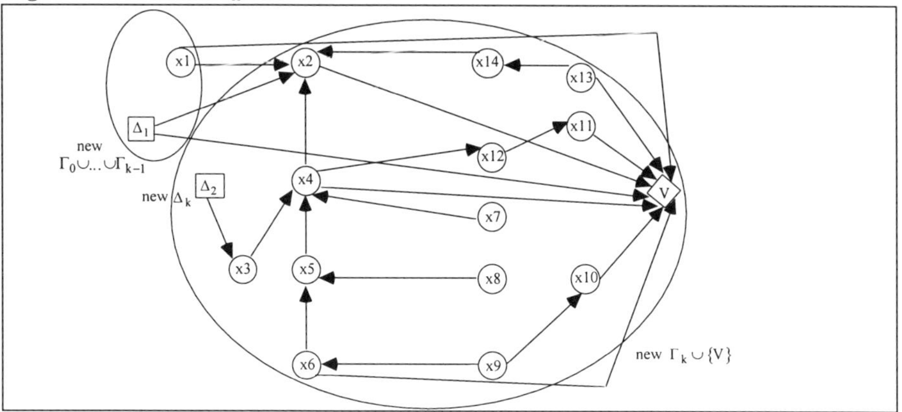

Acknowledgments

This work was partially supported by the National Science Foundation under grants IRI–8902444 and SES–9213558, and by a grant from the General Research Fund at the University of Kansas, grant 3605–XX–0038. I am grateful for comments from my mentors Glenn Shafer and Prakash Shenoy.

REFERENCES

Bellman, R. and Dreyfus, S. (1962) *Applied Dynamic Programming*, Princeton University Press, Princeton, N.J.

Clemen, R.T. (1991) *Making Hard Decisions*, Duxbury Press, Belmont, California.

Gelfand, A. E. and Smith, A. F. M. (1990) Sampling-based approaches to calculating marginal densities, *Journal of the American Statistical Association*, **85**(410), 398-409.

Geman, S. and Geman, D. (1984) Stochastic relaxation, Gibbs distributions, and the Bayesian estoration of images, *IEEE Transactions on Pattern Analysis and Machine Intelligence*, **PAMI-6**, 721-741.

Hastings, W. K. (1970) Monte Carlo sampling methods using Markov chains and their applications, *Biometrika*, **57**(1), 97-109.

Jensen, F. V., Lauritzen, S. L., and Olesen, K. G. (1990) Bayesian updating in causal probabilistic networks by local computations, *Computational Statistics Quarterly* **4**, 269-282.

Ndilikilikesha, P. (1992) Potential influence diagrams, *Working Paper No. 235*, School of Business, University of Kansas, Lawrence, Kansas.

Olmsted, S. M. (1983) On representing and solving decision problems, Ph.D. thesis, Department of Engineering-Economic Systems, Stanford University, Stanford, CA.

Shachter, R. D. (1986) Evaluating influence diagrams, *Operations Research*, **34**(6), 871-882.

Shenoy, P. P. (1992) Valuation-based systems for Bayesian decision analysis, *Operations Research*, **40**(3), 463-484.

Shenoy, P. P. (1993) A new method for representing and solving Bayesian decision problems, in D. J. Hand (ed.), *Artificial Intelligence Frontiers in Statistics: AI and Statistics III*, 119-138, Chapman & Hall, London.

Smith, J. E., Holtzman, S., and Matheson, J. E. (1993) Structuring conditional relationships in influence diagrams, *Operations Research*, **41**(2), 280-297.

7
Modeling and Monitoring Dynamic Systems by Chain Graphs

Alberto Lekuona, Beatriz Lacruz and Pilar Lasala

Departamento de Métodos Estadísticos
Universidad de Zaragoza
50009 Zaragoza (Spain)

ABSTRACT It is widely recognized that probabilistic graphical models provide a good framework for both knowledge representation and probabilistic inference (e.g., see [Cheeseman94], [Whittaker90]). The dynamic behaviour of a system which changes over time requires an implicit or explicit time representation. In this paper, an implicit time representation using dynamic graphical models is proposed. Our goal is to model the state of a system and its evolution over time in a richer and more natural way than other approaches together with a more suitable treatment of the inference on variables of interest.

7.1 Introduction

It is widely recognized that probabilistic graphical models provide a good framework for both knowledge representation and probabilistic inference (e.g., see [Cheeseman94] and [Whittaker90]).

The dynamic behaviour of any specific system which changes over time requires an implicit or explicit time representation. To model such systems is a very important task: the initial structure of the model and its propagation over time, the probabilities attached to the structure, the qualitative and quantitative interrelations among variables in different time slices, etc., need to be taken into account.

In this paper, an implicit time representation using dynamic graphical models is developed.

The goal is to model the state of a system and its evolution over time in a richer and more natural way than in other approaches, together with a more suitable treatment of the inference on variables of interest, according to Bayesian methodology.

In Section 2, a simplified example which will illustrate the ideas developed later is introduced. The qualitative and quantitative issues in the construction of probabilistic dynamic models are described in detail in Section 3. In Section 4, a formulation of the sequential procedure for making inferences over variables is proposed. A brief comment regarding related work is shown in Section 5. Finally, in Section 6, conclusions and future work are discussed.

---

[1] *Learning from Data: AI and Statistics V.* Edited by D. Fisher and H.-J. Lenz. ©1996 Springer-Verlag.

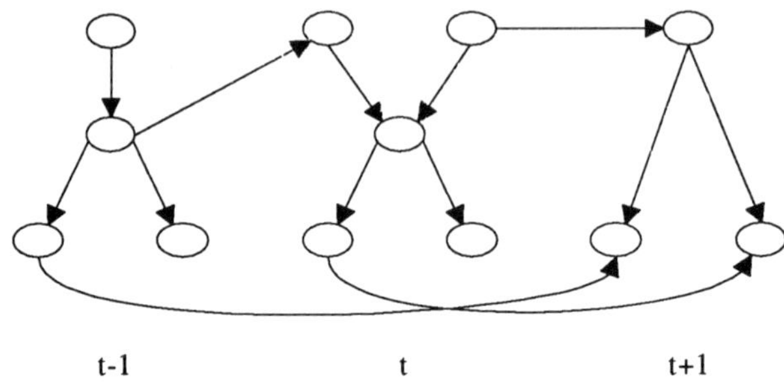

FIGURE 1. Probabilistic dynamic model.

7.2   A medical example

In this section, we aim to describe a simplified example that will be used in the following sections.

In the *Acute Heart Attack Monitoring* (AHAM) is advisable to take into account the possible existence of a Mitral Regurgitation. This failure has a strong influence in forecasting and treatment of Acute Heart Attack. If this Mitral Regurgitation coexists with the Heart Attack, is very important to determine if it is a complication of the Heart Attack or if it was previously present and is due to other causes.

Three evolutionary states in AHAM have been considered and collected in the variable Monitoring (M): Favourable Evolution, Pulmonary Acute Edema and Cardiogenic Shock. The forecasting of these states is given by the values of four parameters, measured in each time unit: Stroke Volume (SV), Mean Blood Pressure (MBP), Pulmonary Wedge Pressure (PWP) and Diuresis (D).

Moreover, two parameters associated with Mitral Regurgitation (MR) are considered: V wave (VW), which indicates the possible existence of Mitral Regurgitation, and Left Atrial Diameter (LAD), which indicates whether the Mitral Regurgitation is a complication of the Heart Attack or not.

The goal is to modelize the evolution of a patient who has suffered an Acute Heart Attack, taking into account a possible Mitral Regurgitation and the information obtained and accumulated in each time slice.

7.3   General issues in the construction of probabilistic dynamic models

Hence, a probabilistic dynamic model will be considered as a sequence of graphs indexed by the time, representing the temporal evolution of a system (see Figure 1). Each graph symbolizes the state of the system and the dependences among its components at a given time. The dynamic behaviour of the components of the system is described by a set of temporal dependences among these components in different time slices.

Furthermore, these dependences are quantified by conditional probability tables associated with the components of the system.

In order to make the management of such models feasible, a set of restrictions must be considered, for both its qualitative and quantitative aspects.

7.3.1 Qualitative issues

The components that describe any dynamic system to be modeled can be split up into observable and non-observable components and they can be depicted by discrete random variables (since the discretization of a continuous variable is always possible, this is not a very restrictive assumption). Observable components collect the evidence of the system measured in each time slice and the main goal will be to make inferences about the present or future state of non-observable components, given a certain accumulated information.

For the purposes of this paper, the above classification is adequate. However, other more specific types of observable variables could be considered, in order to exploit their features: for instance, variables which are independent of time (e.g., sex, chromosomic deficiencies, etc.) and variables which tend to keep the same value (e.g., smoker/non smoker status, economic status, etc.).

In our example, the six variables SV, MBP, PWP, D, VW and LAD are observable, i.e., measurable whereas M and MR are non-observable. We assume that every observable variable will always be measured in every time slice.

Two types of relationships among variables must be considered: those restricted to the same time slice and those between different time slices.

Relationships in the same time slice

The relationships among variables in the same time slice will be displayed according to the following criteria:

- relationships between observable and non-observable variables will be described by directed edges from the former to the latter, meaning that the evidence received in the current time slice modifies the beliefs about the states of the non-observable components. In our example, the main goal of the measurement of the observable variables is either to guarantee the favourable evolution of the patient or to warn of the possibility of complications arising

- relationships between non-observable variables will be described by non-directed edges. There is no reason for assuming either causality or sequentiality because these features cannot be observed in such variables

- relationships between observable variables will not be considered. A parsimonious construction of the model and the assumption, as we have said above, that every observable variable will always be observed, justify not considering these relationships

In Figure 2, a graphic scheme of these criteria applied to our example is represented. Observable variables will be represented by shaded nodes and the remainder will be non-observable variables.

In the first time slice, at the specific moment when the system begins to be observed, the graphical model represents a summary of the behaviour of the system until that moment (which has not been observed) and the structure at that time, together with the intertemporal structure, will determine the evolution of the system.

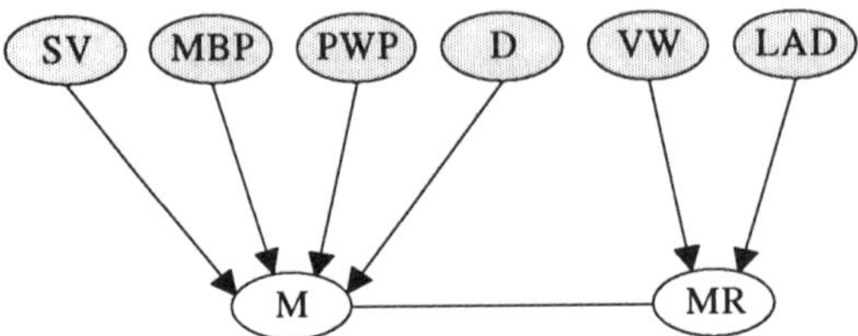

FIGURE 2. Relationships in the same time slice.

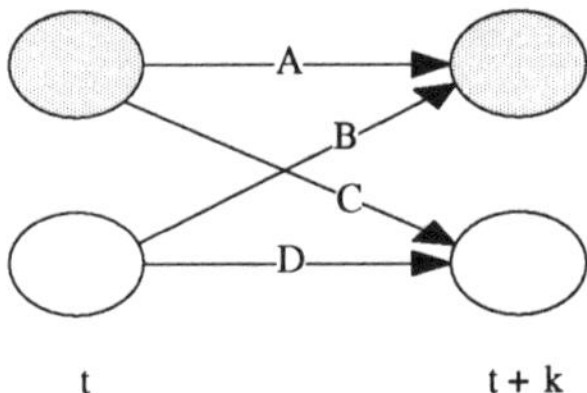

FIGURE 3. Possible relationships between different time slices.

Relationships between different time slices

The relationships among variables in different time slices will be represented according to the following criteria:

- the relationships will be drawn using directed edges, from the older time to the newer time, meaning the natural time sequence

- only relationships among non-observable variables will be considered. On the one hand, the influence of the past in observable variables is irrelevant, because they will be certainly measured sooner or later. This is why relationships A and B (see Figure 3) will never be considered. On the other hand, we claim that the influence of the non-observable variables in the newer time is perfectly described only through the non-observable variables in the older time. Thus, relationship C will not be considered either. In the AHAM example, it is clear that the stored information of the observable variables in several time slices gives information about the future state of the patient, but such information would be irrelevant if the current state of the patient is certainly known. For our example, the relationships between two different time slices are depicted in Figure 4.

From our point of view, the arrows of the directed edges do not necessarily mean cause-effect relationships. They reflect either sequential relationships, inherent to every system evolving over time, or a natural way of reasoning (first, to observe the evidence and subsequently to predict about the non-observable).

Combining directed and undirected edges forces us to consider chain graphs and to use their Markov properties, described in [Frydenberg90].

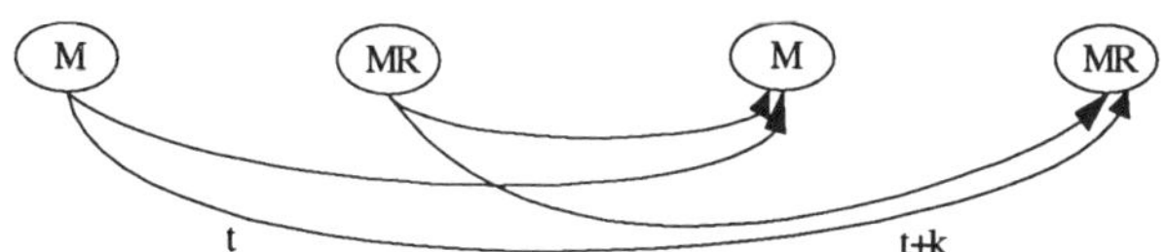

FIGURE 4. Inter-temporal relationships.

7.3.2 Quantitative issues

Let us denote

$E_t = \{$observable variables at time t$\}$

$H_t = \{$non-observable variables at time t$\}$

$S_t = \{E_t, H_t\}$ (describes the whole system at time t)

In order to simplify the model and to make it manageable, a Markovian behaviour is attached to $\{H_t\}_{t \geq 0}$, i.e.,

$$H_{t_n} \perp_P H_{t_0}, ..., H_{t_{n-2}} | H_{t_{n-1}}, \forall t_0 < t_1 < ... < t_n$$

where P is a probability distribution properly defined in some product space. This property assumes all the past history is completely summarized in the more recent past.

The Markovian property is not always justifiable. However, in our example, to assume such a property regarding the evolution of the patient is reasonable because, if the current values of the non-observable variables were certainly known, what had happened previously would not provide aditional information regarding their posterior behaviour.

Taking into account the Markovian condition, arrows from two previous time slices to the same posterior time slice will not simultaneously exist in the graphical representation. Furthermore, all the arrows from the past arise from the immediately preceding time slice. Moreover, it is easy to see that, in the described conditions, we find that $\{S_t\}_{t \geq 0}$ also verifies the Markovian property:

$$S_{t_n} \perp_P S_{t_0}, ..., S_{t_{n-2}} | S_{t_{n-1}}, \forall t_0 < t_1 < ... < t_n$$

Figure 5 shows a schematic summary of the characteristics described above.

Finally, all the structures discussed in Section 3 and the required probabilities must be properly extracted (using, if possible, a database and/or an expert). In this paper, we do not deal with this extraction.

7.4   Inference over the non-observable variables

We describe the sequential procedure which allows inference and forecasting about the present or future states of non-observable variables. Let us represent the accumulated

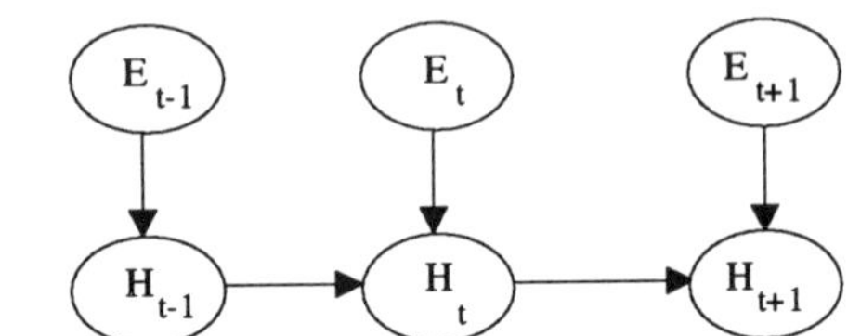

FIGURE 5. Schematic summary of the proposed model.

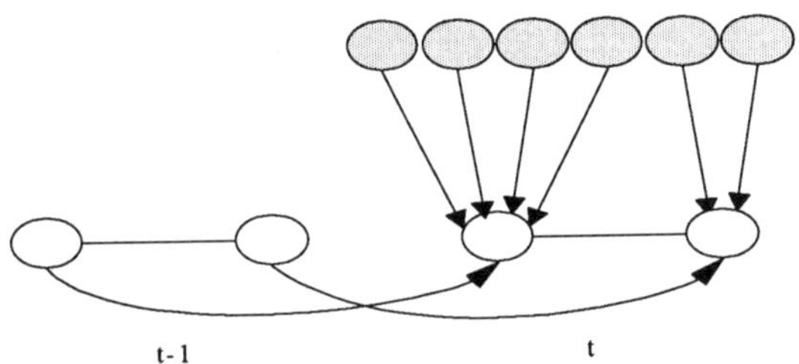

FIGURE 6. Inference process.

evidence up to the present time t by $D_t = \{E_t, D_{t-1}\}$ (with $D_0 = \{E_0\}$). Four steps are considered:

- *Step 1*: Measure the observable variables in time t and, therefore, consider D_t

- *Step 2*: Infer the beliefs of the states of the non-observable variables in the time slice t (see Figure 6). By Bayes:

$$P[H_t|D_t] \propto P[H_t|D_{t-1}] P[E_t|H_t, D_{t-1}]$$

which means a modification of the forecasted beliefs from the previous state. By means of the Bayes theorem we modify the beliefs after receiving new evidence. In the specific case t=0, $P[H_0|D_0]$ is known.

In order to exploit the newly supplied information, a graph in time t must be considered. Provided there is not evidence against the structure used in previous time slices, we propose to maintain it (this is defensible in systems which do not change abruptly). Therefore, D_t will implicitly inherit the same structure of the past.

The meaning of the likelihood $P[E_t|H_t, D_{t-1}]$ is clear: given all the past information and assuming a certain state in the non-observable variables, what behaviour would the observable variables have? The Markov assumption allows the following decomposition:

$$P[E_t|H_t, D_{t-1}] \propto P[S_t|D_{t-1}] = \sum_{H_{t-1}} P[S_t|H_{t-1}] P[H_{t-1}|D_{t-1}]$$

where $P[S_t|H_{t-1}]$ can be factorized according to the chain graph procedure:

$$P[S_t|H_{t-1}] = \prod_j P\left[E_t^j\right] \prod_k P\left[C_t^k|bd\left(C_t^k\right)\right]$$

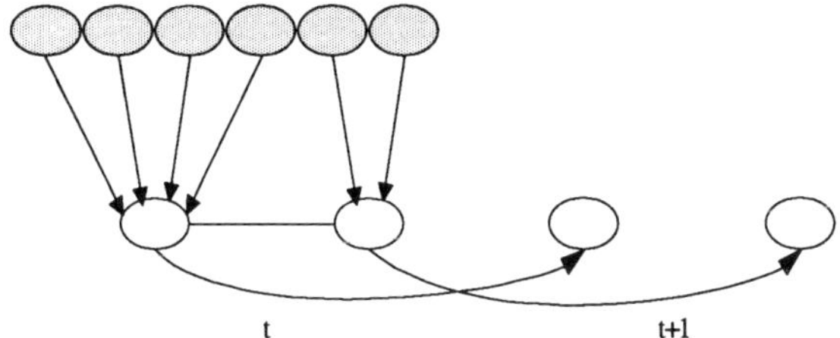

FIGURE 7. Forecasting process.

where $E_t^j \in E_t \forall j$, C_t^k stands for a chain component in H_t and $bd\left(C_t^k\right)$ stands for the boundary of that component, as defined in [Frydenberg90]

- *Step 3*: Forecast the states of the non-observable variables in the time slice t+1, assigning beliefs to them. Using the Markovian property in $\{H_t\}_{t \geq 0}$, it turns out that

$$P\left[H_{t+1}|D_t\right] = \sum_{H_t} P\left[H_{t+1}|H_t\right] P\left[H_t|D_t\right]$$

that is, a weighted mean of transition probabilities associated to the non-observable variables from one period of time to another. The weightings are the actual beliefs in the state of such variables, given all the collected information.

In this forecasting step, the independence of non-observable variables in time t+1, given D_t, must be supposed (the relationships must not appear: we are forecasting the state of the variables in the future but we do not actually know (in time t) if the relationships among non-observable variables will remain in the future) (see Figure 7). Accordingly, we can factorize the previous formula, using the local Markovian property in a chain graph:

$$P\left[H_{t+1}|D_t\right] = \sum_{H_t} \left(\prod_j P\left[H_{t+1}^j|\pi_t\left(H_{t+1}^j\right)\right]\right) P\left[H_t|D_t\right]$$

with $H_{t+1}^j \in H_{t+1}, \forall j$ and $\pi_t\left(H_{t+1}^j\right) = \left\{H_t^k|\exists \text{ arrow } \left(H_t^k, II_{t+1}^j\right)\right\}$

- *Step 4*: t=t+1 and go to Step 1.

7.5   Related literature

Cooper, Horvitz and Heckerman [Cooper88] propose a severe set of restrictions to be considered in temporal probabilistic reasoning but they do not consider stochastic processes for modeling the temporal evolution.

Kjaerulff [Kjaerulff92] proposes a schema for reasoning in dynamic probabilistic networks and he consideres Markov chains for modeling the evolution of the system.

Dagum and Galper [Dagum93a], Dagum, Galper and Horvitz [Dagum92] and Dagum, Galper, Horvitz and Seiver [Dagum93b] simplify the assessment of conditional probabilities in dynamic networks by using simple parametric decompositions. Whenever certain dependence conditions do not hold such assumptions are not easily justifiable.

Berzuini [Berzuini90], Provan [Provan94] and Provan and Clarke [Provan93b] propose the use of semi-Markovian processes. Furthermore, in [Provan94] y [Provan93b], they assume that each variable satisfies the Markov property and they infer the Markovianity of the whole system, only supported under restrictive independence assumptions.

7.6   Conclusions and future work

In this paper, a procedure for modeling dynamic systems has been proposed. We have justified the use of chain graphs instead of directed graphs. The elicitation of the structure of dynamic models using arrows can be very difficult from an expert and almost impossible from a database. Our model only considers arrows for temporal or sequential relationships.

The sequential beliefs updating in our dynamic model is closer to the natural way of thinking of the experts than other approaches that generalize classical time-series analysis or collapse the evidence of the past and present.

Obviously, there are many topics to be deal with in the future. The most important of these are to use databases for acquiring probabilities and learning structures as much as possible and to relax certain restrictions (fixed structures, relationships only between two consecutive times, etc.).

ACKNOWLEDGEMENTS

This research was supported by CICYT grant TIC91-1041. The authors thank Peter Cheeseman for enjoyable and helpful discussions on this material.

7.7   References

[Berzuini90] Berzuini, C. (1990). Modelling temporal processes via belief networks and petri nets, with application to expert systems. *Annals of Mathematics and Artificial Intelligence*, 2, 39-64.

[Cheeseman94] Cheeseman, P.; Olford, R.W. (Eds.) (1994) *Selecting Models from Data. Artificial Intelligence and Statistics IV*. Lecture Notes in Statistics No. 89. Springer-Verlag.

[Cooper88] Cooper, G. F.; Horvitz, E.J.; Heckerman, D.E. (1988) *A method for temporal probabilistic reasoning*. Working paper KSL 88-30. Knowledge Systems Laboratory. Medical Computer Science. Standford University. Standford, California.

[Dagum92] Dagum, P.; Galper, A.; Horvitz, E. (1992) Dynamic network models for forecasting. In Dubois, D.; Wellamn, M.P.; D'Ambrosio, B.; Smets, P. (Eds.) *Uncertainty in Artificial Intelligence. Proceedings of the Eighth Conference*. Morgan Kaufmann, 41-48.

[Dagum93a] Dagum, P.; Galper, A. (1993) Forecasting sleep apnea with dynamic network models. In Hekerman, D.; Mamdani, A. (Eds.) *Uncertainty in Artificial Intelligence. Proceedings of the Ninth Conference*. Morgan Kaufmann, 64-71.

[Dagum93b] Dagum, P.; Galper, A.; Horvitz, E.; Seiver, A. (1993) *Uncertain reasoning and forecasting.* Report KSL-93-47. Knowledge System Laboratory. Standford University. Standford, California.

[Frydenberg90] Frydenberg, M. (1990) The chain graph Markov property. *Scandinavian Journal of Statistics*, vol 7, No. 4, 333-335.

[Kjaerulff92] Kjaerulff, U. (1992) A computational scheme for reasoning in dynamic probabilistic networks. In Dubois, D.; Wellamn, M.P.; D'Ambrosio, B.; Smets, P. (Eds.) *Uncertainty in Artificial Intelligence. Proceedings of the Eighth Conference.* Morgan Kaufmann, 121-129.

[Nicholson92] Nicholson, A. E.; Brady, J.M. (1992) Sensor validation using dynamic belief networks. In Dubois, D.; Wellman, M.P.; D'Ambrosio, B.; Smets, P. (Eds.) *Uncertainty in Artificial Intelligence. Proceedings of the Eighth Conference.* Morgan Kaufmann, 207-214.

[Provan93a] Provan, G.M.A. (1993) Tradeoffs in constructing and evaluating temporal influence diagrams. In Heckerman, D.; Mamdani, A. (Eds.) *Uncertainty in Artificial Intelligence. Proceedings of the Ninth Conference.* Morgan Kaufmann, 40-47.

[Provan93b] Provan, G.M.A.; Clarke, J. R. (1993) Dynamic network construction and updating techniques for the diagnosis of acute abdominal pain. *IEEE Transactions on Pattern Analysis and Machine Intelligence*, vol. 15, No. 3, 299-307.

[Provan94] Provan, G.M.A. (1994) Model selection for diagnosis and treatment using temporal influence diagrams. In [Cheeseman94], 133-142.

[West89] West, M.; Harrison, J. (1989) *Bayesian forecasting and dynamic models.* Springer Verlag.

[Whittaker90] Whittaker, J. (1990) *Graphical models in applied multivariate statistics.* John Wiley and Sons.

8
Propagation of Gaussian belief functions

Liping Liu

School of Business
Albany State College
Albany, GA 31705

ABSTRACT A Gaussian belief function (GBF) can be intuitively described as a Gaussian distribution over a hyperplane, whose parallel sub-hyperplanes are the focal elements. It includes as special cases non-probabilistic linear equations, statistical observations, multivariate Gaussian distributions, and vacuous belief functions. The notion of GBFs was proposed in [Dem 90b], formalized in [Sha 92] and [Liu 95a], and successfully applied in combining independent statistical models in [Liu 95b]. In this paper, we propose a join-tree computation scheme for GBFs. We first represent Dempster's rule for combining GBFs obtained in [Liu 95a] equivalently in terms of matrix sweepings. We then show that the operations of GBFs follow the axioms of Shenoy and Shafer [ShSh 90] and justify the possibility of a join-tree computation scheme for GBFs. The result enriches theory of local computation by extending its application to the combination of statistical models and the integration of knowledge bases. An example is carried out to illustrate how combined inference can be made in accordance with multiple statistical models using graphical belief function models.

8.1 Introduction

Join-tree algorithms, in general, are only applicable to networks with discrete and finite variables. Belief networks with continuous variables can be solved approximately using simulation techniques. However, there are important exceptions. [LaWe 89] extended the join-tree approach to belief networks with mixed variables, some of which are discrete and some conditionally Gaussian. In parallel, this paper proves the axioms of Shenoy and Shafer [ShSh 90] and proposes a local computation scheme for Gaussian belief functions (GBFs).

The notion of GBFs extends Dempster-Shafer theory in representing statistical evidence. In its full generality, a GBF can be intuitively described as a Gaussian distribution over a hyperplane across the members of a partition of the hyperplane into parallel sub-hyperplanes. It includes as special cases non-probabilistic linear equations, statistical observations, multivariate Gaussian distributions, and vacuous belief functions. The notion of GBFs generalizes Bayesian inference of posterior distributions while abandons its most controversial component: improper priors [Sha 76]. It extends, unifies, and clarifies Fisher's fiducial method of posterior reasoning while fills the void of a prior distribution in the logical structure with a vacuous belief function [Dem 90b]. A GBF treats all the components of a statistical model such as observations, model assumptions, and subjective beliefs, not as separate concepts, but as manifestations of a single concept. It allows people to concentrate modeling efforts on recognizing and

---

[1] *Learning from Data: AI and Statistics V*. Edited by D. Fisher and H.-J. Lenz. ©1996 Springer-Verlag.

[2] This paper benefited from many discussions with Glenn Shafer and the foundational work done in [Dem 90b] and [Sha 92]. The research was supported by the Harper Fund.

incorporating independent components of real information. GBFs turn out to have a wide range of real applications. Dempster [Dem 90a, 90b] shows how the Kalman filter can be understood in terms of GBFs. [Liu 95b] further proposes a normative method for combining independent statistical models based on the theory of GBFs. As [Liu 95b] shows, the proposed method has a potential application in automated learning of belief networks from multiple databases that are neither appendable nor joinable [Mai 83]. It generalizes the meta-analysis for integrating independent statistical findings [HuSc 90] and the Bayesian method of estimating common regression coefficients [BoTi 73] in the sense that it can combine models of different kinds that may involve different variables.

Dempster [Dem 90b] sketched how join-tree computation works for belief functions in general. However, it is an open question whether the computation scheme works for GBFs [Sha 92]. The existing work in this area [Kong 86, SSM 87, ShSh 90] applies to finite and to condensable belief functions, but not to GBFs, which are usually continuous but not condensable. Presumably the justification for the finite case can be extended to a justification for the continuous case by a straightforward limiting argument, but this has not been done to date. The primary goal of this paper is to justify the feasibility of a join-tree computation scheme for GBFs. An outline of the rest of this paper is as follows. In 8.2, we introduce the notion of GBFs by drawing on the results in [Liu 95a] and provide appropriate background and motivations to the reader. In 8.3, we represent Dempster's rule of combining GBFs, which is derived in [Liu 95a], alternatively in terms of matrix sweeping operations. In 8.4, we prove the axioms of Shenoy and Shafer [ShSh 90] to justify the possibility of the join-tree computation of GBFs and carry out an example to illustrate how a local propagation scheme may be used to make combined predictions and inferences based on multiple statistical models.

8.2 Gaussian belief functions

A GBF encodes logical and probabilistic knowledge based on a given body of evidence. Logical knowledge is represented by linear equations, which are in turn represented by a hyperplane in a sample space. Probabilistic knowledge is represented by Gaussian distributions across all the members of a partition of the hyperplane into parallel sub-hyperplanes. Less general than an ordinary belief function, whose focal elements may have nonempty intersections, a GBF has the parallel sub-hyperplanes as its mutually exclusive focal elements. Let n, $n - c$, and $n - b$ denote the dimension numbers of the sample space, the hyperplane, and a focal element, respectively. In general, $c \le b \le n$. By appropriately setting up one or two of the dimension numbers c, b, and n, a GBF can be reduced into six nontrivial varieties, which provide building blocks for more complex GBFs. If $b = c = 0$, then the GBF is vacuous and has the sample space as its sole focal element. If $0 < c = b < n$, then the GBF is equivalent to specifying c linear equations. If $c = b = n$, the true point in the sample space is known with certainty, as might occur by direct observation. If $c = 0$ and $b = n$, then the GBF is an ordinary Gaussian probability distribution in the sample space. If $c > 0$ and $b = n$, the GBF is a Gaussian probability distribution over the hyperplane. In the latter two cases, the GBF is Bayesian because its focal elements are singletons with zero dimension. Finally, if $0 = c < b < n$, the GBF is a proper belief function which has a Gaussian distribution for some variables and no-opinions for others.

The above geometric description of GBFs is due to [Dem 90b]. [Liu 995a] elaborates on the idea of [Dem 90b] and [Sha 92] and formally represents a GBF in its full generality as a mathematical construct and a computational object. According to [Liu 95a], if all the variables of interest span a variable space—a finite dimensional vector space whose elements are random variables, then we can consider the sample space to be its dual space—the space of all linear functionals on the variable space. Accordingly, a hyperplane in the sample space is dual to a subspace in the variable space. A wide-sense inner product in the sample space, which specifies the log "density" of a GBF over a hyperplane, is dual to a wide-sense inner product in the variable space, which specifies the covariance of all the variables in a sub-variable space. From this dual correspondence, non-probabilistic linear equations can be represented by a sub-variable space, in which each variable takes on a value with certainty. A multivariate Gaussian distribution over a hyperplane can be represented by a wide sense inner product that specifies the covariance between each pair of variables in the variable space.

Formally, let V be a variable space. A GBF on V is a quintuplet (C, B, L, π, E), where C, B, and L are nested subspaces of V, $C \subseteq B \subseteq L \subseteq V$, π is a wide sense inner product on B with C as its null space, and E is a linear functional on B. We call C *the certainty space*, B *the belief space*, L *the label space*, π *the covariance*, and E *the expectation*. The expectation E and the covariance π define a Gaussian distribution for the variables in B by specifying their means and covariances. This Gaussian distribution is regarded as a full expression of our beliefs, based on a given body of evidence; this item of evidence justifies no beliefs about variables in L going beyond what is implied by the beliefs about the variables in B. (The evidence might justify some further beliefs about variables that are not in L, but these are outside the conversation so far as a belief function with space L is concerned.) The Gaussian distribution assigns zero variance to the variables in C; if X is in C, we are certain that it takes the value $E(X)$ with certainty.

Example 1. Let X, Y, and Z be three variables and x, y, and z be their sample values. A GBF on these variables includes a Gaussian distribution $X + Y \sim N(0.5, 2)$ and a linear equation $x + y + z = 1$. Let L be spanned by X, Y, and Z, C by variable $X + Y + Z$ and B by $X + Y + Z$ and $X + Y$. E and π on B are respectively as follows:

$$E[\alpha_1(X + Y + Z) + \alpha_2(X + Y)] = \alpha_1 + 0.5\alpha_2,$$

$$\pi[\alpha_1(X + Y + Z) + \alpha_2(X + Y), \beta_1(X + Y + Z) + \beta_2(X + Y)] = 2\alpha_2\beta_2.$$

It is easy to see that variable $X + Y + Z$ takes on value 1 with certainty and so C is a one-dimensional certainty space to represent the linear equation $x + y + z = 1$. Therefore, we arrive at GBF $= (C, B, L, \pi, E)$. This GBF has beliefs about each variable in B by giving its mean and variance. Suppose, for example, $Z = (X + Y + Z) - (X + Y)$. Then, $E(Z) = 1 - 0.5 = 0.5$ and $\pi(Z, Z) = 2 \, (-1)(-1) = 2$. However, it has no-opinions for variables in L that are not in B.

The marginalization of a GBF is simply a projection. Suppose (C, B, L, π, E) is a GBF, and M is a subspace of L. Then the marginal of (C, B, L, π, E) on M, denoted by $(C, B, L, \pi, E)^{\downarrow M}$, is another GBF obtained by intersecting certainty space C, belief space B, and label space L with M and restricting the covariance and the expectation to the new belief space:

$$(C, B, L, \pi, E)^{\downarrow M} = (C \cap M, B \cap M, L \cap M, \pi|_{B \cap M}, E|_{B \cap M}).$$

In Example 1, if M is spanned by Z, then $C \cap M = 0$, $B \cap M = L \cap M = \{\alpha Z \mid \alpha \in R\}$, $\pi|_{B \cap M}$ $(\alpha Z, \beta Z) = 2\alpha\beta$, and $E|_{B \cap M} (\alpha Z) = 0.5\alpha$. On the other hand, if M is spanned by X, then $C \cap M$ $= 0$, $B \cap M = 0$, $L \cap M = M$, $\pi|_{B \cap M} (0) = 0$, and $E|_{B \cap M} (0) = 0$. The marginal is vacuous. This is intuitively reasonable because (C, B, L, π, E) carries no knowledge about X.

8.3 Combining Gaussian belief functions

[Liu 95a] derives a rule for combining GBFs by adapting Dempster's rule to the continuous case. In this section, we want to represent it equivalently in terms of matrix sweepings.

Let $\{X_1, X_2, ..., X_n\}$ be a partition of Gaussian random variables with $\text{Cov}(X_i, X_j) = \Sigma_{ij}$ and $E(X_j) = \mu_j$ ($i, j = 1, 2, ..., n$). The shorthand notation

$$M = \begin{pmatrix} \mu_j \\ \Sigma_{ij} \end{pmatrix}_{1-n} \tag{8.1}$$

called an extended matrix, can be used to represent the distribution of the random variables. In (8.1), i means rows, j columns, and 1–n indicates that the variable partition has indices 1, 2, ..., and n. The extended matrix has $(n + 1)$ rows and n columns, where the first row represents the mean vector and the remaining n rows the covariance matrix. Note that the same matrix M can be written differently in the shorthand notation depending on how the variables are partitioned.

Marginalization: Suppose M is an extended matrix for a set of random variables, say X. The marginalization of M to Y, denoted by $M^{\downarrow Y}$, is the submatrix produced by retaining the rows and columns that corresponds to the variables in Y while deleting the rest. Note that Y need not be a subset of X in the definition. Actually, if Y is not a subset of X, $M^{\downarrow Y}$ is the same as $M^{\downarrow Y \cap X}$. Obviously, $M^{\downarrow Y}$ represent the marginal distribution for the variables in $Y \cap X$. Since marginalization is simply a restriction, it is easy to show the following consonance rule:

$$(M^{\downarrow Y})^{\downarrow Z} = M^{\downarrow Y \cap Z} . \tag{8.2}$$

Direct Sum: Suppose M_1 and M_2 are two extended matrices respectively for X and Y. To define their direct sum $M = M_1 \oplus M_2$, we extend M_1 to the set $X \cup Y$ by adding zeros to the elements corresponding to the variables that are not in X. Similarly, we extend M_2 to the set $X \cup Y$ by adding zeros to the elements corresponding to the variables that are not in Y. Then M is simply the sum of the extended M_1 and M_2. It is easy to show the following distributivity rule:

$$(M_1 \oplus M_2)^{\downarrow Y} = (M_1)^{\downarrow Y} \oplus (M_2)^{\downarrow Y}. \tag{8.3}$$

Forward Sweep: Suppose M is an extended matrix for the partition $\{X_1, X_2, ..., X_n\}$. The forward sweep of M on $X_k = x_k$, denoted by $\text{SWP}(X_k = x_k)M$, is defined as follows:

$$\text{SWP}(X_k = x_k)\begin{pmatrix} \mu_j \\ \Sigma_{ij} \end{pmatrix}_{1-n} = \begin{pmatrix} \mu_{j.k} \\ \Sigma_{ij.k} \end{pmatrix}_{1-n},$$

where

$$\mu_{j.k} = \begin{cases} \mu_j - (\mu_k - x_k)(\Sigma_{kk})^{-1}\Sigma_{kj} & j \neq k \\ \mu_k(\Sigma_{kk})^{-1} & j = k \end{cases}$$

$$\Sigma_{ij.k} = \begin{cases} -(\Sigma_{kk})^{-1} & i=j=k \\ \Sigma_{ik}(\Sigma_{kk})^{-1} & j=k\neq i \\ (\Sigma_{kk})^{-1}\Sigma_{kj} & i=k\neq j \\ \Sigma_{ij}-\Sigma_{ik}(\Sigma_{kk})^{-1}\Sigma_{kj} & \text{otherwise} \end{cases}$$

Reverse Sweep: Suppose M is an extended matrix for the partition $\{X_1, X_2, ..., X_n\}$. The reverse sweep of M on $X_k = x_k$, denoted by $RWP(X_k = x_k)M$, is defined as follows:

$$RWP(X_k = x_k)\begin{pmatrix} \mu_j \\ \Sigma_{ij} \end{pmatrix}_{1-n} = \begin{pmatrix} \tilde{\mu}_{j.k} \\ \tilde{\Sigma}_{ij.k} \end{pmatrix}_{1-n},$$

where

$$\tilde{\mu}_{j.k} = \begin{cases} \mu_j-(\mu_k+x_k\Sigma_{kk})(\Sigma_{kk})^{-1}\Sigma_{kj} & j\neq k \\ -\mu_k(\Sigma_{kk})^{-1} & j=k \end{cases}$$

$$\tilde{\Sigma}_{ij.k} = \begin{cases} -(\Sigma_{kk})^{-1} & i=j=k \\ -\Sigma_{ik}(\Sigma_{kk})^{-1} & j=k\neq i \\ -(\Sigma_{kk})^{-1}\Sigma_{kj} & i=k\neq j \\ \Sigma_{ij}-\Sigma_{ik}(\Sigma_{kk})^{-1}\Sigma_{kj} & \text{otherwise} \end{cases}$$

By definition, $\mu_{j.k}$ and $\Sigma_{jj.k}$ are respectively the conditional mean and covariance matrix of X_j given $X_k = x_k$. $\Sigma_{jk.k}$ is the regression coefficient of X_j on X_k. Therefore, if M represents a joint distribution, the marginal of $SWP(X_k = x_k)M$ to X_j is an extended matrix for the conditional distribution of X_j given $X_k = x_k$. The semantics of RWP can be understood from the following equation, which states that $RWP(Y = y)$ and $SWP(Y = y)$ nullify each other.

$$RWP(Y = y)[SWP(Y = y)M] = SWP(Y = y)[RWP(Y = y)M] = M. \tag{8.4}$$

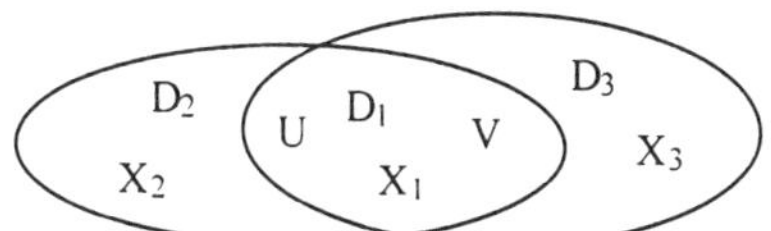

FIGURE 1: The Hypergraph for Bel_1 and Bel_2

We now use the above matrix operations to represent Dempster's rule for combining GBFs [Liu 95a]. Given two GBFs $Bel_1 = (C_1, B_1, L_1, \pi^1, E^1)$ and $Bel_2 = (C_2, B_2, L_2, \pi^2, E^2)$, without loss of generality, assume we can choose a convenient basis $D_1, D_2, D_3, U, V, X_1, X_2, X_3, ...$ such that C_1 is spanned by $\{D_1, D_2, U\}$, C_2 by $\{D_1, D_3, V\}$, B_1 by $\{D_1, D_2, U, V, X_1, X_2\}$, and B_2 by $\{D_1, D_3, V, U, X_1, X_3\}$. In other words, D_1 is a set of deterministic variables and X_1 is a set of random variables commonly borne by both GBFs. U is deterministic in Bel_1 but uncertain in Bel_2. V is otherwise. Bel_1 is certain about D_2 and uncertain about X_2. However, Bel_2 has no opinions about either D_2 or X_2. Similarly, Bel_2 is certain about D_3 and uncertain about X_3 but Bel_1 has no opinions about both D_3 and X_3. The hypergraph representing Bel_1 and Bel_2 is shown in Figure 1. Since D_1 is a common deterministic vector, its value must be the same in both Bel_1 and Bel_2 because otherwise they have not common ground for agreements. Let $D_1 = d_1$ in both Bel_1 and Bel_2. Let $D_2 = d_2$, $U = u$ and $D_3 = d_3$, $V = v$ in Bel_1 and Bel_2, respectively. According to [Liu 95a], $C = C_1 \oplus C_2$ with $D_1 = d_1$, $D_2 = d_2$, $D_3 = d_3$, U

$= u$, $V = v$. Also, $B = B_1 \oplus B_2$ and $L = L_1 \oplus L_2$. Given a basis, a linear functional is specified by the mean vector of the basis and an inner product is specified by the covariance matrix of the basis. Thus, we can represent π and E on B by giving the corresponding extended matrix. By Theorem 1 in [Liu 95a], we can prove the following Theorem 1 (see [Liu 95c]).

Theorem 1. Given any two GBFs

$$\text{Bel}_1 = (C_1, B_1, L_1, \pi^1, E^1): C_1 \text{ is spanned by } \{D_1, D_2, U\} \text{ with } D_1 = d_1, D_2 =$$

$$d_2, U = u, \text{ and } B_1 \text{ by } \{D_1, D_2, U, V, X_1, X_2\} \text{ with } M_1 = \begin{pmatrix} \mu_j^1 \\ \Sigma_{ij}^1 \end{pmatrix}_{0,1,2},$$

and

$$\text{Bel}_2 = (C_2, B_2, L_2, \pi^2, E^2): C_2 \text{ is spanned by } \{D_1, D_3, V\} \text{ with } D_1 = d_1, D_3 =$$

$$d_3, V = v, B_2 \text{ by } \{D_1, D_3, V, U, X_1, X_3\} \text{ with } M_2 = \begin{pmatrix} \mu_j^2 \\ \Sigma_{ij}^2 \end{pmatrix}_{0,1,3},$$

their combination $\text{Bel}_1 \otimes \text{Bel}_2$ is $\text{Bel} = (C, B, L, \pi, E)$:

$$C = C_1 \oplus C_2 \text{ with } D_1 = d_1, D_2 = d_2, D_3 = d_3, U = u, V = v, B = B_1 \oplus B_2,$$

$$L = L_1 \oplus L_2 \text{ with } M = \begin{pmatrix} \mu_j \\ \Sigma_{ij} \end{pmatrix}_{1-3} = \text{RWP}(X_1 = 0)M', \text{ where}$$

$$M' = \text{SWP}(X_1 = 0)[\text{SWP}(V = v)M_1]^{\downarrow[X_1, X_2]} \oplus \text{SWP}(X_1 = 0)[\text{SWP}(U = u)M_2]^{\downarrow[X_1, X_3]}.$$

Example 2. Let Bel_1 denote the following two linear equations involving deterministic variables D and U and random variables X and Y:

$$X = U + 2Y + \varepsilon_X \tag{8.5}$$

$$Y = 1 + 2D + \varepsilon_Y \tag{8.6}$$

where ε_X and ε_Y are independent residual noises with $\varepsilon_X \sim N(0, 4)$ and $\varepsilon_Y \sim N(0, 1)$. Let $D = 1$ and $U = 2$. Then $\text{Bel}_1 = (C_1, B_1, L_1, \pi^1, E^1)$, where C_1 has a basis $\{D, U\}$ with $D = 1$ and $U = 2$, B_1 has a basis $\{D, U, X, Y\}$, the extended matrix for X and Y is

$$M_1 = \begin{pmatrix} 8 & 8 & 2 \\ 3 & 2 & 1 \end{pmatrix}^T.$$

Let Bel_2 be another GBF bearing on random variables U, X, and Z, which represents the following statistical models

$$Z = 1.5U + 0.8 + \varepsilon_Z \tag{8.7}$$

$$X = Z + 5 + \varepsilon_X \tag{8.8}$$

where $U \sim N(1.9, 0.04)$, $\varepsilon_Z \sim N(0, 2)$, and $\varepsilon_X \sim N(0, 6)$ are independent. Then $\text{Bel}_2 = (C_2, B_2, L_2, \pi^2, E^2)$, where $C_2 = 0$, B_2 has a basis $\{U, X, Z\}$, the extended matrix for U, X, and Y is

$$M_2 = \begin{pmatrix} 1.90 & 0.04 & 0.06 & 0.06 \\ 8.65 & 0.06 & 8.09 & 2.09 \\ 3.65 & 0.06 & 2.09 & 2.09 \end{pmatrix}^T.$$

Let $\text{Bel}_1 \otimes \text{Bel}_2 = (C, B, L, \pi, E)$. Then, according to Theorem 1, C is spanned by $\{D, U\}$ and B by $\{D, U, X, Y, Z\}$. U is random in Bel_2 but deterministic in Bel_1. Bel_1 has no random variables that are deterministic in Bel_2. Thus, we only apply the reverse sweep to M_2:

$$[RWP(U=2)M_2]^{\downarrow\{X,Z\}} = \begin{pmatrix} 8.8 & 8 & 2 \\ 3.8 & 2 & 2 \end{pmatrix}^{\mathrm{T}}, \quad SWP(X=0)M_1 = \begin{pmatrix} 1 & -1/8 & 1/4 \\ 1 & 1/4 & 1/2 \end{pmatrix}^{\mathrm{T}},$$

$$SWP(X=0)[SWP(U=2)M_2]^{\downarrow\{X,Z\}} = \begin{pmatrix} 1.1 & -1/8 & 1/4 \\ 1.6 & 1/4 & 3/2 \end{pmatrix}^{\mathrm{T}}.$$

Therefore, applying the direct sum results in

$$SWP(X=0)M_1 \oplus SWP(X=0)[SWP(U=2)M_2]^{\downarrow\{X,Z\}} = \begin{pmatrix} 2.1 & -1/4 & 1/4 & 1/4 \\ 1.0 & 1/4 & 1/2 & 0 \\ 1.6 & 1/4 & 0 & 3/2 \end{pmatrix}^{\mathrm{T}}.$$

Finally, applying $RWP(X=0)$ to the above matrix results in the extended matrix for $Bel_1 \otimes Bel_2$:

$$\begin{pmatrix} 8.4 & 4 & 1 & 1 \\ 3.1 & 1 & 0.75 & 0.25 \\ 3.7 & 1 & 0.25 & 1.75 \end{pmatrix}^{\mathrm{T}}.$$

According to (8.5) and (8.6), the predicted value of X is 8 with variance 8. According to (8.7) and (8.8), the predicted value of X is 8.8 with variance 8. In the combined model, the predicted value of X is 8.4 with variance 4. We note that the variance of X in $Bel_1 \otimes Bel_2$ is smaller than that in both Bel_1 and Bel_2. The variance of Y in $Bel_1 \otimes Bel_2$ is larger than that in Bel_1 and the variance of Z in $Bel_1 \otimes Bel_2$ is larger than that in Bel_2. [Liu 95b] shows that this is generally true. It indicates that the beliefs on a common variable can strengthen each other and the belief on a variable expressed by a single GBF tends to be eroded by another GBF which has no opinions.

8.4 Local propagation in join-trees

A join-tree is a tree-structured graph, where each node is a subset of variables, each pair of neighbors has non-empty intersection, and the intersection of two distinct nodes is contained in every node on the path connecting the two distinct vertices. To obtain a join-tree, we first draw a hypergraph as in Figure 1, where each hyperedge is a set of variables on which a GBF bears. Then we transform the hypergraph into a join-tree by various techniques [Kong 86, Mel 88]. Each node in the resulting join-tree is usually the union of some hyperedges in the hypergraph. We combine the GBFs that bear on the variables of the same join-tree node and load the combined GBF to that node. The join-tree along with the loaded GBFs is then an equivalent representation of the hypergraph and provides a computational structure for efficient reasoning. The basic idea of the join-tree approach is to propagate relevant knowledge by sending and absorbing messages step-by-step in a join-tree. Each step involves sending a message from a node to a neighbor. The complete procedure consists of a series of local computations, each of which involves only a small number of variables that are near each other in the join-tree. The join-tree approach is shown to be applicable to finite and to condensable belief functions [Kong 86, SSM 87, ShSh 90]. Theorem 2 shows that it is also applicable to GBFs, which are neither finite nor condensable. (For the proof, see [Liu 95c]).

Theorem 2. The following axioms of Shenoy and Shafer [ShSh 90] hold for GBFs:

Axiom 1. Combination of GBFs is commutative and associative;

Axiom 2. Let $Bel = (C, B, L, \pi, E)$ and $K \subset M \subset L$. Then $(Bel^{\downarrow M})^{\downarrow K} = Bel^{\downarrow K}$.

Axiom 3. Let $Bel_1 = (C_1, B_1, L_1, \pi_1, E_1)$ and $Bel_2 = (C_2, B_2, L_2, \pi_2, E_2)$. Then
$$(Bel_1 \otimes Bel_2)^{\downarrow L_1} = Bel_1 \otimes (Bel_2)^{\downarrow L_1}.$$

Join-tree algorithms were independently developed in [LaSp 88] for Bayesian expert systems, [Kong 86] and [SSM 88] for belief function expert systems, and slightly improved in [JLO 90]. In the following we adopt the Shafer-Shenoy algorithm to illustrate the propagation of GBFs. The algorithm treats a marginal GBF as a message and repeatedly apply the following rules to propagate messages:

Rule 1. Each node waits to send its message to a given neighbor until it has received messages from all of its other neighbors;

Rule 2. When a node is ready to send its message to a neighbor, it computes the message by collecting all its messages from other neighbors, combining them with its own GBF, and marginalizing the combined GBF to its intersection with the neighbor to whom it is sending.

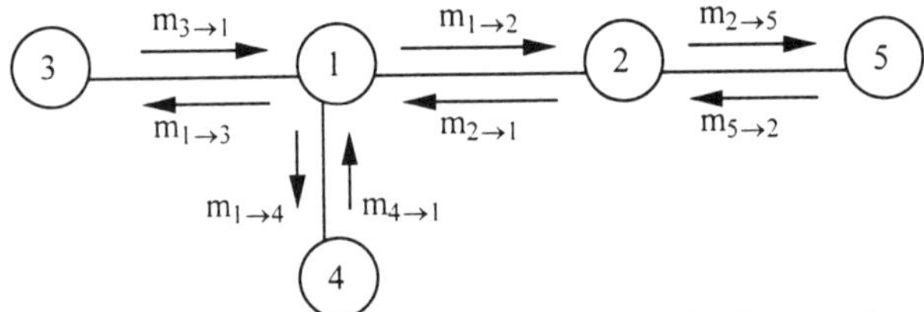

FIGURE 2: The Shafer-Shenoy Architecture for Propagation

Let i, j, k be nodes in a join-tree. Let $m_{i \to j}$ denote the Shafer-Shenoy message to j from neighbor i. Let Bel_j denote the belief function for j. Let $bd(j)$ denote the set of all neighbors of j. Then Rule 2 says that the message from j to neighbor k is given by $m_{j \to k}$

$$m_{j \to k} = \{Bel_j \otimes [m_{i \to j} \mid i \in bd(j)\backslash k]\}^{\downarrow j \cap k}. \tag{8.9}$$

Because of Rule 1, the computation must begin with the leaves of a join-tree. A message from a leave is simply the marginal of its own GBF according to (8.9). In Figure 2, for example, the leaves are 3, 4, and 5. Any of leaves can begin to send its marginal to its unique neighbor. A non-leave node must wait to send its message until it hears from all its other neighbors. In Figure 2, for example, node 1 sends its message to node 2 only after it receives from 3 and 4 and to 3 only after receiving messages from 4 and 2. Similarly, 2 sends its message to 1 after receiving the message from 5 and to 5 after receiving the message from 1.

The GBF in each node does not change during propagation. The message sent to a neighbor is first registered on the communication channel until it is collected in accordance with Rule 2. After all the messages are sent, we have an architecture like Figure 2, where each node stores its original GBF and all the two-way communication channels are filled with messages. Let Bel denote the combination of all the component GBFs in a join-tree. Then,

$$Bel^{\downarrow j} = Bel_j \otimes \{m_{i \to j} \mid i \in bd(j)\}. \tag{8.10}$$

In words, to obtain the marginal of Bel to the variables at node j, we collect all the messages to j and combine them with Bel_j. Therefore, instead of combining all the component GBFs and marginalizing the combination to j, we can compute the same marginal indirectly by local propagation. It is obvious that the latter approach is more efficient because of the locality of its computation. Also, after all the messages are sent, (8.10) can be used to compute the marginal GBFs for all the join-tree nodes.

Example 3. We consider the join-tree computation of five GBFs: Bel_1 and Bel_2 are defined in Example 2, Bel_3 is for deterministic variables D and U with observations D = 1 and U = 2, Bel_4 is for Gaussian variable X with mean 8.5 and variance 2, Bel_5 is for Gaussian variable Z with mean 4 and variance 1. The join-tree representation for these GBFs is shown in Figure 2, where node i has Bel_i, i = 1, 2, ..., 5. Note that node 1 consists of variables D, U, X, and Y and node 2 variables U, X, Z. By (8.9), it is easy to see that $m_{3 \to 1} = (Bel_3)^{\downarrow \{D, U\}} = Bel_3$, $m_{4 \to 1} = (Bel_4)^{\downarrow X} = Bel_4$, and $m_{5 \to 2} = (Bel_5)^{\downarrow Z} = Bel_5$. Then, by (8.9) and Theorem 1, we can compute other messages: $m_{1 \to 2} = \{Bel_1 \otimes m_{3 \to 1} \otimes m_{4 \to 1}\}^{\downarrow 1 \cap 2} = \{Bel_1 \otimes Bel_3 \otimes Bel_4\}^{\downarrow \{U, X\}}$ is GBF with U = 2 and X has mean 8.4 and variance 1.6; $m_{2 \to 5} = \{Bel_2 \otimes m_{1 \to 2}\}^{\downarrow Z}$ is GBF for Z with mean 3.72 and variance 1.58; $m_{1 \to 3} = \{Bel_1 \otimes m_{2 \to 1} \otimes m_{4 \to 1}\}^{\downarrow \{D, U\}}$ is GBF with D = 1 and U = 2; $m_{1 \to 4} = \{Bel_1 \otimes m_{2 \to 1} \otimes m_{3 \to 1}\}^{\downarrow X} = \{Bel_1 \otimes m_{2 \to 1}\}^{\downarrow X}$ is GBF for X with mean 8.07 and variance 3.64; and $m_{2 \to 1} = \{Bel_2 \otimes Bel_5\}^{\downarrow \{U, X\}}$ has extended matrix

$$\begin{pmatrix} 1.91 & 0.039 & 0.019 \\ 8.89 & 0.019 & 6.676 \end{pmatrix}^{T}.$$

After the propagation, we fill all the communication channels with messages. The GBFs stored in all the nodes do not change. Let $Bel = \otimes \{Bel_i \mid i = 1, 2, ..., 5\}$. According to (8.10), we can compute the marginal of Bel for each node in Figure 2. For example, $Bel^{\downarrow X} = Bel_4 \otimes m_{1 \to 4}$ is GBF for X with mean 8.33 and variance 1.29, $Bel^{\downarrow Z} = Bel_5 \otimes m_{2 \to 5}$ is GBF for Z with mean 3.89 and variance 0.61, and $Bel^{\downarrow \{D, U, X, Y\}} = Bel_1 \otimes m_{3 \to 1} \otimes m_{4 \to 1} \otimes m_{2 \to 1}$ is GBF for {D, U, X, Y} with D = 1, U = 2, and X and Y has extended matrix

$$\begin{pmatrix} 8.33 & 1.29 & 0.32 \\ 3.10 & 0.32 & 0.58 \end{pmatrix}^{T}.$$

Variables such as dependent ones in a regression model are sometimes of particular interest. The join-tree technique for GBFs allows us to make a combined prediction for such variables in accordance with all the models available. In Example 2, for instance, the five models have made a combined prediction that the average value of X is 8.33 given D = 1 and U = 2. The propagation scheme also allows us to efficiently absorb evidence and observe its effect throughout the network. For example, suppose we observe that Z = 3 in Figure 2. We can turn Bel_5 into a GBF for deterministic variable Z with Z = 3, propagate the observation through the join-tree as usual, and observe the conditional changes of other variables.

8.5 References

[BoTi 73]   Box, G. E. P. and Tiao, G. C. (1973) *Bayesian Inference in Statistical Analysis*. Addison-Wesley, Reading.

[HuSc 90]   Hunter, J. E. and Schmidt, F. L. (1990) *Methods of Meta-Analysis: Correcting Error and Bias in Research Findings*. Sage, Newbury Park.

[Dem 90a]   Dempster, A. P. (1990a) "Construction and Local Computation Aspects of Network Belief Functions," In *Influence Diagrams, Belief Nets, and Decision Analysis*, R. M. Oliver, J. Q. Smith (eds.), John Wiley and Sons, Chichester.

[Dem 90b]   Dempster, A. P. (1990b) "Normal Belief Functions and the Kalmam Filter," Research Report, Department of Statistics, Harvard University, Cambridge, MA.

[JLO 90]   Jensen, F. V., Lauritzen, S. L., and Olesen, K. G. (1990) "Bayesian Updating in Causal Probabilistic Networks by Local Computations," *Computational Statistics Quarterly*, **4**, 269–282.

[LaSp 88]   Lauritzen, S. L. and Spiegelhalter, D. J. (1988) "Local Computation with Probabilities on Graphical Structures and Their Application to Expert Systems (with discussion)," *Journal of the Royal Statistical Society*, Series B, **50**, 157–224.

[LaWe 89]   Lauritzen, S. L. and Wermuth, N. (1989) "Graphical Models for Associations between Variables, Some of Which are Qualitative and Some Quantitative," *The Annals of Statistics*, **17**, 31–57.

[Liu 95a]   Liu, L. (1995) "A Theory of Gaussian Belief Functions," *International Journal of Approximate Reasoning*, in press.

[Liu 95b]   Liu, L. (1995) "Model Combination Using Gaussian Belief Functions," Research Report, School of Business, University of Kansas, Lawrence, KS.

[Liu 95c]   Liu, L. (1995) "Local Computation of Gaussian Belief Functions," Research Report, School of Business, University of Kansas, Lawrence, KS.

[Kong 86]   Kong, A. (1986) "Multivariate Belief Functions and Graphical Models," Ph.D. Thesis, Department of Statistics, Harvard University, Cambridge, MA.

[Mai 83]   Maier, D. (1983) *The Theory of Relational Databases*. Computer Science Press, Rockville.

[Mel 87]   Mellouli, K. (1987) "On the Propagation of Beliefs in Networks using the Dempster-Shafer Theory of Evidence," Ph.D. Thesis, School of Business, University of Kansas, Lawrence, KS.

[Sha 76]   Shafer, G. (1976) *A Mathematical Theory of Evidence*. Princeton University Press, Princeton.

[Sha 92]   Shafer, G. (1992) "A Note on Dempster's Gaussian Belief Functions," Working paper, School of Business, University of Kansas, Lawrence, KS.

[SSM 87]   Shafer, G., Shenoy, P. P., and Mellouli, K. (1987) "Propagating Belief Functions in Qualitative Markov Trees," *International Journal of Approximate Reasoning*, **3**, 383–411.

[ShSh 90]   Shenoy, P. P. and Shafer, G. (1990) "Axioms for Probability and Belief-Function Propagation," *Uncertainty in Artificial Intelligence*, **4**, 169–198.

9

On Test Selection Strategies for Belief Networks

David Madigan[†] and Russell G. Almond[††]

Department of Statistics, University of Washington and Fred Hutchinson Cancer Research Center, University of Washington[†]

StatSci Division, MathSoft[††]

ABSTRACT

Decision making under uncertainty typically requires an iterative process of information acquisition. At each stage, the decision maker chooses the next best test (or tests) to perform, and re-evaluates the possible decisions. Value-of-information analyses provide a formal strategy for selecting the next test(s). However, the complete decision-theoretic approach is impractical and researchers have sought approximations.

In this paper, we present strategies for both myopic and limited non-myopic (working with known test groups) test selection in the context of belief networks. We focus primarily on utility-free test selection strategies. However, the methods have immediate application to the decision-theoretic framework.

9.1 Introduction

Graphical belief network researchers have developed powerful algorithms to propagate the effects of any piece of information to all the variables in the model in a manner analogous to forward chaining in rule-based expert systems (see, for example, Dawid, 1992). Comparatively little work has been done on the "backward chaining" problem: finding the most cost-effective information sources which will best increase information about a target variable. This is the *Test Selection* problem. There are two parts to the problem: choosing a metric for the value of information, and searching for potential information sources which maximize the metric. We consider both these issues and present workable approaches.

Many authors have recognized the importance of test selection in larger applications. Recent proposed approaches for probabilistic belief networks include: Jensen and Liang (1994), Heckerman, *et al.* (1993), and Almond (1993). In this paper, we propose a semi-automated myopic strategy which synthesizes the critiquing approach of Miller (1983) and Good's idea of a quasi-utility (Good and Card, 1971). We develop search strategies based on these ideas and demonstrate them in the context of a simple imaging application. We also address a limited form of the nonmyopic test selection problem, and propose and demonstrate a Markov chain Monte Carlo solution that generalizes the recent work of Heckerman, *et al.* (1993).

Section 9.2 reviews basic test selection techniques without particular reference to belief

---

[1] *Learning from Data: AI and Statistics V.* Edited by D. Fisher and H.-J. Lenz. ©1996 Springer-Verlag.

[2] *Address for correspondence*: Department of Statistics, GN-22, University of Washington, Seattle, Washington 98195 (madigan@stat.washington.edu).

networks. In particular, it introduces *weights of evidence* and some of their properties, and the concept of "critiquing". Section 9.3 discusses weights of evidence and test selection in the context of belief networks. Section 9.4 addresses the nonmyopic test selection problem and provides a connection with the decision theoretic approach to test selection.

9.2   Basic Test Selection

Before performing a test, we should assess the amount of information that the test will provide. Generally, we should only pursue a test if the information gained from the test is worth the cost of testing. If we have a full decision model (e.g., an influence diagram), then the preferred metric is the *value of information* (Matheson, 1990, provides a review). For a purely probabilistic model, Good and Card (1971) suggest using a *quasi-utility*—a measure of information content which plays the role of a utility. Section 2.1 discusses our suggested quasi-utility, the expected weight of evidence, and Section 2.2 discusses model-independent search strategy issues.

9.2.1   Expected Weight of Evidence

When discriminating between a single hypothesis H and its negation, Good and Card (1971) recommend the *expected weight of evidence* as a quasi-utility measure of the usefulness of a test, T:

$$EW(H:T) = \sum_{j=1}^{n} W(H:t_j)\Pr(t_j \mid H)$$

where $\{t_j, j = 1, ..., n\}$, represent the possible outcomes of the test, T. $W(H:t_j)$ is the weight of evidence for H provided by the evidence $T = t_j$. Specifically:

$$W(H:T_j) = \log \frac{\Pr(T = t_j \mid H)}{\Pr(T = t_j \mid \overline{H})}.$$

Informally, $EW(H:T)$ is the weight of evidence that will be obtained from T "on the average", when the H is true. Glasziou and Hilden (1989) provide a thorough review of the application of quasi-utilities to test selection and justify the choice of the weight of evidence. Good (1985) notes that expected weight of evidence is a generalization of entropy, and regards it as occupying "a central position in human inference." The weight of evidence is closely related to the decision-theoretic "value of perfect information" and Heckerman, Horvitz and Middleton (1993) (hereafter HHM) show how to transform value of information questions into questions involving weights of evidence (see Section 4). See Ben-Basset (1978) and Good and Card (1971) for discussion of alternative entropy-based quasi-utilities for multi-valued hypotheses.

Typically, test selection strategies must account for test costs and misclassification costs. Building on Glasziou and Hilden (1989) and Breiman, *et al.* (1984), Madigan and Almond (1993) proposed a cost-adjusted expected weight of evidence and we refer the interested reader to these references for details.

9.2.2 *Critiquing*

In practice we have found that automated test selection strategies for multi-valued hypotheses (such as medical diagnoses) can exhibit disquieting behavior. Specifically, their line of reasoning may bear no resemblance to typical expert reasoning, moving from indicants relevant to one hypothesis, to indicants relevant to a different hypothesis. The artificial intelligence literature has discussed this problem at length–see, for example, Barr and Feigenbaum (1982, p.82).

To address this problem, we adopt a "critiquing" approach (Miller, 1983). The essential idea of critiquing is to elicit a suggested hypothesis from the user, say $H = h_0$. The system then elicits indicants which are chosen to maximize the probability of quickly accepting or rejecting h_0. Thus, once the user has suggested a hypothesis, the system only elicits indicants that are of direct and usually apparent relevance to that hypothesis. If the hypothesis is rejected, i.e., its probability falls below some defined threshold (Spiegelhalter and Knill-Jones, 1984), the user is prompted for an alternative suggestion, and so on (Miller, 1983, McSherry, 1986). Our experience with this approach in the medical context suggests that clinicians welcome the idea of interacting with the system in this way and that critiquing enhances confidence in the system.

9.3   Myopic Search

The previous section discussed test selection and weights of evidence without regard to the belief network structure. The application to belief networks, while computationally challenging, is straightforward in principle. First the user selects a hypothesis (H) to critique. Next we calculate the expected weight of evidence provided for H by *all* uninstantiated tests. Fortunately, the definition of the expected weight of evidence ensures that these calculations can be carried out in an efficient manner. Exactly two propagations are required: first, H is temporarily instantiated to "true" and $\Pr(T_j^i \mid H)$ is calculated for all uninstantiated tests $T^i, i = 1, \ldots, m$, with states $j = 1, \ldots, n_i$; second, H is temporarily instantiated to "false" and $\Pr(T_j^i \mid \overline{H})$ is again calculated for all i and j. A similar observation is made in Jensen and Liang (1994).

Unfortunately, in very large networks such as Pathfinder (Heckerman *et al.*, 1992), even these calculations are prohibitively expensive. The problem can be framed as a classical search problem looking for the most cost effective test. In the case of belief networks, the structure of the network provides a convenient structure for the search space which can be exploited to find efficient search strategies. In particular, expected weight of evidence is *monotonic* over a class of graphs called Berge Networks[3]—the farther you get from the hypothesis being critiqued, the lower the expected weight of evidence (Section 9.3.1). This property forms the basis of some simple search strategies in the belief network's "junction tree" and related Markov Tree models (Section 9.3.2).

---

[3] A Berge network is a belief network with the property that clique intersections in the associated undirected graph (see Dawid, 1992) contain no more than one node.

9.3.1   Weight of evidence in Berge Networks and Markov Trees

In a Berge network, the expected weight of evidence decreases in a monotone fashion away from the hypothesis, H (Madigan *et al.*, 1994). Therefore, for these networks, the test selection strategy can confine its attention to the immediate neighbors of H. This convenient property derives from the following basic result:

Theorem 1 (Monotonicity): In a belief network with three nodes (A, B and H), if B separates A from H (Figure 1), $EW(H : A) \leq EW(H : B)$.

Proof: see Madigan and Almond (1993).

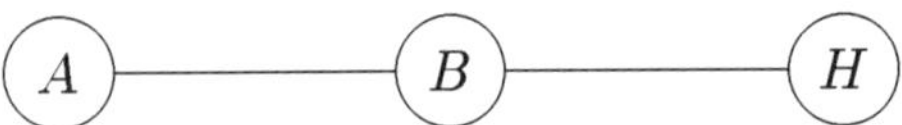

FIGURE 1. *A Simple Berge Network: The weight of evidence A provides for H can be no more than the weight of evidence B provides for H.*

Berge networks have the property that for any pair of connected nodes, the network is collapsible onto a unique "evidence chain" connecting the two nodes. A simple recursion argument extends the above monotonicity property to these chains–see Madigan *et al.* (1994) for details.

For non-Berge networks, no similar property exists. Consider the example of Figure 2. Although $EW(H : B, C) \geq EW(H : A)$, it could be true that $EW(H : A) > EW(H : B)$ and $EW(H : A) > EW(H : C)$. Simply searching all of the neighbors of the target hypothesis is not sufficient for non-Berge networks.

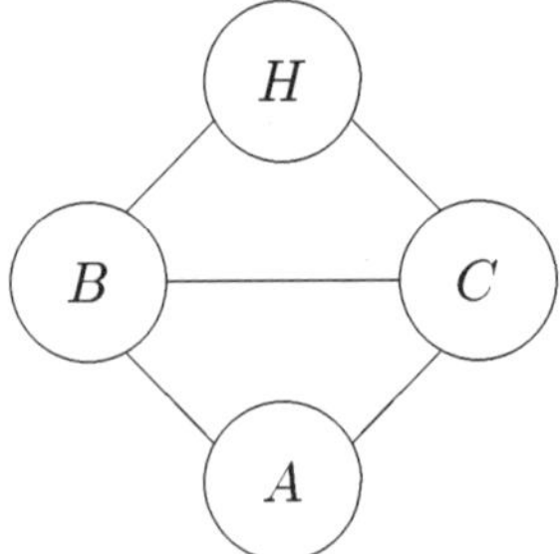

FIGURE 2. *A Simple Non-Berge Network. In this network, the weight of evidence provided by A for H could be greater than that provided by either B or C singly.*

Figure 2 illustrates both the problem and a potential solution. If we cluster the variables B and C, the graph reduces to a simple Berge network (like Figure 1). In a general non-Berge network we can either cluster the variables to form a Berge network, or transform the network into a Markov tree.[4] Many popular algorithms for calculating probabilities in belief networks already build a Markov tree (Dawid, 1992, Almond, 1995). Since Markov

---

[4]A *Markov tree* is a tree whose nodes represent groups of variables and for which all the nodes containing any single variable forms a connected subtree. Both the tree of cliques and the junction tree are Markov trees. See Shenoy and Shafer (1990) or Almond (1995).

trees are always Berge networks, monotonicity holds in the Markov tree model. The next section explores this approach.

9.3.2 Search Strategies in the Markov Tree

Ignoring costs, a simple branch and bound search using the Markov tree structure can find the best test. Almond (1995) recommends constructing the Markov tree by augmenting the tree of cliques or junction tree with nodes representing the individual variables (see Figure 3(b) for an example). Then, starting from the target hypothesis, the algorithm puts all the nodes connected to the hypothesis in the Markov tree on the "search boundary list". Next, it finds the node in the search boundary with the lowest expected weight of evidence for the hypothesis (i.e., the joint weight of evidence of all the variables in the node—see Section 4.1), and expands the node by placing its Markov tree-neighbors on the search boundary. When the node to be expanded represents a single variable corresponding to a test, it follows from the Monotonicity theorem this is the best test.

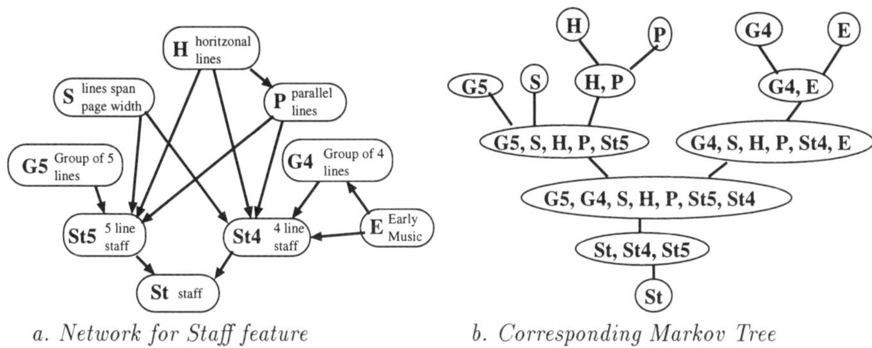

a. Network for Staff feature   *b. Corresponding Markov Tree*

FIGURE 3. A model for determining if a group of lines is a musical staff and its corresponding Markov tree

As an illustration, we show in Figure 3 a model for the task of classifying a group of lines as a musical staff (Almond *et al.*, 1994). The node {St} at the bottom of the Markov tree is the target hypothesis. In this example, lines which span the width of the page are rare, but staff lines are rarely less than a full page wide. Thus, {S} will have high expected weight of evidence. The search might start by looking at the {St, St4, St5} node which trivially has the highest weight of evidence since it contains {St}. Next, the search considers {G5, G4, S, H, P, St5, St4} and then {G5, S, H, P, St5} if it has lower joint expected weight of evidence for {St} than {G4, S, H, P, St4, E}. Next, the search expands {G5, S, H, P, St5}. If the single test variable, {S} has larger expected weight of evidence for {St} than either {G5} or {H, P}, we decide to do that test.

Known background variables play a big role in eliminating unnecessary tests. For example, four line staffs are only relevant in early music. *A priori* knowledge that the music was not early (*E* false) would cause the node {G4, S, H, P, St4, E} to have zero weight of evidence; that branch of the tree would never be searched.

Incorporating test costs requires a little extra care. Branch and bound search will still work if we have a overestimate of the value of the intermediate nodes or an underestimate of the cost. Therefore, we propagate minimum costs towards the target node. For example, the cost associated with the node $\{\mathbf{H}, \mathbf{P}\}$ is the lower of the costs of tests $\{\mathbf{H}\}$ and $\{\mathbf{P}\}$. Now when branch and bound search reaches a node corresponding to a single test variable, there will be no test which yield more expected evidence per unit cost.

9.4   Non-Myopic Search: accounting for feature groups

The test selection strategy described above is *myopic*—it only looks at the effect of extracting one feature at a time. HHM criticize myopic test selection strategies as unrealistic: tests often come bundled in groups. For example, an EKG will typically provide values for a slew of variables; the line detector in Section 9.3.2's example will return a data structure containing the location, direction, length and thickness of the candidate line. Further, a single costly test (such as biopsy) can often be more cost effective than a myriad of cheaper tests, yet a myopic strategy will choose the cheaper test (HHM provide an example from Pathfinder). Essentially, a myopic strategy continually shops around for bargains, when spending a little more freely may ultimately prove more cost effective (Glasziou and Hilden, 1989).

There are two challenges in going from myopic to non-myopic search. First, we must be able to calculate the joint expected weight of evidence of the test group, which gets computationally expensive as the size of the group increases (Section 9.4.1). Second, in the case where the application does not define the test groups, we must *find* the sets of variables with the largest joint expected weight of evidence.

9.4.1   Joint Expected Weight of Evidence

Suppose $T^1, T^2, \ldots, T^p$ form a test group, and we wish to calculate their joint expected weight of evidence:

$$EW(H : T^1, T^2, \ldots, T^p) = \sum_{\mathcal{T}} \Pr(T^1_{i_1}, T^2_{i_2}, \ldots, T^p_{i_p} \mid H) \log \frac{\Pr(T^1_{i_1}, T^2_{i_2}, \ldots, T^p_{i_p} \mid H)}{\Pr(T^1_{i_1}, T^2_{i_2}, \ldots, T^p_{i_p} \mid \overline{H})} \quad (9.1)$$

where the summation is over $\mathcal{T}$, all the possible values of the tests in the group. However, as HHM and Jensen and Liang indicate, the number of terms in the summation above can lead to computational difficulties with even a modest group size.

Simulation based methods provide a workable solution to this problem. The key point is that (9.1) above is precisely the expectation of:

$$\log \frac{\Pr(T^1, T^2, \ldots, T^p \mid H)}{\Pr(T^1, T^2, \ldots, T^p \mid \overline{H})} \quad (9.2)$$

with respect to $\Pr(T^1, T^2, \ldots, T^p \mid H)$. Both simple Monte Carlo in the Markov tree and Markov chain Monte Carlo (Kong, 1991, Neal, 1993, Buntine, 1994) provide methods for approximating this expectation, although we only describe the latter. We construct an irreducible Markov chain $\{T(t) = T^1(t), T^2(t), \ldots, T^p(t)\}$, for $t = 1, 2, \ldots$ with state space $\mathcal{T}$ which has equilibrium distribution $\Pr(T^1, T^2, \ldots, T^p \mid H)$. Then for any well-behaved function $g(T(t))$ defined on $\mathcal{T}$, if we simulate this Markov chain for $t = 1, \ldots, N$, the average:

$$\hat{g} = \frac{1}{N} \sum_{t=1}^{N} g(T(t)) \tag{9.3}$$

converges with probability one to $E(g(T))$ as N goes to infinity. To compute (9.1) in this fashion we set $g(T(t)) = \log \frac{\Pr(T^1(t), T^2(t), \ldots, T^p(t) | H)}{\Pr(T^1(t), T^2(t), \ldots, T^p(t) | \overline{H})}$.

To implement the Markov chain we define a neighborhood $\mathrm{nbd}(T)$ for each $T \in \mathcal{T}$ which is the set of elements of $\mathcal{T}$ which differ from T in just one of $T^1, T^2, \ldots, T^p$ (larger neighborhoods are also possible). Define a transition matrix q by setting $q(T \to T') = 0$ for all $T' \notin \mathrm{nbd}(T)$ and $q(T \to T')$ constant for all $T' \in \mathrm{nbd}(T)$. If the chain is currently in state T, we proceed by drawing T' from $q(T \to T')$. We accept it with probability:

$$\min\left\{ 1, \frac{\mathrm{pr}(T' \mid H)}{\mathrm{pr}(T \mid H)} \right\}. \tag{9.4}$$

Otherwise the chain stays in state T. Diagnostics exist for assessing how many cycles are needed and how many should be discarded (Raftery and Lewis, 1992), for assessing convergence (Geyer, 1993), and for overcoming difficulties with multimodal discrete distributions (Lin, 1992).

As an illustration, we consider the coronary artery disease study of Detrano, *et al.* (1989). The study attempted to find clinical variables to predict the presence of coronary artery disease without an intrusive angiograph, and measured a number of clinical and test variables for 303 patients referred for to the Cleveland Clinic for coronary angiography. The data are available through the Murphy and Aha (1992) repository. The variables are as follows (the number in parentheses is the number of states for each variable):

Clinical Data: *Age* (3), *Sex* (2), *Rest-Bp* (3; Systolic blood pressure at rest), *Chest-Pain* (4).

Routine Test Data: *Chol* (4; Serum cholesterol in mg/dl), *Fast-Bsug* (2; Fasting Blood Sugar in mg/dl), *Rest-Ecg*(3; Electrocardiographic results at rest).

Exercise Data: *Max-Heart-Rate* (3), *Exer-Angina* (2; Exercise induced angina), *Old-Peak* (3; ST depression induced by exercise relative to rest), *Slope-Peak* (3; The slope of the peak exercise ST segment), *Exer-Thal-Defects* (3; Exercise thallium scintigraphic defects).

Experimental Non-invasive Test: *Colored-Floro* (4; Number of blood vessels colored by fluoroscopy).

Outcome Variables: *Health-State* (5), *Healthy?* (2).

Almond and Madigan (1993) selected a belief network model for these data and we show this model in Figure 4. Treating *Healthy?* as the target variable, we used the Markov chain Monte Carlo (MCMC) algorithm to compute the expected weight of evidence (EW) for a number of test groups, and we show the results in Table 9.1. This example is sufficiently small that we can also calculate the exact expected weight of evidence for each group. The results show that within 1,000 samples, the MCMC algorithm provides a reasonable approximation to the exact expected weight of evidence. Frequently the rank order of the expected weights of evidence for different test groups will be of primary importance. In that case, as few as 100 samples will often prove adequate.

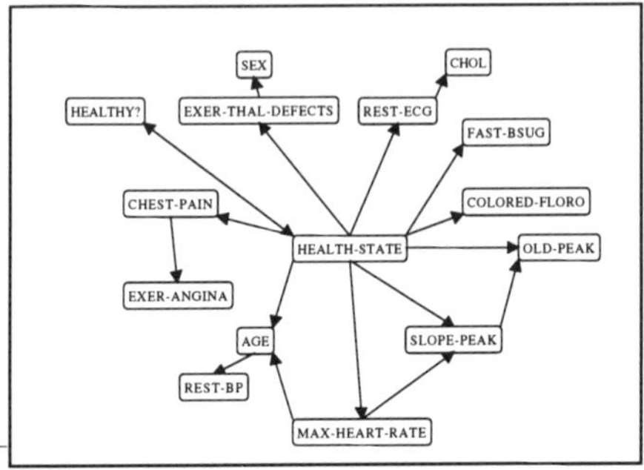

FIGURE 4. Belief network for the Heart model

TABLE 9.1. Markov chain Monte Carlo computation of the expected weight of evidence for test groups in the coronary artery disease example

Indicant Group	*Exact EW*	*EW at 500 samples*	*EW at 1,000 samples*	*EW at 10,000 samples*
Routine Test Data	0.15	0.57	0.14	0.18
Exercise Data	45.87	52.81	45.89	42.86
Clinical Data	24.96	26.98	26.67	24.92

9.4.2 Relationship to Decision-Theoretic Test Selection

The simulation approach outlined above provides a general solution to the problem addressed by HHM in the context of a specific class of graphs (graphs where the tests in the group are conditionally independent given the diseases or form a Markov chain structure in the graph). They show that the calculation of the value of perfect information for a test group, reduces to the assessment of a number of inequalities like:

$$\Pr\left(\log\frac{\Pr(T^1, T^2, \ldots, T^p \mid H)}{\Pr(T^1, T^2, \ldots, T^p \mid \overline{H})}\right) > W^* \tag{9.5}$$

where W^* depends on the actual utilities. To compute this using the Markov chain, set:

$$g(T) = \mathrm{I}\left(\log\frac{\Pr(T^1, T^2, \ldots, T^p \mid H)}{\Pr(T^1, T^2, \ldots, T^p \mid \overline{H})} > W^*\right), \tag{9.6}$$

where I is an indicator function. Unlike the HHM method, our approach does not require any constraints on the graph.

9.4.3 Searching for Groups

A general solution to the nonmyopic test selection problem requires that we *find* the groups of variables which are most cost effective relative to some target hypothesis. Unfortunately, the individual test variables might be widely scattered through the Markov tree model and we do not have a fully satisfactory solution to the combinatorial problem that arises. HHM (Section VII) provide some suggestions.

References

Almond, R.G. [1993]. "Lack of Information Based Control in Expert Systems." In Hand, D.J (ed). *Artificial Intelligence Frontiers in Statistics: AI and Statistics III*, Chapman and Hall, pp 82–89.

Almond, R.G. [1995]. *Graphical Belief Modelling.* Chapman and Hall, in press.

Almond, R.G., M.Y. Jaisimha, E. Arbogast, and S.F. Elston[1994]. "Intelligent Image Browsing and Feature Extraction." *StatSci Research Report No. 25,* 1700 Westlake Ave, N. Suite 500, Seattle, WA 98117.

Almond, R.G., and D. Madigan[1993]. "Using GRAPHICAL-BELIEF to Predict Risk for Coronary Artery Disease." *StatSci Research Report No. 19,* 1700 Westlake Ave, N. Suite 500, Seattle, WA 98117.

Barr, A. and Feigenbaum, E. [1982]. *Handbook of Artificial Intelligence,* volume 2. Kaufmann, Los Altos.

Ben-Basset, M. [1978]. "Myopic policies in sequential classification." *IEEE Transactions on Computing,* (**27**, 170–174.

Breiman, L., J. Friedman, and C. Stone[1984]. *Classification and Regression Trees.* Wadsworth International.

Buntine, W.[1994]. "Learning with Graphical Models." *Journal of Artificial Intelligence Research,,* (**2**, 159–225.

Dawid, A. P. [1992]. "Applications of a general propagation algorithm for probabilistic expert systems." *Statistics and Computing,* (**2**), 25–36.

Detrano, R., A. Janosi, W. Steinbrunn, M. Pfisterer, J-J. Schmid, S. Sandhu, K.H. Guppy, S. Lee, and V. Froelicher [1989]. "International Application of a New Probability Algorithm for the Diagnosis of Coronary Artery Disease." *American Journal of Cardiology,* (**64**) 304–310.

Geyer, C.J. [1993]. "Practical Markov chain Monte Carlo." *Statistical Science,* **8,** .

Glasziou, P. and J. Hilden[1989]. "Test Selection Measures." *Medical Decision Making* (**9**), 133-141.

Good, I.J. [1985]. "Weight of Evidence : a brief survey." In : *Bayesian Statistics 2,* Bernardo, J.M., DeGroot, M.H., Lindley, D.V., and Smith, A.F.M., eds, North Holland : New York, 249–269.

Good, I. J. and Card, W. [1971]. "The diagnostic process with special reference to errors." *Method of Inferential Medicine.* (**10**), 176–188.

Heckerman, D., E. Horvitz, and B. Middleton[1993]. "An Approximate Nonmyopic Computation for Value of Information." *IEEE Transaction of Pattern Analysis and Machine Intelligence* (**15**), 292–298.

Heckerman, D., E. Horvitz, and B.N. Nathwani[1992]. "Toward normative expert systems: Part I. The Pathfinder Project." *Methods of Information in Medicine* (**31**), 90–105.

Jensen, F.V. and J. Liang[1994]. "drHugin: A system for hypothesis driven data request." In: *Bayesian Belief Networks and Probabilistic Reasoning*, Gammerman, A., ed., UNICOM : London, to appear.

Kong, A.[1991]. "Efficient Methods for Computing Linkage Likelihoods of Recessive Diseases in Inbred Pedigrees." *Genetic Epidemiology*, 8, 81–103.

Lin, S. [1992]. "On the performance of Markov chain Monte Carlo methods on pedigree data and a new algorithm." *Technical Report 231*, Department of Statistics, University of Washington.

Madigan, D., Mosurski, K., and Almond, R.G. [1994]. "Explanation in Belief Networks." Submitted for publication.

Madigan, D. and Almond, R.G.[1993]. "Test Selection Strategies for Belief Networks" StatSci Research Report 20.

McSherry, D.M.G.[1986]. "Intelligent dialogue based on statistical models of clinical decision making." *Statistics in Medicine*, 5,497–502.

Matheson, J.E.[1990]. "Using Influence Diagrams to Value Information and Control." *Influence Diagrams, Belief Nets and Decision Analysis,* Oliver, Robert M. and Smith James Q. (ed.) John Wiley & Sons.

Miller, P. [1983]. "ATTENDING: Critiquing a physician's management plan." *IEEE Transactions on Pattern Analysis and Machine Intelligence,* (5, 449–461.

Murphy, P.M. and Aha, D.W.[1992]. *UCI Repository of Machine Learning Databases.* Online database maintained at the Department of Information and Computer Science, University of California, Irvine, CA.

Neal, R.M. [1993]. "Probabilistic inference using Markov chain Monte Carlo methods." *Technical Report CRG-TR-93-1*, Department of Computer Science, University of Toronto.

Raftery, A.E. and Lewis, S.L. [1992]. "How many iterations in the Gibbs sampler?" In : *Bayesian Statistics 4*, Bernardo, J.M., Berger, J.O., Dawid, A.P. and Smith, A.F.M., eds, Oxford University Press : Oxford, 763–773.

Shenoy, P.P. and Shafer, G. [1990]. "Axioms for Probability and Belief-Function Propagation." in *Uncertainty in Artificial Intelligence,* 4, 169-198.

Spiegelhalter, D.J. and Knill-Jones, R.P. [1984]. "Statistical and knowledge based approaches to clinical decision support systems, with an application in gastroenterology (with discussion)." *Journal of the Royal Statistical Society (Series A)* (**147**), 35–77.

10

Representing and Solving Asymmetric Decision Problems Using Valuation Networks

Prakash P. Shenoy

School of Business
University of Kansas
Summerfield Hall
Lawrence, KS 66045-2003 USA.
p-shenoy@ukans.edu

ABSTRACT This paper deals with asymmetric decision problems. We describe a generalization of the valuation network representation and solution technique to enable efficient representation and solution of asymmetric decision problems. The generalization includes the concepts of indicator valuations and effective frames. We illustrate our technique by solving Raiffa's oil wildcatter's problem in complete detail.

10.1 Introduction

This paper deals with asymmetric decision problems. An asymmetric decision problem can be defined most easily using its decision tree representation. In a decision tree, a path from the root node to a leaf node is called a *scenario*. We say a decision problem is asymmetric if there exists a decision tree representation of it such that not all scenarios include all variables in the problem. In asymmetric decision problems, some scenarios may exclude either some chance variables, or some decision variables, or both. The main goal of this paper is to describe a valuation network representation and solution of asymmetric decision problems.

Influence diagrams and valuation networks as originally conceived were designed for symmetric decision problems. For asymmetric decision problems, these techniques makes an asymmetric problem symmetric by adding variables and dummy configurations to scenarios. In doing so, we increase the computational burden of solving the problem. For this reason, representing and solving asymmetric problems has been the subject of several studies in recent years.

In the influence diagram literature, four techniques have been proposed by Call and Miller [1990], Smith *et al.* [1993], Fung and Shachter [1990], and Covaliu and Oliver [1996], to deal with asymmetric decision problems. Each of these four techniques is a hybrid of influence diagram and decision tree techniques. In essence, influence diagram representation is used to capture the uncertainty information, and decision tree representation is used to capture the structural asymmetry information.

In this paper, we investigate the use of valuation networks to represent and solve asymmetric decision problems. The structural asymmetry information is represented by indicator valuations. An indicator valuation is a special type of a probability valuation whose values are

---

[1] *Learning from Data: AI and Statistics V*. Edited by D. Fisher and H.-J. Lenz. © 1996 Springer-Verlag.

restricted to either 0 or 1. Indicator valuations enable us to reduce the domain of probability valuations and this contributes greatly to improving the computational efficiency of the solution technique. We use indicator valuations to define effective frames as subsets of frames of variables. All numeric information is specified only for effective frames. The solution technique is mostly the same as in the symmetric case. The main difference is that all computations are done on the effective frames of variables. This contributes to the increased efficiency of the solution technique. Also, when restricted to effective frames, the values of indicator valuations are identically one, and therefore indicator valuations can be handled implicitly and this contributes further to the increased efficiency of the solution technique.

An outline of the remainder of the paper is as follows. In Section 2, we give a verbal statement of the oil wildcatter's problem [Raiffa 1968]. This is an asymmetric decision problem. In Section 3, we describe the valuation network representation method for asymmetric decision problems and illustrate it using the oil wildcatter's problem. In Section 4, we sketch a fusion algorithm for solving valuation network representations. Finally, in Section 5, we summarize and conclude.

10.2 The Oil Wildcatter's Problem

The oil wildcatter's (OW) problem is reproduced with minor modifications from Raiffa [1968].

An oil wildcatter must decide either to drill (d) or not drill ($\sim d$). He is uncertain whether the hole is dry (dr), wet (we) or soaking (so).

Table I gives his monetary payoffs and his subjective probabilities of the various states. The cost of drilling is \$70,000. The net return associated with the d-we pair is \$50,000 which is interpreted as a return of \$120,000 less the \$70,000 cost of drilling. Similarly the \$200,000 associated with the d-so pair is a net return (a return of \$270,000 less the \$70,000 cost of drilling).

At a cost of \$10,000, the wildcatter could take seismic soundings which will help determine the geological structure at the site. The soundings will disclose whether the terrain below has no structure (ns)—that's bad, or open structure (os)—that's so-so, or closed structure (cs)—that's really hopeful. The experts have provided us with Table II which shows the prob-

TABLE I. The utility matrix for the OW problem.

	Wildcatter's profit, \$ (υ)		Act		Probability of state
			drill (d)	not drill ($\sim d$)	
State	Dry	(dr)	−70,000	0	0.500
	Wet	(we)	50,000	0	0.300
	Soaking	(so)	200,000	0	0.200

TABLE II. Probabilities of seismic test results conditional on the amount of oil.

	P(R | O)		Seismic Test Results (R)		
			No Structure (ns)	Open Structure (os)	Closed Structure (cs)
Amount	Dry	(dr)	0.600	0.300	0.100
of	Wet	(we)	0.300	0.400	0.300
Oil (O)	Soaking	(so)	0.100	0.400	0.500

FIGURE 1. Decision tree representation and solution of the OW problem.

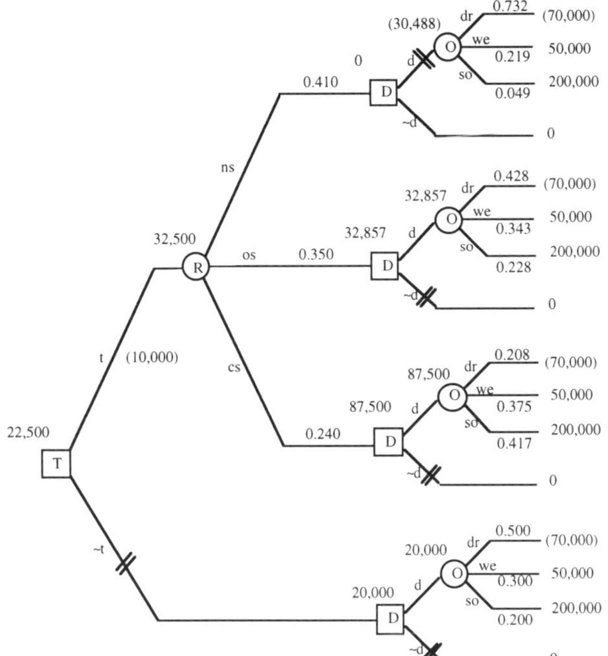

abilities of seismic test results conditioned on the amount of oil.

Figure 1 shows a decision tree representation and solution of this problem. The optimal strategy is to do a seismic test; not drill if seismic test reveals no structure, and drill if the seismic test reveals either open or closed structure. The expected profit associated with this strategy is $22,500.

Notice that the OW problem is asymmetric. This problem has 16 scenarios. Of these, 9 scenarios include all four variables, 3 scenarios include only variables T, R, and D, 3 scenarios include only variables T, D, O, and one scenario includes only variables T and D.

10.3 Valuation Network Representation

In this section, we describe the valuation network representation technique and illustrate it using the oil wildcatter's (OW) problem.

A valuation network representation is specified at three levels — graphical, dependence, and numeric. This is somewhat analogous to Howard and Matheson's [1981] relational, functional, and numerical levels of specification of influence diagrams. The graphical and dependence levels have qualitative (or symbolic) knowledge, whereas the numeric level has quantitative knowledge.

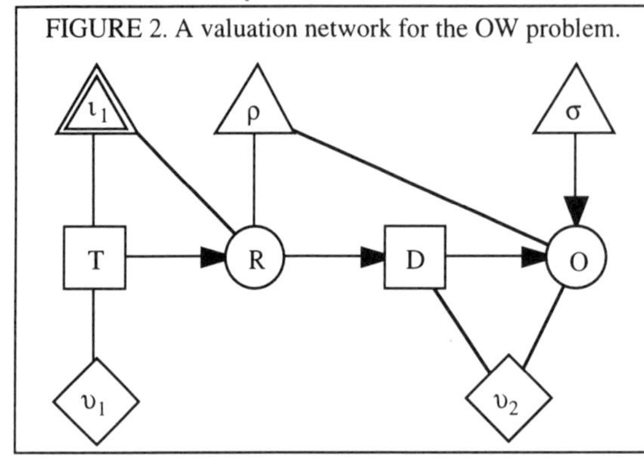

FIGURE 2. A valuation network for the OW problem.

10.3.1 Graphical Level

At the graphical level, a valuation network representation consists of a graph called a *valuation network*. Figure 2 shows a valuation network for the OW problem. A valuation network consists of two types of nodes — variable and valuation. Variables are further classified as either decision or chance, and valuations are further classified as either indicator, probability, or utility. Thus in a valuation network, there are in all five different types of nodes — decision, chance, indicator, probability, and utility.

Decision Nodes. Decision nodes correspond to decision variables and are depicted by rectangles. In the OW problem, there are two decision nodes labeled T, and D. T represents the seismic test decision, and D represents the drill decision.

Chance Nodes. Chance nodes correspond to chance variables and are depicted by circles. In the OW problem, there are two chance nodes labeled R and O. R represents the seismic test result, and O represents the amount of oil.

Let X_D denote the set of all decision variables, let X_R denote the set of all chance variables, and let X denote $X_D \cup X_R$.

Indicator Valuations. Indicator valuations represent qualitative constraints on the joint frames of decision and chance variables and are depicted by double-triangular nodes. The set of variables directly connected to an indicator valuation by undirected edges constitutes the domain of the indicator valuation. In the OW problem, there is one indicator valuation labeled ι_1. ι_1's domain is $\{T, R\}$. ι_1 represents the constraint that seismic test result is not available if the oil wildcatter decides not to do the seismic test.

Utility Valuations. Utility valuations represent factors of the joint utility function and are depicted by diamond-shaped nodes. The set of variables directly connected to a utility valuation constitutes the domain of the utility valuation. Depending on whether the utility function decomposes additively or multiplicatively, the factors are additive or multiplicative (or perhaps some combination of the two). In the OW problem, there are two additive utility valuations labeled υ_1, and υ_2. υ_1's domain is $\{T\}$, and υ_2's domain is $\{D, O\}$. υ_1 represents the profit from the seismic test decision, and υ_2 represents the profit from the drill decision.

Probability Valuations. Probability valuations represent multiplicative factors of the family of joint probability distributions of the chance variables in the problem, and are depicted by triangular nodes. The set of all variables directly connected to a probability valuation constitutes the domain of the probability valuation. In the OW problem, there are two probability valuations labeled σ, and ρ. σ's domain is $\{O\}$, and ρ's domain is $\{R, O\}$.

Information Constraints. The specification of the valuation network at the graphical level includes directed arcs between pairs of distinct variables. These directed arcs represent information constraints. Suppose R is a chance variable and suppose D is a decision variable. An arc R→D means that the true value of R is known to the decision maker (DM) at the time the DM has to choose an alternative from D's frame, and, conversely, an arc from D$\rightarrow$R means that the true value of R is not known to the DM at the time the DM has to choose an alternative from D's frame.

10.3.2 Dependence Level

Next, we specify valuation network representation at the dependence level. Like the graphical level, the dependence level involves only qualitative (or symbolic) knowledge.

Frames. Associated with each variable X is a *frame* w_X. We assume that all variables have finite frames. In the OW problem, $w_T = \{t, \sim t\}$, where t denotes do seismic test, and $\sim t$ denotes not do seismic test; $w_R = \{ns, os, cs, nr\}$, where ns denotes no structure, os denotes open structure, cs denotes closed structure, and nr denotes no result; $w_D = \{d, \sim d\}$, where d denotes drill, and $\sim d$ denotes not drill; $w_O = \{dr, we, so\}$ where dr denotes dry, we denotes wet, and so denotes soaking.

Configurations. We often deal with non-empty subsets of variables in x. Given a non-empty subset h of x, let w_h denote the Cartesian product of w_X for X in h, i.e., $w_h = \times\{w_X | X \in h\}$. We can think of w_h as the set of possible values of the joint variable h. Accordingly, we call w_h the *frame for h*. Also, we refer to elements of w_h as *configurations of h*. We use this terminology even when h consists of a single variable, say X. Thus we refer to elements of w_X as *configurations of X*.

Indicator Valuations. Suppose s is a subset of variables. An *indicator valuation for s* is a function $\iota: w_s \rightarrow \{0, 1\}$. The values of indicator valuations represent probabilities. The only values assumed by an indicator valuation are 0 and 1, hence the term indicator valuation. An efficient way of representing an indicator valuation is simply to describe the elements of the frame that have value 1, i.e., we represent ι by Ω_ι where $\Omega_\iota = \{x \in w_s | \iota(x) = 1\}$. Obviously, $\Omega_\iota \subseteq w_s$. To minimize jargon, we also call Ω_ι an indicator valuation for s.

In the OW problem, we have one indicator valuation—ι_1 (or Ω_{ι_1}) with domain $\{T, R\}$. This indicator valuation is specified as follows:

$$\Omega_{\iota_1} = \{(t, ns), (t, os), (t, cs), (\sim t, nr)\}.$$

ι_1 represents the constraint that the seismic test result is not available if the oil wildcatter decides not to do the seismic test.

Projection of Configurations. *Projection* of configurations simply means dropping extra coordinates; if (t, ns, d, dr) is a configuration of $\{T, R, D, O\}$, for example, then the projection of (t, ns, d, dr) to $\{T, R\}$ is simply (t, ns), which is a configuration of $\{T, R\}$.

If g and h are sets of variables, $h \subseteq g$, and x is a configuration of g, then let $x^{\downarrow h}$ denote the projection of x to h.

Marginalization of Indicator Valuations. Suppose Ω_{ι_a} is an indicator valuation for a, and suppose $b \subseteq a$. The marginalization of Ω_{ι_a} to b, denoted by $\Omega_{\iota_a}{}^{\downarrow b}$, is an indicator valuation for b given by

$$\Omega_{\iota_a}{}^{\downarrow b} = \{x \in w_b \mid (x, y) \in \Omega_{\iota_a} \text{ for some } y \in w_{a-b}\}.$$

To illustrate this definition, consider the indicator valuation Ω_{ι_1} for $\{T, R\}$ in the OW problem. The marginal of Ω_{ι_1} for $\{T\}$ is given by the indicator valuation

$$\Omega_{\iota_1}{}^{\downarrow T} = \{t, \sim t\}.$$

Combination of Indicator Valuations. Suppose Ω_{ι_a} is an indicator valuation for a, and suppose Ω_{ι_b} is an indicator valuation for b. The combination of Ω_{ι_a} and Ω_{ι_b}, denoted by $\Omega_{\iota_a} \otimes \Omega_{\iota_b}$, is an indicator valuation for a$\cup$b given by $\Omega_{\iota_a} \otimes \Omega_{\iota_b} =$

$$\{\mathbf{x} \in W_{a \cup b} \mid \mathbf{x}^{\downarrow a} \in \Omega_{\iota_a} \text{ and } \mathbf{x}^{\downarrow b} \in \Omega_{\iota_b}\}.$$

Effective Frames. Suppose $\{\Omega_{\iota_1}, \ldots, \Omega_{\iota_p}\}$ is the set of indicator valuations in a given problem such that Ω_{ι_j} is an indicator valuation for s_j, $j = 1, \ldots, p$. Without loss of generality, assume that $s_1 \cup \ldots \cup s_p = \mathsf{x}$. (If a variable, say X, is not included in the domain of some indicator valuation, include the vacuous indicator valuation Ω_ι for $\{X\}$, i.e., $\Omega_\iota = w_X$.) Suppose s is a subset of variables. The effective frame for s, denoted by Ω_s, is given by

$$\Omega_s = \{\otimes\{\Omega_{\iota_k} \mid s_k \cap s \neq \varnothing\}^{\downarrow s}.$$

In words, the effective frame for s is defined in two steps as follows. First we combine indicator valuations whose domains include a variable in s. Second, we marginalize the resulting combination to eliminate variables not in s.

To illustrate this definition, consider the indicator valuations Ω_{ι_1} for $\{T, R\}$. Let Ω_{ι_2} denote the vacuous indicator valuation for $\{O\}$. Then, for example, the effective frame for $\{R, O\}$ is given by $\Omega_{\{R, O\}} = (\Omega_{\iota_1} \otimes \Omega_{\iota_2})^{\downarrow\{R, O\}} =$

$\{(ns, dr), (ns, we), (ns, so), (os, dr), (os, we), (os, so), (cs, dr), (cs, we), (cs, so), (nr, dr), (nr, we), (nr, so)\}$.

Notice that the definitions of combination and marginalization of indicator valuations satisfy the three axioms needed for local computation [Shenoy and Shafer 1990]. Thus, we can compute effective frames using local computation [Shenoy 1994]. Thus, e.g., to compute the effective frame for $\{R, O\}$, by definition, $\Omega_{\{R, O\}} = (\Omega_{\iota_1} \otimes \Omega_{\iota_2})^{\downarrow\{R, O\}}$. However, if we use local computation, we can compute $\Omega_{\{R, O\}} = \Omega_{\iota_1}^{\downarrow R} \otimes \Omega_{\iota_2}$. Notice that the combination in $(\Omega_{\iota_1} \otimes \Omega_{\iota_2})^{\downarrow\{R, O\}}$ is on the frame of $\{T, R, O\}$ whereas the combination in $\Omega_{\iota_1}^{\downarrow R} \otimes \Omega_{\iota_2}$ is only on the frame of $\{R, O\}$.

As we will see shortly, all the numeric information in probability and utility valuations are specified on effective frames only. Thus, the definitions of marginalization and combination of indicator valuations allow us to compute effective frames using local computation.

10.3.3 Numeric Level

Finally, we specify a valuation network at the numeric level. At this level, we specify the details of the utility and probability valuations.

Utility Valuations. Suppose $u \subseteq \mathsf{x}$. A utility valuation υ for u is a function $\upsilon \colon \Omega_u \to R$, where R is the set of real numbers. The values of υ are utilities. If υ is a utility valuation for u, we say u is the *domain* of υ.

In the OW problem, there are two utility valuations υ_1 for $\{T\}$, and υ_2 for $\{D, O\}$. Table III shows the details of these utility valuations.

Probability Valuations. Suppose $p \subseteq \mathsf{x}$. A *probability valuation* π for p is a function $\pi \colon \Omega_p \to [0, 1]$. The values of π are probabilities. If π is a valuation for p, then we say p is the *domain* of π.

In the OW problem, there are two probability valuations, σ for $\{O\}$, and ρ for $\{O, R\}$. σ represents the prior probability for O, and ρ represents the conditional probability of R given

TABLE III. Utility valuations in the OW problem.

$\Omega_{\{D, O\}}$		υ_2
d	dr	$-70,000$
d	we	$50,000$
d	so	$200,000$
$\sim d$	dr	0
$\sim d$	we	0
$\sim d$	so	0

Ω_T	υ_1
t	$-10,000$
$\sim t$	0

TABLE IV. Probability valuations in the OW problem.

Ω_O	σ
dr	.500
we	.300
so	.200

$\Omega_{\{O, R\}}$		ρ
dr	ns	.600
dr	os	.300
dr	cs	.100
dr	nr	1
we	ns	.300
we	os	.400
we	cs	.300
we	nr	1
so	ns	.100
so	os	.400
so	cs	.500
so	nr	1

O and the fact that the values of R are not ruled out by structural constraints (see [Shenoy 1993b] for more details). Table IV shows the details of these probability valuations.

We have now completely defined a valuation network representation of a decision problem. In summary, a valuation network representation of a decision problem Δ consists of decision variables, chance variables, indicator valuations, probability valuations, utility valuations, and information constraints, $\Delta = \{x_D, x_R, \{\iota_1, ..., \iota_p\}, \{\upsilon_1, ..., \upsilon_m\}, \{\rho_1, ..., \rho_n\}, \rightarrow\}$.

10.4 A Fusion Algorithm

In this section, we sketch a fusion algorithm for solving valuation network representations of decision problems.

The fusion algorithm is essentially the same as in the symmetric case [Shenoy 1992]. The main difference is in how indicator valuations are handled. Since indicator valuations are identically one on effective frames, there are no computations involved in combining indicator valuations. This contributes to the efficiency of the solution technique. Indicator valuations do contribute domain information and cannot be totally ignored.

Fusion with respect to a decision variable D is defined as follows. All utility and indicator valuations that include D in their domain are combined together, and the resulting utility valuation υ is marginalized such that D is eliminated from its domain. A new indicator valuation ζ_D for h corresponding to the decision function for D is created. The utility valuations that do not include D in their domain remain unchanged. All probability and indicator valuations that include D in their domain are combined together and the resulting probability valuation ρ is combined with ζ_D and the result is marginalized so that D is eliminated from its domain. The probability and indicator valuations that do not include D in their domains remain unchanged.

Fusion with respect to a chance variable C is defined as follows. The utility, probability, and indicator valuations whose domains do not include C remain unchanged. A new probability valuation, say ρ, is created by combining all probability and indicator valuations whose domain include C and marginalizing C out of the combination. Finally, we combine all probability and indicator valuations whose domains include C, divide the resulting probability valuation by the new probability valuation ρ that was created, combine the resulting probability valuation with the utility valuations whose domains include C, and finally marginalize the resulting utility valuation such that C is eliminated from its domain.

The details of the fusion algorithm are given in [Shenoy 1993b]. Figure 3 depicts the fusion algorithm graphically for the OW problem. The numerical details are shown in Tables V–IX.

FIGURE 3. The fusion algorithm for the OW problem.

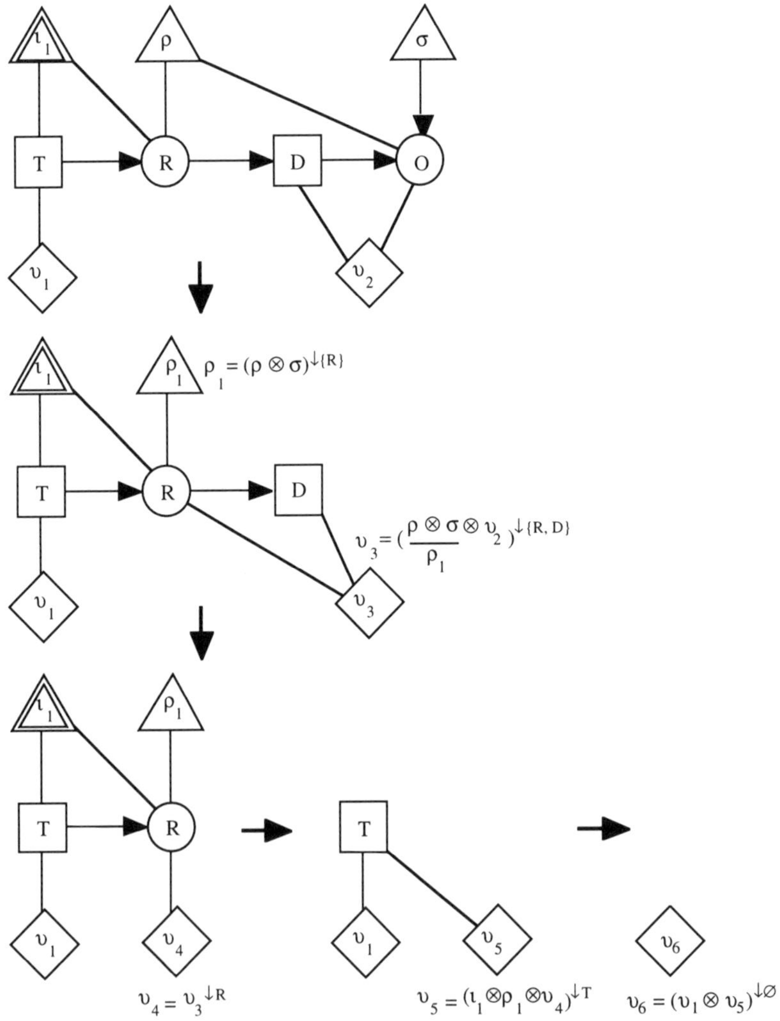

TABLE V. Fusion with respect to O.

$\Omega_{\{R, O\}}$		ρ	σ	$\rho \otimes \sigma$ $= \rho'$	$\rho'^{\downarrow R}$ $= \rho_1$	ρ'/ρ_1 $= \rho''$
ns	dr	.600	.500	.300	.410	.732
ns	we	.300	.300	.090		.219
ns	so	.100	.200	.020		.049
os	dr	.300	.500	.150	.350	.428
os	we	.400	.300	.120		.343
os	so	.400	.200	.080		.228
cs	dr	.100	.500	.050	.240	.208
cs	we	.300	.300	.090		.375
cs	so	.500	.200	.100		.417
nr	dr	1	.500	.500	1	.500
nr	we	1	.300	.300		.300
nr	so	1	.200	.200		.200

TABLE VI. Fusion with respect to O (continued from Table V).

$\Omega_{\{R, D, O\}}$			ρ''	υ_2	$\rho'' \otimes \upsilon_2$ $= \upsilon'$	$\upsilon'^{\downarrow \{R, D\}}$ $= \upsilon_3$
ns	d	dr	.732	−70	−51.24	−30.49
ns	d	we	.219	50	10.95	
ns	d	so	.049	200	9.80	
ns	~d	dr	.732	0	0	0
ns	~d	we	.219	0	0	
ns	~d	so	.049	0	0	
os	d	dr	.428	−70	−29.96	32.86
os	d	we	.343	50	17.15	
os	d	so	.228	200	45.60	
os	~d	dr	.428	0	0	0
os	~d	we	.343	0	0	
os	~d	so	.228	0	0	
cs	d	dr	.208	−70	−14.56	87.5
cs	d	we	.375	50	18.75	
cs	d	so	.417	200	83.4	
cs	~d	dr	.208	0	0	0
cs	~d	we	.375	0	0	
cs	~d	so	.417	0	0	
nr	d	dr	.500	−70	−35	20
nr	d	we	.300	50	15	
nr	d	so	.200	200	40	
nr	~d	dr	.500	0	0	0
nr	~d	we	.300	0	0	
nr	~d	so	.200	0	0	

TABLE VII. Fusion with respect to D.

$\Omega_{\{R, D\}}$		υ_3	$\upsilon_3^{\downarrow R}$ $= \upsilon_4$	ξ_D
ns	d	−30.49		
ns	~d	0	0	~d
os	d	32.86	32.86	d
os	~d	0		
cs	d	87.5	87.5	d
cs	~d	0		
nr	d	20	20	d
nr	~d	0		

TABLE VIII. Fusion with respect to R.

$\Omega_{\{T, R\}}$		$\iota_1 \otimes \rho_1$	υ_4	$\iota_1 \otimes \rho_1 \otimes \upsilon_4$	$(\iota_1 \otimes \rho_2 \otimes \upsilon_4)^{\downarrow T}$ $= \upsilon_5$
t	ns	.410	0	0	32.50
t	os	.350	32.86	11.50	
t	cs	.240	87.5	21	
~t	nr	1	20	20	20

TABLE IX. Fusion with respect to T.

Ω_T	υ_1	υ_5	$\upsilon_1 \otimes \upsilon_5$	$(\upsilon_1 \otimes \upsilon_5)^{\downarrow \varnothing}(\blacklozenge)$ $= \upsilon_6(\blacklozenge)$	$\xi_T(\blacklozenge)$
t	−10	32.50	22.50	22.50	t
~t	0	20	20		

10.5 Summary and Conclusion

The main contribution of this paper is a generalization of the valuation network technique for representing and solving asymmetric decision problems. The structural asymmetry in a decision problem is represented by indicator valuations. An indicator valuation is a special type of a probability valuation. Indicator valuations allow us to reduce the domain of probability valuations. This contributes to the efficiency of the solution technique. Also, indicator valuations are used to define effective frames. An effective frame is a subset of a frame. All computations are done on effective frames, and this contributes also to the efficiency of the solution technique.

In [Shenoy 1993b], we compare the asymmetric valuation network representation and solution technique with the symmetric valuation network technique described in [Shenoy 1992], and with the influence diagram-based technique of Smith *et al.* [1993].

10.6 Acknowledgments

This work is based upon work supported in part by the National Science Foundation under Grant No. SES-9213558. I am grateful to Concha Bielza, Rui Guo, Liping Liu, and an anonymous referee for comments and discussions. Comments from Concha Bielza have resulted in removal of several errors from [Shenoy 1993b] from which this article is derived.

10.7 References

Call, H. J. and W. A. Miller (1990), "A comparison of approaches and implementations for automating decision analysis," *Reliability Engineering and System Safety*, **30**, 115–162.

Covaliu, Z. and R. M. Oliver (1996), "Formulation and solution of decision problems using sequential decision diagrams," *Management Science*, to appear.

Fung, R. M. and R. D. Shachter (1990), "Contingent influence diagrams," Working Paper, Advanced Decision Systems, Mountain View, CA.

Howard, R. A. and J. E. Matheson (1981), "Influence diagrams," in R. A. Howard and J. E. Matheson (eds.) (1984), *The Principles and Applications of Decision Analysis*, **2**, 719–762, Strategic Decisions Group, Menlo Park, CA.

Raiffa, H. (1968), *Decision Analysis*, Addison-Wesley, Reading, MA.

Shenoy, P. P. (1992), "Valuation-based systems for Bayesian decision analysis," *Operations Research*, **40**(3), 463–484.

Shenoy, P. P. (1993a), "A new method for representing and solving Bayesian decision problems," in D. J. Hand (ed.), *Artificial Intelligence Frontiers in Statistics: AI and Statistics III*, 119–138, Chapman & Hall, London.

Shenoy, P. P. (1993b), "Valuation network representation and solution of asymmetric decision problems," Working Paper No. 246, School of Business, University of Kansas, Lawrence, KS.

Shenoy, P. P. (1994), "Consistency in valuation-based systems," *ORSA Journal on Computing*, **6**(3), 281–291.

Shenoy, P. P. and G. Shafer (1990), "Axioms for probability and belief-function propagation," in R. D. Shachter, T. S. Levitt, J. F. Lemmer and L. N. Kanal (eds.), *Uncertainty in Artificial Intelligence*, *4*, 169-198, North-Holland, Amsterdam.

Smith, J. E., S. Holtzman and J. E. Matheson (1993), "Structuring conditional relationships in influence diagrams," *Operations Research*, **41**(2), 280–297.

11
A Hill-Climbing Approach for Optimizing Classification Trees

Xiaorong Sun, Steve Y. Chiu and Louis Anthony Cox

U S WEST Advanced Technology
Boulder, CO 80303, USA
email: {*xsun,schiu,tony*}*@advtech.uswest.com*

ABSTRACT We consider the problem of minimizing the expected cost of determining the correct value of a binary-valued function when it is costly to inspect the values of its arguments. This type of problem arises in distributed computing, in the design of diagnostic expert systems, in reliability analysis of multi-component systems, and in many other applications. Any feasible solution to the problem can be described by a sequential inspection procedure which is usually represented by a binary classification tree. In this paper, we propose an efficient hill-climbing algorithm to search for the optimal or near-optimal classification trees. Computational results show that the hill-climbing approach was able to find optimal solutions for 95% of the cases tested.

11.1 Introduction

The following minimum expected-cost classification problem arises in many logical inference, reliability testing, pattern recognition, and statistical classification applications. Suppose that $X = (x_1, \cdots, x_n)$ is a binary pattern vector with n components and that $\phi(X)$ is a function mapping pattern vectors into $\{0, 1\}$. If pattern vectors are generated randomly according to a probability n-vector $(p_1, \cdots, p_n)$, where p_i is the probability that $x_i = 1$ and $1 - p_i$ is the probability that $x_i = 0$; and if component x_i can be inspected at a cost c_i to determine its true value (0 or 1), then the minimum expected-cost classification problem is to find an adaptive sequential inspection strategy that minimizes the expected cost of determining $\phi(X)$.

An adaptive sequential inspection strategy can be described as a binary classification tree or decision tree. The leaf nodes of the tree represent the results of the inspection and are labeled by the corresponding values of the $\phi(\cdot)$ function (instead of "0" or "1", we shall use "F" or "W" to label the leaf nodes to avoid confusion with the component indices). The non-leaf nodes represent inspection actions and are labeled by indices of the component to be inspected. Each non-leaf node v has exactly two sons called 0-son and 1-son of v, and v is called the father of its sons. Suppose node v is labeled i, its 0-son v_0 has label j and its 1-son v_1 has label k, then component j (or k) will be inspected next if component i is found to be equal to 0 (or 1). The first component to be inspected is the root node of the tree.

The expected cost of a tree can be calculated recursively in the following way. The costs of leaf nodes are 0. The cost $C(v)$ of a non-leaf node v is the sum of c_i (the cost to inspect

---

component i) and the expected cost of its sons, i.e., $C(v) = c_i + p_i C(v_1) + (1 - p_i)C(v_0)$. The cost of the tree is defined to be the cost of its root.

FIGURE 1 gives an example of a classification tree where x_4 is to be inspected first. If $x_4 = 0$, then x_2 will be inspected next; otherwise, x_1 will be inspected next. The rest of the tree is followed similarly until the value of $\phi(X)$ can be determined. The numbers on the right side of each node are the cost of that node when $p_1 = 0.2$, $p_2 = 0.4$, $p_3 = 0.1$, $p_4 = 0.6$, and $c_1 = 8$, $c_2 = 2$, $c_3 = c_4 = 1$. The expected cost of the tree is 7.192.

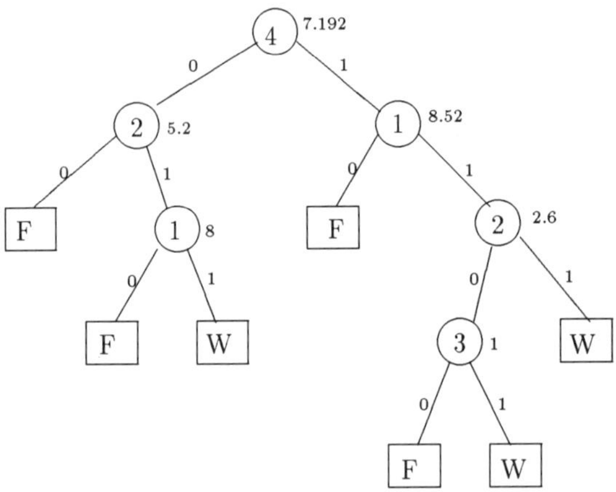

FIGURE 1. An example of a decision tree

Prominent applications of the problem include the following:

Efficient computation in a distributed database environment. Suppose that a processor in a computer network is assigned the task of computing $\phi(X)$, but that the component of X are stored at different remote locations in the network. If x_i is stored at a node having a distance (measured in a communication cost or accessing time) c_i from the processor computing ϕ, then the problem of sequentially selecting components of X to retrieve over the network so as to minimizing the expected cost (or time, etc.) to compute $\phi(X)$ can be formulated as this classification problem.

Optimal design of expert systems. If the components of X represent the true values of n propositions in an expert system and if $\phi(X)$ is a function determining a conclusion from the true values of these propositions, then the problem of efficient inference (deriving correct conclusion at least expected cost) can be formulated as a classification problem.

Efficient diagnostic procedure. If $\phi(X)$ is a biostatistical discriminant of classification function with independent variables $x_1, \cdots, x_n$ that are costly to observe, e.g., because they require intrusive test upon the patient being classified, then the problem is to adaptively select variables for observation so as to minimize the expected cost of computing $\phi(X)$.

Reliability analysis of multi-component systems. If the n components of X represent the states of n components in a newly manufactured multicomponent system, and if $\phi(X)$ is the structure function having value 1 or "W" (standing for "working") if the system operates, and 0 or "F" (standing for "failed") if otherwise, then the problem is to determine whether the system will operate with minimum expected cost. A well-known structure is the linear threshold system which is defined by

$$\phi(X) = \begin{cases} 1 & \text{if } w_1 x_1 + w_2 x_2 + \cdots + w_n x_n \geq k, \\ 0 & \text{otherwise.} \end{cases}$$

When $w_1 = \cdots = w_n = 1$, the system is called a *k-out-of-n* system. When $w_1 = \cdots = w_n = 1$ and $k = 1$, it is called a *parallel system*. When $w_1 = \cdots = w_n = 1$ and $k = n$, it is called a *series system*.

The classification problem has been proved to be NP-hard[Cox89]. The only known polynomial-time solvable case is k-out-of-n system[Ben-Dov81, Chang90], which includes series and parallel systems.

It is obvious that an optimal decision strategy does not test any component twice, which implies for any leaf of a classification tree, the path from the root to the leaf contains no pair of nodes which have the same label. So if we already observed the values of some components (which form a *partial tree*), then we need only consider the sub-problem restricted by these values. To our knowledge, all existing heuristics for solving this problem focus on procedures to construct a good decision tree starting from the root. (The IDL approach in [Velde90] only finds a tree with minimal "size". The costs and probabilities associated with component variables are not considered). No local improvement method has been proposed since it is not obvious one can move from one tree to another in the space of all decision trees This paper is the first attempt to develop such a method. We will construct a delete/add (DA) move and show that any tree in the solution space can be obtained from any other tree by a series of DA moves. Section 11.2 contains the necessary notations to describe the algorithm. Section 11.3 describes the DA move in detail. Section 11.4 shows the optimality of our heuristic solutions for the parallel and series systems. Section 11.5 reports the computational results.

11.2   Notations

Consider a function $\phi(X)$ mapping a pattern vector X to $\{0,1\}$. The ith component is *irrelevant* to ϕ if $\phi(x_1, \cdots, x_{i-1}, 1, x_{i+1}, \cdots, x_n) = \phi(x_1, \cdots, x_{i-1}, 0, x_{i+1}, \cdots, x_n)$ for any 0,1-vector $(x_1, \cdots, x_{i-1}, x_{i+1}, \cdots, x_n)$. Otherwise, the ith component is *relevant* to ϕ. If $E = \{1, 2, \cdots, n\}$ is a set of components and the function ϕ satisfies:

(i) Monotonicity: If $X \leq Y$ (i.e., $x_i \leq y_i$ for $i = 1, \cdots, n$), then $\phi(X) \leq \phi(Y)$.

(ii) Relevancy: All components are relevant.

then (E, ϕ) is called a *coherent system*. For such a system, each component can have only one state, i.e., 0 (failed) or 1 (working). So an n-dimensional 0,1-vector X is also called the *state vector* of the system. (See [Barlow75] and [Ben-Dov81] for background). Note that we assume the data set is consistent here, i.e., no identical patterns have different ϕ values.

For a coherent system (E, ϕ), a *1-set* is defined as any set of component values for a state vector X that guarantees $\phi(X) = 1$. A *minimal 1-set* is a 1-set that ceases to be one if any of the component values in it are removed. Similarly, a *0-set* is defined as any set of component values for a state vector X that guarantees $\phi(X) = 0$, and a *minimal 0-set* is a 0-set containing no proper subsets that are also 0-sets. By the monotonicity of coherent system, a minimal 1-set contains no component of value 0, and a minimal 0-set contains no component of value 1. The components of value 0 in a 1-set, or the components of value 1 in a 0-set, are called *inessential* components.

Now consider a decision tree T, and a path P from the root to a leaf labeled "W". It can be seen that the set of the component values determined by the path P form a 1-set of the system. Similarly, the component values determined by a path from the root to a leaf labeled "F" form a 0-set. A path P from the root to a leaf of the tree T is called a *minimal path* of the tree if the corresponding component values form a minimal 1-set or 0-set. The following lemma can easily be seen to be true for a coherent system.

Lemma 11.2.1 *Suppose T is a decision tree and P is a path from the root to a leaf labeled by "F" ("W"). If P has a component i which has value 1 (0) in path P, then P is not minimal.*

Visually, for any decision tree drawn on the plane, if we place 1-sons under the right branch and 0-sons under the left branch of their fathers, then any path except the extreme left and right ones is not minimal. By definition, the set of component values determined by any non-minimal path P from the root to a leaf contains an inessential component i (which will also be called an inessential component of the path P). Thus, it is unnecessary to test the component i if the actual system state (i.e., working or failed) is determined by the path P. In this way, we can build a new tree $T_{P,i}$, with possibly a lower expected cost, in which component i is deleted from path P. The detail is described in the next Section.

Since no path from root to leaf may contain the same component more than once, a path can be uniquely identified by all of its labels in the order from the root to leaf. Thus, we will also use the component label to identify a node in a path in the following sections.

11.3   The DA move

For a path P with an inessential component i of a decision tree T, if component i is deleted from P, we may add i to other paths of T, and use subtrees of T to build a new tree $T_{P,i}$. This is the main idea of the DA move, i.e., **Delete** a component from a path, and **Add** it to other paths if necessary.

We now give a formal definition of the DA move. Consider a decision tree T and a path $P = v_1, v_2, \cdots, v_p$ specified by the labels of its nodes such that v_1 is the label of the root, and v_p is the label of a leaf (P is shown by the dotted line in FIGURE 2). Assume P is not minimal with an inessential component v_i. For any j such that $i \leq j < p$, denote by t_j the value of the component v_j determined by the path P, and let $s_j = 1 - t_j$. Denote by T_j the subtree of T rooted at the s_j-son of the node labeled v_j.

Since v_i is inessential in path P, we may build a new tree $T_{P,i}$ in which node v_i will be deleted from P. For any j such that $i < j < p$, the node labeled v_j in the path P of tree T will remain as a node in the new tree with the same label. Since we want P remain as a path of the new tree, the t_j-son of the node labeled v_j is still v_{j+1}. However, the s_j-son of v_j in the new tree will be a new node u_j with label v_i (see FIGURE 2). Obviously, we

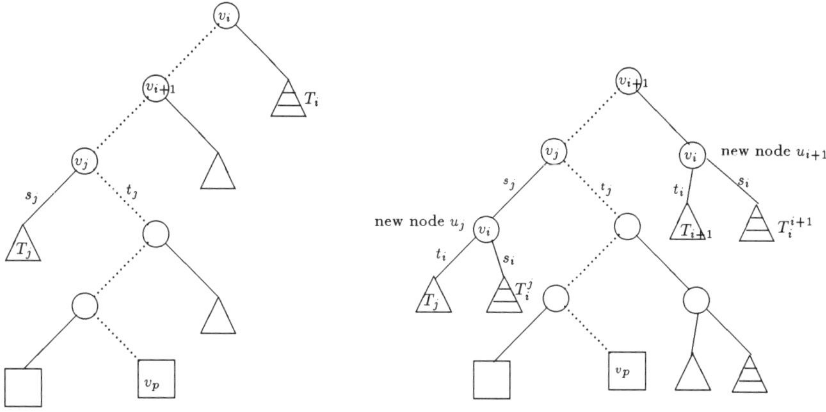

Tree T and a path P Tree $T_{P,i}$ obtained by a single DA move from T

FIGURE 2. Illustration of DA move

may still use T_j as the t_i-son of u_j (strictly speaking, as the subtree rooted at the t_i-son of node u_j). The s_i-son T_i^j of u_j will be *derived* from the subtree T_i described as follows.

We can not simply use T_i as the s_i son of u_j since otherwise some paths may contain some components from $\{v_{i+1}, \cdots, v_j\}$ twice.

Now compare the subtree T_i of tree T, and the subtree T_i^{i+1} rooted at the s_i-son of the new node u_{i+1} in the new tree $T_{P,i}$ which will be built. We can see that both subtrees assume exactly the same values of components $v_1, \cdots, v_i$. The difference between them is that T_i^{i+1} assumes that the value of component v_{i+1} is also known (i.e., s_{i+1}). Therefore, we can use the same inspection strategy specified by T_i to test the sub-system restricted by the component values of $v_1, \cdots, v_{i+1}$, only remembering that we do not have to test component v_{i+1}. In other word, the subtree T_i^{i+1} can be derived from T_i by replacing all subtrees rooted at those nodes which have label v_{i+1} by the corresponding subtrees rooted at its s_{i+1}-son. In this case, we shall say that T_i^{i+1} is *derived* from T_i restricted by $x_{v_{i+1}} = s_{i+1}$.

In general, for any j such that $i < j < p$, we can build a subtree T_i^j from T_i as the s_i-son of u_j in the new tree recursively by maintaining a tree S which is initialized to be T_i and do the following:

For $j = i + 1, \cdots, p - 1$
begin
    Let T_i^j be the tree derived from S restricted by $x_{v_j} = s_j$;
    Replace S by the tree derived from S restricted by $x_{v_j} = t_j$;
end.

Using $T_i^{i+1}, \cdots, T_i^{p-1}$, we can define a new tree $T_{P,i}$ from a given tree T by the following rules.

(a) If v_i is not the label of the root of T, then replace the t_i-son of v_i by v_{i+1};

Else replace the root of T by v_{i+1}.

(b) For each node $v_j (i < j < p)$ of P:

If the set of component values defined by $x_{v_j} = s_j$ and $x_{v_k} = t_k$ for every k such that $1 \le k < i$ form a 1-set (0-set), then add a s_j-leaf to $T_{P,i}$ with label "W" ("F");

Else add a new node u_j with label v_i, and let T_i^j, T_j be its s_i-son and t_i-son respectively.

(c) All other nodes will be unchanged.

The following lemma holds:

Lemma 11.3.1 *The new tree $T_{P,i}$ is also a decision tree for the same coherent system.*

Proof: The definitions of $T_i^{i+1}, \cdots, T_i^{p-1}$, together with rules (a) and (b) guarantee that no path of the tree contains the same component more than once. Rules (b) and (c) guarantee that the leaf labels of $T_{P,i}$ identify the states of the system. $\square$

In the space of all decision trees, the move from T to the new decision tree $T_{P,i}$ constitutes a *DA move*. A given decision tree can be improved by a hill-climbing heuristics using DA moves. One (greedy) approach is to always move to the next tree which has the maximum amount of improvement. Another method is to move to the first direction in which an improvement is found. The details are described as follows, where $C(T)$ is the cost of tree T.

ALGORITHM: DA-climb heuristics.
INPUT: A coherent system (E, ϕ) and a decision tree T.
OUTPUT: An improved decision tree T.

Step 1. Find a path P which is not minimal with an inessential component i such that $C(T_{P,i}) < C(T)$.

If there is no such path, then stop and output T; Else go to Step 2.

Step 2. Let $T = T_{P,i}$, and go to Step 1.

ALGORITHM: Greedy DA-climb heuristic.
INPUT and OUTPUT are the same as above algorithm.

Step 1. Find a path P which is not minimal with an inessential component i such that $C(T) - C(T_{P,i})$ is maximized.

If the the maximum is 0, then stop and output T; Else go to Step 2.

Step 2. Let $T = T_{P,i}$, and go to Step 1.

The following theorem theoretically guarantees the quality of the solutions found by the heuristics.

Theorem 11.3.1 *Let T_1 and T_2 be any two decision trees of the same coherent system. Then T_1 can be always transformed to T_2 by a series of DA moves.*

Proof: Let i_1 be the label of the root r_1 of T_1, and i_2 be that of the root r_2 of T_2, Since all components are relevant, T_1 must have a node v with label i_2. If $i_1 \neq i_2$, then node v of T_1 must have a father u. By Lemma 11.2.1, there is a path P in T_1 such that u is inessential for P. It can be seen that in the tree $T_{P,u}$ obtained by a single DA move, the depth of v which has a label i_2 is decreased by 1. After a series of such DA moves, eventually we will have a tree T' whose root has the same label i_2 as T_2.

Now the two subtrees of the root of T' define the same sub-systems as that of T_2, respectively. The theorem then follows by induction on the number of the components in the coherent system. $\Box$

11.4   Series and parallel systems

For series and parallel systems, the decision tree corresponds to a permutation of all components of the system. Therefore, we have

Lemma 11.4.1 *For series and parallel systems, one DA move is equivalent to switching the order of two adjacent non-leaf nodes of the permutation tree.*

Proposition 11.4.1 *Both heuristics DA-climb and Greedy DA-climb give the optimum decision tree for series and parallel systems.*

Proof: This is a direct conclusion of Lemma 11.4.1 and the known fact that [Ben-Dov81] the optimal test order for series system is the permutation $\pi_1, \cdots, \pi_n$ satisfying

$$\frac{c_{\pi_1}}{q_{\pi_1}} \leq \cdots \leq \frac{c_{\pi_n}}{q_{\pi_n}}.$$

The proof for parallel system is similar. $\Box$

11.5   Computational results

Both hill-climbing heuristics were coded in C, and run on a Sparc 10 workstation. The optimal decision tree is found by a dynamic programming recursion [Cox89].

Heuristic *DA-climb* uses a *depth-first search* [Tarjan72] method to find if there is a path P which is not minimal with an inessential component i such that $C(T_{P,i}) < C(T)$. If such a path P and i are found, then we immediately replace T by $T_{P,i}$. *Greedy DA-climb* uses the same search method to exhaust all paths of current tree, and finds the maximum defined in Step 1.

Our heuristics have been tested on linear threshold systems, which are an important class of reliability systems [Sheng69]. The structure functions are generated using linear inequalities $w_1 x_1 + \cdots + w_n x_n \geq k$ whose coefficients are uniformly distributed in the interval $[1, 99]$, and k is set to the largest integer less than $\beta \sum_{i=1}^{n} w_i$. (It can be seen that any linear threshold system satisfies the monotonicity condition of coherent systems.

Although it may not satisfy the relevancy condition, both DA-climb and greedy DA-climb heuristics can still be applied).

From lemma 11.2.1, a minimal 1-set (0-set) can not contain components with value 0 (1). To check if a subset of components with value 1's form a minimal 1-set, we need only check if the sum of the coefficients w_i of these components is larger than or equal to k, and if any proper subset with one less component also has such property. Similarly, to check if a subset of components with value 0's form a minimal 0-set, we need only check if the sum of the coefficients w_i of the components not in the subset is larger than or equal to k, and if any subset with one more component also has such property.

The probabilities $p_1, \cdots, p_n$ are uniformly generated in the interval $(0.01, 0.95)$. The costs $c_1, \cdots, c_n$ are real numbers uniformly distributed in the interval $(1.00, 99.00)$.

The initial decision tree is constructed based on the strategy which examines all components in the order $\pi_1, \cdots, \pi_n$ such that $w_{\pi_1} \geq \cdots \geq w_{\pi_n}$.

For each $n = 5, 6, 7$, both heuristics were run for 100 problems. In TABLE 11.1, $\beta = 0.3$. The left and right column under Opt(%) show the percentage of times heuristic *DA-climb* and *Greedy DA-climb* obtain the optimal solutions. The columns under Ave.Err(%) show the average errors of the two heuristics relative to the optimal solutions. The columns under Larg.Err(%) show the largest error relative to the optimal solutions.

n	Opt(%)		Ave.Err(%)		Larg.Err(%)	
5	98	99	0.01	0.00	0.94	0.06
6	98	97	0.01	0.01	0.60	1.07
7	97	96	0.01	0.04	0.48	2.84

TABLE 11.1. $\beta = 0.3$

TABLE 11.2 shows similar results for 100 randomly generated problems with $\beta = 0.5$.

n	Opt(%)		Ave.Err(%)		Larg.Err(%)	
5	99	99	0.03	0.03	2.69	2.69
6	95	93	0.01	0.01	0.38	0.38
7	95	95	0.02	0.02	1.43	1.43

TABLE 11.2. $\beta = 0.5$

Since we can find the exact optimal tree for k-out-of-n system for any n [Ben-Dov81], we compared our heuristic solutions with the optimal solution for larger n. In TABLE 11.3, we can see that for all problems we tested, both heuristics find optimal solutions. The columns under Ave.CPU show the average CPU time in seconds. The columns under Ave.Iter show the number of steps each heuristic took to reach a local optimum. Since *Greedy DA-climb* needs to search the whole tree it is not surprising to see that it takes more CPU time than *DA-climb*, but takes fewer steps to reach a local optimum.

11.6   Conclusion

We have developed a heuristic procedure method to search the space of binary decision trees based on hill climbing ideas. Computational results show that our heuristic is able to

n	Opt(%)		Ave.CPU		Ave.Iter	
5	100	100	0.01	0.02	8.6	5.0
6	100	100	0.14	0.31	22.1	11.7
7	100	100	0.61	1.39	39.6	19.2
8	100	100	1.91	5.40	64.8	30.4
9	100	100	9.15	31.27	102.3	46.6
10	100	100	23.69	78.11	163.2	69.5

TABLE 11.3. k-*out-of-n* system, $\beta = 0.8$

find optimal solutions in most cases. While hill-climbing is ubiquitously used on standard optimization problems, the possibility of applying it to classification trees has not been previously recognized.

In contrast to the dynamic programming or any other exact solution procedures, which may take a long time without presenting any feasible solution, our hill-climbing heuristic always maintains a feasible solution. Although the heuristic may also take a long time to run because of the size of the classification tree, it can be terminated at any time and still delivers a feasible solution.

11.7 REFERENCES

[Barlow75] R.E. Barlow and F. Proschan, *Statistical theory of reliability and life testing*, Holt, Rinehart and Winston, New York, 1975.

[Ben-Dov81] Y. Ben-Dov, Optimal testing procedures for special structures of coherent system, *Management Science* 27, No. 12 (1981), 1410-1420.

[Chang90] M.F. Chang, W. Shi and W.K. Fuchs, Optimal diagnosis procedures for k-out-of-n system, *IEEE Trans. on Computers* 39(4) (1990) 559-564.

[Cox89] L.A. Cox, Y. Qiu and W. Kuehner, Heuristic least-cost computation of discrete classification functions with uncertain argument values, *Annals of Operations research*, 21(1989) 1-30.

[Sheng69] C.L. Sheng, *Threshold Logic*, Academic Press, New York, 1969.

[Tarjan72] R.E. Tarjan, Depth-first search and linear graph algorithms, *SIAM J. Computing* 1 (1972) 146-160.

[Velde90] W. Van de Velde, Incremental induction of topologically minimal trees, *Proc. of the Seventh International Conference on Machine Learning* (1990) 66-74.

Part III

Search Control in Model Hunting

12
Learning Bayesian Networks is NP-Complete

David Maxwell Chickering

Computer Science Department
University of California at Los Angeles
dmax@cs.ucla.edu

ABSTRACT

Algorithms for learning Bayesian networks from data have two components: a scoring metric and a search procedure. The scoring metric computes a score reflecting the goodness-of-fit of the structure to the data. The search procedure tries to identify network structures with high scores. Heckerman et al. (1995) introduce a Bayesian metric, called the BDe metric, that computes the relative posterior probability of a network structure given data. In this paper, we show that the search problem of identifying a Bayesian network—among those where each node has at most K parents—that has a relative posterior probability greater than a given constant is NP-complete, when the BDe metric is used.

12.1 Introduction

Recently, many researchers have begun to investigate methods for learning Bayesian networks. Many of these approaches have the same basic components: a scoring metric and a search procedure. The scoring metric takes a database of observed cases D and a network structure B_S, and returns a score reflecting the goodness-of-fit of the data to the structure. A search procedure generates networks for evaluation by the scoring metric. These approaches use the two components to identify a network structure or set of structures that can be used to predict future events or infer causal relationships.

Cooper and Herskovits (1992)—herein referred to as CH—derive a Bayesian metric, which we call the BD metric, from a set of reasonable assumptions about learning Bayesian networks containing only discrete variables. Heckerman et al. (1995)—herein referred to as HGC—expand upon the work of CH to derive a new metric, which we call the BDe metric, which has the desirable property of *likelihood equivalence*. Likelihood equivalence says that the data cannot help to discriminate equivalent structures.

We now present the BD metric derived by CH. We use B_S^h to denote the hypothesis that B_S is an I-map of the distribution that generated the database.[2] Given a belief-network structure B_S, we use Π_i to denote the parents of x_i. We use r_i to denote the number of states of variable x_i, and $q_i = \prod_{x_l \in \Pi_i} r_l$ to denote the number of instances of Π_i. We use the integer j to index these instances. That is, we write $\Pi_i = j$ to denote the observation of the jth instance of the parents of x_i.

---

[1] *Learning from Data: AI and Statistics V.* Edited by D. Fisher and H.-J. Lenz. ©1996 Springer-Verlag.
[2] There is an alternative causal interpretation of network structures not discussed here. See HGC for details.

Using reasonable assumptions, CH derive the following Bayesian scoring metric:

$$p(D, B_S^h|\xi) = p(B_S^h|\xi) \cdot \prod_{i=1}^{n} \prod_{j=1}^{q_i} \frac{\Gamma(N_{ij}')}{\Gamma(N_{ij}' + N_{ij})} \cdot \prod_{k=1}^{r_i} \frac{\Gamma(N_{ijk}' + N_{ijk})}{\Gamma(N_{ijk}')} \tag{12.1}$$

where ξ is used to summerize all background information, N_{ijk} is the number of cases in D where $x_i = k$ and $\Pi_i = j$, $N_{ij} = \sum_{k=1}^{r_i} N_{ijk}$, $N_{ij}' = \sum_{k=1}^{r_i} N_{ijk}'$, and $\Gamma(\cdot)$ is the *Gamma* function. The parameters N_{ijk}' characterize our prior knowledge of the domain. We call this expression or any expression proportional to it the BD (*Bayesian Dirichlet*) metric.

HGC derive a special case of the BD metric that follows from likelihood equivalence. The resulting metric is the BD metric with the prior parameters constrained by the relation

$$N_{ijk}' = N' \cdot p(x_i = k, \Pi_i = j|B_{S_C}^h, \xi) \tag{12.2}$$

where N' is the user's *equivalent sample size* for the domain, and $B_{S_C}^h$ is the hypothesis corresponding to the complete network structure. HGC note that the probabilities in Equation 12.2 may be computed from a *prior network*: a Bayesian network encoding the probability of the first case to be seen.

HGC discuss situations when a restricted version of the BDe metric should be used. They argue that in these cases, the metric should have the property of *prior equivalence*, which states that $p(B_{S1}^h|\xi) = p(B_{S2}^h|\xi)$ whenever B_{S1} and B_{S2} are equivalent.

HGC show that the search problem of finding the l network structures with the highest score among those structure where each node has at most one parent is polynomial whenever a decomposable metric is used. In this paper, we examine the general case of search, as described in the following decision problem:

K-LEARN
INSTANCE: Set of variables U, database $D = \{C_1, \ldots, C_m\}$, where each C_i is an instance of all variables in U, scoring metric $M(D, B_S)$ and real value p.
QUESTION: Does there exist a network structure B_S defined over the variables in U, where each node in B_S has at most K parents, such that $M(D, B_S) \geq p$?

Höffgen (1993) shows that a similar problem for PAC learning is NP-complete. His results can be translated easily to show that K-LEARN is NP-complete for $k > 1$ when the BD metric is used. In this paper, we show that K-LEARN is NP-complete, even when we use the BDe metric and the constraint of prior equivalence.

12.2 K-LEARN is NP-Complete

In this section, we show that K-LEARN is NP-complete, even when we use the likelihood-equivalent BDe metric and the constraint of prior equivalence.

The inputs to K-LEARN are (1) a set of variables U, (2) a database D, (3) the relative prior probabilities of all network structures where each node has no more than K parents, (4) parameters N_{ijk}' and N_{ij}' for some node–parent pairs and some values of i, j, and k, and (5) a value p.

The input need only include enough parameters N_{ijk}' and N_{ij}' so that the metric score can be computed for all network structures where each node has no more than K parents.

Consequently, we do not need the N'_{ijk} and N'_{ij} parameters for nodes having more than K parents, nodes with parent configurations that always have zero prior probabilities, and values of i, j, and k for which there is no corresponding data in the database. Also, we emphasize that the parameters N'_{ijk} must be derivable from some joint probability distribution using Equation 12.2.

Given these inputs, we see from Equation 12.1 that the BDe metric for any given network structure and database can be computed in polynomial time. Consequently, K-LEARN is in NP. In the following sections, we show that K-LEARN is NP-hard. In Section 12.2.1, we give a polynomial time reduction from a known NP-complete problem to 2-LEARN. In Section 12.2.2, we show that 2-LEARN is NP-hard using the reduction from Section 12.2.1, and then show that K-LEARN for $K > 2$ is NP-hard by reducing 2-LEARN to K-LEARN. In this discussion, we omit conditioning on background information ξ to simplify the notation.

12.2.1   Reduction from DBFAS to 2-LEARN

In this section we provide a polynomial time reduction from a restricted version of the feedback arc set problem to 2-LEARN. The general feedback arc set problem is stated in Garey and Johnson (1979) as follows:

FEEDBACK ARC SET
INSTANCE: Directed graph $G = (V, A)$, positive integer $K \leq |A|$.
QUESTION: Is there a subset $A' \subset A$ with $|A'| \leq K$ such that A' contains at least one arc from every directed cycle in G?

It is shown in Garvill (1977) that **FEEDBACK ARC SET** remains NP-complete for directed graphs in which no vertex has a total in-degree and out-degree more than three. We refer to this restricted version as **DEGREE BOUNDED FEEDBACK ARC SET**, or DBFAS for short.

Given an instance of **DBFAS** consisting of $G = (V, A)$ and K, our task is to specify, in polynomial time, the five components of an instance of 2-LEARN. To simplify discussion, we assume that in the instance of DBFAS, no vertex has in-degree or out-degree of zero. If any such vertex exists, none of the incident edges can participate in a cycle and we can remove the vertex from the graph without changing the answer to the decision problem.

To help distinguish between the instance of **DBFAS** and the instance of 2-LEARN, we adopt the following convention. We use the term *arc* to refer to a directed edge in the instance of **DBFAS**, and the term *edge* to refer to a directed edge in the instance of 2-LEARN.

We construct the variable set U as follows. For each node v_i in V, we include a corresponding binary variable v_i in U. We use $\mathcal{V}$ to denote the subset of U that corresponds to V. For each arc $a_i \in A$, we include five additional binary variables $a_{i1}, \ldots, a_{i5}$ in U. We use $\mathcal{A}_i$ to denote the subset of U containing these five variables, and define $\mathcal{A}$ to be $\mathcal{A}_1 \cup \ldots \cup \mathcal{A}_{|A|}$. We include no other variables in U.

The database D consists of a single case $C_1 = \{1, \ldots, 1\}$.

The relative prior probability of every network structure is one. This assignment satisfies our constraint of prior equivalence. From Equation 12.1 with database $D = C_1$

and relative prior probabilities equal to one, the BDe metric—denoted $M_{BDe}(D, B_S)$—becomes

$$M_{BDe}(C_1, B_S) = \prod_i \frac{N'_{ijk}}{N'_{ij}} \qquad (12.3)$$

where k is the state of x_i equal to one, and j is the instance of Π_i such that the state of each variable in Π_i is equal to one. The reduction to this point is polynomial.

To specify the necessary N'_{ijk} and N'_{ij} parameters, we specify a prior network and then compute the parameters using Equation 12.2, assuming an arbitrary equivalent sample size of one.[3] From Equation 12.3, we have

$$M_{BDe}(C_1, B_S) = \prod_i p(x_i = 1 | \Pi_i = 1, \ldots, 1, B^h_{S_C}) \qquad (12.4)$$

To demonstrate that the reduction is polynomial, we show that the prior network can be constructed in polynomial time. In Section 12.2.3 (Theorem 12), we show that each probability in Equation 12.4 can be inferred from the prior network in constant time due to the special structure of the network.

We denote the prior Bayesian network $\mathcal{B} = (\mathcal{B}_S, \mathcal{B}_P)$. The prior network $\mathcal{B}$ contains both *hidden* nodes, which do not appear in U, and *visible* nodes which do appear in U. Every variable x_i in U has a corresponding visible node in $\mathcal{B}$ which is also denoted by x_i. There are no other visible nodes in $\mathcal{B}$. For every arc a_k from v_i to v_j in the given instance of DBFAS, $\mathcal{B}$ contains ten hidden binary nodes and the directed edges as shown in Figure 1 at the end of this subsection.

In the given instance of DBFAS, we know that each node v_i in V is adjacent to either two or three nodes. For every node v_i in V which is adjacent to exactly two other nodes in G, there is a hidden node h_i in $\mathcal{B}$ and an edge from h_i to x_i. There are no other edges or hidden nodes in $\mathcal{B}$.

We use h_{ij} to denote the hidden node parent common to visible nodes x_i and x_j. We create the parameters $\mathcal{B}_P$ as follows. For every hidden node h_{ij} we set

$$p(h_{ij} = 0) = p(h_{ij} = 1) = \frac{1}{2}$$

Each visible node in $\mathcal{B}$ is one of two types. The type of a node is defined by its conditional probability distribution. Every node a_{i5} in $\mathcal{B}$ (corresponding to the fifth variable created in U for the ith arc in the instance of DBFAS) is a *type II* node, and all other nodes are *type I* nodes. A type I node has the conditional probability distribution shown in Table 12.1.

We say that two variables in U are *prior siblings* if the corresponding nodes in the prior network $\mathcal{B}$ share a common hidden parent. We use S_{x_i} to denote the set of all variables in U which are prior siblings of x_i.

For each type II node a_{i5}, we define the *distinguished siblings* as the set $D_{a_{i5}} = \{a_{i3}, a_{i4}\} \subset S_{a_{i5}}$. Table 12.2 shows the conditional probability distribution of a type II node x_i with distinguished siblings $\{x_j, x_k\}$.

---

[3]Because there is only one case in the database, only the ratios $\frac{N'_{ijk}}{N'_{ij}}$ are needed (see Equation 12.3), and from Equation 12.2 the equivalent sample size is irrelevant. In general the equivalent sample size will need to be specified to uniquely determine the parameter values.

TABLE 12.1. Conditional probability distribution for a type I node.

h_{ij}	h_{ik}	h_{il}	$p(x_i = 1 \mid h_{ij}, h_{ik}, h_{il})$
0	0	0	0
0	0	1	1
0	1	0	1
0	1	1	0
1	0	0	1
1	0	1	0
1	1	0	0
1	1	1	0

TABLE 12.2. Conditional probability distribution for a type II node x_i with $D_{x_i} = \{x_j, x_k\}$.

h_{ij}	h_{ik}	h_{il}	$p(x_i = 1 \mid h_{ij}, h_{ik}, h_{il})$
0	0	0	$\frac{1}{3}$
0	0	1	1
0	1	0	$\frac{2}{3}$
0	1	1	0
1	0	0	$\frac{2}{3}$
1	0	1	0
1	1	0	$\frac{1}{3}$
1	1	1	0

There are $|V| + 5|A|$ visible nodes in $\mathcal{B}$, each visible node has at most three hidden node parents, and each probability table has constant size. Thus, the construction of $\mathcal{B}$ takes time polynomial in the size of the instance of DBFAS.

We now derive the value for p. From Equation 12.4, we obtain

$$
\begin{aligned}
M_{BDe}(C_1, B_S) \quad &= \prod_i p(x_i = 1 \mid \Pi_i = 1, \ldots, 1, B^h_{S_C}) \\
&= \prod_i \delta^{3 - |\Pi_i \cap S_{x_i}|} \cdot \frac{p(x_i = 1 \mid \Pi_i = 1, \ldots, 1, B^h_{S_C})}{\delta^{3 - |\Pi_i \cap S_{x_i}|}} \\
&= \delta^{\left(3n - \sum_i |\Pi_i \cap S_{x_i}|\right)} \prod_i s'(x_i \mid \Pi_i, S_i)
\end{aligned}
\tag{12.5}
$$

where $\delta < 1$ is a positive constant that we shall fix to be $15/16$ for the remainder of the paper.

Let σ be the total number of prior sibling pairs as defined by $\mathcal{B}$, and let γ be the number of prior sibling pairs which are not adjacent in B_S. The sum $\sum_i |\Pi_i \cap S_{x_i}|$ is the number of edges in B_S which connect prior sibling pairs and is therefore equal to $\sigma - \gamma$. Rewriting Equation 12.5, we get

$$
M_{BDe}(C_1, B_S) = \delta^{(3n - (\sigma - \gamma))} \prod_i s'(x_i \mid \Pi_i, S_i) = c' \cdot \delta^\gamma \prod_i s'(x_i \mid \Pi_i, S_i)
\tag{12.6}
$$

We now state 3 lemmas, postponing their proofs to Section 12.2.3. A network structure B_S is a *prior sibling graph* if all pairs of adjacent nodes are prior siblings. (Not all pairs of prior siblings in a prior sibling graph, however, need be adjacent.)

Lemma 1 *Let B_S be a network structure, and let $B_{S'}$ be the prior sibling graph created by removing every edge in B_S which does not connect a pair of prior siblings. Then it follows that $M_{BDe}(C_1, B_{S'}) \geq M_{BDe}(C_1, B_S)$*

Throughput the remainder of the paper, the symbol α stands for the constant $24/25$.

Lemma 2 *If B_S is a prior sibling graph, then for every type I node x_i in B_S, if Π_i contains at least one element, then $s'(x_i \mid \Pi_i, S_i)$ is maximized and is equal to $m_1 = 64/135$. If $\Pi_i = \emptyset$, then $s'(x_i \mid \Pi_i, S_i) = \alpha \cdot m_1$.*

Lemma 3 *If B_S is a prior sibling graph, then for every type II node x_i in B_S, if $\Pi_i = D_{x_i}$, where D_{x_i} is the set of two distinguished siblings of x_i, then $s'(x_i|\Pi_i, S_i)$ is maximized and is equal to $m_2 = 40/81$. If $\Pi_i \neq D_{x_i}$ then $s'(x_i|\Pi_i, S_i) \leq \alpha \cdot m_2$.*

Finally, we define p in the instance of 2-LEARN as

$$p = c' m_1^{|V|} \left(m_1^4 m_2\right)^{|A|} \alpha^K \tag{12.7}$$

where m_1 and m_2 are defined by Lemma 2 and 3 respectively, and c' is the constant from Equation 12.6.

The value for p can be derived in polynomial time. Consequently, the entire reduction is polynomial.

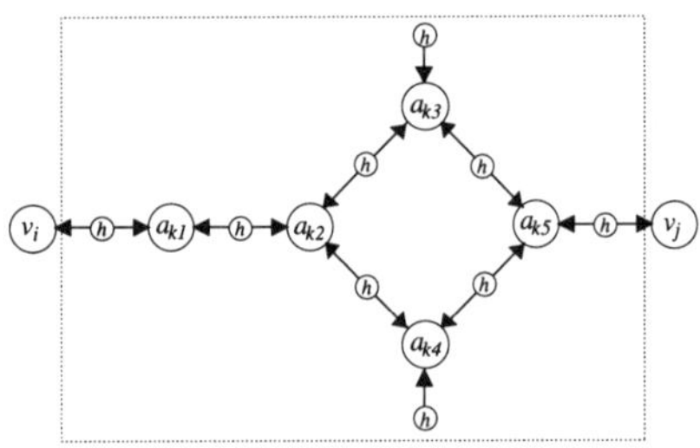

FIGURE 1. Subgraph of the prior net $\mathcal{B}$ corresponding to the *kth* arc in A from v_i to v_j.

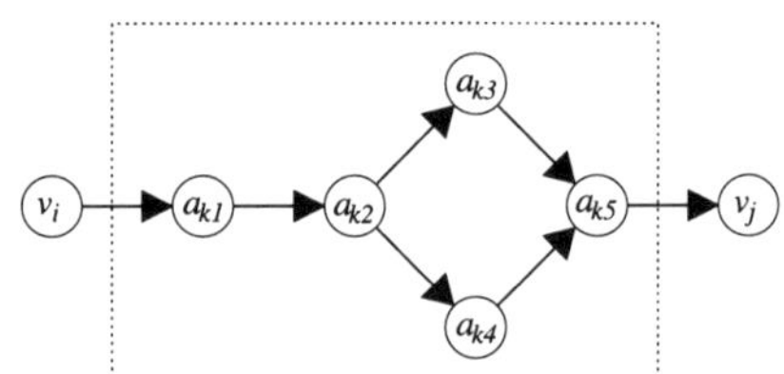

FIGURE 2. Optimal configuration of the edges incident to the nodes in $\mathcal{A}_k$ corresponding to the arc from v_i to v_j.

12.2.2 Proof of NP-Hardness

In this section, we first prove that 2-LEARN is NP-hard using the reduction from the previous section. Then, we prove that K-LEARN is NP-hard for all $k > 1$, using a reduction from 2-LEARN.

The following lemma explains the selection of p made in Equation 12.7, which in turn facilitates the proof that 2-LEARN is NP-hard. Let γ_k be the number of prior sibling pairs $\{x_i, x_j\}$ which are not adjacent in B_S, where at least one of $\{x_i, x_j\}$ is in $\mathcal{A}_k$. It follows that $\sum_k \gamma_k = \gamma$, and we can express Equation 12.6 as

$$M_{BDe}(C_1, B_S) \;=\; c' \left[\prod_{x_i \in V} s'(x_i|\Pi_i, S_i)\right] \left[\prod_j t(\mathcal{A}_j, \gamma_j)\right] \tag{12.8}$$

where $t(\mathcal{A}_j, \gamma_j) = \delta^{\gamma_j} \prod_{x_i \in A_j} s'(x_i|\Pi_i, S_i)$.

Lemma 4 *Let B_S be a prior sibling graph. If each node in $\mathcal{A}_k$ is adjacent to all of its prior siblings, and the orientation of the connecting edges are as shown in Figure 2, then $t(\mathcal{A}_k, \gamma_k)$ is maximized and is equal to $m_1^4 \cdot m_2$. Otherwise, $t(\mathcal{A}_k, \gamma_k) \leq \alpha \cdot m_1^4 \cdot m_2$.*

Proof: In Figure 2, every type I node in $\mathcal{A}_k$ has at least one prior sibling as a parent, and the single type II node has its distinguished siblings as parents. Thus, by Lemmas 2 and

3, the score $s'(x_i|\Pi_i, S_i)$ for each node $x_i \in \mathcal{A}_k$ is maximized. Furthermore, every pair of prior siblings are adjacent. Thus, we have

$$t(\mathcal{A}_j, \gamma_j) \;=\; \delta^{\gamma_j} \prod_{x_i \in A_j} s'(x_i|\Pi_i, S_i) = \delta^0 \cdot m_1 \cdot m_1 \cdot m_1 \cdot m_1 \cdot m_2$$

Suppose there exists another orientation of edges incident to the nodes in $\mathcal{A}_k$ such that that $t(\mathcal{A}_k, \gamma_k) > \alpha \cdot m_1^4 \cdot m_2$. Because $\delta < \alpha$ ($\frac{15}{16} < \frac{24}{25}$), every pair of prior siblings must be adjacent in this hypothetical configuration. Furthermore, every node in $\mathcal{A}_k$ must achieve its maximum score, else the total score will be bounded above by $\alpha \cdot m_1^4 \cdot m_2$. From Lemma 3 and Lemma 2, it follows that the resulting configuration must be identical to Figure 2
$\square$

The next two theorems prove that 2-LEARN is NP-hard.

Theorem 5 *There exists a solution to the 2-LEARN instance constructed in Section 12.2.1 with $M_{BDe}(C_1, B_S) \geq p$ if there exists a solution to the given DBFAS problem with $A' \leq K$.*

Proof: Given a solution to DBFAS, create the solution to 2-LEARN as follows: For every arc $a_k = (v_i, v_j) \in A$ such that $a_k \notin A'$, insert the edges in B_S between the corresponding nodes in $\mathcal{A}_k \cup v_i \cup v_j$ as shown in Figure 2. For every arc $a_k = (v_i, v_j) \in A'$, insert the edges in B_S between the corresponding nodes in $\mathcal{A}_k \cup v_i \cup v_j$ as shown in Figure 2, except for the edge between a_{k1} and a_{k2} which is reversed and therefore oriented from a_{k2} to a_{k1}.

To complete the proof, we must first show that B_S is a solution to the 2-LEARN instance, and then show that $M_{BDe}(C_1, B_S)$ is greater than or equal to p. Because each node in B_S has at most two parents, we know B_S is a solution as long as it is acyclic. By construction, B_S cannot contain a cycle unless there is a cycle in G for which none of the edges are contained in A'. Because G is a solution to DBFAS, this implies B_S is acyclic. We now derive $M_{BDe}(C_1, B_S)$. Let $\mathcal{A}^{opt}$ be the subset of $\mathcal{A}_k$ sets which correspond to the arcs in $A \setminus A'$. Rewriting Equation 12.8 we get

$$M_{BDe}(C_1, B_S) \;=\; c' \left[\prod_{x_i \in V} s'(x_i|\Pi_i, S_i) \right] \cdot \left[\prod_{\mathcal{A}_j \in \mathcal{A}^{opt}} t(\mathcal{A}_j, \gamma_j) \right] \cdot \left[\prod_{\mathcal{A}_k \in A \setminus \mathcal{A}^{opt}} t(\mathcal{A}_k, \gamma_k) \right]$$

Every node $x_i \in V$ has at least one prior sibling node as a parent because each node in the instance of DBFAS has an in-degree of at least one. Furthermore, Lemma 4 guarantees that for every $\mathcal{A}_k$ in $\mathcal{A}^{opt}$, $t(\mathcal{A}_j, \gamma_j)$ equals $m_1^4 \cdot m_2$. Now consider any $\mathcal{A}_k$ in $A \setminus \mathcal{A}^{opt}$. All prior sibling pairs for which at least one node is in this set are adjacent in B_S, so γ_k is zero. Furthermore, every node in this set attains a maximum score, except for the type I node a_{k2} which by Lemma 2 attains a score of $\alpha \cdot m_1$. Plugging into Equation 12.9 we have

$$\begin{aligned}
M_{BDe}(C_1, B_S) \;&=\; c' \left[m_1^{|V|} \right] \cdot \left[(m_1^4 \cdot m_2)^{|\mathcal{A}^{opt}|} \right] \cdot \left[(m_1^4 \cdot m_2 \cdot \alpha)^{|A \setminus \mathcal{A}^{opt}|} \right] \\
&=\; c' m_1^{|V|} \left[m_1^4 \cdot m_2 \right]^{|A|} \alpha^{|A'|}
\end{aligned}$$

Because $\alpha < 1$ and $|A'| \leq K$ we conclude that $M_{BDe}(C_1, B_S) \geq p$. $\square$

Theorem 6 *There exists a solution to the given DBFAS problem with $A' \leq K$ if there exists a solution to the 2-LEARN instance constructed in Section 12.2.1 with $M_{BDe}(C_1, B_S) \geq p$.*

Proof: Given the solution B_S to the instance of 2-LEARN, remove any edges in B_S which do not connect prior siblings. Lemma 1 guarantees that the BDe score does not decrease due to this transformation.

Now create the solution to DBFAS as follows. Recall that each set of nodes $\mathcal{A}_k$ corresponds to an arc $a_k = (v_i, v_j)$ in the instance of DBFAS. Define the solution arc set A' to be the set of arcs corresponding to those sets $\mathcal{A}_k$ for which the edges incident to the nodes in $\mathcal{A}_k$ are *not* configured as shown in Figure 2.

To complete the proof, we first show that A' is a solution to DBFAS, and then show that $|A'| \leq K$. Suppose that A' is not a solution to DBFAS. This means that there exists a cycle in G that does not pass through an arc in A'. For every arc (v_i, v_j) in this cycle, there is a corresponding directed path from v_i to v_j in B_S (see Figure 2). But this implies there is a cycle in B_S which contradicts the fact that we have a solution to 2-LEARN. ¿From Lemma 4 we know that each set $\mathcal{A}_k$ that corresponds to an arc in A' has $t(\mathcal{A}_k, \gamma_k)$ bounded above by $\alpha \cdot m_1^4 \cdot m_2$. Because $M_{BDe}(C_1, B_S) \geq p$, we conclude from Equation 12.8 that there can be at most K such arcs. $\square$

Theorem 7 *K-LEARN with $M_{BDe}(D, B_S)$ satisfying prior equivalence is NP-hard for every integer $K > 1$.*

Proof: Because 2-LEARN is NP-hard, we establish the theorem by showing that any 2-LEARN problem can be solved using an instance of K-LEARN.

Given an instance of 2-LEARN, an equivalent instance of K-LEARN is identical to the instance of 2-LEARN, except that the relative prior probability is zero for any structure that contains a node with more than two parents[4]. It remains to be shown that this assignment satisfies prior equivalence. We can establish this fact by showing that no structure containing a node with more than two parents is equivalent to a structure for which no node contains more than two parents.

Chickering (1995) shows that for any two equivalent structures B_{S_1} and B_{S_2}, there exists a finite sequence of arc reversals in B_{S_1} such that (1) after each reversal B_{S_1} remains equivalent to B_{S_2}, (2) after all reversals $B_{S_1} = B_{S_2}$, and (3) if the edge $v_i \rightarrow v_j$ is the next edge to be reversed, then v_i and v_j have the same parents with the exception that v_i is also a parent of v_j. It follows that after each reversal, v_i has the same number of parents as v_j did before the reversal, and v_j has the same number of parents as v_i did before the reversal. Thus, if there exists a node with l parents in some structure B_S, then there exists a node with l parents in any structure that is equivalent to B_S. $\square$

12.2.3   Proof of Lemmas

To prove Lemmas 1 through 3, we derive $s'(x_i | \Pi_i, S_{x_i})$ for every pair $\{x_i, \Pi_i\}$. Let x_i be any node. The set Π_i must satisfy one of the following mutually exclusive and collectively exhaustive assertions:

---

[4]Note that no new parameters need be specified.

Assertion 1 For every node x_j which is both a parent of x_i and a prior sibling of x_i (i.e. $x_j \in \Pi_i \cap S_{x_i}$), there is no prior sibling of x_j which is also a parent of x_i.

Assertion 2 There exists a node x_j which is both a parent of x_i and a prior sibling of x_i, such that one of the prior siblings of x_j is also a parent of x_i.

The following theorem shows that to derive $s'(x_i|\Pi_i, S_{x_i})$ for any pair $\{x_i, \Pi_i\}$ for which Π_i satisfies Assertion 1, we need only compute the cases for which $\Pi_i \subseteq S_{x_i}$.

Theorem 8 *Let x_i be any node in B_S. If Π_i satisfies Assertion 1, then $s'(x_i|\Pi_i, S_{x_i}) = s'(x_i|\Pi_i \cap S_{x_i}, S_{x_i})$.*

Proof: From Equation 12.5, we have

$$s'(x_i|\Pi_i, S_{x_i}) = \frac{p(x_i|\Pi_i, B_{S_C}^e)}{\delta^{3-|\Pi_i \cap S_{x_i}|}} \tag{12.9}$$

Because Π_i satisfies Assertion 1, it follows by construction of $\mathcal{B}$ that x_i is d-separated from all parents that are not prior siblings once the values of $\Pi_i \cap S_{x_i}$ are known. $\square$

For the next two theorems, we use the following equalities.[5]

$$p(h_{ij}, h_{ik}, h_{il}) = p(h_{ij})p(h_{ik})p(h_{il}) \tag{12.10}$$

$$p(h_{ij}, h_{ik}, h_{il}|x_j) = p(h_{ij}|x_j)p(h_{ik})p(h_{il}) \tag{12.11}$$

$$p(h_{ij}, h_{ik}, h_{il}|x_j, x_k) = p(h_{ij}|x_j)p(h_{ik}|x_k)p(h_{il}) \tag{12.12}$$

$$p(h_{ij} = 0|x_i = 1) = \frac{2}{3} \tag{12.13}$$

Equation 12.10 follows because each hidden node is a root in $\mathcal{B}$. Equation 12.11 follows because any path from x_j to either h_{ik} or h_{il} must pass through some node $x \neq x_j$ which is a sink. Equation 12.12 follows from a similar argument, noting from the topology of $\mathcal{B}$ that $x \notin \{x_j, x_k\}$. Equation 12.13 follows from Tables 1 and 2, using the fact that $p(h_{ij} = 0)$ equals $1/2$.

Theorem 9 *Let x_i be any type I node in B_S for which Π_i satisfies Assertion 1. If $|\Pi_i \cap S_{x_i}| = 0$ then $s'(x_i|\Pi_i, S_{x_i}) = \alpha \cdot m_1$. If $|\Pi_i \cap S_{x_i}| = 1$ then $s'(x_i|\Pi_i, S_{x_i}) = m_1$. If $|\Pi_i \cap S_{x_i}| = 2$ then $s'(x_i|\Pi_i, S_{x_i}) = m_1$.*

Proof: Follows by solving Equation 12.9, using Equations 12.10 through 12.13 and the probabilities given in Table 12.1. $\square$

Theorem 10 *Let x_i be any type II node in B_S for which Π_i satisfies assertion 1. If $|\Pi_i \cap S_{x_i}| = 0$ then $s'(x_i|\Pi_i) = \alpha^2 \cdot m_2$. If $|\Pi_i \cap S_{x_i}| = 1$ then $s'(x_i|\Pi_i) = \alpha \cdot m_2$. If $|\Pi_i \cap S_{x_i}| = 2$ and $\Pi_i \neq D_{x_i}$ then $s'(x_i|\Pi_i) = \alpha \cdot m_2$. If $\Pi_i = D_{x_i}$ then $s'(x_i|\Pi_i) = m_2$.*

---

[5]We drop the conditioning event $B_{S_C}^e$ to simplify notation.

Proof: Follows by solving Equation 12.9, using Equations 12.10 through 12.13 and the probabilities given in Table 12.2. $\square$

Now we show that if Assertion 2 holds for the parents of some node, then we can remove the edge from the parent which is not a sibling without decreasing the score. Once this theorem is established, the lemmas follow.

Theorem 11 *Let x_i be any node. If $\Pi_i = \{x_j, x_k\}$, where $x_j \in S_{x_i}$ and $x_k \in S_{x_j}$, then $s'(x_i|x_j) \geq s'(x_i|x_j, x_k)$.*

Proof: For any node we have

$$p(x_i = 1|x_j = 1, x_k = 1) \;\; = \;\; \frac{p(x_i = 1)p(x_k = 1|x_i = 1)p(x_j = 1|x_i = 1, x_k = 1)}{p(x_k = 1)p(x_j = 1|x_k = 1)}$$

Because x_i and x_k are not prior siblings, it follows that $p(x_k|x_i) = p(x_k)$. Expressing the resulting equality in terms of $s'(x_i|\Pi_i, S_{x_i})$, noting that x_i has only one prior sibling as a parent, and canceling terms of δ, we obtain

$$s'(x_i|\{x_j, x_k\}, S_{x_i}) = s'(x_i|\emptyset, S_{x_i}) \frac{s'(x_j|\{x_i, x_k\}, S_{x_j})}{s'(x_j|\{x_k\}, S_{x_j})} \tag{12.14}$$

If x_j is a type I node, or if x_j is a type II node and x_i and x_k are *not* its distinguished siblings, then $s'(x_j|\{x_i, x_k\}, S_{x_j})$ equals $s'(x_j|\{x_k\}, S_{x_j})$, which implies that we can improve the local score of x_i by removing the edge from x_k. If x_j is a type II node, and $D_{x_j} = \{x_i, x_k\}$, then $s'(x_j|\{x_i, x_k\}, S_{x_j})$ equals $(1/\alpha) \cdot s'(x_j|\{x_k\}, S_{x_j})$, which implies we can remove the edge from x_k without affecting the score of x_i. $\square$

The preceding arguments also demonstrate the following theorem.

Theorem 12 *For any pair $\{x_i, \Pi_i\}$, where $|\Pi_i| \leq 2$, the value $p(x_i = 1|\Pi_i)$ can be computed from $\mathcal{B}$ in constant time when the state of each of the variable in Π_i is equal to one.*

12.3 REFERENCES

[Chickering, 1995] Chickering, D. M. (1995). A Transformational characterization of Bayesian network structures. In *Proceedings of Eleventh Conference on Uncertainty in Artificial Intelligence,* Montreal, QU. Morgan Kaufman.

[Cooper and Herskovits, 1992] Cooper, G. and Herskovits, E. (1992). A Bayesian method for the induction of probabilistic networks from data. *Machine Learning,* 9:309–347.

[Garey and Johnson, 1979] Garey, M. and Johnson, D. (1979). *Computers and intractability: A guide to the theory of NP-completeness.* W.H. Freeman.

[Garvil, 1977] Garvil, F. (1977). Some NP-complete problems on graphs. In *Proc. 11th Conf. on Information Sciences and Systems,* Johns Hopkins University, pages 91–95. Baltimore, MD.

[Heckerman et al., 1995] Heckerman, D., Geiger, D., and Chickering, D. (1995). Learning discrete Bayesian networks. *Machine Learning,* 20:197-243.

[Höffgen, 1993] Höffgen, K. (revised 1993). Learning and robust learning of product distributions. Technical Report 464, Fachbereich Informatik, Universität Dortmund.

13
Heuristic Search for Model Structure: the Benefits of Restraining Greed

John F. Elder IV

Computational & Applied Mathematics Dept.
 & Center for Research on Parallel Computation
Rice University
Houston Texas 77251-1892
elder@rice.edu

ABSTRACT Inductive modeling or "machine learning" algorithms are able to discover structure in high-dimensional data in a nearly automated fashion. These adaptive statistical methods – including decision trees, polynomial networks, projection pursuit models, and additive networks – repeatedly search for, and add on, the model component judged best at that state. Because of the huge model space of possible components, the choice is typically *greedy*; that is, optimal only in the very short term. In fact, it is usual for the analyst and algorithm to be greedy at three levels: when choosing a 1) term within a model, 2) model within a family, and 3) family within a wide collection of methods. It is better, we argue, to "take a longer view" in each stage. For the first stage (term selection) examples are presented for classification using decision trees and estimation using regression. To improve the third stage (method selection) we propose *fusing* information from disparate models to make a combined model more robust. (Fused models merge their output estimates but also share information on, for example, variables to employ and cases to ignore.) Benefits of fusing are demonstrated on a challenging classification dataset, where the task is to infer the species of a bat from its chirps.

13.1   Introduction

Modern inductive modeling or "machine learning" techniques can often discover useful patterns in high-dimensional data in a nearly automated fashion. These adaptive statistical algorithms – including decision trees, polynomial networks, projection pursuit models, and additive networks – incrementally build up the model structure (inputs, interconnections, and terms) as well as set parameter values. That is, they sequentially add the component, from the set of candidates, which works best with the existing collection. This model expansion typically ceases when the structure is judged to optimally trade off 1) training accuracy and 2) simplicity (of model form or function surface). Complexity is regulated in order to protect against the serious danger of overfit, and thus lead to greater success on new data.

The heuristic search procedures employed by these methods are all *greedy* to some degree, as a consequence of making workable choices in an open or combinatorially huge domain of possible models. This procedure basically works well in practice, though it is known that, theoretically, greedy searches can result in arbitrarily worse models than optimal ones (when such are possible) on finite data sets. Some real and artificial examples

---

[1] *Learning from Data: AI and Statistics V.* Edited by D. Fisher and H.-J. Lenz. ©1996 Springer-Verlag.

of this will be shown for regression subset selection and decision tree construction.

Modeling algorithms typically return for use the single best model found, though many alternatives with similar predicted performance are identified. If variables are selected by that model, the analyst may conclude more from their membership than is warranted, since alternative models can employ very different variable sets with essentially the same potential. Lastly, even analysts who try several modeling methods, to explore the effect of different basis functions, usually seek the "best" method for the problem, when a more useful focus might be to find the best estimates for the outputs.

Three stages of *greediness* are thus usually a part of the inductive modeling process, in the selection of the best 1) term, 2) model, and 3) method. It is useful then, to explore what restraining greed, or "broadening the scoring horizon", can do for each stage. For stage one (term selection), a recent decision tree algorithm, *Texas 2-Step*, is described which looks two steps ahead to choose threshold variables and values, rather than one. (That is, it judges a split not by the purity of the resulting child nodes, but how the grandchildren turn out.) Similar examples are given for regression, where a method recently introduced to identify severe collinearity can be used to expand the situations in which optimal subset techniques are valid. Potential improvements to stage two (model selection) are briefly addressed. Lastly, several inductive and nonparametric techniques are compared on a challenging field application: identifying a bat's species by its chirps. It is demonstrated that stage three (method selection) can be improved by merging disparate models; *i.e.*, employing a weighted sum of their outputs – an idea that is relatively well known though little practiced. Yet, it is further demonstrated that even more performance can be extracted if the sharing of information between methods (variables selected, cases discounted, etc.) is more thorough – a process we call *model fusion*.

13.2 Automated Induction

Inductive algorithms are, at one level, "black boxes" for developing classification, estimation, or control models from sample data. They automatically search a vast space of potential models for the best inputs, structure (terms and interconnections), and parameter values. The models are pieced together in a stepwise manner into a feed-forward network (*e.g.*, tree) of simple nodes. The better methods also *prune* unnecessary terms or nodes from the model, thereby regulating complexity to reduce the chance of *overfit*. Overfit models are over-specialized to the training data and generalize poorly (fail on new data). This is widely held to be the chief danger of using inductive methods.

Complexity is primarily regulated either through

1. *term penalties*, using model selection criteria such as Cp (Mallows, 1973) or *Minimum Description Length*, MDL (Rissanen, 1978),

2. *weight penalties* which encourage "shrinkage" of the parameters towards smaller (less influential) values,

3. *roughness penalties* (integrated second derivatives of the estimation surface), or

4. *tests on withheld data* (*e.g.*, V-fold cross-validation).

The penalties add to an error measure, and models having the lowest combined score are judged the best candidates for use. For example, *Stepwise regression* can be considered a low-level automated induction algorithm. Though the set of possible models (linear combinations of subsets of original inputs) is quite constrained, the procedure does identify which variables among the candidates to employ.

Another, more crude, regulatory method is also commonly employed with *artificial neural networks* (ANNs): halting parameter adjustment before convergence. Since their structure is fixed, ANNs are not explicitly inductive; however, they may implicitly be adaptive as many nodes start out having little effect on the response surface. This is because most random node weights lead to essentially linear functions (nodes operating in either the middle or extreme of the sigmoidal transfer function) and such linear functions are absorbed by subsequent layers of nodes. Only as nodes get pushed into the curved part of the sigmoids during training do parameters become active, increasing the danger of overfit as training continues. If much of the typical ANN structure is underutilized, it could explain a couple of puzzles about ANNs at first glance: 1) they often are less over-parameterized than they appear, and 2) their performance can be surprisingly robust with respect to changes in the network structure. (Oddly then, improving the ANN search – typically, the iterative, local, gradient-based *backpropagation* procedure – without simplification of the model structure, may result in better training but worse out-of-sample performance!).[2]

Leading automated induction methods, using "building blocks" consisting of logistic functions, splines, polynomials, planes, non-parametric smoothes of weighted sums, etc. – are briefly described in (Elder, 1993; and Elder & Pregibon, 1995) along with their chief strengths and weaknesses. Here, we focus on one of the weaknesses: greediness, and look briefly at its effect on regression and decision trees.

13.3 Subset Selection in Regression

Due to the combinatorial explosion of a trial-and-error search process (the methods are at least polynomial in the inputs and often exponential), a greedy heuristic is often employed: models are constructed in stages, and only the current step is optimized at a given time. *Forward selection* finds the single best term, then adds to it the term which works best with the first, then the one which best assists the pair, and so on. (Note that this is very much more useful than a "first impression" model, which ranks the candidate terms according to their individual performance and employs the top K.) *Reverse elimination* begins with a "full" model and sequentially removes the least useful term.

A combined method, *stepwise selection* (*e.g.*, Draper & Smith, 1966) considers removing variables after each new variable is introduced. The standard selection mechanism, checking *F-to-enter* and *F-to-exit* significance values, is a kind of heuristic term penalty method, but not a correct use of F-tests. (The static significance measure is invalid in the dynamic modeling situation and can lead to highly inflated confidences in the resulting parameter values (Miller, 1990).)

---

[2]Note that removing small terms or nodes may not help over-parameterization, where useless terms can appear significant though they collectively cancel. (The dangers of collinear variables in regression are analogous.)

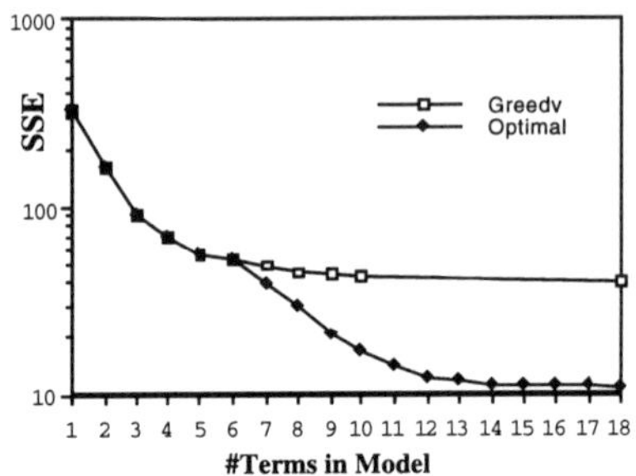

FIGURE 1. Greedy vs. Optimal Subset Selection.

This greedy growth strategy makes the search feasible and often discovers useful features, but can miss "reachable" structure in the data; that is, within the form of the basis functions employed. For example, given $Y = \{1, 1, 1, 1\}$, $X_1 = \{1, 1, 1, 0\}$, $X_2 = \{1, 1, 0, 0\}$, $X_3 = \{0, 0, 1, 1\}$, a stepwise procedure would first choose X_1 with which to estimate Y, and then seek to add another X. However, an exact model, $Y = X_2 + X_3$, would not include that single best input. Surprisingly, even if there is agreement between the forward and backward procedures on the best model of each size, they can differ by an arbitrarily large amount from some of the best subsets (Berk, 1978).

For example, Desroachers & Mohseni (1984) presented a purportedly optimal algorithm for model selection, and demonstrated it on a problem of estimating rocket engine temperature (from Lloyd & Lipow, 1962), where their small set results agreed with earlier analyses by Draper & Smith (1966). However, the approach turned out to be a version of forward selection. To compare these models with optimal subsets (of the candidate set defined by Desroachers & Mohseni), a new technique for term elimination had to be developed (Elder, 1990). Figure 1 shows the SSE of the greedy and optimal models of each size. The greedy errors leveled off at a limit of 40, while the optimal subset models were able to reach nearly the minimum error possible for the data (approximated by the Y axis base). Clearly, greedy methods can be improved upon significantly, in training, on real applications.

But the degree to which greediness generally hurts performance in practice, on new data, is an open question. Berk (1978) sounded a slightly cautionary note in the case of regression subset selection. Using nine well-studied data sets (having from 4 to 15 predictors, 13 to 541 cases, and often more analysts!), he noted the maximum training error difference between all-subsets (optimal) models and both 1) forward selection and 2) reverse elimination models. An improvement of up to 29% in SSE was observed. Then, the sample distributions of each data set were employed to generate synthetic data with known population characteristics, and the study again performed for this new evaluation data. Figure 2 plots the training vs. evaluation data differences for the forward and reverse models from the (Berk, 1978) study. Most evaluation differences were smaller and in a tighter range (-2 to 7%, with one exception). Also, in two cases, a greedy method won on the evaluation data by a slight margin.

Note that the differences are somewhat exaggerated, as the maximum disagreement between methods is shown, not that at some automated stopping point. For instance, the

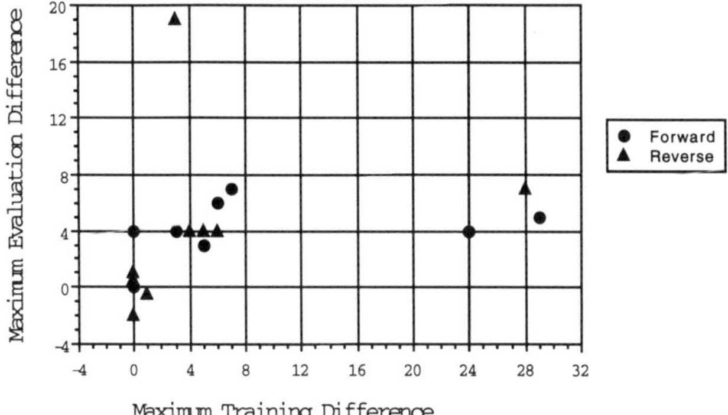

FIGURE 2. Ex. Training vs. Evaluation Improvement of Optimal over both Forward and Reverse Greedy Methods.

two worst reverse values (one training, one evaluation), are for models of size 1 and 2 – where the forward method would clearly be preferable. Still, the greedy training and evaluation under-performances are correlated, and it seems that regression differences on new data, while usually less dramatic than on training data, are still likely to be significant.

For regression model building, a logical improvement to the stage 1 greedy growth strategy (while stopping short of the hope of "optimal" models) is to add *chunks* of terms at a time, rather than just one. This is the heart the approach taken in GMDH-like techniques, such as ASPN (Algorithm for the Synthesis of Polynomial Networks; Elder, 1985; Elder & Brown, 1992). There, sets of several terms, employing a few independent variables, are considered for inclusion simultaneously, then pared down by reverse elimination. Nodes of such equations are built up until the added complexity cannot be justified, according to a penalty criterion – either Predicted Squared Error (A. Barron, 1984) or MDL. An ASPN regression network, such as that shown in Figure 3, can have multiple layers of diverse nodes, each with several terms, resulting in a flexible compound function form.

Extensive comparison with more greedy algorithms has yet to be performed, but several researchers have successfully employed such regression networks on applications which had proven very difficult by other methods, including automatic pipe inspection (Mucciardi, 1982), fish stock classification (Prager, 1988), reconfigurable flight control (Elder & Barron, 1988), tactical weapon guidance (Barron & Abbott, 1988), and temperature distribution forecasting (Fulcher & Brown, 1991). Though several areas of possible improvement have been identified (Elder & Brown, 1992), its success suggests that taking complex, rather than simple, steps might improve other constructive algorithms for induction, such as those used to build decision trees.

13.4   Constructing Decision Trees

Though there are other and earlier decision tree algorithms (*e.g.*, ID3 and CHAID), CART (Classification and Regression Trees, Breiman, Friedman, Olshen & Stone, 1984) is perhaps the best known and, arguably, most powerful. Some of its nicer features include built-in cross-validation, the ability to handle categorical variables and missing data, and

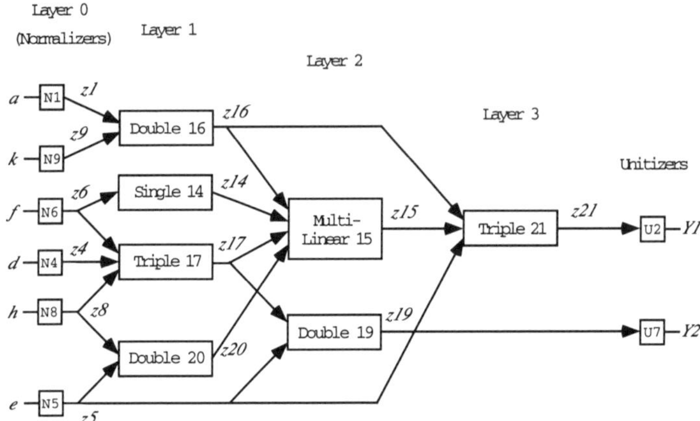

FIGURE 3. Sample Regression Network.

a good presentation of the output. (Versions are also appearing which tie into commercial statistical packages and improve the interface.) Still, the basic classification algorithm is very simple: try to discriminate between classes by recursively bifurcating the data until the resulting groups are as pure as can be sustained. That is, start with all the training data and choose the univariate threshold split ($e.g.$, $x_3 < 1.14$) which divides the sample into two maximally pure parts ($i.e.$, minimizes the sample variance of the sum).[3] Then, continue with each of the parts (child nodes) until either no splits are possible, or the leaves (terminal nodes of the tree) are pure (represent only one class) or have some minimum size. Then, CART prunes back (simplifies) the tree, typically using cross-validation, to avoid overfit. This over-training followed by pruning was found by CART's authors to lead to better trees than under the competing method of trying to select the growth stopping point. For estimation, the leaves are set to the mean or median value of the cases contained, forming a piecewise-constant surface.

This simple splitting approach is nevertheless powerful, as a sequence of threshold questions quickly conditions an individual case. Each path down the tree can have its own important variables, and outliers have no special influence. Also, as with other methods which implicitly select variables, a user is freed to try many more candidate variables than otherwise, since CART will sift through them unfettered by concerns about multicollinearity, a bane of regression methods. However, if the candidate variables are jointly useful, relatively independent, and not beset by many outliers, other methods of discrimination can outperform CART.

Here, we wonder simply if CART's strategy of choosing the greedy split cannot be improved. As a motivating example, consider the XOR-like data of Table 13.1. CART forms the approximation tree of Figure 4(a) (using a leaf size limit of ≤ 2 cases). Its greedy search does not find the simpler, exact tree of Figure 4(b). To explore whether an

---

[3]Multi-linear splits ($e.g.$, $x_1 + 2x2 < 3$) are possible, but do not work well in CART in practice, perhaps because of a poor internal search routine. Newer algorithms ($e.g.$, OC1 by Murthy $et\ al.$, 1994) claim to do better at this.

TABLE 13.1. Greedy Counter-Example for CART.

Y	a	b	c
0	0	0	0
0	1	0	0
1	0	0	1
1	1	0	1
1	0	1	0
1	1	1	0
0	1	1	1
0	1	1	1

extension of the horizon to two steps ahead would be beneficial, a decision tree algorithm called *Texas Two-Step* was written.

13.5   Texas Two-Step (TX2step)

The algorithm TX2step is a slimmed-down version of CART for classification which is not able to handle missing data, perform internal cross-validation, set misclassification costs, or adjust priors, and so on. Yet it can look two steps ahead to choose the current split, and thereby finds the tree of Figure 4(b) given the data of Table 13.1. TX2step has one other new feature: given more than one split which results in the same score, it uses that with the largest relative gap between border training cases. That is, the tie-breaker to choose the dimension d of the split depends on

$$gap[d] = 0.5\frac{(\min(\text{Right } Xd) - \max(\text{Left } Xd))}{\max(Xd) - \min(Xd)}$$

The algorithm can optionally be greedy as well; in that mode, and ignoring gaps, it was validated on several test problems to reproduce the same tree as CART without cross-validation. Therefore, to focus solely on the greediness issue, TX2step-1 (with gap measurement) was actually run in place of CART on the example application shown next. Training was performed until all nodes were pure, but those leaves with a majority class having fewer than 3 cases were pruned back (*i.e.*, reabsorbed into their parent node).

13.6   Application Example: Identifying Bat Species

Researchers from the University of Illinois, Urbana/Champaign[4] measured bat echolocation calls and extracted a few dozen time-frequency features from the signals, such as low frequency at the 3db level, relative time position of the signal peak, and amplitude ratio of the first and second harmonics. This work is toward developing an automated classifying system to track species of bats – especially those considered endangered. After visualization of projections of the data by the author, and analysis of correlations, multicollinearity, redundancy, and outliers (for suggested techniques see *e.g.*, Elder, 1993),

---

[4]Biologists Ken White, Curtis Condon, & Al Feng, and Electrical Engineers Oliver Kaefer & Doug Jones.

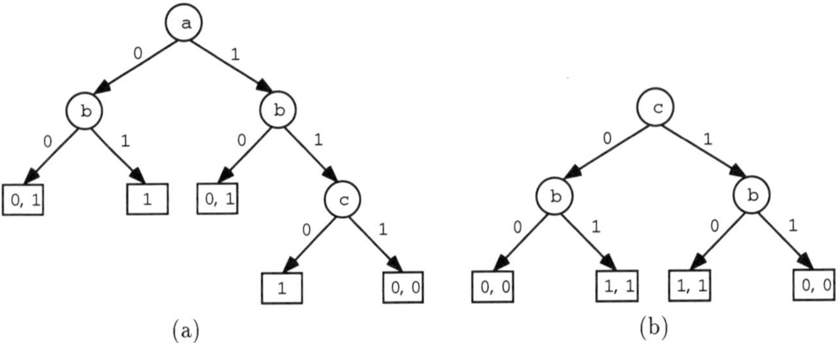

(a)   (b)

FIGURE 4. Inaccurate Greedy Decision Tree (a) and Correct Decision Tree (b).

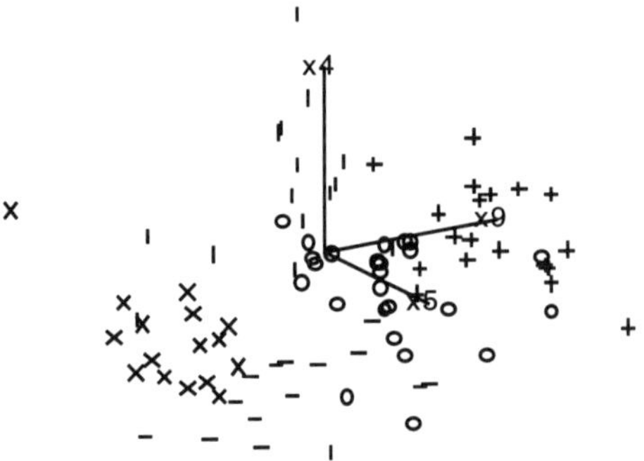

FIGURE 5. Example Projection of Bat Classes.

some variables were eliminated and other new ones tried at UIUC, resulting in a database of 93 cases, each with 15 candidate input features, representing 5 different species (classes) of bats.[5] An informative projection of the data is shown in Figure 5, where the classes are noted by different symbols. Note that the groups do tend to cluster but that a fair amount of overlap is evident in this (and all low-D) views.

Trained on all the data, the 1-step tree, shown in Figure 6(a), had 5 splits (17 prior to pruning) and made 13 training errors.[6] (In the trees, "Yes" answers travel to the left child; "No" to the right.) The 2-step tree of Figure 6(b) started out simpler, with 14 splits, but pruned less, ending with 10 splits and only 5 training errors. The best root node split happened to be greedy but several other splits were not. For example, the data in the right child node of the root, shown in Figures 7 (a) and (b), are those 58 of 93 cases where $x_5 > 101.5$. The greedy tree was drawn to split first on $x_{20} < 3.59$, then on

---

[5] A single bat from a sixth *Long-Eared* species contributed 5 signals originally, but was removed as it could be easily distinguished by its low-frequency signals, and since only one representative does not allow proper testing.

[6] It takes under a second on a SPARC-2 to run the 1-step algorithm on this problem, but 75 seconds for 2.

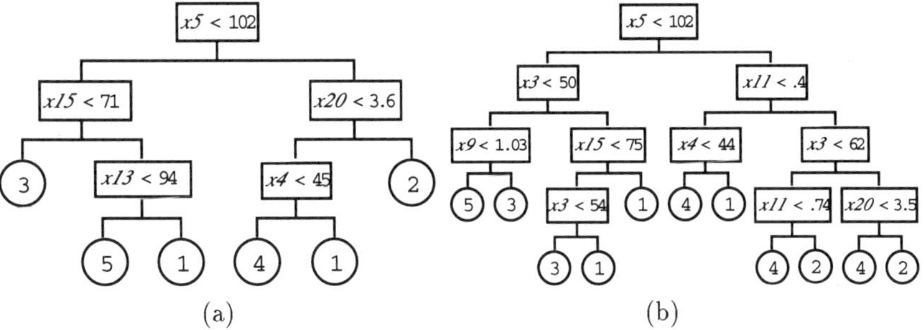

FIGURE 6. CART (1-step) Tree (a) and TX2step Tree (b).

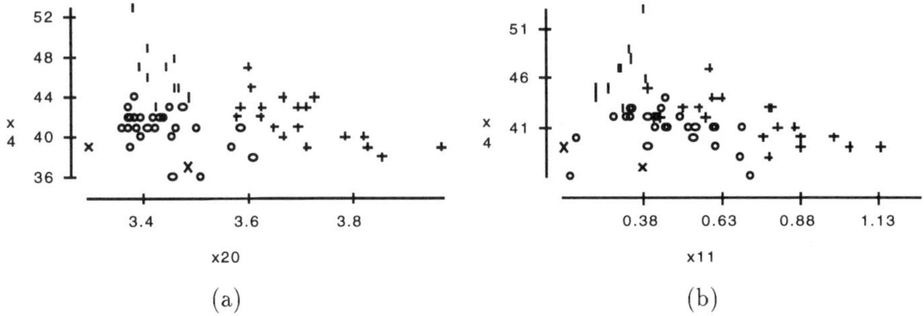

FIGURE 7. View at Right Node for CART (a) and TX2step (b).

$x_4 < 44.5$, and it was incorrect for 6 cases on that branch. The 2-step tree instead first chose $x_{11} < 0.39$ – a seemingly worse split, but when followed by $x_4 < 43.5$ on one branch, one which allowed it to correctly classify 4 more cases. (The difficulties this split caused its sibling branch were cleared up by subsequent splits.) Importantly, the 2-step cuts were often more appealing visually; that is, they accorded more with what an analyst would do when viewing two dimensions of data simultaneously, rather than one.

As expected, the less greedy algorithm performed better on training data. But, the best test, of course, involves new data. Since there were not many cases, a cross-validation evaluation was performed, where all 3-8 signals for each bat, in turn, were held out of training and independently run down the tree for testing (18 runs for each method). The CART-like 1-step tree gets 43 of 93 signals correct (46%), TX2step 54 (58%), and the ANN (results are courtesy of Oliver Kaefer & Doug Jones of UIUC) performs best with 64 (69%). The difference in accuracy for the tree methods appears more critical when using a *voting scheme*, where several different signals from a single bat are classified and the majority class is assigned. Then, CART misses 11 of the 18 bats but TX2step only 6. (The voting ANN misses just 4.)

In this experiment (counter to our usual experience), the tree methods were outperformed by an ANN. However, the variable selection performed by CART and TX2step proved helpful to the ANN; a network trained on all 35 original data features got only 52% correct in batwise cross-validation, one trained on 17 variables (those given as candidates to the tree methods) was 63% correct, and the ANN using only the 8 tree-selected variables did best with 69%. Here, simpler ANNs performed better on new data. (Clearly, techniques for adapting ANN structure to the data – more flexibly than through stepwise methods as say, *cascade correlation*, are worthy of careful attention.)

The data characteristics of the problem – filtered features, lack of outliers, clustered classes – which helped the neural network perform well, should also be agreeable to exemplar-based statistical techniques, such as *kernels* and *nearest neighbors*. Accordingly, we tried the latter out, along with *polynomial networks*, an estimative technique capable of finding nonlinear decision boundaries when employed for classification. For the latter, five indicator (0/1) variables took the place of the output, and the maximum one was taken as the estimated class. This scheme results in 62% accuracy for the polynomial network algorithm (ASPN) on new cross-validated (CV) data.

For simple K-nearest neighbors (K-NN), a new case is labeled with the class of the majority of the closest K known cases, generally using Euclidean distance as the metric. This simple nonparametric method is very flexible; however, moreso than with methods which select variables, it is sensitive to the dimensions employed, as any unnecessary one will contribute to the distance calculations and thereby obscure the true relationships. So, the 8 dimensions found useful by the tree methods were employed, and a 3-NN algorithm got 70% CV accuracy. Further, when all subsets of these 8 variables were exhaustively explored, a set of 5 were found with 80% accuracy, indicating the importance of dimension selection for the algorithm.[7]

13.7 Model Fusion

Individually, the methods can be aligned according to ascending CV accuracy in the order: Trees (46% for 1-step, 58% for 2), Polynomial Networks (62%), Neural Networks (52% for 35 variables, 63% for 17, and 69% for 8), Nearest Neighbors (70% for the same 8 variables, up to 80% for the best 5). (For other problems, with other data characteristics, these rankings could be completely opposite.) Generally, the number of bats correctly classified by voting (taking the majority class of a set of signals when multiple observations are made of one bat) was proportional to the overall accuracy, with one important exception: polynomial networks (PNs) were able to get the most bats right when voting (15 of 18) *if obvious outliers were ignored*. Since PNs do not limit their outputs between 0 and 1 like the other estimative technique (ANNs) used here, over-extrapolated outputs can be obvious. When these are removed, the voting improved.

It is an ancient statistical principle[8] that merging a set of independent estimates can reduce the variance of the output. The suite of techniques studied seem different enough to warrant this: not only do they employ quite different basis functions and methods of estimation and complexity regulation, but similarly performing methods make mistakes on different cases. For example, the 8-variable ANN and 8-variable 3-NN methods are only 1% off in overall performance, but they classify 35% of the cases differently.

Accordingly, the four techniques were averaged, and the CV performance improved

---

[7]It's not valid to use this performance number, as the CV accuracy was maximized through a search; another level of CV would be required to get an honest estimate of performance on truly new data.

[8]Proverbs 24:6b: "... in a multitude of counselors there is safety."

to 74%. When the weights were made proportional to their individual performance, this increased to 78%. Then, when the polynomial network was allowed to identify and remove outliers for all four techniques, the best performance (82%) was obtained.

13.8   Concluding Remarks

For stage 1 (term selection) we have seen that regression subsets and decision trees can be suboptimal if the single best step is always taken. This is true in other venues as well. Cover (1974) showed an investigation in which greed hurts, where: If only one experiment is allowed, E_1 provides the most information, but if two are possible, then independent versions of the "worse" experiment E_2 are better. This was also shown to be the case on new data for regression and for decision trees, where a version of CART was outperformed on an example problem by TX2step, which looks ahead an additional step when selecting a threshold for the current node.

Stage 2 (model selection) was only lightly addressed; yet clearly, merging (retaining, and averaging the outputs of) several good candidate models *within* a family can improve on the robustness of their estimates, and better indicate the suite of useful variables. This can be a very straightforward (if internal) change for some algorithms, such as polynomial networks, which explicitly consider, and retain until the end, numerous candidate models built up in layers of nodes.

Improvements to stage 3 (method selection) were addressed here, and a challenging field classification task employed to demonstrate the potential of *fusing* diverse methods together to get the best model. This fusion incorporates sharing weighted output estimates, where the weights plausibly come from the independent method's relative performance, but also implies more interaction. Here, nonparametric methods worked best on the data, but variable selection from both the decision tree and polynomial network methods, and case selection from the latter, were able to improve the results on new data significantly further. In summary, these explorations suggest that reducing greed at each stage of the inductive modeling process – by lengthening the scoring horizon, or sharing information between competing methods – can be expected to improve the performance of induced models on new cases.

9. REFERENCES

Barron, A. R. (1984). Predicted Squared Error: A Criterion for Automatic Model Selection. Ch. 4 of (Farlow, 1984)

Barron, R. L. & D. Abbott (1988). User of Polynomial Networks in Optimum, Real-time, Two-Point Boundary Value Guidance of Tactical Weapons, *Proc. Military Comp. Conf.*, Anaheim, CA, May 3 - 5.

Berk, K. N. (1978). Comparing Subset Regression Procedures, *Technometrics, 20*, no. 1: 1-6.

Breiman, L., J. H. Friedman, R. A. Olshen, & C. J. Stone (1984). *Classification and Regression Trees*. Wadsworth & Brooks, Pacific Grove, CA.

Cover, T. M. (1974). The Best Two Independent Measurements Are Not the Two Best. *IEEE Trans. Systems, Man & Cybernetics, 4*.

Desroachers, A. & S. Mohseni (1984). On Determining the Structure of a Non-Linear System, *International Journal of Control, 40*: 923-938.

Draper, N. R. & H. Smith (1966). *Applied Regression Analysis.* Wiley, New York.

Elder, J. F. IV (1985). *User's Manual: ASPN: Algorithm for Synthesis of Polynomial Networks* (4th Ed., 1988). Barron Assoc. Inc., Stanardsville, VA.

Elder, J. F. IV (1990). Feature Elimination Using High-Order Correlation, *Proc. Aerospace Applications of Artificial Intelligence*, Dayton, OH, Oct. 29-31: 65-72.

Elder, J. F. IV (1993). Assisting Inductive Modeling through Visualization, *Proc. Joint Statistical Mtg.*, San Francisco, CA, Aug. 7-11.

Elder, J. F. IV & R. L. Barron (1988). Automated Design of Continuously-Adaptive Control: The "Super-Controller" Strategy for Reconfigurable Systems, *Proc. American Control Conf.*, Atlanta, GA, June 15-17.

Elder, J. F. IV & D. E. Brown (1992). Induction and Polynomial Networks, Univ. VA Tech. Report IPC-TR-92-9. (Forthcoming in 1995 as Chapter 3 in *Advances in Control Networks and Large Scale Parallel Distributed Processing Models, Vol. 2.* Ablex, Norwood, NJ.

Elder, J. F. IV, & D. Pregibon (1995, in press) A Statistical Perspective on Knowledge Discovery in Databases, Chapter 4 in *Advances in Knowledge Discovery and Data Mining*, eds. U. M. Fayyad, G. Piatetsky-Shapiro, P. Smyth, & R. Uthurusamy, AAAI/MIT Press.

Farlow, S. J. (1984), Ed. *Self-Organizing Methods in Modeling: GMDH Type Algorithms.* Marcel Dekker.

Fulcher, G. E. & D. E. Brown (1991). *A Polynomial Network for Predicting Temperature Distributions*, Institute for Parallel Computation Tech. Report 91-008, Univ. VA.

Ivakhnenko, A. G. (1968). The Group Method of Data Handling – A Rival of the Method of Stochastic Approximation, *Soviet Automatic Control, 3.*

Lloyd, D. K., & M. Lipow (1962). *Reliability: Management, Methods, and Mathematics.* Prentice Hall, Englewood Cliffs: 360.

Mallows, C. L. (1973). Some Comments on Cp, *Technometrics, 15*: 661-675.

Miller, A. J. (1990). *Subset Selection in Regression.* Chapman and Hall, NY.

Mucciardi, A. N. (1982). *ALN 4000 Ultrasonic Pipe Inspection System. Nondestructive Evaluation Program: Progress in 1981*, EPRI Report NP-2088-SR, Jan.

Murthy, S. K., S. Kasif, & S. Salzberg (1994). A System for Induction of Oblique Decision Trees, *Journal of Artificial Intelligence, 2*: 1-32.

Prager, M. H. (1988). Group Method of Data Handling: A New Method for Stock Identification. *Trans. American Fisheries Society, 117*: 290-296.

Rissanen, J. (1978). Modeling by Shortest Data Description, *Automatica, 14*: 465-471.

14
Learning Possibilistic Networks from Data

Jörg Gebhardt and Rudolf Kruse

Department of Mathematics and Computer Science
University of Braunschweig
D-38106 Braunschweig, Germany
Email: gebhardt@ibr.cs.tu-bs.de

ABSTRACT We introduce a method for inducing the structure of (causal) *possibilistic* networks from databases of sample cases. In comparison to the construction of Bayesian belief networks, the proposed framework has some advantages, namely the explicit consideration of *imprecise (set-valued) data*, and the realization of a controlled form of *information compression* in order to increase the efficiency of the learning strategy as well as approximate reasoning using local propagation techniques. Our learning method has been applied to reconstruct a non-singly connected network of 22 nodes and 24 arcs without the need of any a priori supplied node ordering.

14.1   Introduction

Bayesian networks provide a well-founded normative framework for knowledge representation and reasoning with *uncertain*, but *precise* data. Extending pure probabilistic settings to the treatment of *imprecise (set-valued)* information usually restricts the computational tractability of the corresponding inference mechanisms. It is therefore near at hand to consider alternative uncertainty calculi that provide a justified form of *information compression* in order to support efficient reasoning in the presence of imprecise and uncertain data without affecting the expressive power and correctness of decision making.

Such a modelling approach is appropriate for systems that accept *approximate* instead of crisp reasoning due to a non-significant sensitivity concerning slight changes of information. Possibility theory [Zadeh78, Dubois88] seems to be a promising framework for this purpose.

In this paper we focus our interest on the concept of a *possibilistic causal network*, which is a directed acyclic graph (DAG) and a family of (conditional) possibility distributions. Since the covering of all aspects of possibilistic reasoning is beyond the scope of this paper, we will confine to the problem of inducing the structure of a possibilistic causal network from data.

In Section 14.2 we introduce a possibilistic interpretation of databases of set-valued samples. Based on this semantic background, Sections 14.3 and 14.4 deal with possibilistic networks and the structure induction method, respectively. In Section 14.5, we mention some basic ideas and important results including an example of the successful application of our approach.

---

[1] *Learning from Data: AI and Statistics V.* Edited by D. Fisher and H.-J. Lenz. ©1996 Springer-Verlag.
[2] This work has partially been funded by CEC-ESPRIT III Basic Research Project 6156 (DRUMS II)

14.2   Possibilistic Interpretation of Sample Databases

Let $\mathrm{Obj}(X_1,\ldots,X_n)$ be an *object type* of interest, which is characterized by a set $V = \{X_1,\ldots,X_n\}$ of *variables (attributes)* with finite domains $\Omega^{(i)} = \mathrm{Dom}(X_i)$, $i = 1,\ldots,n$.

The *precise specification* of a current object state of this type is then formalized as a tuple $\omega_0 = (\omega_0^{(1)},\ldots,\omega_0^{(n)})$, taken from the *universe of discourse* $\Omega = \Omega^{(1)} \times \ldots \times \Omega^{(n)}$. Any subset $R \subseteq \Omega$ can be used as a *set-valued specification* of ω_0, which consists of all states that are possible candidates for ω_0. R is therefore called *correct* for ω_0, if and only if $\omega_0 \in R$. R is called *imprecise*, iff $|R| > 1$, *precise*, iff $|R| = 1$, and *contradictory*, iff $|R| = 0$.

Suppose that general knowledge about dependencies among the variables is available in form of a database $\mathcal{D} = (D_j)_{j=1}^m$ of sample cases. Each case D_j is interpreted as a (set-valued) correct specification of a previously observed representative object state $\omega_j = (\omega_j^{(1)},\ldots,\omega_j^{(n)})$.

Supporting imprecision (non-specificity) consists in stating $D_j = D_j^{(1)} \times \ldots \times D_j^{(n)}$, where $D_j^{(i)}$ denotes a nonempty subset of $\Omega^{(i)}$. We assume that $\omega_j^{(i)} \in D_j^{(i)}$ is satisfied, but no further information about any preferences among the elements in $D_j^{(i)}$ is given. When the cases in $\mathcal{D}$ are applied as an imperfect specification of the current object state ω_0, then *uncertainty* concerning ω_0 occurs in the way that the underlying frame conditions (here named as *contexts*, and denoted by c_j), in which the sample states ω_j have been observed, may only for some of the cases coincide with the context on which the observation of ω_0 is based. The complete description of context c_j depends on the physical frame conditions of ω_j, but is also influenced by the frame conditions of observing ω_j by a human expert, a sensor, or any other observation unit. For the following consideration, we make some assumptions on the relationships between contexts and context-dependent specifications of object states. In particular, we suppose that our knowledge about ω_0 can be represented by an *imperfect specification* $\Gamma = (\gamma, P_C)$, $C = \{c_1,\ldots,c_m\}$, $\gamma : C \to \mathfrak{P}(\Omega)$, $\gamma(c_j) = D_j$, $j = 1,\ldots,m$, with C denoting the set of contexts, $\gamma(c_j)$ the context-dependent set-valued specification of ω_j, P_C a probability measure on C, and $\mathfrak{P}(\Omega)$ the power set of Ω. $P_C(\{c\})$ quantifies the probability of occurrence of context $c \in C$. If all contexts are in the same way representative and thus equally likely, then P_C should be the uniform distribution on C.

We suppose that C can be formalized as a subset of a Boolean algebra of propositions. The mapping $\gamma : C \to \mathfrak{P}(\Omega)$ indicates the assumption that there is a functional dependency of the sample cases from the underlying contexts, so that each context c_j uniquely determines its set-valued specification $\gamma(c_j) = D_j$ of ω_j. It is reasonable to state that $\gamma(c_j)$ is correct for ω_j (i.e.: $\omega_j \in \gamma(c_j)$) and of *maximum specificity*, which means that no proper subset of $\gamma(c_j)$ is guaranteed to be correct for ω_j with respect to context c_j. Related to the current object state of interest, specified by the (unknown) value $\omega_0 \in R$, and observed in a new context c_0, any c_j in C is adequate for delivering a set-valued specification of ω_0, if c_0 and c_j, formalized as logical propositions, are not contradicting. Intersecting the context-dependent set-valued specifications $\gamma(c_j)$ of all contexts c_j that do not contradict c_0, we obtain the most specific correct set-valued specification of ω_0 with respect to γ.

The idea of using set-valued mappings on probability fields in order to treat uncertain and imprecise data refers to similar random-set-like approaches that were suggested, for instance, in [Strassen64], [Dempster68], and [Kampe82]. But note that for operating on imperfect specifications in the field of knowledge-based systems, it is important to provide adequate semantics. We addressed this topic in more detail elsewhere [Gebhardt93a, Gebhardt93b].

When we are only given a database $\mathcal{D} = (D_j)_{j=1}^m$ of sample cases, where $D_j \subseteq \Omega$ is assumed to be a context-dependent most specific specification of ω_j, we are normally not in the position to fully describe the contexts c_j in the form of propositions that are taken from an appropriate underlying Boolean algebra of propositions. For this reason it is convenient to carry out an information compression by paying attention to the context-dependent specifications rather than to the contexts themselves. We do not directly refer to $\Gamma = (\gamma, P_C)$, but to its degree of α-*correctness* w.r.t. ω_0, which is defined as the total mass of all contexts c_j that yield a correct context-dependent specification $\gamma(c_j)$ of ω_0. If we are given any $\omega \in \Omega$, then $\Gamma = (\gamma, P_C)$ is called α-*correct* w.r.t. ω, iff

$$P_C \left(\{ c \in C \mid \omega \in \gamma(c) \} \right) \geq \alpha, \qquad 0 \leq \alpha \leq 1.$$

Note that, although the application of a probability measure P_C suggests disjoint contexts, we do not make any assumptions about the interrelation of contexts. With respect to the frame conditions that cause the observations, the contexts may be identical, partially corresponding, or disjoint. We add their weights, because disjoint contexts are the "worst case" in which we can not restrict the total weight to a smaller value without losing correctness. In this manner, a possibility degree is the upper bound for the total weight of the combined contexts.

Suppose that our only information about ω_0 is the α-correctness of Γ w.r.t. ω_0, without having any knowledge of the description of the contexts in C. Under these restrictions, we are searching for the most specific set-valued specification $A_\alpha \subseteq \Omega$ of ω_0, namely the largest subset of Ω such that α-correctness of Γ w.r.t. ω is satisfied for all $\omega \in A_\alpha$. It easily turns out that the family $(A_\alpha)_{\alpha \in [0,1]}$ consists of all α-*cuts* $[\pi_\Gamma]_\alpha$ of the induced *possibility distribution*

$$\begin{aligned}
\pi_\Gamma : \Omega &\rightarrow [0,1], \\
\pi_\Gamma(\omega) &= P_C \left(\{ c \in C \mid \omega \in \gamma(c) \} \right),
\end{aligned}$$

where for any π, taken from the set $\text{POSS}(\Omega)$ of all possibility distributions that can be induced from imperfect specifications w.r.t. Ω, the α-cut $[\pi]_\alpha$ is defined as follows:

$$\begin{aligned}
[\pi]_\alpha &= \{ \omega \in \Omega \mid \pi(\omega) \geq \alpha \}, \qquad 0 < \alpha \leq 1, \\
[\pi]_0 &= \Omega.
\end{aligned}$$

Note that $\pi_\Gamma(\omega)$ can in fact be viewed as a degree of possibility for the truth of "$\omega = \omega_0$": If $\pi_\Gamma(\omega) = 1$, then $\omega \in \gamma(c)$ holds for all contexts $c \in C$, which means that $\omega = \omega_j$ is possible for all sample object states ω_j, $j = 1, \ldots, m$, so that $\omega_0 = \omega$ should be possible without any restriction.

If $\pi_\Gamma(\omega) = 0$, then $\omega_j = \omega$ has been rejected for ω_j, $j = 1, \ldots, m$, since $\omega \notin \gamma(c_j)$ is true for the set-valued specifications $\gamma(c_j)$ of ω_j. This entails the impossibility of $\omega_0 = \omega$, if the description of the context c_0 for the specification of ω_0 is assumed to be a conjunction of the descriptions of any contexts in C.

If $0 < \pi_\Gamma(\omega) < 1$, then there are contexts that support $\omega_0 = \omega$ as well as contexts that contradict $\omega_0 = \omega$. The quantity $\pi_\Gamma(\omega)$ reflects the maximum possible total mass of contexts that support $\omega_0 = \omega$.

Note that in the recent years several proposals for the semantics of a theory of possibility as a framework for reasoning with uncertain and imprecise data have been made. Among the numerical approaches, we like to mention the epistemic interpretation of fuzzy sets [Zadeh78], the axiomatic view of possibility theory using possibility measures [Dubois88], one-point coverages of random sets [Nguyen78], contour functions of consonant belief functions [Shafer76], falling shadows in set-valued statistics [Wang83], Spohn's theory of epistemic states [Spohn90], and possibility theory based on likelihoods [Dubois93]. Ignoring the underlying interpretation of contexts, π_Γ formally coincides with the one-point coverage of Γ, when it is interpreted as a (not necessarily nested) random set. From a semantics point of view, operating on possibility distributions in our setting may better be strongly oriented at the concept of α-correctness. For an extensive presentation of this background of possibility theory, we refer to [Gebhardt93b, Gebhardt93c]. Special aspects of possibility measures for decision making in this framework have been considered in [Gebhardt94].

14.3   The Concept of a Possibilistic Network

For the following investigations, we suppose that all relevant dependencies among the variables in V can be represented in a qualitative way with the aid of a *dependency hypergraph* $H = (V, E)$ [Berge76].

¿From a quantitative point of view, a dependency hypergraph $H = (V, \mathcal{E})$, when applied to a relation $R \subseteq \Omega$, induces a *constraint network* $\mathcal{N}_H(R)$ over V, which is defined as the family of nonempty relations

$$R_E \stackrel{\text{def}}{=} \Pi_E^V(R),$$

with Π_E^V denoting the pointwise projection from Ω^V onto Ω^E. In this connection, let $\Omega^{\{v\}}$ be the domain of the variable $v \in V$. If $W \subseteq V$ is an arbitrary subset of variables, then

$$\Omega^W \stackrel{\text{def}}{=} \begin{cases} \bigtimes_{v \in W} \Omega^{\{v\}}, & \text{if } W \neq \emptyset \\ \{\varepsilon\}, & \text{if } W = \emptyset \end{cases}$$

is defined as the product of their domains, where the empty tuple ε is the only element of $\Omega^\emptyset$.

Since $\mathcal{N}_H(R)$ specifies local dependencies among the values of the variables in $E \in \mathcal{E}$, $\mathcal{N}_H(R)$ may be less informative than R. More particularly, defining

$$\text{rel}(\mathcal{N}_H(R)) \stackrel{\text{def}}{=} \{\omega \in \Omega \mid \forall E \in \mathcal{E} : \Pi_E^V(\omega) \in R_E\}$$

as the set of all global dependencies in Ω that can be derived from $\mathcal{N}_H(R)$, we obtain

$$R \subseteq \text{rel}(\mathcal{N}_H(R)).$$

Structure identification of relation R is the task of finding a dependency hypergraph $H = (V, \mathcal{E})$ such that

$$R = \text{rel}(\mathcal{N}_H(R))$$

holds. The induced constraint network $\mathcal{N}_H(R)$ is then called a *lossless join decomposition* of R which *describes* or *represents* R.

Given a dependency hypergraph $H = (V, \mathcal{E})$, any most specific set-valued specification R of the current object state ω_0 is assumed to have a lossless join decomposition $(R_E)_{E \in \mathcal{E}}$. This property has an important influence on the interpretation of a sample database $\mathcal{D}$, caused by the decomposability of $\mathcal{D}$ into a family $(\mathcal{D}_E)_{E \in \mathcal{E}}$ of databases, where each $\mathcal{D}_E$ provides sample cases of observed dependencies among the variables contained in the hyperedge E. The database $\mathcal{D}_E = (D_j^E)_{j=1}^m$ consists of the set-valued specifications $D_j^E = \Pi_E^V(D_j)$ of the dependencies $\Pi_E^V(\omega_j)$ that are part of the sample object states ω_j. Given the dependency hypergraph H and the database $\mathcal{D}$, the pair $(\mathcal{D}, H)$ may be applied in order to imperfectly specify ω_0. Our knowledge about ω_0 can be represented with the aid of a family $(\Gamma_E)_{E \in \mathcal{E}}$ of imperfect specifications $\Gamma_E = (\gamma_E, P_E)$, where $\gamma_E : C_E \to \mathfrak{P}(\Omega^E)$ is defined on the set $C_E = \{c_1^E, \ldots, c_m^E\}$ of those contexts c_j^E that reflect the frame conditions for specifying the sample dependency $\Pi_E^V(\omega_j)$, i.e., $\gamma_E(c_j^E) \overset{\text{def}}{=} D_j^E$.

Under the assumption that the contexts c_j^E in C^E are equally likely, we choose $P_E(c_j^E) \overset{\text{def}}{=} \frac{1}{m}$ for all $j = 1, \ldots, m$. The family $\mathcal{N}_H(\mathcal{D}) = (\pi_{\Gamma_E})_{E \in \mathcal{E}}$ of the induced possibility distributions π_{Γ_E} is called a *possibilistic (constraint) network* over V.

Stating α_E-correctness of Γ_E w.r.t. $\Pi_E^V(\omega_0)$, we obtain $R_E = [\pi_{\Gamma_E}]_{\alpha_E}$ as the most specific correct set-valued specification of $\Pi_E^V(\omega_0)$ that follows from the interpretation of $\mathcal{D}$ and H. The specifications R_E can be combined to the constraint network $\mathcal{N} = (R_E)_{E \in \mathcal{E}}$ as a lossless join decomposition of $\text{rel}(\mathcal{N})$, which is the resulting most specific correct set-valued specification of ω_0.

The following definition introduces an alternative view of a possibilistic network, which incorporates causality aspects in the way that it deals with a directed acyclic graph of qualitative dependencies, and a family of (conditional) possibility distributions.

Definition 1.   Let $V = \{X_1, \ldots, X_n\}$ denote a set of variables, $\Omega^{(i)} = \text{Dom}(X_i)$, $i = 1, \ldots, n$, their attached finite domains, and $\Omega = \Omega^{(1)} \times \ldots \times \Omega^{(n)}$ their common universe of discourse. Furthermore, let $\mathcal{D} = (D_j)_{j=1}^m$ be a database of (set-valued) sample cases $D_j = D_j^{(1)} \times \ldots \times D_j^{(n)}$ with $\emptyset \neq D_j^{(i)} \subseteq \Omega^{(i)}$ for $j = 1, \ldots, m$.

Let $\Gamma = (\gamma, P_C)$, determined by $\gamma : C \to \mathfrak{P}(\Omega)$, $C = \{c_1, \ldots, c_m\}$, and $\gamma(c_j) = D_j$, $j = 1, \ldots, m$, be an imperfect specification of the current object state $\omega_0 \in \Omega$ of interest.

Let $G = (V, E)$ be a directed acyclic graph. For any $W \subseteq V$ and $X \in V \backslash W$, define

$$\varphi[\mathcal{D}; W \to V] : \quad \Omega \quad \to \quad \text{POSS}(\Omega^{\{X\}}),$$
$$\varphi[\mathcal{D}; W \to V](\omega) \quad \overset{\text{def}}{=} \quad \pi_{\mathcal{D}}(X \mid W = \omega), \text{ where}$$
$$\pi_{\mathcal{D}}(X \mid W = \omega)(\omega') \quad \overset{\text{def}}{=} \quad P_C\left(\left\{c \in C \mid \omega \in \Pi_W^V(\gamma(c)) \text{ and } \omega' \in \Pi_{\{X\}}^V(\gamma(c))\right\}\right).$$

For any $X \in V$, let

$$\text{par}_G(X) \overset{\text{def}}{=} \{Y \in V \mid (Y, X) \in E\}$$

denote the *set of all parent nodes of* X. Then, the family

$$\mathcal{N}_G(\mathcal{D}) \overset{\text{def}}{=} (\varphi[\mathcal{D}; \text{par}_G(X) \to X])_{X \in V}$$

is called a *possibilistic causal network* for ω_0, induced by $\mathcal{D}$ and G.

Note that $[\pi_{\mathcal{D}}(X \mid W = \omega)]_\alpha$ is the most specific correct set-valued specification of $\Pi^V_{\{X\}}(\omega_0)$ that follows from $\mathcal{D}$, given the instantiation of the variables in W ($\Pi^V_W(\omega_0) = \omega$) and the α-correctness of $\Gamma_{W \cup \{X\}}$ w.r.t. $\Pi^V_{W \cup \{X\}}(\omega_0)$.

The possibilistic causal network $\mathcal{N}_G(D)$ is an information-compressed representation of our knowledge about ω_0, given the database $\mathcal{D}$ and a DAG $G = (V, E)$ of causal dependencies among the variables in V. The DAG G induces the dependency hypergraph

$$H(G) = (V, E_G), \qquad E_G \stackrel{\text{def}}{=} \{\{X\} \cup \text{par}_G(X) \mid X \in V\},$$

and, incorporating the database $\mathcal{D}$, the possibilistic constraint network $\mathcal{N}_{H(G)}(D)$.

14.4   Inducing Possibilistic Networks from Data

In this section we present a new method for inducing a possibilistic causal network $\mathcal{N}_G(D)$ and therefore its attached possibilistic constraint network $\mathcal{N}_{H(G)}(D)$ from a database $\mathcal{D}$ of sample cases. This method is based on viewing a DAG as an abstract deductive reasoning scheme. More particularly, let $<$ be a topological ordering of V, i.e. all nodes $v, v' \in V$ with $v' < v$ satisfy the condition $(v, v') \notin E$. Suppose that $v_1 < v_2 \cdots < v_n$ reflects this ordering. For any $\omega \in \Omega$, we either want to identify ω as the current object state ($\omega = \omega_0$) or to verify $\omega \neq \omega_0$ by marking those variables v_i, $i = 1, \ldots, n$, for which $\Pi^V_{\{v_i\}}(\omega)$ turns out to be inconsistent with the available general knowledge about ω_0. This knowledge is encoded by $\mathcal{N}_G(D)$ and the total mass α_Z of all contexts in C_Z whose description does not contradict the description of the context c_0^Z for the specification of the dependency $\Pi^V_Z(\omega_0)$ within the current object state ω_0. Let $\omega' = \Pi^V_{\text{par}_G(X)}(\omega)$ and $\omega'' = \Pi^V_{\{X\}}(\omega)$. Since we assume that the descriptions of the addressed contexts are not available, we only know that $0 \leq \alpha_Z \leq \alpha(\omega', \omega'', X)$ holds, where

$$\alpha(\omega', \omega'', X) \stackrel{\text{def}}{=} \pi_{\mathcal{D}}(X \mid \text{par}_G(X) = \omega')(\omega'')$$

is the maximum possible correctness degree of Γ_Z w.r.t. $\Pi^V_Z(\omega)$. Applied to any $\omega \in \Omega$, the reasoning scheme works as follows:

Scheme 2.
Pseudocode of a deductive reasoning scheme for identifying or rejecting any $\omega \in \Omega$ as the current object state ω_0, given a possibilistic causal network $\mathcal{N}_G(\mathcal{D})$.

for $i := 1$ **to** n **do begin**

    assign $X := v_i$ and calculate $\pi_i := \pi_{\mathcal{D}}(X \mid \text{par}_G(X) = \Pi^V_{\text{par}_G(X)}(\omega))$;

    if $\pi_i \not\equiv 0$

    then determine the total mass α_Z of all contexts in C_Z, $Z = \text{par}_G(X) \cup \{X\}$, whose description does not contradict the description of the context c_0^Z for the specification of $\Pi^V_Z(\omega_0)$; note that Γ_Z is α_Z-correct w.r.t. $\Pi^V_Z(\omega_0)$;

            provide further information such that either $\Pi^V_{\{X\}}(\omega_0)$ is identified within the set $\Phi_i := [\pi_i]_{\alpha_Z}$ of all the remaining possible instantiations of X, or $\Pi^V_{\{X\}}(\omega) \neq \Pi^V_{\{X\}}(\omega_0)$ is recognized.

if $(\pi_i \equiv 0)$ **or** $(\Pi^V_{\{X\}}(\omega) \neq \Pi^V_{\{X\}}(\omega_0))$

then mark X as a variable where an erraneous instantiation has been detected $(\omega \neq \omega_0)$

end

The amount of information that has to be added in order to identify $\Pi^V_{\{X\}}(\omega_0)$ within the set Φ_i of possible alternatives can be quantified by $H(\Phi_i)$, where H denotes the *Hartley measure of information* [Hartley28].

For any nonempty finite set A,

$$H(A) \stackrel{\text{def}}{=} \log_2 |A|$$

is the number of elementary propositions (measured in bits), whose truth values must be determined for the specification and thus the identification of a single element in the reference set A.

Note that in the case $\Phi_i = \emptyset$, there is no need for adding further information, since no possible alternative instantiations of v_i exist. We therefore define $H(\emptyset) \stackrel{\text{def}}{=} 0$.

Based on Hartley information, we measure the nonspecificity of $\mathcal{N}_G(D)$ with respect to $\omega \in \Omega$, namely the total amount of additional information (beyond D) that is necessary either to identify ω as the current object state ($\omega = \omega_0$) by carrying out inference in the presented deductive reasoning scheme, or to mark all variables for which the reasoning process provides a contradiction ($\omega \neq \omega_0$). We assume that all choices of α_Z-correctness degrees in the interval $[0, \alpha(\omega', \omega'', X)]$ are equally likely, and that in our a priori state of knowledge, we are indifferent concerning the value of ω_0. Summarizing these ideas, we obtain the following definition.

Definition 3.   Let $G = (V, E)$ be a DAG, $\mathcal{D}$ a database of sample cases, and $\mathcal{N}_G(\mathcal{D}) = (\varphi[\mathcal{D}; \text{par}_G(X) \to X])_{X \in V}$ their induced possibilistic causal network. Then, for any $\omega \in \Omega$ and any family $\Lambda(\omega) = (\alpha_i(\omega))_{i=1}^n$ of correctness degrees that satisfy $0 \leq \alpha_i(\omega) \leq \alpha(\Pi^V_{\text{par}_G(X_i)}(\omega), \Pi^V_{\{X_i\}}(\omega), X_i)$, the quantity

$$\text{Nonspec}[\mathcal{N}_G(\mathcal{D})](\omega, \Lambda(\omega)) = \sum_{i=1}^n H\left([\varphi[\mathcal{D}; \text{par}_G(X_i) \to X_i](\omega)]_{\alpha_i(\omega)}\right)$$

is called the *nonspecificity of* $\mathcal{N}_G(\mathcal{D})$ *w.r.t.* ω *and* $\Lambda(\omega)$.

Assuming uniform distributions on Ω and all $\Lambda(\omega)$, $\omega \in \Omega$, then

$$E(\text{Nonspec}[\mathcal{N}_G(\mathcal{D})]) = \frac{1}{|\Omega|} \sum_{\omega \in \Omega} \int_{\Lambda(\omega)} \text{Nonspec}[\mathcal{N}_G(\mathcal{D})](\omega, \Lambda(\omega)) dF_{\Lambda(\omega)}$$

is called the *expected nonspecificity of* $\mathcal{N}_G(\mathcal{D})$.

The concept of the expected nonspecificity of a causal constraint network is helpful for inducing an optimal DAG G that minimizes $E(\text{Nonspec}[\mathcal{N}_G(D)])$ relative to $\mathcal{D}$ and a chosen class of DAGs. The following Theorem shows another representation of this quantity, which is more convenient for computational aspects.

Theorem 4.   *For any $X \in V$ and any $W \subseteq V$, $X \notin W$, define*

$$m(W, X) \stackrel{\text{def}}{=} \frac{1}{|\Omega^W| \cdot |\Omega^{\{X\}}|} \cdot$$

$$\sum_{\substack{\omega' \in \Omega^W, \omega'' \in \Omega^{\{X\}}: \\ \alpha(\omega', \omega'', X) > 0}} \int_0^{\alpha(\omega', \omega'', X)} \frac{1}{\alpha(\omega', \omega'', X)} H\left([\pi_D(X \mid W = \omega')]_\alpha\right) d\alpha.$$

Then,

$$E(\mathrm{Nonspec}[\mathcal{N}_G(\mathcal{D})]) = \sum_{X \in V} m(\mathrm{par}_G(X), X).$$

14.5   Results and Concluding Remarks

- Let $\mathcal{G}_k(V)$, $k \geq 2$, denote the class of all directed acyclic graphs w.r.t. V that satisfy the condition $|\mathrm{par}(X)| \leq k - 1$ for all $X \in V$. Based on Theorem 14.4, we developed an *Algorithm G1* for determining a DAG $G \in \mathcal{G}_k(V)$ that minimizes $E(\mathrm{Nonspec}[\mathcal{N}_G(D)])$ among all DAGs in $\mathcal{G}_k(V)$ that satisfy the node ordering constraints of G. Algorithm G1 has a time complexity of $O(mn^k r^k)$, where r is defined as $r = \max\{|\Omega^{(i)}| \, | \, i = 1, \ldots, n\}$. It does not need any presupposed node ordering, and although it is only optimal w.r.t. a subclass of $\mathcal{G}_k(V)$, it nevertheless tends to deliver a good choice w.r.t. $\mathcal{G}_k(V)$.

- Eliminating arcs of a DAG $G \in \mathcal{G}_k(V)$ in order to get a more simple DAG $G' \in \mathcal{G}_{k-1}(V)$, is connected with a loss of information which is quantified by the corresponding increasement of $E(\mathrm{Nonspec}[\mathcal{N}_{G'}(D)])$ in comparison to the expectation $E(\mathrm{Nonspec}[\mathcal{N}_G(D)])$.

- Algorithm G1 can be made more efficient with the aid of a Greedy search method, starting with $\mathcal{G}_2(V)$ and stepwise extending the optimal output graph with respect to those arcs that reflect the strongest causal dependencies (i.e. the smallest degree of nonspecificity) to the class $\mathcal{G}_k(V)$. The resulting *Algorithm G2* has a time complexity of $O(mn^2 r^k)$.

- From a graph theoretical point of view, independence in the DAG of a possibilistic causal network is represented in the same way as independence in Bayesian networks. Due to the different uncertainty calculus, independence in our approach turns out to basically coincide with the concept of *non-interactivity* well-known from possibility theory. Non-interactivity satisfies the basic properties of independence as proposed in [Pearl88], with the exception of the intersection axiom.

- From database theory it is well-known that given a relation R and any hypergraph H, deciding whether $\mathrm{rel}(\mathcal{N}_H(R)) = R$ is NP-hard. Constructing a lossless join decomposition of a relation within a class of dependency hypergraphs is presumably intractable even in cases where each individual member of the class is tractable [Dechter92].

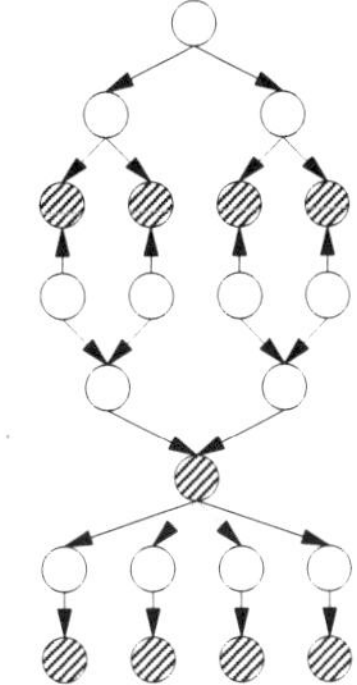

Fig.1 : The blood–type example network

Since the problem of structure identification in relational data can be viewed as a special case of the corresponding problem in the possibilistic framework, it is out of reach to get a general, non-heuristic, efficient algorithm for possibilistic learning. One of the tasks of our future research in this field therefore consists in working out, how far Algorithms G1 and G2 deliver tight approximations of optimal (i.e. maximum specificity preserving) decompositions.

- Algorithms G1 and G2 have successfully been applied for reconstructing the network shown in Figure 1. The underlying application refers to a Bayesian approach, implemented with HUGIN [Andersen89] for daily use in Denmark. It deals with the determination of the genotype and verifying the parentage in the F-blood group system of Danish Jersey Cattle [Rasmussen92]. The application is supported by a real database of 747 sample cases for the 9 attributes marked as grey nodes in Figure 1, including lots of missing values. There is also additional expert knowledge regarding the quantitative dependencies among other attributes. Using this information we extended the database to an artificial database for all attributes. Running Algorithm G2 on this database, the network could be efficiently reconstructed in the possibilistic setting without erraneous links, except from those dependencies, where a unique directing of arcs is not possible, since not expressable in a database. We will include our learning strategy in the system POSSINFER [Kruse94], a software tool for *possibilistic inference* that we develop in cooperation with Deutsche Aerospace in the field of data fusion problems.

Acknowledgements

We would like to thank S.L. Lauritzen and L.K. Rasmussen for supporting us with the bloodtype determination example.

14.6 REFERENCES

[Andersen89] S.K. Andersen, K.G. Olesen, F.V. Jensen, and F. Jensen. HUGIN — A Shell for Building Bayesian Belief Universes for Expert Systems, *Proc. 11th Int. Joint Conf. on AI*, 1080–1085, 1989.

[Berge76] C. Berge. *Graphs and Hypergraphs.* North Holland, New York, 1976.

[Dechter92] R. Dechter and J. Pearl. Structure Identification in Relational Data, *Artificial Intelligence*, 58:237–270, 1992.

[Dempster68] A.P. Dempster. Upper and Lower Probabilities Generated by a Random Closed Interval, *Ann. Math. Stat.*, 39:957–966, 1968.

[Dubois88] D. Dubois and H. Prade. *Possibility Theory.* Plenum Press, New York, 1988.

[Dubois93] D. Dubois, S. Moral, and H. Prade. A Semantics for Possibility Theory Based on Likelihoods, CEC–ESPRIT III BRA 6156 DRUMS II, Annual Report, 1993.

[Gebhardt93a] J. Gebhardt and R. Kruse. The Context Model — An Integrating View of Vagueness and Uncertainty, *Int. J. of Approximate Reasoning*, 9:283–314, 1993.

[Gebhardt93b] J. Gebhardt and R. Kruse. A Comparative Discussion of Combination Rules in Numerical Settings, CEC–ESPRIT III BRA 6156 DRUMS II, Annual Report, 1993.

[Gebhardt93c] J. Gebhardt and R. Kruse. A New Approach to Semantic Aspects of Possibilistic Reasoning, In: M. Clarke, R. Kruse, and S. Moral, editors. *Symbolic and Quantitative Approaches to Reasoning and Uncertainty*, Lecture Notes in Computer Science, 747, 151–160, Springer, Berlin, 1993.

[Gebhardt94] J. Gebhardt and R. Kruse. On an Information Compression View of Possibility Theory, *Proc. 3rd IEEE Int. Conf. on Fuzzy Systems*, Orlando, 1994.

[Hartley28] R.V.L. Hartley. Transmission of Information, *The Bell Systems Technical J.*, 7:535–563, 1928.

[Kampe82] J. Kampé de Fériet. Interpretation of Membership Functions of Fuzzy Sets in Terms of Plausibility and Belief. In: M.M. Gupta and E. Sanchez, editors. *Fuzzy Information and Decision Processes*, North–Holland, 13–98, 1982.

[Kruse94] R. Kruse, J. Gebhardt, and F. Klawonn. *Foundations of Fuzzy Systems*, Wiley, Chichester, 1994.

[Nguyen78] H.T. Nguyen. On Random Sets and Belief Functions, *J. of Mathematical Analysis and Applications*, 65:531–542, 1978.

[Pearl88] J. Pearl. *Probabilistic Reasoning in Intelligent Systems — Networks of Plausible Inference.* Morgan Kaufman, San Mateo 1988.

[Rasmussen92] L.K. Rasmussen. Blood Group Determination of Danish Jersey Cattle in the F-blood Group System, *Dina Research Report 8*, Dina Foulum, 8830 Tjele, Denmark, November 1992.

[Shafer76] G. Shafer. *A Mathematical Theory of Evidence*, Princeton University Press, Princeton, 1976.

[Spohn90] W. Spohn. A General Non–Probabilistic Theory of Inductive Reasoning, In: R.D. Shachter, T.S. Levitt, L.N. Kanal. and J.F Lemmer, editors, *Uncertainty in Artificial Intelligence*, 149–158, North Holland, Amsterdam, 1990.

[Strassen64] V. Strassen. Meßfehler und Information, *Zeitschrift Wahrscheinlichkeitstheorie und verwandte Gebiete*, 2:273–305, 1964.

[Wang83] P.Z. Wang. From the Fuzzy Statistics to the Falling Random Subsets, In: P.P. Wang, editor. *Advances in Fuzzy Sets, Possibility and Applications*, 81–96, Plenum Press, New York, 1983.

[Zadeh78] L.A. Zadeh. Fuzzy Sets as a Basis for a Theory of Possibility, *Fuzzy Sets and Systems*, 1:3–28, 1978.

15
Detecting Imperfect Patterns in Event Streams Using Local Search

Adele E. Howe

Computer Science Department
Colorado State University
Fort Collins, CO 80523
USA
howe@cs.colostate.edu

ABSTRACT Recurring patterns in event streams may indicate causal influences. Such patterns have been used as the basis of software debugging. One technique, dependency detection, for finding these patterns requires exhaustive search and perfect patterns, two characteristics that are unrealistic for event streams extracted from software executions. This paper presents an enhanced version of dependency detection that uses a more flexible pattern matching scheme to extend the types of patterns detected and local search to reduce the computational demands. The new version was tested on real and artificial data to determine whether local search is effective for detecting strong patterns.

15.1   Introduction

The problem of inferring causality from empirical observations has been well studied. Several approaches are notable for efficiently constructing complex causal models of the inter-relationships between variables (e.g., [8, 3, 2]). These approaches tend to rely on correlations and co-variances among the variables as the basis for inferring causality. However, for some applications, the available data are categorical observations over time (e.g., event streams or execution traces of programs); for example, patterns in execution traces form the basis of several methods of debugging software (e.g., [1, 4]). These applications are less amenable to solution by methods based on correlation and co-variance.

An alternative approach, called *Dependency Detection*, searches the event streams for often recurring sequences [6]. The set of recurring sequences (called *dependencies*) indicates events that commonly co-occur and forms a weak model of causality.

Although promising, dependency detection is limited in several ways. First, the underlying search is exhaustive, looking for all possible dependencies in the data. As a consequence, the computational complexity increases exponentially with the length of the sequences. Second, the sequences are rigid, which means that only exact matches count and that any noise (i.e., insertion of some other unrelated event into the stream) will not count as an example of the sequence. Third, the technique considers only a single stream rather than multiple streams of parallel events; Oates et al. have developed a technique for multi-stream dependency detection [7]. This paper describes how local search with

---

[1] *Learning from Data: AI and Statistics V.* Edited by D. Fisher and H.-J. Lenz. ©1996 Springer-Verlag.

flexible matching has been used to overcome the first two limitations.

15.2   Finding Significant Sequences

The goal is to find sequences that repeat with unusual frequency in event streams. This section describes the basis for finding these sequences (dependency detection) and enhancements to the basic method that find longer sequences that are insensitive to minor variations (*local search dependency detection*).

15.2.1   Dependency Detection

Dependency Detection applies contingency table analysis to build a set of simple models of causes of a particular event's occurrence. It was designed to be the first step in a partially automated debugging process, which was originally developed for the Phoenix planning system [5]. Consequently, the algorithm and the examples refer to failure recovery actions of the Phoenix planner and the failures that occurred during the planner's execution. It has been used as well for detecting sources of search inefficiency in another planning system.

Dependency detection searches Phoenix execution traces for statistically significant patterns: co-occurrences of particular precursors (events or combinations of events) leading to a target event (i.e., a failure)[2]. The execution traces are sequences of events: alternating failures and the recovery actions that repaired them, e.g.,

$$F_a \rightarrow R_{aa} \rightarrow F_b \rightarrow R_{zz} \rightarrow F_a \rightarrow R_{aa} \rightarrow F_b.$$

The idea is that the appearance of repetitive sequences (which will be called *patterns* to distinguish potential models from the data), such as $R_{aa} \rightarrow F_b$, may designate early warnings of severe failures or indicate partial causes of the failures.

Dependency detection exhaustively tested the significance of a well-defined set of patterns using contingency table analysis. The set of patterns is crucial because the patterns form the hypothesized models of causality for the data; a model that is not in the set will not be tested. The pattern set defines the size of the search space. To test each pattern, the algorithm needs a method of matching the pattern to the data and, to reduce redundancy and potentially conflicting models, needs a method of reconciling similar models. Dependency detection exploits characteristics of the Phoenix data; the set of patterns included all combinations of a precursor followed immediately by a failure; the precursor could be a single failure, a single recovery action or a two event sequence of a failure and a recovery action. (The precursor-failure patterns are written as [F F], [R F] and [FR F], respectively.) While [R F] and [FR F] match the event streams exactly, the [F F] pattern matches by ignoring the value of the intervening recovery action. Similar patterns are reconciled by exploiting a property of the underlying statistic.

The dependency detection algorithm has three steps.

1. Gather execution traces.

2. For each defined pattern of precursor and failure, a contingency table is built and

---

[2]As terminology, a pattern is composed of a precursor and a failure.

	F_b	$F_{\overline{b}}$
R_{aa}	5	10
$R_{\overline{aa}}$	10	100

TABLE 15.1. Sample contingency table from Dependency Detection for sequence $[R_{aa}F_b]$

tested for the appearance of the pattern in the execution traces.

3. Each significant pattern is tested for overlap with other similar significant patterns.

First, the algorithm gathers the data, which is alternating failures and recovery actions. Second, it steps through all possible patterns. For each pattern, it arranges four counts from the event streams in a 2x2 contingency table and applies a G-test[3] to the table. Figure 15.1 shows an example contingency table for the sequence $[R_{aa}F_b]$ (which indicates that the precursor is a recovery action of type aa and the failure is of type b). The four counts are: the number of times the precursor is followed by the target failure (upper left cell) and not followed by the target failure (upper right cell), as well as when other precursors are followed by the failure (lower left cell) and not followed by it (lower right cell). A G-test of the example table produces $G = 6.79, p < .009$, suggesting a dependency between the two events.

For each significant pattern ($\alpha < .05$), the third step prunes out similar (overlapping) patterns using a Homogeneity G-test. Two patterns overlap when one pattern contains either a particular failure or a particular recovery action as precursor (i.e., are either [R F] or [F F]) and the other contains both (i.e., [FR F] where one of the events in the precursor is the same as in the singleton). The process exploits the additivity property of the G-test by treating the data for the [FR F] pattern as a partition of the [F F] or [R F] (whichever is being compared) data and results in a list of all significant patterns found in the execution traces. (Details of the pruning process can be found in [6].)

Computationally, dependency detection requires a complete sweep of the execution traces to test each pattern. The number of patterns equals the number of types of failures squared (for FF) plus the number of failures squared times the number of recovery actions (for FRF) plus the number of recovery actions times the number of failures (for RF). In total, this is roughly $M * N^3$ where M is the length of the event traces, N is the number of types of events in the traces (i.e., the total number of types of recovery actions and failures). N is cubed to subsume the FRF case in which the precursor has both a failure and a recovery action; if we wished to consider longer patterns of events, then the 3 would be replaced by the length of the desired pattern.

15.2.2 *Local Search Dependency Detection*

Dependency detection is limited primarily by its reliance on exhaustive search of possible patterns. This was feasible initially because the search space included only all possible immediately preceding influences on failure; the traces contained at most eleven types of

---

[3]The G-test is similar to chi-square in its use and composition; its primary difference is that the results are additive, meaning that the summation of G results for partitions of a sample is equal to G for the entire sample.

failure and eight recovery actions, producing, at most, 1177 possible combinations.

The original algorithm was modified to include more flexible matching and a non-exhaustive, but computationally more manageable search through local search. More flexible matching of patterns to data means that the event streams may be composed of many types of events, which need not alternate as before[4]. The events can be of different values for several types. Any set of events with time stamps for ordering is a legal event stream. In this view, the event streams for the Phoenix data are described as having two types of events, failures and recovery actions, which include eleven and eight values respectively. The data is passed to the program in a file with three columns: time stamp, event type and event value.

Improving the Matching

The original dependency detection required an exact match of pattern constituents to events observed in the traces. Many domains may not produce uninterrupted sequences, e.g., ones in which parallel processes are incorporated in a single event trace or regularly occurring, but uninteresting events may disrupt patterns.

To make the matching insensitive to the insertion of random events, the contingency table construction was changed to match a *relative order* pattern within a window. As before, the pattern consists of a precursor and a final target event. For example, we could use a precursor of any four recovery actions and a target of a particular failure. For the construction of a contingency table, we need to determine whether the precursor matches (thus, we add one to a count in the top row of the contingency table) and whether the target event matches (in which case we add one to a count in the left column of the table).

In this algorithm, the precursor is a relative order of a specified number of events. A *window* designates an area over which the algorithm will match the precursor. A pattern matches a given window position when the target event is at the last position and when the elements of the precursor appear in the window in the same *order* as they do in the pattern. If the window is exactly the length of the precursor, then the match must be exact as in the original dependency detection algorithm; if the window length is greater than that of the precursor, then the match checks the relative order of the constituents of the precursor. For example, if the precursor is [A B] and the window is of length three, then the precursor will match a portion of the event streams in which A is followed immediately by B or A and B are separated by a single event.

The algorithm for constructing the contingency table is:

1. Initially position the match window at the first position in the event streams in which the window's last position (the target) is over an *event of the same type* and at least $W - 1$ events proceed it.

2. Repeatedly re-position the window as follows until the end of the event stream is reached:

    (a) Position the window with its last position over the next event of the same type as the target event.

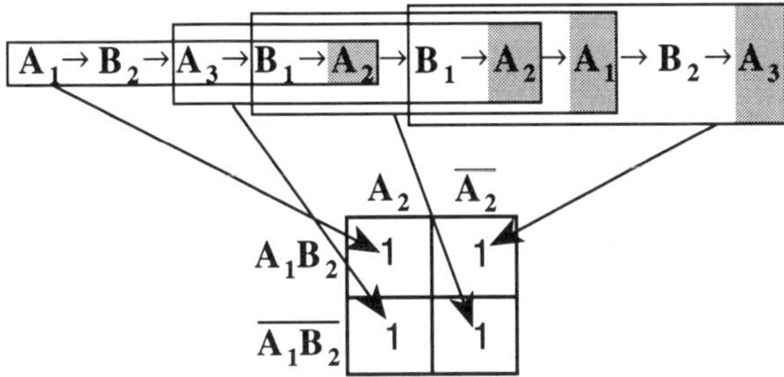

FIGURE 1. Sliding a window to construct a contingency table.

(b) If the target event matches the last position, then prepare to add to a cell in the left column; otherwise, prepare to add to the right column.

(c) If the precursor matches within the window, then add one to the top row and column indicated by the last step of the contingency table; otherwise, add to the bottom row and column indicated as before.

For a given pattern, incidence counts are gathered by sliding an imaginary match window over the event streams. The window defines an area of length W. By default, this length is $2P + 1$, which indicates that twice as many elements can be found in the window as are expected to be part of the precursor $(2P)$ and 1 accommodates the target event in the pattern. In other words, we can allow P spurious events to slip between salient events. The window is re-positioned by moving its last position (the target) to the next event of the same type as the target. As in the original formulation, four counts are tallied for the contingency table: precursor with target, precursor without target, not precursor with target and not precursor without target. Each window positioning contributes to one and only one of the four positions in the contingency table. Figure 1 shows four positionings of the sliding window and their contribution to the contingency table for the $[A_1 \ B_2, \ A_2]$ pattern (the target is of type A and has the value A_2). After sliding the window through all event streams, a G-test is run on the resulting contingency table.

Reducing the Search

The purpose of the search is to test models of co-occurrence and find those that seem to represent significant dependencies between events in time. Because of the computational problems of increasing the number and character of the potential models, the dependency detection algorithm also was modified to apply local instead of exhaustive search. In this case, local search refers to a process of starting with a randomly selected initial pattern (a precursor and target event combination) and applying operators to find the best pattern from that one; the search continues until no better pattern can be found through operator application. Multiple initial starts of local search are done to collect different dependencies and explore different parts of the search space.

Local search has several advantages over exhaustive search for this application. First, we can control how much time is devoted to search through the operators applied and the number of trials. Second, we can be assured of getting the strongest dependencies; local search prunes the number of dependencies to be considered and usually prunes spurious dependencies based on few data.

The algorithm incorporates local search as follows:

1. Gather event streams of salient events.

2. Select parameters for the pattern: a specific target event, a length for the precursor (P) and a length for the window (W).

3. For each of T trials

   (a) Construct an initial pattern by randomly selecting P events in some order.

   (b) Find the best pattern by searching from the initial pattern:
       i. Apply the permutation operator to the best pattern found so far.
       ii. For each permutation, test its significance by constructing a contingency table for the relative order matching and applying the G test to it.
       iii. Select the best permutation.
       iv. If the permutation is no better than the previous best, then halt search and return the best that was found; otherwise, repeat from step i.

4. Return a list of the best patterns found in each trial.

For the second step, as described in the previous section, patterns may be relative order and of a specified length with a specific target event. The window length indicates the number of allowed spurious events. For the third step, the search finds the best related pattern: the one with the highest G value for its contingency table. For each of the T trials, the initial pattern is composed of a precursor of length P with the elements chosen randomly plus the target event at the end.

New patterns are generated by applying a local search operator to the current best. Several local search operators were considered that permute the pattern by reordering and replacing pattern elements, but the best operator, in terms of efficiency and efficacy, is 1-OPT. 1-OPT optimizes one element in the pattern at a time. Progressing through the pattern positions in order, this operator considers all available elements as replacements; the operator replaces the element originally in the position with the best element for that position by testing the possibilities.

The algorithm continues to apply 1-OPT to subsequent best patterns until the resulting G value cannot be improved. The G values are obtained by constructing contingency tables and applying the G test as was described in the last subsection. The output is a list of the patterns returned in each of the T trials.

15.3   Results

Dependency detection, in its original form, has already been shown to be effective at finding interesting patterns in the Phoenix data. The two modifications described here extend the types of patterns found and search process for identifying significant patterns. The utility of the pattern extension remains to be seen; we are currently exploring it in another project on debugging different planners. The change in search process will only be an improvement if significant dependencies are surrounded by basins of attraction for

Failure	1	2	3	4	5	6	7	8
# unique	14	10	5	6	8	8	10	14
# $p < .01$	9	2	3	3	8	1	3	7
Avg. depth	5.7	4.8	6.8	5.3	5.8	5.5	5.0	4.8

TABLE 15.2. Results of local search for 8 target failures with precursors of length 3, window of length 7 on a Phoenix data set. The first two rows represent the number of trials out of 100 which found unique patterns and highly significant patterns. The last row indicates the average number of applications of the local search operator during a trial.

directing the search. If the topology of the search space is relatively flat, then the number of random starts required to find significant dependencies will need to be large, thus reducing or eliminating any computational advantage over brute force exhaustive search.

Tests of the original dependency detection showed that similar patterns tended all to be significant. In other words, it was likely that if one pattern was significant, another pattern that shared all but one of the same elements would also be significant. In a sense, a strongly significant pattern is a basin of attraction in the search space; this characteristic is exploited by local search. To confirm whether such basins of attraction exist and can be exploited by the algorithm, the implementation was tested on execution traces from Phoenix and on synthetic data.

15.3.1 Finding Dependencies in Phoenix Data

The efficacy of local search was tested on some data sets from Phoenix. The results were similar in each data set and are reported for a data set with 946 events (alternating failures and recovery actions) in its event streams. For each of eight target failures (those that appeared in the event streams at least five times), the algorithm started with 100 randomly generated precursors of length three (which makes a pattern of length 4) using a window size of seven and searched for the highest rated neighbor that could be reached from each. Each precursor included either failures or recovery actions for 18 possible values in each precursor position. The searches found a significant dependency in about one-third of the trials, and half were highly significant ($p < .01$). As would be expected, many of the dependencies were found in more than one trial and significant dependencies tended to be related to other significant dependencies. Table 15.2 summarizes the results for each target failure (designated by number) by number of unique dependencies found, number of dependencies found with $p < .01$, and the average search depth to find the dependencies. The search depth suggests a strong relationship between significant patterns and indicates the number of times the operator was applied. The 100 trials on the eight failures required three hours of CPU time on a SPARC IPX with the code written in Lucid Common Lisp.

As shown in Table 15.3, searching for length four precursors exhibited similar, if slightly less successful results. Highly significant dependencies are found in, on average, 18 out of 200 trials. In this case, the search tends to proceed much further, but this is balanced out by many of the initial starts not appearing in the data and many of their neighbors not appearing in the data either. Not too surprisingly, the 200 trials on the eight failures required twice as much computation (six hours).

As would be expected, the dependencies found exhibited a high degree of overlap.

Failure	1	2	3	4	5	6	7	8
# unique	22	19	18	12	16	20	9	29
# $p < .01$	14	10	12	7	11	4	4	15
Avg. depth	8.5	7.3	7.6	7.5	8.1	7.0	6.9	6.8

TABLE 15.3. Results for precursors of length 4, window of length 9

As described earlier, a shorter pattern overlaps with a longer pattern when the longer pattern is the shorter pattern with events added to one or the other end. Over 50% of the length three precursors overlapped (i.e., were substrings) of the length four precursors. Additionally, comparisons to exhaustive search of longer dependencies showed overlap as well: nearly all the relative order dependencies were found within the absolute order dependencies.

15.3.2 *Finding Dependencies in Synthetic Data*

In addition to the Phoenix data, the local search algorithm was tested on finding dependencies in synthetic data. The synthetic data was used to probe a concern: Can local search find dependencies even when there is no relationship between similar patterns?

The synthetic data was produced by generating token streams with biases. For this test, the bias was three patterns that were selected in advance. For each pattern, the precursor would appear at each position in the event streams with a likelihood of .05; if the precursor was inserted, then the target event followed it with likelihood of .80. Twenty seven synthetic data sets were generated, which included three numbers of tokens (15, 25, 35), three stream lengths (100, 500 1000), and three pattern lengths (2, 3, 4). Local search found all the patterns of length 2 and 3 with 15 or 25 tokens and at least 500 events in the streams, but none of the patterns of length 4 with 15 or 35 tokens. In addition, few of the patterns were found when there were 35 tokens. These results suggest that local search can detect relative order dependencies even when the neighborhoods have less meaning. However, increasing numbers of tokens and length of precursor means that local search is less likely to effectively search the space.

15.4 Conclusions

Although computational problems have been overcome, testing and collecting more dependencies exacts a cost in terms of certainty and model deluge. The algorithm requires serial testing of hypotheses: contingency table analysis is run on similar patterns for the same data repeatedly in the course of local search. Consequently, the probability threshold of .05 is no longer an accurate figure. We are currently investigating methods of adjusting α down to compensate for the serial testing inherent in this algorithm. Given the fairly shallow depth of the search (number of tests to be run during a search) and the low values of P found in the most significant dependencies, it appears that adjusting α will enhance the certainty of the results without unduly dropping the number of dependencies detected in the data.

From the point of view of someone interpreting dependencies, it is easy to be overwhelmed with the number found. Local search addresses this problem in two ways: First, only the most significant dependencies are likely to be found. Although the algorithm is doing multiple hill-climbs, it will likely find repeats in subsequent trials of the same pattern at the top of a large hill. Second, the search neighborhood in local search provides a natural clustering: the basins of attraction around most locally optimal patterns. These basins ensure that only one of the set of similar dependencies will be returned, and they can be exploited to cluster dependencies that contain the same subsequences.

Subsequences that are repeated across dependencies are probably themselves significant or may indicate cases in which several events exert a similar influence. For example, in the Phoenix data, some of the failures were variants on the same problem (e.g., two of them were failures in calculating paths), and some of the recovery actions were variants on the same method (e.g., two of them replanned, but from different points in the plan). We would like to cluster related dependencies to find commonalities and to reduce the burden of looking over a long list.

Clusters can be formed by finding locally optimal patterns and then backing out of the basin at individual positions to find other highly significant related patterns. The algorithm was augmented to "back out" of local maxima and gather patterns from the neighborhood that were also significant. If we examine patterns local to the maxima found in the test with precursors of length three and window of length seven, then we find that at each position in the pattern at least one and as many as six out of sixteen other patterns are significant. With local search, we can find clusters of somewhat less significant but highly similar composition.

Local search dependency detection is data directed, finds highly significant dependencies and provides a convenient method of clustering. As demonstrated on the Phoenix data, the local search method is effective at detecting long dependencies in data with considerable overlap in the elements of the dependencies: it finds a large number of the most significant dependencies in a manageable amount of computation time. However, not too surprisingly, when tested on synthetic data, it does not perform nearly as well when there is little relationship between subsequences and the long sequences that form dependencies.

Acknowledgments: This research was supported by a Colorado State University Diversity Career Enhancement grant and ARPA/Rome Laboratory contract F30602-93-C-0100. The US Government is authorized to reproduce and distribute reprints for governmental purposes notwithstanding any copyright notation hereon. I wish to thank Darrell Whitley for his advice on the application of local search, Aaron Fuegi for his programming, Steve Leathard for running tests, Tim Oates for the synthetic data generator, and Douglas Fisher and anonymous reviewers for their patient and careful reading and suggestions for improvement.

15.5 REFERENCES

[1] Peter C. Bates and Jack C. Wileden. High-level debugging of distributed systems: The behavioral abstraction approach. Department of Computer and Information Science 83-29, University of Massachusetts, Amherst, MA, March 1983.

[2] Paul R. Cohen, Lisa A. Ballesteros, Dawn E. Gregory, and Robert St. Amant. Regression can build predictive causal models. Technical Report 94-15, Dept. of Computer Science, University of Massachusetts/Amherst., 1994.

[3] C. Glymour, R. Scheines, P. Spirtes, and K. Kelly. *Discovering Causal Structure.* Academic Press, 1987.

[4] N.K. Gupta and R.E. Seviora. An expert system approach to real time system debugging. In *Proceedings of the IEEE Computer Society Conference on AI Applications*, pages 336–343, Denver, CO, December 5-7 1984.

[5] Adele E. Howe. Analyzing failure recovery to improve planner design. In *Proceedings of the Tenth National Conference on Artificial Intelligence*, pages 387–393, July 1992.

[6] Adele E. Howe and Paul R. Cohen. Detecting and explaining dependencies in execution traces. In P. Cheeseman and R.W. Oldford, editors, *Selecting Models from Data; Artificial Intelligence and Statistics IV*, volume 89 of *Lecture Notes in Statistics*, chapter 8, pages 71–78. Springer-Verlag, NY,NY, 1994.

[7] Tim Oates, Dawn Gregory, and Paul R. Cohen. Detecting complex dependencies in categorical data. In *Preliminary Papers of the Fifthe International Workshop on Artificial Intelligence and Statistics*, January 1995.

[8] Judea Pearl and T. S. Verma. A theory of inferred causation. In J.A. Allen, R. Fikes, and E. Sandewall, editors, *Principles of Knowledge Representation and Reasoning: Proceedings of the Second International Conference*, pages 441–452, San Mateo, CA, April 1991. Morgan Kaufmann.

16
Structure Learning of Bayesian Networks by Hybrid Genetic Algorithms

Pedro Larrañaga, Roberto Murga, Mikel Poza and Cindy Kuijpers

Department of Computer Science and Artificial Intelligence
University of the Basque Country,
P.O. Box 649, 20080 San Sebastián, Spain.

ABSTRACT This paper demonstrates how genetic algorithms can be used to discover the structure of a Bayesian network from a given database with cases. The results presented, were obtained by applying four different types of genetic algorithms – SSGA (Steady State Genetic Algorithm), GAeλ (Genetic Algorithm elistist of degree λ), hSSGA (hybrid Steady State Genetic Algorithm) and the hGAeλ (hybrid Genetic Algorithm elitist of degree λ) – to simulations of the ALARM Network. The behaviour of these algorithms is studied as their parameters are varied.

16.1   Introduction

In recent years, the search for the structure of a Bayesian network able to reflect all existing relations of interdependence in a database of cases has constituted a research topic of fundamental importance. Although the first algorithms were related to tree and polytree structures (e.g. [Chow68, Rebane89]), research has been concentrated upon multiple connected structures [Fung90, Herskovits90, Cooper92, Bouckaert93, Wedelin93, Lauritzen93, Chickering95, Bouckaert94]. For an introduction on Bayesian networks, see [Pearl88].

In this article we propose to obtain Bayesian network structures with the help of an intelligent search process based on genetic algorithms.

Evolutionary algorithms are probabilistic search algorithms which simulate natural evolution. They were proposed about 30 years ago. Their application to combinatorial optimization problems has, however, only recently become an actual research topic. Three different types of evolutionary algorithms exist: genetic algorithms [Holland75, Goldberg89, Davis91], evolution strategies [Schwefel67] and evolutionary programming [Fogel62]. This paper, however, focusses upon genetic algorithms (GAs). GAs are search algorithms based on the mechanics of natural selection and genetics. They combine "survival of the fittest" among string structures with a structured yet randomized information exchange to form a search algorithm which, under certain conditions, evolves to the optimum with probability 1 [Eiben90, Chakraborty93, Rudolph94].

In GAs the search space of a problem is represented as a collection of individuals. The individuals are represented by character strings, which are often referred to as *chromo-*

---

[1] *Learning from Data: AI and Statistics V*. Edited by D. Fisher and H.-J. Lenz. ©1996 Springer-Verlag.
[2] We thank Gregory F. Cooper for providing his simulation of the ALARM Network.
This work was supported by the Diputación Foral de Gipuzkoa, under grant OF 94/0762, by the Fondo de Investigación Sanitaria, Ministerio de Sanidad y Consumo, under grant 94/1370, and by the grant PI94/78 from the Gobierno Vasco - Departamento de Educación, Universidades e Investigación.

somes. The purpose of a GA is to find the individual from the search space with the best "genetic material". The quality of an individual is measured with an objective function. The part of the search space to be examined in each iteration is called the *population*.

A genetic algorithm works approximately as follows. First, the initial population is chosen at random, and the quality of each of its individual is determined. Next, in every iteration parents are selected from the population. These parents produce children, which are added to the population. For all newly created individuals of the resulting population a probability near zero exits that they "mutate", i.e. that they change their hereditary distinctions. Later, some individuals are removed from the population according to a selection criterion in order to reduce the population to its initial size. One iteration of the algorithm is referred to as a *generation*.

The operators which define the child production process and the mutation process are called the *crossover operator* and the *mutation operator*, respectively. Mutation and crossover play different roles in the GA. Mutation is needed to explore new states and helps the algorithm to avoid being trapped on local optima. Crossover should increase the average quality of the population. By choosing adequate crossover and mutation operators, as well as a reduction mechanism, the probability that the GA results in a near-optimal solution in a reasonable number of iterations is enlarged.

In Figure 1 we show the basic structure of a genetic algorithm (GA).

BEGIN GA
    Obtain the initial population at random.
    WHILE NOT stop **DO**
      BEGIN
        Select parents from the population.
        Produce children from the selected parents.
        Mutate the children.
        Add the children to the population.
        Reduce the population to its original size.
      END
END GA

FIGURE 1. Pseudo code of a genetic algorithm (GA).

16.2   Proposed Approach

We represent a Bayesian network structure by a *connectivity matrix* $C = (c_{ij})_{i,j=1,...,n}$, where

$$c_{ij} = \begin{cases} 1 & \text{if } (j \text{ is a parent node of } i) \text{ and } (i > j), \\ 0 & \text{otherwise.} \end{cases}$$

The inequality $i > j$ originates in the assumed ancestral order between the variables. Because of the inequality the crossover and mutation operators to be used are closed operators.

We consider four different genetic algorithms to which we refer as the SSGA (Steady State Genetic Algorithm), the GAeλ (Genetic Algorithm elitist of degree λ), the hSSGA (hybrid Steady State Genetic Algorithm) and the hGAeλ (hybrid Genetic Algorithm elitist of degree λ). Here the word *hybrid* indicates the addition of a *local optimizer* to the algorithm.

In an iteration of the SSGA and the hSSGA only one new individual is created, while in the GAeλ and the hGAeλ the generation replacement has a global character. In all algorithms, the reduction criterion is elitist. In the SSGA and the hSSGA, the created individual is compared with the worst existing individual at the time of creation. In the GAeλ, and the hGAeλ, however, the population at time $t + 1$ consists of the λ best individuals drawn from the union of two sets: (1) the set of the λ individuals existing at time t, and (2) the λ children created from the individuals at time t.

The behaviour of all algorithms is studied with the help of three different *population sizes* λ ($\lambda = 10$, $\lambda = 50$, $\lambda = 100$).

The *objective function* used to evaluate the quality of a structure, is based on the formula proposed by Cooper and Herskovits [Cooper92], for a joint probability $P(B_S, D)$ of a Bayesian network structure B, and a database D, expressed in terms of the natural logarithm. Therefore, our aim is to find the structure with the highest joint probability.

The *selection function* is based on the rank of the objective function. If we denote by I_t^j the j-th individual of the population at time t, and by $\mathrm{rank}(g(I_t^j))$ the rank of its objective function, the probability $p_{j,t}$ that individual I_t^j is selected to be a parent is equal to

$$p_{j,t} = \frac{\mathrm{rank}(g(I_t^j))}{\lambda(\lambda + 1)/2}.$$

The *reproduction function* to be used is the so-called 1-point crossover operator. Following the selection of two parents, the probability that these parents are crossed is 1. This probability makes it feasible to compare the algorithms. The *mutation operator* consists of the probabilistic alteration of the bits, which represent the connectivity matrix. This alteration is performed with a probability near to zero. We consider two different mutation probabilities p_m, namely $p_m = 0.001$ and $p_m = 0.01$.

The algorithms *stop* when either, 10,000 structures have been evaluated or when in 1000 successive iterations of the algorithm, the evaluation of the best structure in the current population corresponds with the average evaluation over the structures in that population.

The initial population of network structures is generated at random, subject to the restriction that a node in a network structure never has more than m parent nodes (in our case, $m = 4$).

After applying the crossover and mutation operators, the created structures do not necessarily satisfy the restriction that each node of a network structure has at most m parent nodes. To enforce this restriction, the SSGA and GAeλ algorithms select n parent nodes at random, ($0 \leq n \leq m$), for each node of a created network structure. This approach will give poor results. Therefore, we try a second approach that is the hybridization of each of SSGA and GAeλ algorithms with the help of a *local optimizer*. This optimizer selects the best subset of at most m parent nodes for each node in a network structure. The process of generating structures and the application of the local optimizer, is repeated in every iteration of the algorithm.

TABLE 16.1. Evaluation of the ALARM Network structure.

Number of cases ALARM Network	$log P(D|B_S)$
100	-6.3860e02
200	-1.1413e03
500	-2.6461e03
1000	-5.0345e03
2000	-9.7291e03
3000	-1.4412e04
10,000	-4.7086e04

16.3   Results

We describe the results of an experiment in which a database generated by simulation of a Bayesian network is used to search for the structure which has a maximal joint probability. This joint probability is compared with the corresponding value of the structure of the initial network structure. Also the Hamming distance between both structures, and the number of evaluations needed to obtain convergence are considered. All results were obtained with a SPARCserver 1000 under operating system Solaris 2.3.

We applied the algorithms to a database of 10,000 cases generated with the ALARM Network, which was constructed by Beinlinch et $al.$ [Beinlich89] as a prototype to model potential anesthesia problems in the operating room. The simulation of the 10,000 cases of this network has been achieved with the help of a Monte Carlo technique developed for Bayesian networks by Henrion [Henrion88]. It corresponds with the first 10,000 cases generated by Herskovits [Herskovits91]. We have considered different subsets consisting of the first 100, 200, 500, 1000, 2000, 3000, and 10,000 cases from the original database. The evaluations of the initial structures for the different databases can be seen in Table 16.1.

All algorithms are evaluated with respect to the population size and the mutation rate. For every possible combination of parameters 10 executions were carried out. Therefore, the total number of evaluations performed for every database of cases was $2\times2\times3\times2\times10 = 240$.

In Table 16.2 and Table 16.3 the results are presented. BOF refers to the best value found of the objective function and AOF is the average value of the objective function. HD refers to the Hamming distance between the ALARM Network structure and the one with the best objective function, while the number between parentheses is the average Hamming distance.

By comparing Table 16.2 with Table 16.3, we see the importance of the local optimizer. Table 16.2 indicates that the evaluation function of the ALARM Network was improved for small population sizes (λ=10) only, but all results of the hybrid algorithms hSSGA and hGAeλ (see Table 16.3) were better than the ones presented in Table 16.1.

Another remarkable point is the small variability in the results found by the hybrid algorithms (see Table 16.3) with respect to the ones obtained by the SSGA and the GAeλ (see Table 16.2).

TABLE 16.2. Results obtained with the SSGA and the GAeλ.

		SSGA			GAeλ		
		$\lambda = 10$	$\lambda = 50$	$\lambda = 100$	$\lambda = 10$	$\lambda = 50$	$\lambda = 100$
	BOF	-6.2014e02	-6.2363e02	-6.3223e02	-6.2090e02	-6.2650e02	-6.4073e02
100	AOF	-6.3952e02	-6.4680e02	-6.5899e02	-6.3993e02	-6.5037e02	-6.6497e02
	HD	31 (48.2)	32 (51.7)	34 (56.5)	37 (49.4)	38 (55.0)	51 (65.7)
	BOF	-1.1256e03	-1.1307e03	-1.1432e03	-1.1253e03	-1.1348e03	-1.1587e03
200	AOF	-1.1556e03	-1.1673e03	-1.1911e03	-1.1543e03	-1.1755e03	-1.2038e03
	HD	19 (38.5)	19 (42.4)	34 (52.9)	17 (38.1)	22 (45.5)	37 (57.9)
	BOF	-2.6350e03	-2.6438e03	-2.6741e03	-2.6354e03	-2.6492e03	-2.7069e03
500	AOF	-2.6918e03	-2.7186e03	-2.7740e03	-2.6908e03	-2.7262e03	-2.7946e03
	HD	11 (34.8)	18 (41.2)	29 (52.9)	12 (33.2)	22 (45.1)	41 (58.8)
	BOF	-5.0279e03	-5.0404e03	-5.0799e03	-5.0286e03	-5.0611e03	-5.1533e03
1000	AOF	-5.1117e03	-5.1741e03	-5.2491e03	-5.1137e03	-5.2026e03	-5.3185e03
	HD	4 (30.2)	15 (40.8)	28 (53.7)	2 (30.2)	20 (45.3)	38 (60.5)
	BOF	-9.7200e03	-9.7440e03	-9.8159e03	-9.7200e03	-9.7538e03	-9.8769e03
2000	AOF	-9.8504e03	-9.8890e03	-9.9481e03	-9.8436e03	-9.9188e03	-10.0068e03
	HD	3 (30.7)	11 (41.4)	28 (54.9)	3 (30.1)	16 (47.6)	43 (66.1)
	BOF	-1.4404e04	-1.4425e04	-1.4485e04	-1.4405e04	-1.4450e04	-1.4649e04
3000	AOF	-1.4578e04	-1.4722e04	-1.4957e04	-1.4580e04	-1.4810e04	-1.5173e04
	HD	1 (29.0)	12 (41.4)	29 (56.2)	2 (30.3)	22 (50.5)	48 (68.8)
	BOF	-4.7079e04	-4.7118e04	-4.7279e04	-4.7079e04	-4.7163e04	-4.7362e04
10,000	AOF	-4.7462e04	-4.8019e04	-4.8506e04	-4.7531e04	-4.8264e04	-4.9083e04
	HD	2 (31.1)	15 (46.4)	34 (61.2)	2 (32.5)	26 (56.6)	42 (72.3)

Because of the mentioned considerations, we decided to analyze the algorithms SSGA and GAeλ separately from the hybrid algorithms.

The analysis of the 1680 (240 $\times$ 7) runs has been carried out using the Kruskal-Wallis test, which looks for differences statistically significants.

16.3.1   Analysis of the results of the SSGA and the GAeλ

Objective Function

The average behaviour of the SSGA is similar to the GAeλ. There are no statistically significant differences in any of the 7 databases. However, statistically significant differences exist with respect to the population size, obtaining the best performance with $\lambda=10$. The results found with the mutation rate $p_m=0.001$ are significantly better than the ones found with $p_m=0.01$.

Number of evaluations needed until convergence

For the population sizes $\lambda=50$ and $\lambda=100$, none of the algorithms was able to converge under the stop criterion earlier described. For $\lambda=10$, the convergence velocity, both of the SSGA as well as of the GAeλ, is significantly larger for $p_m=0.001$ than for $p_m=0.01$ for

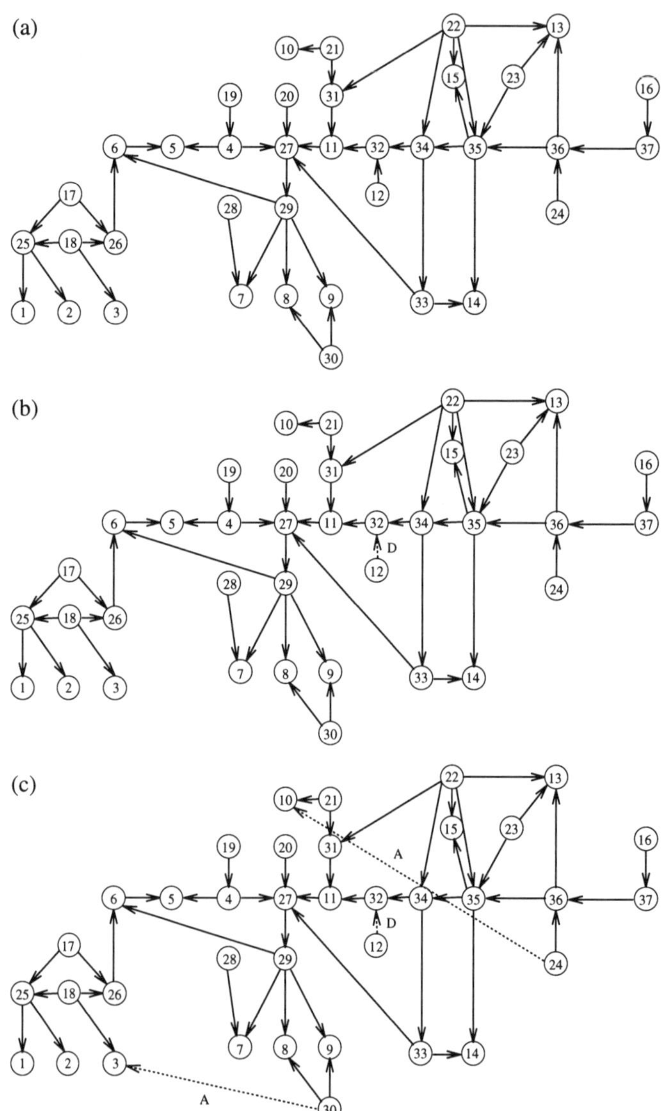

FIGURE 2. (a) The ALARM network structure. (b) The network structure learned by the hSSGA and the hGAeλ algorithms from a 10000-case and 3000-case database generated from the ALARM network. (c) The network structure learned by the hSSGA and the hGAeλ algorithms from a 2000-case database generated from the ALARM network. Arcs that are added or deleted with respect to the ALARM network are indicated with A and D respectively.

TABLE 16.3. Results obtained with the hSSGA and the hGAeλ.

		hSSGA			hGAeλ		
		$\lambda = 10$	$\lambda = 50$	$\lambda = 100$	$\lambda = 10$	$\lambda = 50$	$\lambda = 100$
	BOF	-6.1901e02	-6.1901e02	-6.1901e02	-6.1901e02	-6.1901e02	-6.1901e02
100	AOF	-6.1928e02	-6.1932e02	-6.1921e02	-6.1945e02	-6.1927e02	-6.1920e02
	HD	34 (35.6)	34 (35.6)	33 (35.6)	34 (35.7)	34 (35.6)	32 (35.3)
	BOF	-1.1249e03	-1.1249e03	-1.1249e03	-1.1249e03	-1.1249e03	-1.1249e03
200	AOF	-1.1249e03	-1.1253e03	-1.1249e03	-1.1249e03	-1.1249e03	-1.1250e03
	HD	19 (19.9)	19 (20.0)	20 (20.0)	19 (19.9)	18 (19.6)	17 (19.4)
	BOF	-2.6350e03	-2.6350e03	-2.6350e03	-2.6350e03	-2.6350e03	-2.6350e03
500	AOF	-2.6360e03	-2.6353e03	-2.6350e03	-2.6350e03	-2.6350e03	-2.6350e03
	HD	12 (13.0)	11 (12.5)	12 (12.3)	12 (12.6)	12 (12.1)	11 (12.1)
	BOF	-5.0279e03	-5.0279e03	-5.0279e03	-5.0279e03	-5.0279e03	-5.0279e03
1000	AOF	-5.0279e03	-5.0279e03	-5.0279e03	-5.0281e03	-5.0303e03	-5.0279e03
	HD	4 (4.0)	4 (4.0)	4 (4.0)	4 (4.1)	3 (4.1)	4 (4.1)
	BOF	-9.7200e03	-9.7200e03	-9.7200e03	-9.7200e03	-9.7200e03	-9.7200e03
2000	AOF	-9.7206e03	-9.7200e03	-9.7200e03	-9.7205e03	-9.7200e03	-9.7200e03
	HD	3 (3.1)	3 (3.0)	3 (3.0)	3 (3.1)	3 (3.0)	3 (3.0)
	BOF	-1.4404e04	-1.4404e04	-1.4404e04	-1.4404e04	-1.4404e04	-1.4404e04
3000	AOF	-1.4404e04	-1.4404e04	-1.4404e04	-1.4404e04	-1.4404e04	-1.4407e04
	HD	1 (1.0)	1 (1.0)	1 (1.0)	1 (1.0)	1 (1.0)	1 (1.2)
	BOF	-4.7078e04	-4.7078e04	-4.7078e04	-4.7078e04	-4.7078e04	-4.7078e04
10,000	AOF	-4.7078e04	-4.7082e04	-4.7078e04	-4.7078e04	-4.7083e04	-4.7078e04
	HD	1 (1.0)	1 (1.1)	1 (1.0)	1 (1.0)	1 (1.1)	1 (1.0)

all databases considered.

The poor results found with this first approach we attribute to the "blind" parent selection process used for maintaining the restriction on the maximum number of parent nodes.

16.3.2   Analysis of the results of the hSSGA and the hGAeλ

Objective Function

For the small databases (100, 200 and 500 cases) we found statistically significant differences for the mutation rate, obtaining the best performance for p_m=0.01. Differences due to population size were only significant for the 500-case database, where performance improved as the population size became larger. The large databases (1000, 2000, 3000, 10,000 cases) did not give statistically significant differences with respect to any of the three parameters considered (the type of the GA, the population size and the mutation rate).

Number of evaluations needed until convergence

The stop criterion was sufficient for guaranteeing the convergence of the hybrid algorithms. We found, for all databases, that the hSSGA converges significantly faster than the hGAeλ. Moreover, the algorithms converged faster as the population size became smaller. Finally a mutation rate equal to 0.01 resulted in a faster convergence than a mutation rate of 0.001.

The best structure obtained by the hybrid algorithms coincided for both the algortihms and was found with both the 3000-case database as well as the 10,000-case database. If we compare this structure (see Figure 2(b)) with the ALARM Network (see Figure 2(a)), we see that the only difference between the two structures is the arc from node 12 to node 32, which is missing in the best structure found by the hybrid algorithms. The best structure found by the hybrid algorithms with the 2000-case database is shown in Figure 2(c). This structure has, in comparison with the ALARM Network, two additional arcs (the arc from node 24 to node 10, and the arc from node 30 to node 3) and one missing arc (the arc from node 12 to node 32).

The obtained improvements using the local optimizer, we interprete as an empirical demonstration of the validity of our hybrid approach. The local search related to every node involves that unimportant parts of the search space are not examined.

16.4   Conclusions and Future Research

We have illustrated how the genetic approach can be used in the Structure Learning of Bayesian networks from a database of cases.

First, we have tried an approach in which a "blind" selection process was used to "repair" created structures with nodes which have too many parent nodes. Second, we have followed a hybrid approach. The results of the latter approach are far better than the results of the former approach. In this case the results are independent of the generation gap and, in outline, also of the mutation rate and the population size.

In the future we plan to tackle the more general problem in which the assumption of the ancestral ordering between the variables is not assumed. Other potential research is related to the use of different evaluation functions, some of which appear in [Bouckaert93, Bouckaert94, Chickering95].

We expect that our approach can also be applied to dynamical Bayesian networks. Also, it would be interesting to investigate the use of other heuristical search methods, like i.e. tabu search.

16.5   REFERENCES

[Beinlich89] Beinlich, I.A., Suermondt, H.J., Chavez, R.M., & Cooper, G.F. (1989). The ALARM monitoring system: A case study with two probabilistic inferences techniques for belief networks. *Proceedings of the Second European Conference on Artificial Intelligence in Medicine* (pp. 247–256).

[Bouckaert93] Bouckaert, R.R. (1993). Probabilistic network construction using the minimum description length principle. In M. Clarke, R. Kruse & S. Moral (Eds.) *Symbolic and Quantitative Approaches to Reasoning and Uncertainty – ECSQARU-93*, No. 747, Lectures Notes in Computing Science, pp. 41-48, Springer-Verlag.

[Bouckaert94] Bouckaert, R.R. (1994). Properties of Bayesian belief network learning algorithms. *Uncertainty in Artificial Intelligence, Tenth Annual Conference* (pp. 102–

109). San Francisco, CA: Morgan Kaufmann.

[Chakraborty93] Chakraborty, U.K., & Dastidar, D.G. (1993). Using reliability analysis to estimate the number of generations to convergence in genetic algorithms. *Information Processing Letters, 46*, 199–209.

[Chickering95] Chickering, D.M., Geiger, D., & Heckerman, D. (1995). Learning Bayesian networks: Search methods and experimental results. *Preliminary Papers of the Fifth International Workshop on Artificial Intelligence and Statistics* (pp. 112–128).

[Chow68] Chow, C.K., & Liu, C.N. (1968). Approximating discrete probability distributions with dependence trees. *IEEE Transactions on Information Theory, 14*, 462–467.

[Cooper92] Cooper, G.F., & Herskovits, E.H. (1992). A Bayesian method for the induction of probabilistic networks from data. *Machine Learning, 9*, 309–347.

[Davis91] Davis, L. (Ed.) (1991). *Handbook of Genetic Algorithms*. New York: Van Nostrand Reinhold.

[Eiben90] Eiben, A.E., Aarts, E.H.L., & Van Hee, K.M. (1990). *Global convergence of genetic algorithms: An infinite Markov chain analysis.* Computing Science Notes, Eindhoven University of Technology, The Netherlands.

[Fogel62] Fogel, L.J. (1962). Atonomous automata. *Ind. Res., 4*, 14–19.

[Fung90] Fung, R.M., & Crawford, S.L. (1990). CONSTRUCTOR: A system for the induction of probabilistic models. *Proceedings of AAAI* (pp. 762–769).

[Goldberg89] Goldberg, D.E. (1989). *Genetic Algorithms in Search, Optimization and Machine Learning.* Reading, MA: Addison-Wesley.

[Henrion88] Henrion, M. (1988). Propagating uncertainty in Bayesian networks by probabilistic logic sampling. In J. Lemmer & L. Kanal (Eds.) *Uncertainty in Artificial Intelligence, 2*, pp. 149–163, North-Holland.

[Herskovits91] Herskovits, E. H. (1991). *Computer based probabilistic-network construction.* Doctoral dissertation, Medical Information Sciences, Stanford University.

[Herskovits90] Herskovits, E. H., & Cooper, G.F. (1990). *Kutató: An entropy-driven system for construction of probabilistic expert systems from databases.* Report KSL-90-22, Knowledge Systems Laboratory, Medical Computer Science, Stanford University.

[Holland75] Holland, J.H. (1975). *Adaptation in Natural and Artificial Systems.* The University of Michigan Press.

[Lauritzen93] Lauritzen, S.L., Thiesson, B., & Spiegelhalter, D.J. (1993). Diagnostic systems created by model selection methods-A case study. *Fourth International Workshop on Artificial Intelligence and Statistics* (pp. 93–105).

[Pearl88] Pearl, J. (1988). *Probabilistic Reasoning in Intelligent Systems.* San Mateo, CA: Morgan Kaufmann.

[Rebane89] Rebane, G., & Pearl, J. (1989). The recovery of causal polytrees from statistical data. In L. Kanal, T. Levitt & J. Lemmer (Eds.) *Uncertainty in Artificial Intelligence, 3*, pp. 175–182, North-Holland.

[Rudolph94] Rudolph, G. (1994). Convergence analysis of canonical genetic algoritms. *IEEE Transactions on Neural Networks, vol. 5, no. 1*, 96–101.

[Schwefel67] Schwefel, H.-P. (1967). *Numerische Optimierung von Computer-Modellen mittels der Evolutionsstrategie.* Basel: Birkhäuser.

[Wedelin93] Wedelin, D. (1993). *Efficient algorithms for probabilistic inference combinatorial optimization and the discovery of causal structure from data.* Doctoral dissertation, Chalmers University of Technology, Göteborg.

17
An Axiomatization of Loglinear Models with an Application to the Model-Search Problem

Francesco M. Malvestuto

Department of Electrical Engineering
University of L'Aquila
Rojo Poggio, 67040 (AQ)
Italy
malvestuto@vaxaq.cc.univaq.it

ABSTRACT A good strategy to save computational time in a model-search problem consists in endowing the search procedure with a mechanism of logical inference, which sometimes allows a loglinear model to be accepted or rejected on logical grounds, without resorting to the numeric test. In principle, the best inferential mechanism should based on a complete axiomatization of loglinear models. We present a (probably incomplete) axiomatization, which can be translated into a graphical inference procedure working with directed acyclic graphs, and show how it can be applied to find an efficient solution to the model-search problem.

17.1 Introduction

Loglinear models are widely employed in analysis of categorical data [1, 8] and in design of probabilistic expert systems [15]. The model-search problem consists in discovering what loglinear models fit a given probability distribution that describes a certain phenomenon of interest. Now, the evaluation of a single loglinear model requires the execution of a time-consuming numeric routine (*validity test*) and, since all possible loglinear models are in number exponential in the number of variables, their evaluation is very expensive even for a moderate number of variables.

A strategy to save computational time consists in divising an *informed* search procedure that includes an inferential mechanism which at each step, on the basis of the knowledge of decisions taken on previously examined models, attempts to infer on logical grounds the acceptance or the rejection of models that are still to be examined. The *efficiency* of an informed search procedure for a given probability distribution may be measured by the quantity

$$1 - \text{(no. of tests executed)}/\text{(size of the search space)} .$$

Examples of informed procedures can be found in [2, 3, 6, 7, 10]. Of course, the efficiency of an informed search procedure depends on the extent that the procedure is informed of the logical properties possessed by the class of loglinear models the search space is composed of. In the best case, the inferential routine of an

---

informed search procedure manages to decide the acceptance or the rejection of the model under examination if and only if that decision is a logical consequence of the decisions taken on models that have been previously examined. Unfortunately, an arbitrary class of loglinear models need not be such that *all* the consequences of a given set of (accepted) models from that class are derivable by using inference rules or, in equivalent terms, an arbitrary class of loglinear models need not admit a complete axiomatization; for example, it is not likely that the whole class of loglinear models be (completely) axiomatizable. However, saturated binary models (i.e., fixed-context independences and conditional independences) admit a sound and complete axiomatization [4, 5, 10].

In this paper, we present an axiomatization for arbitrary loglinear models from which an informed search procedure can be easily obtained. Moreover, we provide a graphical translation of the axiomatization proposed: each axiom is translated into a formal rule for manipulating directed acyclic graphs, here called *derivation DAGs*, and a model that has not yet been examined, will be accepted without being tested if one manages to construct a derivation DAG whose leaves represent the generators of the model.

The paper is organized as follows. Section 2 contains basic definitions. In Section 3 we present a sound axiomatization of loglinear models, which in Section 4 is translated in a graphical inference mechanism. In Section 5 we state some results which show the power of the axiomatization proposed. Section 6 closes with some open problems.

17.2 Terminology

Let the universe of discourse be defined by a finite set U of variables with associated finite (variation) domains and V a nonempty subset of U; a V-tuple, denoted by v, is an element of the Cartesian product of the domains associated with the variables in V. Let $p(\mathbf{u})$ be a probability distribution over U, $p(v)$ its marginal with respect to V and $\mathbf{S} = \{V_1, \ldots, V_n\}$ ($n \geq 1$) a set covering of V; if there exist real functions $\psi_1, \ldots, \psi_n$ respectively of $V_1, \ldots, V_n$ such that, for all v

$$p(v) = \psi_1(v_1) \times \ldots \times \psi_n(v_n) \, ,$$

then we say that $p(\mathbf{u})$ satisfies the (n-ary) *loglinear model* (*model*, for short) *generated* by $\mathbf{S}$, denoted by $[\mathbf{S}]$, and we call the sets $V_1, \ldots, V_n$ the *generators* of the model. If $V = U$, we speak of *universal* (also called "full" or "saturated") *models*; otherwise, we speak of *marginal models*.

Let $\mathbf{P}$ be the set of all possible (non-negative) probability distributions over U. The subset of $\mathbf{P}$ consisting of all probability distributions satisfying a model α will be denoted by $\mathbf{P}(\alpha)$. Moreover, if Σ is a set of models, by $\mathbf{P}(\Sigma)$ we denote the set of all probability distributions satisfying all models in Σ, that is, $\mathbf{P}(\Sigma) = \cap_{\alpha \in \Sigma} \mathbf{P}(\alpha)$.

A unary model $[\{V\}]$ will be also called a *trivial* model since it is satisfied by any probability distribution from **P**.

Models can be regarded as logical sentences so that the relations of implication and equivalence between models can be stated as follows. Given a (possibly empty) set Σ of models and a single model α, we say that Σ *(logically) implies* α or, equivalently, α is a *(logical) consequence* of Σ, denoted by $\Sigma \Rightarrow \alpha$, if $P(\Sigma)$ is a subset of $P(\alpha)$. Notice that a trivial model is a consequence of any (possibly empty) set of models. Moreover, Σ and α are *(logically) equivalent* if $P(\Sigma) = P(\alpha)$. Finally, the *logical closure* of Σ is the set

$$\Sigma^+ = \{\alpha : \Sigma \Rightarrow \alpha\} .$$

Of course, Σ is a subset of Σ^+ and we say that Σ is *logically closed* if $\Sigma = \Sigma^+$.

Given a class $\mathbb{M}$ of models (e.g., the class of universal models), there may exist inference rules that describe the structure of any closed set of models from $\mathbb{M}$. A typical inference rule is one that asserts that if certain models hold, then so must others; an inference rule is expressed by the statement "**if** Σ **then** α", where $\Sigma \cup \{\alpha\}$ is a suitable set of models from $\mathbb{M}$.

Let $\Sigma \cup \{\alpha\}$ be a set of models from $\mathbb{M}$ and let $\mathcal{A}$ be a set of inference rules; we say that α is *derivable* from Σ, denoted by $\Sigma \to \alpha$, if we can apply the rules in $\mathcal{A}$ to models in Σ in such a way to obtain α. The set $\mathcal{A}$ is an *axiomatization* of $\mathbb{M}$ if, for every subset $\Sigma \cup \{\alpha\}$ of $\mathbb{M}$, the set of models derivable from Σ is a subset of the logical closure of Σ; that is, $\Sigma \Rightarrow \alpha$ whenever $\Sigma \to \alpha$; furthermore, $\mathcal{A}$ is a *complete axiomatization* of $\mathbb{M}$ if, for every subset $\Sigma \cup \{\alpha\}$ of $\mathbb{M}$, the set of models derivable from Σ coincides with the logical closure of Σ.

We close this section by introducing some definitions related to set-theoretical aspects of the generating class of a model.

Let **S** be a set covering of a nonempty subset V of the universe of discourse U. A *path* in **S** is a sequence $<V_1, ..., V_k>$ $(k \geq 1)$ of sets from **S** such that, if $k > 1$, then $V_i \cap V_{i+1} \neq \emptyset$ for $i = 1, ..., k\text{-}1$. Let X be a subset of U, not necessarily a subset of V. A path $<V_1, ..., V_k>$ in **S** *passes through* X if $k > 1$ and there exists an index i $(1 \leq i \leq k\text{-}1)$ such that $V_i \cap V_{i+1} \subseteq X$; moreover, two sets Y and Z from **S** are *separated* by X if $Y \neq Z$ and every path $<V_1, ..., V_k>$ from Y to Z (that is, with $V_1 = Y$ and $V_k = Z$) passes through X. Two sets in **S** are *X-nonseparable* if they are not separated by X. The relation of X-nonseparability is an equivalence relation on **S** and the equivalence classes are called the *X-components* of **S**. The *boundary* of an X-component C of **S** is the subset of X made up of the variables that occur in C; the set class formed by the boundaries of the X-components of **S** is denoted by $\partial_X \mathbf{S}$ and called the *derivative* of **S** with respect to X.

Example 1. Let $S = \{ADF, BCE, DE\}$ and $X = ABCDG$. The X-components of S are $C_1 = \{ADF\}$ and $C_2 = \{BCE, DE\}$ and the derivative of S with respect to X is $\partial_X S = \{AD, BCD\}$.

17.3 Decomposable Models

Now, let us consider the case of binary models. Observe that the assumption that p satisfies the model $[\{V_1, V_2\}]$ can be paraphrased by saying that in p the two variable sets $Y = V_1 \backslash V_2$ and $Z = V_2 \backslash V_1$ are *independent given* $X = V_1 \cap V_2$ (if $X = \varnothing$, then Y and Z are independent). The notion of conditional independence attracted the interest of information scientists since it seems to translate the common-sense concepts of "sufficiency" or "irrelevance": no information about Y is contained in Z over and above that contained in X or, equivalently, the amount of information on Y resulting from learning Z is null after learning X. Furthermore, from a computational point of view, the problem of testing if a conditional independence holds for a given probability distribution is an easy task. This motivated the attempts of finding efficient search procedures for selecting all valid conditional independences for a given probability distribution. Since the number of universal conditional independences over n variables is equal to $(3^n - 2^{n+1} + 1)/2$, the usefulness of an inferential mechanism which could reduce the number of validity tests to execute is evident; fortunately, universal conditional independences admit a complete axiomatization [5, 10] and efficient procedures can be worked out [10].

Decomposable models form the largest class of models that enjoy the above-mentioned desirable properties of conditional independences on account of the following facts: (*i*) a model is (logically) equivalent to a set of conditional independences if and only if it is decomposable [9, 13], (*ii*) the computational problem of testing if a decomposable model holds for a given probability distribution is an easy task [1], and (*iii*) universal decomposable models admit a complete axiomatization [14].

Formally, a model with generating class S is said to be *decomposable* if there is an ordering $<V_1, \dots, V_n>$ of the sets in S (the generators of the model $[S]$) such that if $n > 1$ then for each i, $2 \leq i \leq n$, there exists a $j_i < i$ for which

$$(V_1 \cup \dots \cup V_{i-1}) \cap V_i = V_{j_i} \cap V_i .$$

Such an ordering of the sets in S, to be called a *running-intersection ordering*, allows us to graphically represent S by a forest whose vertices represent the sets in S and whose edges link the pairs of vertices representing the pairs of sets (V_i, V_{j_i}).

Example 2. Consider the decomposable model with generating class $S = \{ABC, ABD, ACE, BCF\}$. By making use of any running-intersection ordering we can represent S by the tree shown in Figure 1.

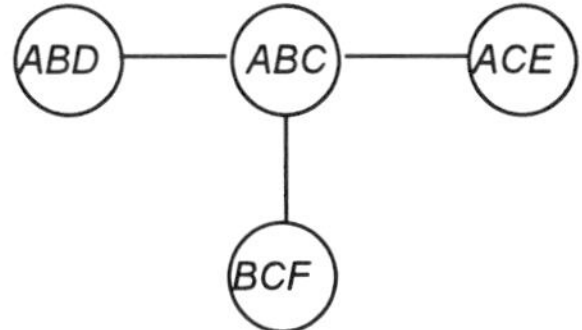

Figure 1. *A tree representing the generating class of a decomposable model*

Remark 1. Let [S] be a decomposable model and F a forest that represents S according to some running-intersection ordering. If X and X' are two generators of [S] that are adjacent in F, and C is the X-component of S containing X', then the boundary of C equals $X \cap X'$. It follows that each set in $\partial_X(\mathbf{S})\backslash\{X\}$ is a subset of some generator of [S] (distinct from X).

17.4 An Axiomatization of Unrestricted Models

In this section, we present an axiomatization of the whole class of unrestricted (universal and marginal) models.

Now, consider the following set of inference rules, where by $V(\mathbf{S})$ we denote the set of variables that occur in set class **S**:

A0: "**if** $\varnothing$ **then** $[\{X\}]$",   for $X \subseteq U$.

A1: "**if** [S] **then** $[\mathbf{S} \cup \{X\}]$",   for $X \subseteq V(\mathbf{S})$.

A2: "**if** $[\mathbf{S} \cup \{X, Y\}]$ **then** $[\mathbf{S} \cup \{XY\}]$".

A3: "**if** $[\mathbf{S} \cup \{XA\}]$ **then** $[\mathbf{S} \cup \{X\}]$",   for $A \notin V(\mathbf{S})$.

A4: "**if** $\{[\mathbf{S} \cup \{X\}], [\mathbf{R}]\}$ **then** $[\mathbf{S} \cup \mathbf{R} \cup \partial_X \mathbf{S}]$",   for $X = V(\mathbf{R})$.

Remark 2. In [11] it is proved that, given two set classes **R** and **S** with $V(\mathbf{R}) \subseteq V(\mathbf{S})$, [R] is a logical consequence of [S] if and only if each generator of [R] is a superset of some set in $\partial_{V(\mathbf{R})}\mathbf{S}$. On the other hand, it is not difficult to see that any model over set of variables V such that each of its generators are supersets of some set in $\partial_V \mathbf{S}$ is derivable from [S] using axioms A1, A2 and A3. Therefore, logical closures of single models can be computed using A1, A2 and A3, only.

In [12] we proved that the set of the four inference rules above is an axiomatization; however, it is not very likely that this axiomatization is complete.

In the rest of this section we show that the axioms above can be translated into graphical rules in such a way that a model α turns out to be derivable from a given set of models Σ if and only if there exists a directed acyclic graph, we call a *derivation DAG* for Σ, whose leaves are exactly the generators of α. Consequently, the task of inferring a model reduces to that of finding a suitable DAG with leaves the generators of the model.

Let Σ be a set of models. Assume without loss of generality that for each nonempty subset X of the universe of discourse U, the trivial model $[\{X\}]$ belongs to Σ. Then the *derivation DAGs for* Σ are defined recursively as follows.

(0) If $[S] \in \Sigma$ and $S = \{V_1, ..., V_n\}$, then the DAG with root $V(S)$ and children $V_1, ..., V_n$, is a derivation DAG for Σ.

(1) If a derivation DAG for Σ has leaves $V_1, ..., V_n$ and X is a subset of the variable set $\cup_i V_i$, then the DAG formed by making X the child of any non-leaf vertex, is a derivation DAG for Σ.

(2) If X and Y are leaves of a derivation DAG for Σ, then the DAG formed by adding XY as a child of both X and Y is a derivation DAG for Σ.

(3) If a derivation DAG for Σ has leaves $V_1, ..., V_n, XA$, and $A \notin V_i$ for $i = 1, ..., n$, then the DAG formed by adding a vertex X as a child of XA is a derivation DAG for Σ.

(4) If G, G' and G'' are derivation DAGs for Σ such that

> the leaves of G are $V_1, ..., V_n, X$,
> the leaves of G' are the boundaries of the X-components of the set class
> $\{V_1, ..., V_n, X\}$, and
> the leaves of G'' form a set covering of X,

then the DAG formed by putting each leaf of G' distinct from X and each leaf of G'' as a child of both the leaves of G and G' labeled by X, is a derivation DAG for Σ.

It is important to note that parts (1)-(4) of the above definition correspond exactly to the axioms A1-A4. In fact, derivation DAGs and the axioms are equivalent, which can be proved by structural induction.

Example 3. Let $S = \{CE, DE\}$, $X = ABCD$, $R = \{AC, BD\}$ and $Q = \{AC, BD, CD, CE, DE\}$. Let $\Sigma = \{[S \cup \{X\}], [R]\}$. In order to prove that $\Sigma \Rightarrow [Q]$ we show that $[Q]$ is derivable from Σ by constructing a derivation DAG for Σ whose leaves are the generators of $[Q]$.

STEP 1. {*a DAG with leaves the generators of* $[S \cup \{X\}]$ *is constructed*}
By applying the graphical rule (0) to $[S \cup \{X\}]$, we construct the DAG G (see Figure 2) with root $ABCDE$ and children ABC, CE and DE.

STEP 2. {*a DAG with leaves the sets composing* $\partial_X(S \cup \{X\})$ *is constructed*}
Starting from a copy of G, we construct the DAG G' (see Figure 2) with root $ABCDE$ and children $ABCD$ and CD, obtained by applying the graphical rule (2) to the

children *CE* and *DE* of *ABCDE* and, hence, the graphical rule (3) to the resulting vertex *CDE*.

STEP 3. {*a DAG with leaves the generators of* [R] *is constructed*}
By applying the graphical rule (0) to [R], construct the DAG *G"* (see Figure 2) with root *ABCD* and children *AC* and *BD*.

STEP 4. {*a DAG with leaves the generators of* [Q] *is constructed*}
The DAGs *G*, *G'* and *G"* are merged into one DAG (see Figure 3) by applying rule (4). The resulting DAG has leaves *AC*, *BD*, *CD*, *CE* and *DE*, that is, the generators of Q.

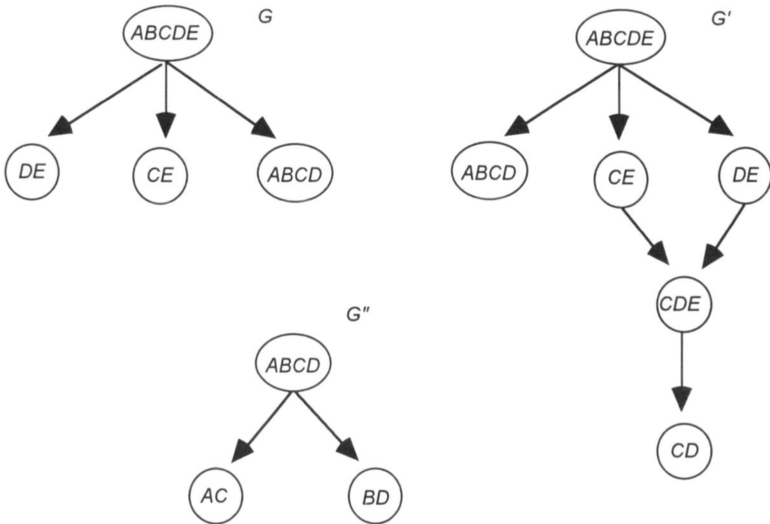

Figure 2. *Three derivation DAGs*

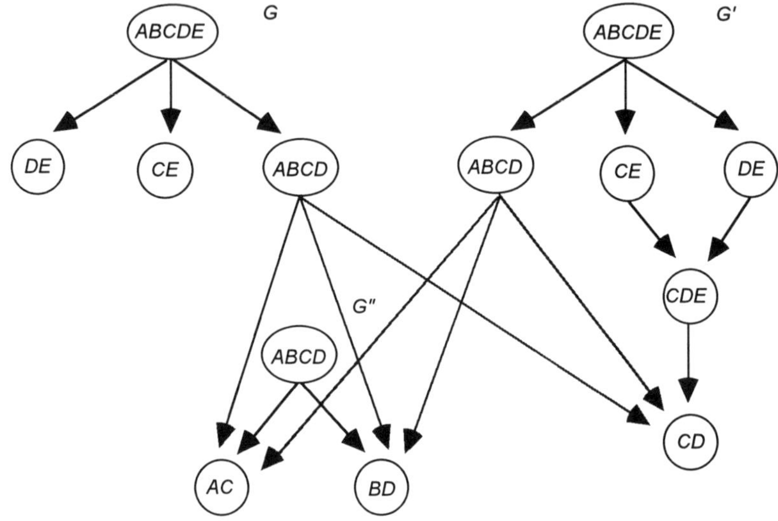

Figure 3. *The derivation DAG obtained by merging the derivation DAGs of Figure 2*

17.5 An Axiomatization of Universal Models

From the axiomatization of unrestricted models presented in Section 4 we can easily derive the following axiomization of universal models.

A0:   "**if** $\varnothing$ **then** $[\{U\}]$".
A1:   "**if** $[S]$ **then** $[S \cup \{X\}]$",   for $X \subseteq U.$
A2:   "**if** $[S \cup \{X, Y\}]$ **then** $[S \cup \{XY\}]$".
A3*:   "**if** $\{[S \cup \{X\}], [R]\}$ **then** $[S \cup \partial_X R \cup \partial_X S]$" .

Notice that the rule A3* is easily derivable from A3 and A4, so that the set $\{A0, A1, A2, A3^*\}$ is a proper axiomatization of universal models. furthermore, the complete axiomatizations of universal binary models and universal decomposable models derive both from the set $\{A0, A1, A2, A3^*\}$ [14].

On the basis of the axiom set $\{A1, A2, A3^*\}$, it is not difficult in principle to work out an inferential routine.

Suppose that we are given a probability distribution p and we want to determine the set of all loglinear models satisfied by p. We make use of two set variables $\mathbb{Y}$ and $\mathbb{N}$ that will contain the sets of models evaluated as they are accepted and rejected, respectively. Initially, both of them are empty and the validity test is executed on a model; so, after the first test either $\mathbb{Y}$ or $\mathbb{N}$ is nonempty. During the execution of the search procedure, $\mathbb{Y}$ and $\mathbb{N}$ will change their contents on consequence of the results of the test or of the inferential routine. After each run of the validation test, we will execute the inferential routine according to the following procedure scheme.

Case 1: *the test resulted in the acceptance of model* α.
STEP 1. By applying axioms A1 and A2, determine $\{\alpha\}^+$.

STEP 2. Determine the set Σ of models that can be inferred by applying axiom A3 to α and to some model in $\mathbb{Y}$ (if any), that is,

$$\Sigma := \{\sigma: \exists\,\beta \in \mathbb{Y}\ A3(\alpha, \beta) = \sigma\}.$$

STEP 3. Set $\mathbb{Y} := \cup_{\sigma \in \Sigma} \{\sigma\}^+ \cup \{\alpha\}^+ \cup \mathbb{Y}$.

Case 2: *the numeric routine resulted in the rejection of model* α.
STEP 1. By applying axioms A1 and A2, determine the set Σ of models that imply α.

STEP 2. Set $\mathbb{N} := \Sigma \cup \mathbb{N}$.

17.6 Open Problems

After closing, we wish to mention an open question.

– Is there an algorithm that, given a set $\Sigma \cup \{\alpha\}$ of universal loglinear models, allows us to decide whether α is or is not implied by Σ?

Acknowledgement.

This work was supported by the Research Funds of the Italian University & Research Ministry.

17.7 References

1.   Bishop, Y.M., Fienberg, S. and Holland, P., *Discrete Multivariate Analysis*. M.I.T. Press, 1975.
2.   Edwards, D., and Havranek, T., A fast procedure for model search in multidimensional contingency tables, *Biometrika* 72 (1985), 339-351.
3.   Edwards, D., and Havranek, T., A fast model selection procedure for large families of models, *J. of the American Statistical Association* 82 (1987), 205-213.
4.   Geiger, D., Paz, A. and Pearl, J., Axioms and algorithms for inferences involving probabilistic independence, *Information and Computation* 91 (1991), 128-141.
5.   Geiger, D. and Pearl, J., Logical and algorithmic properties of conditional independence and graphical models, *The Annals of Statistics* 21 (1993), 2001-2021.
6.   Havranek, T., On application of statistical model search techniques in constructing a probabilistic knowledge base, *Trans. 11th Prague Conf. on "Information Theory, Statistical Decision Functions and   Random   Processes"* (1990).

7. Havranek, T., On model methods, *Proc. 8th Symposium on "Computational Statistics"* (1990).

8. Havranek, T., On model methods, *Proc. Con. on "Symbolic-Numeric Data Analysis and Learning"* (1991).

9. Hill, J.R., Comment, *Statistical Science* 8 (1993), 258-261.

10. Malvestuto, F.M., A unique formal system for binary decompositions of database relations, probability distributions, and graphs, *Information Sciences* 59 (1992), 21-52.

11. Malvestuto, F.M., Testing implication of hierarchical log-linear models for probability distributions, to appear in *Statistics and Computing*.

12. Malvestuto, F.M., Formal treatment of loglinear models for probability distributions, *Proc. 3rd Workshop on "Uncertainty Processing in Expert Systems"* (1994).

13. Malvestuto, F.M., Statistical versus relational join dependencies, *Proc. 7th Int. Conf. on "Scientific & Statistical Database Management"* (1994).

14. Malvestuto, F.M., Formal theories of probabilistic dependency models, *Proc. World Conf. on "Fundamentals of Artificial Intelligence"* (1995).

15. Pearl, J., *Probabilistic Reasoning in Intelligent Systems*. Morgan Kaufman Pub., 1988.

18
Detecting Complex Dependencies in Categorical Data

Tim Oates, Matthew D. Schmill, Dawn E. Gregory and Paul R. Cohen

Computer Science Department, LGRC
University of Massachusetts
Box 34610
Amherst, MA 01003-4610

ABSTRACT Locating and evaluating relationships among values in multiple streams of data is a difficult and important task. Consider the data flowing from monitors in an intensive care unit. Readings from various subsets of the monitors are indicative and predictive of certain aspects of the patient's state. We present an algorithm that facilitates discovery and assessment of the strength of such predictive relationships called *Multi-stream Dependency Detection* (MSDD). We use heuristic search to guide our exploration of the space of potentially interesting dependencies to uncover those that are significant. We begin by reviewing the dependency detection technique described in [3], and extend it to the multiple stream case, describing in detail our heuristic search over the space of possible dependencies. Quantitative evidence for the utility of our approach is provided through a series of experiments with artificially-generated data. In addition, we present results from the application of our algorithm to two real problem domains: feature-based classification and prediction of pathologies in a simulated shipping network.

18.1 Dependency Detection

Consider a network of seaports, with ships carrying cargo between the ports according to a complex schedule. Unforeseen occurrences at a single port or a group of ports, such as severe weather or mechanical failures, can impact the schedule at adjacent ports. If the current state of the network can be used to predict future states of the network, it may be possible to adjust the schedule to minimize adverse effects of such unforeseen occurrences. Similarly, it would be very useful to determine how the future state of a patient, as indicated by various monitors in an Intensive Care Unit (ICU), depends on the current state of the patient. Note that data in the form of time series is not required for the discovery and exploitation of such predictive relationships. Machine learning algorithms that perform feature-based classification determine how a class label depends on various subsets of a feature vector. In all of the above examples, the goal is to determine whether one set of features can be used to predict another set of features. The two sets of features may be taken from the same source at different times (e.g. ICU monitors), or they may be taken from different sources with no notion of time (e.g. a feature vector and a class label). We will revisit both the shipping network and classification examples later in this paper.

A *dependency* is an unexpectedly frequent or infrequent co-occurrence of events over

---

time. Our goal is to find dependencies between tokens contained in multiple streams. A stream is a sequence of values produced over time, and a token is one of the finite set of values that a stream can produce. Dependencies across multiple streams may take many forms: perhaps token A in stream 1 predicts token B in stream 2, or perhaps token A in stream 1 *and* token C in stream 2 predict token B in stream 2. In general, if stream j contains t_j distinct tokens, there are $[\prod_{j=1}^{n} t_j + 1]^2$ possible dependencies between two items.

The dependency detection technique in [3] uses contingency tables to assess the significance of dependencies in a *single* stream of data. Let (t_p, t_s, δ) denote a dependency. Each dependency *rule* states that when the precursor token, t_p, occurs at time step i in the stream, the successor token, t_s, will occur at time step $i + \delta$ in the stream with some probability. When this probability is high, the dependency is strong.

Consider the stream ACBABACCBAABACBBACBA. Of all 19 pairs of tokens at lag 1 (e.g. AC, CB, BA, ...) 7 pairs have B as the precursor; 6 of those have A as the successor, and one has something other than A (denoted $\overline{A}$), as the successor. The following contingency table represents this information:

$$Table(\text{B},\text{A},1) = \begin{array}{c|ccc} & \text{A} & \overline{\text{A}} & total \\ \hline \text{B} & 6 & 1 & 7 \\ \overline{\text{B}} & 1 & 11 & 12 \\ \hline total & 7 & 12 & 19 \end{array}$$

It appears that A depends strongly on B because it almost always follows B and almost never follows anything else ($\overline{B}$). We can determine the significance of each dependency by computing a G statistic for its contingency table:

$$G \begin{pmatrix} n_1 & n_2 & r_1 \\ n_3 & n_4 & r_2 \\ c_1 & c_2 & t \end{pmatrix} = 2 \left[n_1 \log \frac{n_1 t}{r_1 c_1} + n_2 \log \frac{n_2 t}{r_1 c_2} + n_3 \log \frac{n_3 t}{r_2 c_1} + n_4 \log \frac{n_4 t}{r_2 c_2} \right]$$

For example, the contingency table shown above has a G value of 12.38, which is significant at the .001 level, so we reject the null hypothesis that A and B are independent and conclude that (B,A,1) is a real dependency.

We extend this technique to the multiple stream case by introducing the concept of a *multi-token*. A multi-token represents the value of any or all streams at any given time i. For a series with n streams, all multi-tokens will have the form $< x_1, \ldots, x_n >$, where x_j indicates the value in stream j. In order to support the "any or all" requirement, we add a special wildcard symbol, *, to the set of values that may appear in each stream. Thus we can indicate a "don't care" condition by placing an * in the appropriate stream.

For a multi-stream example, consider the following streams:

$$\text{AC}\textbf{BABA}\text{CCBAABAC}\textbf{BA}\text{C}\textbf{BA}\text{C}$$
$$\text{BA}\textbf{CACA}\text{BACBABAB}\textbf{CA}\text{B}\textbf{CA}\text{B}$$

The dependency ($<$B,C$>$,$<$A,A$>$,1) indicated in boldface is significant at the .01 level with a G value of 7.21. The corresponding contingency table is:

$$Table(\text{B},\text{A},1) = \begin{array}{c|ccc} & <\text{A,A}> & \overline{<\text{A,A}>} & total \\ \hline <\text{B,C}> & 4 & 1 & 5 \\ \overline{<\text{B,C}>} & 2 & 12 & 14 \\ \hline total & 6 & 13 & 19 \end{array}$$

We now have both syntax and semantics for multi-stream dependencies. Syntactically, a dependency can be expressed as a triple containing two multi-tokens (a precursor and a successor) and an integer (the lag). For each of the n streams, the multi-tokens contain either a token that may appear in the stream or a wildcard. Dependencies can also be expressed in the form $x \rightarrow_\delta y$ where x and y are multi-tokens. Semantically, this says the occurrence of x indicates or predicts the occurrence of y, δ time steps in the future.

18.2   Searching for Dependencies

The problem of finding significant two-item dependencies can be framed in terms of search. A node in the search space consists of a precursor/successor pair – a predictive rule. The goal is to find predictive rules that are "good" in the sense that they apply often and are accurate. The root of the search space is a pair of multi-tokens with the wildcard in all n positions. For $n = 2$, the root is $< *, * > \rightarrow < *, * >$. The children of a node are generated by replacing (instantiating) a single wildcard in the parent, in either the precursor or successor, with a token that may appear in the appropriate stream. For example, the node $< A, * > \rightarrow < *, X >$ has both $< A, Y > \rightarrow < *, X >$ and $< A, * > \rightarrow < B, X >$ as children.

The rule corresponding to a node is always more specific than the rules of its ancestors and less specific than any of its descendants. This fact can be exploited in the search process by noting that as we move down any branch in the search space, the value in the top left cell of the contingency table (n_1) can only remain the same or get smaller. This leads to a powerful pruning heuristic. Since rules based on infrequently co-occurring pairs of multi-tokens (those with small n_1) are likely to be spurious, we can establish a minimum size for n_1 and prune the search space at any node for which n_1 falls below that cutoff. In practice, this heuristic dramatically reduces the size of the search space that needs to be considered.

Our implementation of the search process makes use of *best first search* with a heuristic evaluation function. That function strikes a tunable balance between the expected numbers of hits and false positives for the predictive rules when they are applied to previously unseen data from the same source. We define *aggressiveness* as a parameter, $0 \leq a \leq 1$, that specifies the value assigned to hits relative to the cost associated with false positives. For a given node (rule) and its contingency table, let n_1 be the size of the top left cell, let n_2 be the size of the top right cell, and let t_S be the number of non-wildcards in the successor multi-token. The value assigned to each node in the search space is $S = t_S(an_1 - (1-a)n_2)$. High values of aggressiveness favor large n_1 and thus maximize hits without regard to false positives. Low aggressiveness favors small n_2 and thus minimizes false positives with a potential loss of hits.

Since the size of the search space is enormous, we typically impose a limit on the number of nodes expanded. The output of the search is simply a list of the nodes, and thus predictive rules, generated.

18.3   Empirical Evaluation

In this section we evaluate the performance of the algorithm on artificially-generated data sets. The goal is to answer a variety of questions regarding the behavior of the algorithm over its domain of applications. Artificial data simplifies this task since the "real" dependencies are known, providing means for distinguishing structure in the data from noise.

Artificial data sets are generated by random sampling and applying a set of probabilistic *structure rules*: $R = \{(P, Pr_P, S, Pr_S)\}$. Each series is initialized by generating n streams of length l, sampled randomly from the token set T. Values for n, l, T, and R are determined by the experiment protocol. Default values are $n = 5$, $l = 100$, $T = \{\text{A},\text{B},\text{C},\text{D},\text{E}\}$, and $R = \{(< \text{A},\text{A},*,*,* >,.1,< \text{C},\text{D},\text{D},*,* >,.8),(< * \text{ C},\text{C},*,* >,.1,< *,\text{A},\text{A},\text{B},* > ,.8),(< *,*,\text{D},\text{D},* >,.1,< *,*,\text{D},\text{C},\text{B} >,.8)\}$.

Structure is then introduced into this random series in two phases: first, seed the precursors P into each time-slice with probability Pr_P; then, whenever a time-slice i matches the precursor of a rule r, insert the successor into time-slice $i + \delta$ with probability $Pr_S(r)$. For analysis, we can partition the resulting series into noise and structure by determining which components are predicted by the dependency rules $(P(r), S(r), \delta)$ for each structure rule $r \in R$.

In each experiment, we run one or more iterations of the search algorithm for each experiment condition. Unless different values are specified by the experiment protocol, we gather 5000 predictive rules with aggressiveness set to 0.5. These rules are post-processed as described below, and used to make *predictions* in ten new data sets generated from the same structure rules. The results are evaluated with respect to two factors: *predictive power* (the total number of predictions made) and *accuracy* (the percentage of the predictions that were correct). These factors are considered separately for the structure and noise portions of the data set.

18.3.1   Selecting the Best Dependency Rules

The MSDD search algorithm generates a large set of dependencies, from which we would like to select the most accurate and predictive rules. Since all our experiments depend on the quality of this selection process, the first question we wish to answer is, "what post-processing strategy will select the best predictive rules?" Although more sophisticated techniques may be needed to resolve redundancy, the simplest approach is to *filter* and *sort* the rules, first discarding rules that do not conform to certain criteria, and then ranking them according to some precedence function.

In this experiment, four different filter criteria are combined with six different sort functions for a total of 24 experiment conditions. The filter options discard rules under the following conditions: (1) never; (2) G not significant at the 0.05 level ($G < 3.84$); (3) $n_1 < 5$; and (4) $n_1 < n_2$. The remaining rules are then sorted according to one of these six functions: (1) randomly; (2) the G statistic (computed over the training data); (3) the number of true instances n_1; (4) the approximate number of true predictions $n_1 \times t_S$; (5) the percentage of instances that are true $n_1/(n_1+n_2)$; and (6) the approximate percentage of predictions that are true, $n_1 \times t_S/(n_1 + n_2)$. Note that for a given rule, t_S tells us the number tokens values that will be predicted when the precursor matches. By multiplying sorting functions (3) and (5) by t_S to yield sorting functions (4) and (6) respectively, we are biasing the rule selection process toward those rules that make many rather than few predictions about actual stream contents. The intuition is that, all other things (including accuracy) being equal, we prefer rules that tell us more about what we will see on the next time step.

We ran five iterations of each condition on data sets with default structure. The results indicate that the highest predictive power and accuracy are achieved when discarding rules with fewer than 5 true instances (filter condition 3), and sorting them according to the G statistic (sort condition 2). This result is as expected: the rules that remain are unlikely to be spurious dependencies, and they are applied in order of their significance.

18.3.2 Comparison of Search Heuristics

Now that we know how to effectively use the output of MSDD, we can address important issues regarding the performance of the algorithm. In this experiment, we compare the performance of the S heuristic (defined in Section 18.2) to other heuristics and across different levels of aggressiveness.

All the search heuristics used in this experiment are based on contingency table analysis of the dependency rules. In addition to the S heuristic, we also use:

1. A normalized S value $\frac{S}{(n_1+n_2)(n_1+n_3)}$, where S is normalized by its *expected count*.

2. The aggressiveness-weighted ratio of hits to false-positives, $\frac{an_1}{(1-a)n_2}$.

3. The aggressiveness-weighted fraction of the instances that are hits, $\frac{an_1}{n_1+n_2}$.

The results (which are not included here due to space constraints) confirm that S is the best of these heuristics: it produces good accuracy and predictive power while allowing the user to tune the performance with the aggressiveness parameter; the other heuristics are not affected by tuning. As expected, high aggressiveness favors predictive power while low values favor accuracy.

18.3.3 Effects of Inherent Structure

Perhaps the most important question to be resolved is: How strong must a dependency be in order for it to be found by the algorithm? In practical terms, this involves two issues: how frequently a dependency occurs and how often the precursor multitoken appears but the successor multitoken does not. In this experiment, we generated 243 data sets of default size, with 1, 3, or 5 structure rules spanning all combinations of: precursor size $t_P \in \{1,3,5\}$, precursor probability $Pr_P \in \{0.1,0.2,0.3\}$, successor size $t_S \in \{1,3,5\}$, successor probability $Pr_S \in \{0.1,0.5,0.9\}$.

The results of this experiment are very encouraging. They indicate that the successor probability is the only limitation on the accuracy of the algorithm, even though the number of rules, the size and probability of the precursor patterns determine the amount of structure that is available to be predicted. Further exploration is required to confirm these results.

18.3.4 Effects of Problem Size

The final issue to be resolved is the influence of the problem size on the performance of the algorithm. In this experiment, we are primarily concerned with the level of performance attained for a given number of predictive rules as the problem size increases. Ideally, we can bound performance as a polynomial function of the input size.

In this experiment, we generate 27 data sets spanning all combinations of: number of streams $n \in \{5, 10, 20\}$, stream length $l \in \{100, 1000, 5000\}$, and number of tokens $|T| \in \{5, 10, 20\}$. For each data set, we let MSDD generate 1000, 5000, 10000, and 20000 predictive rules, with aggressiveness set to 0.5.

This experiment has several interesting results. First, performance actually *improves* as the number of different tokens increases; intuitively, this is due to the probability of each individual token decreasing as the variety of tokens increases. Second, the accuracy of the algorithm is basically constant as the stream length increases. This is due to the probability distributions remaining constant as the length increases. The time requirement of the algorithm does increase with stream length. Finally, it appears that MSDD need only generate $n \times 1000$ search nodes to discover the significant dependencies; this is a very strong claim that needs to be supported by further experimentation.

18.4   Applications

Recall two of the example applications from the introduction that were used to motivate our discussion of MSDD: feature-based classification and predicting the state of a shipping network. In this section we discuss the performance of MSDD on both of those tasks.

18.4.1   Feature-Based Classification

In the interest of generality, we applied MSDD to a task for which it was not explicitly designed: feature-based classification. We present results for thirteen datasets from the UC Irvine collection. Twelve of those datasets were selected from a list of thirteen presented in [9] as being a minimal representative set that covers several important features that distinguish problem domains. For a dataset with n attributes and a single class label, the precursor multi-tokens were n-ary feature vectors, and the successor "multi-tokens" contained only the class label. The actual contents of all precursors (i.e. instantiated attributes and their values) and successors (i.e. classification) were determined by MSDD during its search. The results are presented below in Table 18.1. The accuracy shown in the table is the mean obtained over ten trials where the data was randomly split on each trial into a training set containing 2/3 of the instances and a test set containing the remaining 1/3. The exceptions are NetTalk (training data was generated from a list of the 1000 most common English words, and accuracy was tested on the full 20,008 word corpus), Monks-2 (a single trial with 169 training instances and 432 test instances to facilitate comparison with results contained in [7]), Soybean (a single trial with 307 training instances and 376 test instances), and Mushroom (500 training instances and 7624 test instances). We compared MSDD's performance with other published results for each dataset [2, 4, 7, 8]. On ten datasets for which we had multiple published results, MSDD performance exceeds half of the reported results on six datasets. Only on the Soybean dataset did MSDD perform badly. Nearly all of the 20,000 search nodes generated for that dataset were devoted to predicting a single majority class. An unusually large number of highly accurate rules for predicting that class exist, and were therefore found and expanded by the search algorithm. Due to the high branching factor of the Soybean dataset (it contains 35 attributes), the node limit on the search tree was quickly reached. We are currently exploring solutions to this problem. For a more complete comparison than that shown in Table 18.1, refer to [5].

Data Set	Mean Accuracy	Search Nodes	Other Results from Literature
Breast Cancer	95.15%	10,000	1-nearest neighbor 93.7%
Diabetes	71.33%	10,000	ADAP 76%
Heart Disease	79.21%	20,000	ID3 71.2%; C4 75.8%; back prop 80.6%
Hepatitis	80.77%	10,000	CN2 80.1%; C4 81.2%; Bayes 84.0%
LED-7	70.54%	5,000	CART 71%; C4 72.6%; Bayes 74%
LED-24	71.28%	5,000	CART 70%; NTgrowth+ 71.5%; Bayes 74%
Lymphography	78.16%	15,000	Assistant-86 76%; CN2 82%; Bayes 83%
NetTalk	70.11%	50,000	NetTalk 77%
Monks-2	79.17%	5,000	CN2 69.0%; ID3 69.1%; back prop 100%
Mushroom	99.49%	30,000	GINI 98.6%; Info Gain 98.6%; C4 100%
Soybean	13.83%	20,000	IWN 97.1%
Thyroid	95.46%	20,000	
Waveform-40	73.02%	15,000	Nearest neighbor 38%; CART 72%; Bayes 86%

TABLE 18.1. Performance of MSDD as a feature-based classifier on thirteen datasets from the UC Irvine collection.

18.4.2 Pathology Prediction

We applied MSDD to the task of predicting pathologies in a simulated shipping network called TransSim. When several ships attempt to dock at a single port at the same time, most will be queued to await a free dock, resulting in a *bottleneck*. We built a pathology demon that predicts the potential for bottlenecks before they actually form, and we built an agent that modifies the shipping schedule in an effort to keep predicted pathologies from materializing. Using the demon as an oracle, we gathered data from a single run of the simulator and used MSDD to generate rules to predict bottlenecks. To assess the utility of the previously generated rules, we ran ten simulations in each of two conditions; one with the existing demon and another with the demon replaced by the rules. We used t tests to determine whether mean costs associated with each simulation were lower in the demon condition as compared to the rule condition. The results are presented below in Table 18.2. Note that the number of pathologies predicted (PP) by the demon is almost twice the number predicted by the rules and, therefore, the agent made about twice as many schedule modifications (SM). However, of our five cost measures — average docking queue length (QL), the amount of time that cargo spends sitting idle on docks or in queued ships (IC), the amount of time that cargo spends in transit (CT), the degree to which the carrying capacity of each ship is utilized (SU), and the number of simulated days required to deliver all cargo (SD) — only one, SD, was significantly lower in the demon condition when compared to the rule condition. That is, even though the agent is taking a much more active role, performance is not significantly better. Inspection of execution traces shows that the demon is much more likely than the rule set to predict short-lived pathologies. The rules are good at forecasting substantial pathologies, ones that will not go away of their own accord, but miss the more fleeting pathologies. Said differently, MSDD rules are not misled by small, noisy fluctuations in the state of the simulation. This behavior is beneficial when we view disruption to the original schedule as a cost that we want to minimize.

Cost	Demon Mean	Rule Mean	p Value
PP	184.2	94.6	0.0001
CT	2289.3	2377.9	0.0689
IC	1149.8	1202.1	0.1844
QL	637.7	640.5	0.9177
SD	131.1	141.6	0.0019
SU	188.8	202.2	0.3475
SM	21.6	9.2	0.0001

TABLE 18.2. Comparison of simulation costs using demon and MSDD rules for pathology prediction.

This experiment suggests that MSDD can discover indicators of pathological states in TransSim from high level domain information. MSDD can identify relevant state information to emulate the objective function of an external oracle. One limitation of this approach, as compared with the demon, is that an initial run of the simulator is required to gather data to drive the rule generation process. However, the domain knowledge supplied to the MSDD algorithm was minimal in comparison to the demon.

18.5   Work in Progress

We developed and are currently evaluating an incremental version of the MSDD algorithm called IMSDD. MSDD is a batch algorithm; all data from the streams must be present when the search begins. That may not be possible or optimal in cases where learning about the environment must be interleaved with acting in it, or when structure in the environment changes over time. Likewise, it may not be possible for agents with bounded computational resources to process all data in all streams. IMSDD sacrifices the ability to examine and recount data in batches in order to gain adaptable, any-time behavior.

IMSDD takes a data-driven, bottom-up approach to forming rules. As IMSDD receives input, it stores multitoken pairs as fully instantiated words. Tokens in streams that exhibit no contingent structure are *generalized* as a move towards representing the data's true structure. IMSDD follows a *predict* $\rightarrow$ *verify* $\rightarrow$ *generalize* $\rightarrow$ *update* loop. Based on the current input multitoken, IMSDD predicts the next multitoken, evaluates that prediction, uses the input to form new generalizations, and then updates its internal data structures. The data structure maintained by IMSDD is called a *precursor tree*. Figure 1 shows an example of such a tree. To update the counts for an input word (a precursor/successor multi-token pair), the precursor is parsed through the tree to a leaf. Each leaf contains a successor table that maintains the counts of all tokens that have followed the precursor. For example, the input word $< c, b > < b, a >$ would be parsed to the leftmost successor table shown Figure 1, causing the count for b in row 1 to be increased to 3 and the count for a in row 2 to be increased to 4. Precursors are generalized when paths exist in the tree that differ in exactly one position, e.g. $< c, b >$ and $< b, b >$ would generalize to $< *, b >$. The successor table for the new path is created by summing the contents of the successor tables on the paths that were generalized over (see Figure 1). Predictions are made by parsing the current input multi-token along all applicable paths (multiple paths are possible with wildcards), gathering the set of most likely successor token values from each successor table, and choosing the one that maximizes a score similar to the MSDD

S value. Two additional mechanisms were implemented to control the size of the data structure and the search space. First, during IMSDD's selection process to find the best successor to predict, a pruning component selects a fixed number of the worst rated rules to be excised. Second, rather than parsing a precursor down all possible paths, wildcard paths are taken with some probability. In effect, we randomly sample from the set of possible paths.

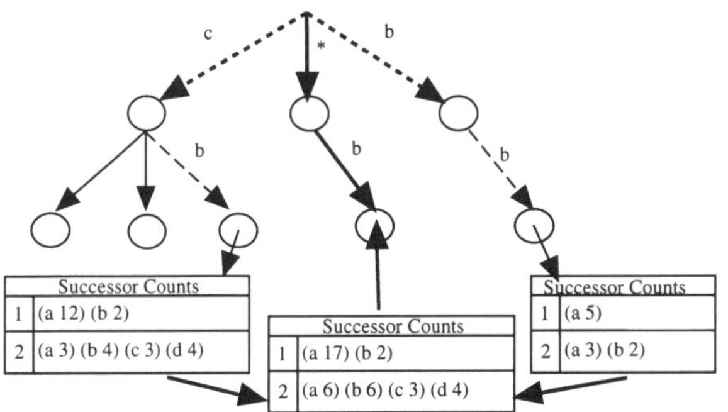

FIGURE 1. An example IMSDD precursor tree.

To evaluate IMSDD we defined the metrics *adjusted hit rate (ahr)* to be the number of correct token predictions divided by the number of tokens seeded in the dataset, and *fp-rate (fpr)* to be the number of incorrectly predicted tokens divided by the total number of tokens. The first mechanisms we examined were the pruning strategy and the sampling policy for precursor generalization. We recorded the learning curves generated by IMSDD as we varied the sampling rate and turned pruning on and off. Figure 2a suggests that both pruning and sampling have small effects. We next explored the effects of increasing n, the number of streams. Figure 2b shows the effect of increasing n from 5 to 9. First, note that the overall shape of the learning curve is similar for all three values of n. Second, both the onset of the learning curve and the point at which IMSDD can account for 100% of the structure differ are delayed as n increases. The first result suggests that the learning algorithm, when scaled up, might exhibit the same facility for learning rules and accounting for structure. The second result implies that due to the larger stream size, there is some degree of difficulty learning good initial generalizations given the higher dimensional search space.

18.6   Conclusion

In this paper we described how the problem of finding significant dependencies between the tokens in multiple streams of data can be framed in terms of search. The notion of dependencies between pairs of tokens introduced in [3] was extended to pairs of multi-tokens, where a multi-token describes the contents of several streams rather than just one. We introduced the Multi-Stream Dependency Detection (MSDD) algorithm that performs a general-to-specific best-first search over the exponentially sized space of possible dependencies between multi-tokens. The search heuristic employed by MSDD strikes a tunable balance between the expected number of hits and false positives for the dependencies

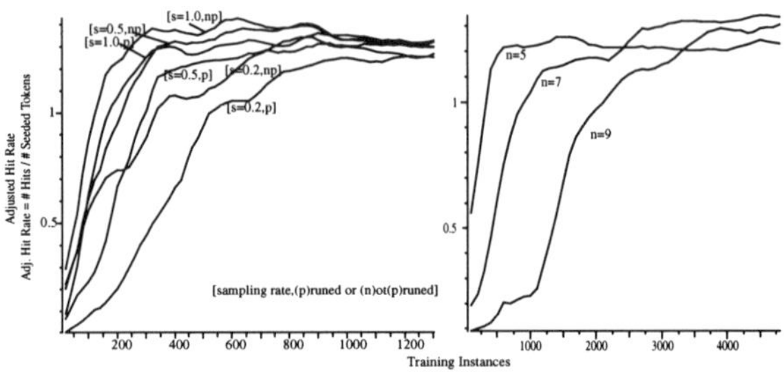

FIGURE 2. (a) Effects of sampling and pruning on IMSDD learning curves. (b) IMSDD learning curves for 5, 7, and 9 streams.

discovered when they are applied as predictive rules to previously unseen data from the same source. We presented results from an empirical evaluation of MSDD's performance over a wide range of artificially generated data. In addition, we applied MSDD to the task of pathology prediction in a simulated shipping network and to a number of classification problems from the UC Irvine collection. The results that we obtained are very encouraging.

We are continuing to evaluate both MSDD and IMSDD. We are particularly interested in making quantitative statements about performance bounds of the algorithms as a function of input characteristics. Also, we are working to remove the need for a fixed sized multi-token and a fixed time interval between multi-tokens.

Acknowledgments

This work is supported by ARPA/Rome Laboratory under contract #'s F30602-91-C-0076 and F30602-93-C-0010. The U.S. government is authorized to reproduce and distribute reprints for governmental purposes notwithstanding any copyright notation hereon.

18.7 REFERENCES

[1] Bennett, K. P. and Mangasarian, O. L. Robust linear programming discrimination of two linearly inseparable sets. In *Optimization Methods and Software* 1, 1992, 23-34 (Gordon and Breach Science Publishers).

[2] Holte, Robert C. Very simple classification rules perform well on most commonly used datasets. In *Machine Learning*, (11), pp. 63-91, 1993.

[3] Howe, Adele E. and Cohen, Paul R. Understanding Planner Behavior. To appear in *AI Journal*, Winter 1995.

[4] Murphy, P. M., and Aha, D. W. *UCI Repository of machine learning databases* [Machine-readable data repository]. Irvine, CA: University of California, Department of Information and Computer Science, 1994.

[5] Oates, Tim. MSDD as a Tool for Classification. Memo 94-29, Experimental Knowledge Systems Laboratory, Department of Computer Science, University of Mas-

sachusetts, Amherst, 1994.

[6] Oates, Tim and Cohen, Paul R. Toward a plan steering agent: Experiments with schedule maintenance. In *Proceedings of the Second International Conference on Artificial Intelligence Planning Systems*, pp. 134-139, 1994.

[7] Thrun, S.B. The MONK's problems: A performance comparison of different learning algorithms. Carnegie Mellon University, CMU-CS-91-197, 1991.

[8] Wirth, J. and Catlett, J. Experiments on the costs and benefits of windowing in ID3. In *Proceedings of the Fifth International Conference on Machine Learning*, pp. 87-99, 1988.

[9] Zheng, Zijian. A benchmark for classifier learning. Basser Department of Computer Science, University of Sydney, NSW.

Part IV
Classification

19

A Comparative Evaluation of Sequential Feature Selection Algorithms

David W. Aha[†] and Richard L. Bankert[††]

Navy Center for Applied Research in AI[†]
Naval Research Laboratory
Washington, D.C. 20375-5337
aha@aic.nrl.navy.mil

Marine Meteorology Division[††]
Naval Research Laboratory
Monterey, CA 93943
bankert@nrlmry.navy.mil

ABSTRACT Several recent machine learning publications demonstrate the utility of using feature selection algorithms in supervised learning tasks. Among these, *sequential feature selection* algorithms are receiving attention. The most frequently studied variants of these algorithms are *forward* and *backward sequential selection*. Many studies on supervised learning with sequential feature selection report applications of these algorithms, but do not consider variants of them that might be more appropriate for some performance tasks. This paper reports positive empirical results on such variants, and argues for their serious consideration in similar learning tasks.

19.1 Motivation

Feature selection algorithms attempt to reduce the number of dimensions considered in a task so as to improve performance on some dependent measures. In this paper, we restrict our attention to supervised learning tasks, where our dependent variables are classification accuracy, size of feature subset, and computational efficiency.

Feature selection has been studied for several decades (e.g., Fu, 1968; Mucciardi & Gose, 1971; Cover & van Campenhout, 1977). Several publications have reported performance improvements for such measures when feature selection algorithms are used (e.g., Almuallim & Dietterich, 1991; Doak, 1992; Kononenko, 1994; Caruana & Freitag, 1994; Skalak, 1994; Moore & Lee, 1994; Aha & Bankert, 1994; Townsend-Weber & Kibler, 1994; Langley & Sage, 1994). Feature selection algorithms are typically composed of the following three components:

1. **Search algorithm** This searches the space of feature subsets, which has size 2^d, where d is the number of features.

2. **Evaluation function** This inputs a feature subset and outputs a numeric evaluation. The search algorithm's goal is to maximize this function.

3. **Performance function** The performance task studied in this paper is classification.

---

[1]*Learning from Data: AI and Statistics V.* Edited by D. Fisher and H.-J. Lenz. ©1996 Springer-Verlag.

>Given the subset found to perform best by the evaluation function, the classifier is
>used to classify instances in the dataset.

Doak (1992) identified three categories of search algorithms: exponential, randomized, and sequential. Exponential algorithms (e.g., branch and bound, exhaustive) have exponential complexity in the number of features and are frequently prohibitively expensive to use (i.e., they have complexity $O(2^d)$, where d is the number of features). Randomized algorithms include genetic and simulated annealing search methods. These algorithms attain high accuracies (Doak, 1992; Vafaie & De Jong, 1993; Skalak, 1994), but they require biases to yield small subsets. How this is best done remains an open issue. Sequential search algorithms have polynomial complexity (i.e., $O(d^2)$); they add or subtract features and use a hill-climbing search strategy. We are investigating frequently used variants of these algorithms (Aha & Bankert, 1994).

The most common sequential search algorithms for feature selection are *forward sequential selection* (FSS) and *backward sequential selection* (BSS), and we focus on these algorithms in this paper. FSS begins with zero attributes, evaluates all feature subsets with exactly one feature, and selects the one with the best performance. It then adds to this subset the feature that yields the best performance for subsets of the next larger size. This cycle repeats until no improvement is obtained from extending the current subset. BSS instead begins with all features and repeatedly removes a feature whose removal yields the maximal performance improvement.

Doak (1992) reported that BSS frequently outperformed FSS, perhaps because BSS evaluates the contribution of a given feature in the context of all other features. In contrast, FSS can evaluate the utility of a single feature only in the limited context of previously selected features. Caruana and Freitag (1994) note this problem with FSS, but their results favor neither algorithm. Based on these observations, it is not clear whether FSS will outperform BSS on a given dataset with an unknown amount of feature interactions. Therefore, we study both in this paper.

Doak (1992) evaluated a beam search variant of FSS and noted that it frequently outperformed FSS. He predicted that a similar variant of BSS would outperform BSS. We investigate both variants in this paper.

This paper focuses on variants of BSS and FSS and their application to a sparse data set of particular interest to the Naval Research Laboratory's Marine Meteorology Division. We define a space of parameterized algorithms in which these algorithms are two of many possible instantiations. We show that feature selection improves the performance of a case-based classifier on this dataset, provide evidence for John, Kohavi, and Pfleger's (1994) conjecture that *wrapper* models outperform *filter* models (described in Section 19.3), and provide evidence against Doak's (1992) implication that BSS is always preferable to FSS. Finally, we show that variants of BSS and FSS can frequently outperform them, and argue that similar variants should always be tested when possible.

19.2   Cloud Pattern Classification Task

Feature selection algorithms hold the potential to improve classification performance for sparse datasets with many features and for datatsets where the features used to describe cases are not known to be highly relevant to the performance function (i.e., here, the classifier). Both of these situations are true for the dataset we examined.

The cloud pattern classification dataset was provided by the Marine Meteorology Division of the Naval Research Laboratory[2] It is sparse; its number of cases is small (69) and they are each described by a comparatively large number (98) of features. Furthermore, these features were chosen under the assumption that they *may* provide useful information for classifying cloud patterns. However, no verification was performed to ensure that the chosen features are predictive of particular cloud patterns. Some of the features are known to be relevant for discriminating some cloud pattern classes, but their relevance for distinguishing all cloud patterns is unknown.

Thus, we expected this to be an ideal dataset for testing sequential selection variants that use classifiers which perform poorly in the presence of many features with potentially low relevance. It is well known that nearest neighbor algorithms perform poorly under such situations (e.g., Aha, 1992). However, such non-parametric classifiers are excellent choices for evaluation functions since manual parameter-tuning is not feasible whenever a large number of (different) feature subsets must be evaluated. Therefore, we chose IB1 (Aha, 1992) as the classifier; it is an implementation of the nearest neighbor classifier.

The features used to describe each instance consist of shape, size, spectral, and textural measures for each cloud pattern region. Examples of each group include such features as compactness, perimeter size, maximum pixel value, and cluster prominence, respectively. All features are defined over a continuous range of values. Bankert (1994a; 1994b) describes these features in more detail.

19.3   Framework

In our experiments, we investigated using FSS and BSS as the search algorithm, and we used IB1 as the classifier. We used both IB1 and the Calinski-Harabasz separability index measure as the evaluation function.[3] Our selection of these two evaluation functions was motivated by the hypothesis that *wrapper* models, which use the classifier itself as the evaluation function, outperform *filter* models, which do not. This was conjectured by John, Kohavi, and Pfleger (1994). Doak (1992) cited some informal evidence for this conjecture but did not describe a detailed analysis.

Three additional search-related variables were varied in our experiments that allowed us to test beam searching variants of FSS and BSS. These variables constrain how a beam search is conducted. Figure 19.3 describes our parameterizable framework, which we refer to as BEAM. Briefly, BEAM implements beam-search variants of the given search algorithm. A *queue* of *states* is maintained, where a state is a subset of features and an indication as to which subsets immediately derivable from it (i.e., the state's children) have not yet been evaluated. The evaluation of each state is maintained with it on the queue. Initialize_queue initializes the queue with either the complete set of features (for BSS) or the empty set (for FSS). Its evaluation is computed, and this is recorded as the initial *best* state. BEAM then loops until the queue is empty. The first step of

---

[2] Bankert (1994b) describes feature selection experiments with a variant of this dataset, but the differences in these datasets prevent a direct comparison.

[3] Milligan and Cooper (1985) found that the Calinski-Harabasz separability index performed best among 30 separability indices in their experiments.

```
Inputs:
 S: Search algorithm
 E: Evaluation function
 q: Size of the queue
 m: Number of states evaluated per iteration
 n: Number of subsets evaluated per state
Partial Key:
 queue: An ordered queue of states
 best: The best-performing state


BEAM(S, E, q, m, n)
1. queue = Initialize_queue(S, E)
2. best = Initialize_best(queue)
3. While (queue is not empty) do:
   A. states = Select_states(queue, m)
   B. evaluations = Evaluate_states(S, E, states, n)
   C. queue = Update_queue(states, evaluations, queue, q)
   D. best = Update_best(best, queue)
4. Output: best
```

FIGURE 1. BEAM: A Framework for Sequential Feature Selection

the loop stochastically selects m states from the queue, where states higher in the queue have a higher probability of being selected.[4] The next step evaluates n child states (i.e., subsets for BSS, supersets for FSS) derivable by search algorithm S from each of these m states using evaluation function E. The queue is then updated; states that have been exhaustively searched (i.e., all its children have been evaluated) by S are dropped, and newly-evaluated states that are among the current top q states are integrated into their appropriately ordered locations in the queue. Finally, the best state located so far is maintained in *best*. After the queue has been exhausted, this best state is output for use by the classifier. The user must provide the five inputs listed in Figure 1 (i.e., S, E, q, m, and n).

19.4 Empirical Comparisons

All experiments were run using a leave-one-out strategy. We report averages for classification accuracy, size of selected feature subset, and number of subsets evaluated. Averages were computed over ten runs for all the experiments in which IB1 was used as the evaluation function. Our implementation of the Calinski-Harabasz separability index measure is deterministic because ties rarely occur when it evaluates states (i.e., when used as an evaluation function, it outputs a continuous value). Ties occur more frequently when us-

---

[4]More specifically, this function selects ordered state i with probability $P(i) = (q - i + 1)/\sum_{j=1}^{q} j$, where q is the size of the queue.

TABLE 19.1. Best and Averages (Accuracy, Size of Selected Subset, and Number of Subsets Evaluated) for IB1 When Using FSS and BSS (with a Queue Size of One) for Two Different Evaluation Functions

Search	EvalFn	Best			Averages		
		Acc	Sz	#Eval	Acc	Sz	#Eval
IB1		62.3%	98	1	-	-	-
FSS	IB1	89.9%	6	4,852	88.1%	13.1	4,852.0
BSS	IB1	84.1%	18	4,699	79.9%	30.8	4,343.8
FSS	Index	66.7%	20	4,852	-	-	-
BSS	Index	68.1%	45	4,849	-	-	-

ing IB1 since BEAM uses it to evaluate states in a leave-one-out test on classification accuracy. Since tied states are randomly ordered (among themselves) on the queue, this can cause IB1 to behave differently on each run. Thus, Table 1 shows averages when IB1 is used as the evaluation function, but not otherwise.

The following subsections summarize our investigations of four hypotheses. Sections 4.1, 4.2, and 4.3 discuss results with the standard FSS and BSS algorithms, whereas Section 4.4 concerns their beam search variants.

19.4.1   Feature selection is appropriate for this dataset

Since this dataset is sparse (i.e., 98 features, only 69 instances) and several features have unknown and possibly low classification relevance, we expected feature selection to improve IB1's performance. Table 19.1 displays the results for IB1 and the average results for four feature selection algorithms (i.e., the FSS and BSS search algorithms were tested using both IB1 and the Calinski-Harabasz separability index as the evaluation function). Feature selection consistently increased accuracy and reduced feature set size. These results support this hypothesis.

19.4.2   The wrapper model is superior to the filter model

Two evaluation functions were compared: IB1 and the Calinski-Harabasz separability index. We expected that, since IB1 was also used as the classifier, using it as the evaluation function would yield superior results.

As shown in Table 19.1, the performance of the standard FSS and BSS algorithms (i.e., $q = 1$, $m = 1$, $n =$ all) is much lower when using the separability index measure rather than IB1 as the evaluation function. This supports our hypothesis. We believe the reason for this is that the separability index measure's bias is different than IB1's bias. Therefore, the index measure will not necessarily select a good-performing subset for IB1.

19.4.3   BSS outperforms FSS

Doak (1992) found that BSS often outperformed FSS in his studies, but he examined smaller numbers of features in his task. We were unsure which algorithm would yield better results.

The average accuracies for FSS were higher than for BSS, and their differences in accuracy were significant according to a one-tailed t-test ($p < 0.05$). FSS also found

TABLE 19.2. Best and Averages when using IB1 as the Evaluation Function and a Queue Size of 25

Search	m	n	Best			Averages		
			Acc	Sz	#Eval	Acc	Sz	#Eval
FSS	1	1	91.3%	20	57,122	87.8%	21.8	56,387.4
FSS	1	all	94.2%	22	110,058	89.3%	19.7	117,015.1
FSS	all	all	95.7%	16	114,850	94.2%	13.7	114,854.9
BSS	1	1	89.9%	12	2,871	86.8%	10.0	3,918.9
BSS	1	all	91.3%	10	8,031	86.5%	10.6	10,764.5
BSS	all	all	87.0%	11	110,881	84.1%	23.6	104,837.3

significantly ($p < 0.05$) smaller subset sizes but required significantly more evaluations ($p < 0.1$).

Table 19.2 summarizes the results for the two search algorithms when we varied the number of states selected per evaluation (m) and the number of subsets evaluated per selected state (n) while using a queue size (q) of 25. Although FSS's average accuracy is higher for all three variants, it is significantly higher only for the third variant ($p < 0.01$). FSS also yields significantly larger-sized subsets only for the first variant ($p < 0.1$), and evaluated significantly more subsets ($p < 0.05$) in all three cases.

The accuracy results differ from the majority of results found by Doak (1992), thus contradicting this hypothesis. We suspect that BSS is more easily confused by large numbers of features because the deletion of a single feature, though perhaps relevant, can have less effect in the presence of a larger number of features. We plan to explore this hypothesis further in systematic experimentation with artificially generated data. However, the general pattern found here suggests that FSS is preferred when the optimal number of selected features is small while BSS is preferred if otherwise. This also suggests that BSS's performance can be enhanced in these situations when it is biased towards using small-sized subsets of features. We show evidence for this in (Aha & Bankert, 1994). Similarly, in situations when most features are relevant, we expect that FSS' performance can be enhanced by biasing its search to start with large-sized subsets of features.

19.4.4   Beam search variants outperform standard FSS and BSS

Caruana and Freitag (1994) report evidence for this hypothesis when using combinations of these algorithms, but not for simple extensions of them. Doak (1992) found that a beam search version of FSS was preferable to it, but did not evaluate such extensions of BSS.

The accuracies for the beam searching variants in Table 19.2, for which the queue size was 25, were all higher than when the queue has size one (Table 19.1). These differences were significant for the third variant of FSS and all variants of BSS ($p < 0.05$). Furthermore, the BSS beam-search variants also significantly reduced the size of the selected feature subsets ($p < 0.1$). However, these sizes increased significantly for two of the FSS variants ($p < 0.05$). Finally, the number of evaluated subsets significantly increased for five of the six variants tested ($p < 0.005$), though this decreased significantly for BSS ($m = 1, n = 1$) ($p < 0.1$).

In summary, many beam search variants of the standard sequential feature selection approaches can significantly increase classification accuracies on this task. Simultaneously, some can also significantly reduce the size of the subsets located, but they tend to significantly increase search requirements (i.e., the number of subsets evaluated).

19.5   Conclusions

This paper examines variants of forward and backward sequential feature selection algorithms for a cloud pattern classification task defined by a sparse dataset with numerous features. When using a nearest neighbor algorithm as the classifier, our results show that (1) feature selection improves accuracy on this task, (2) using the classifier as the evaluation function yields better performance than when using a separability index, (3) BSS does not always outperform FSS (contrary to some claims), and (4) beam search variants of these algorithms can improve accuracy, at least on this task.

We found similar results with other datasets and plan to discuss them in extensions of this paper. Additional studies showed that using random initial subset selection with a comparatively small number of features drastically reduces the amount of search performed by BSS without sacrificing accuracy.

Acknowledgements

Thanks to John Grefenstette, Paul Tag, Karl Branting, Pat Langley, Diana Gordon, Doug Fisher, and the reviewers for their comments and suggestions.

19.6   REFERENCES

Aha, D. W. (1992). Generalizing from case studies: A case study. In *Proceedings of the Ninth International Conference on Machine Learning* (pp. 1–10). Aberdeen, Scotland: Morgan Kaufmann.

Aha, D. W., & Bankert, R. L. (1994). Feature selection for case-based classification of cloud types: An empirical comparison. In D. W. Aha (Ed.) *Case-Based Reasoning: Papers from the 1994 Workshop* (Technical Report WS-94-01). Menlo Park, CA: AAAI Press.

Almuallim, H., & Dietterich, T. G. (1991). Learning with many irrelevant features. In *Proceedings of the Ninth National Conference on Artificial Intelligence* (pp. 547–552). Menlo Park, CA: AAAI Press.

Bankert, R. L. (1994a). Cloud classification of AVHRR imagery in maritime regions using a probabilistic neural network. *Journal of Applied Meteorology, 33*, 909–918.

Bankert, R., L. (1994b). Cloud pattern identification as part of an automated image analysis. *Proceedings of the Seventh Conference on Satellite Meteorology and Oceanography* (pp. 441–443). Boston, MA: American Meteorological Society.

Caruana, R., & Freitag, D. (1994). Greedy attribute selection. In *Proceedings of the Eleventh International Machine Learning Conference* (pp. 28–36). New Brunswick, NJ: Morgan Kaufmann.

Cover, T. M., & van Campenhout, J. M. (1977). On the possible orderings in the measurement selection problem. *IEEE Transactions on Systems, Man, and Cybernetics*, 7, 657–661.

Doak, J. (1992). *An evaluation of feature selection methods and their application to computer security* (Technical Report CSE-92-18). Davis, CA: University of California, Department of Computer Science.

Fu, K. S. (1968). *Sequential methods in pattern recognition and machine learning.* New York: Academic Press.

John, G., Kohavi, R., & Pfleger, K. (1994). Irrelevant features and the subset selection problem. In *Proceedings of the Eleventh International Machine Learning Conference* (pp. 121–129). New Brunswick, NJ: Morgan Kaufmann.

Kononenko, I. (1994). Estimating attributes: Analysis and extensions of RELIEF. In *Proceedings of the 1994 European Conference on Machine Learning* (pp. 171–182). Catania, Italy: Springer Verlag.

Langley, P., & Sage, S. (1994). Oblivious decision trees and abstract cases. In D. W. Aha (Ed.), *Case-Based Reasoning: Papers from the 1994 Workshop* (Technical Report WS-94-01). Menlo Park, CA: AAAI Press.

Milligan, G. W., & Cooper, M. C. (1985). An examination of procedures for determining the number of clusters in a data set. *Psychometrika, 50*, 159–179.

Moore, A. W., & Lee, M. S. (1994). Efficient algorithms for minimizing cross validation error. In *Proceedings of the Eleventh International Conference on Machine Learning* (pp. 190–198). New Brunswick, NJ: Morgan Kaufmann.

Mucciardi, A. N., & Gose, E. E. (1971). A comparison of seven techniques for choosing subsets of pattern recognition properties. *IEEE Transaction on Computers, 20*, 1023–1031.

Skalak, D. (1994). Prototype and feature selection by sampling and random mutation hill climbing algorithms. In *Proceedings of the Eleventh International Machine Learning Conference* (pp. 293–301). New Brunswick, NJ: Morgan Kaufmann.

Townsend-Weber, T., & Kibler, D. (1994). Instance-based prediction of continuous values. In D. W. Aha (Ed.), *Case-Based Reasoning: Papers from the 1994 Workshop* (Technical Report WS-94-01). Menlo Park, CA: AAAI Press.

Vafaie, H., & De Jong, K. (1993). Robust feature selection algorithms. In *Proceedings of the Fifth Conference on Tools for Artificial Intelligence* (pp. 356–363). Boston, MA: IEEE Computer Society Press.

20

Classification Using Bayes Averaging of Multiple, Relational Rule-based Models

Kamal Ali and Michael Pazzani

Department of Information and Computer Science,
University of California, Irvine, CA, 92717, USA.
ali@ics.uci.edu, pazzani@ics.uci.edu

ABSTRACT We present a way of approximating the posterior probability of a rule-set model that is comprised of a set of class descriptions. Each class description, in turn, consists of a set of relational rules. The ability to compute this posterior and to learn many models from the same training set allows us to approximate the expectation that an example to be classified belongs to some class. The example is assigned to the class maximizing the expectation. By assuming a uniform prior distribution of models, the posterior of the model does not depend on the structure of the model: it only depends on how the training examples are partitioned by the rules of the rule-set model. This uniform distribution assumption allows us to compute the posterior for models containing relational and recursive rules. Our approximation to the posterior probability yields significant improvements in accuracy as measured on four relational data sets and four attribute-value data sets from the UCI repository. We also provide evidence that learning multiple models helps most in data sets in which there are many, apparently equally good rules to learn.

20.1   Introduction

There has been much work in learning relational models of data in recent years (e.g. FOIL: [Quinlan90]; FOCL: [Pazzani-Kibler91]; GOLEM: [3], CLINT: [2], ML-SMART: [1]). Here we present results that apply Bayesian probability theory (e.g. [Berger85]) to learning and classification using class descriptions consisting of first-order (relational) rules. According to this theory, it is necessary to learn several models of the data. In our work, each model consists of one class description per class in the data so learning multiple models implies learning many descriptions for each class. Although there has been prior work in applying Bayesian probability theory for learning and combining evidence from multiple decision trees ([Buntine90]) there has been no such work for relational rules or for rule-set models. A rule-set model is defined to consist of a set of class descriptions, one description per class in the data. Each description, in turn, consists of a set of rules that all conclude for the same class (Table 20.1).

Our core algorithm is based on FOIL which learns a disjoint covering for a class. That is, it learns rules for a class that cover disjoint subsets of the training examples of that class. Therefore, ideally, each learned rule should correspond to a disjunct or subclass in the class. By applying FOIL stochastically many times for each class, we are able to model each disjunct more than once. This in turn leads to a better overall model of each subclass and hence of the class. Our goal is to demonstrate that classifications obtained

---

TABLE 20.1. HYDRA-MM learns many class descriptions for each class. For each class it combines evidence from descriptions of that class to calculate the expectation that a test example belongs to that class. The test example is assigned to the class that maximizes that expectation.

	1st model of the data	2nd model of the data
Description for class a	class-a(X,Y) :- b(X), c(Y). class-a(X,Y) :- d(X,Z), e(Z,Y).	class-a(X,Y) :- b(X), c(Y). class-a(X,Y) :- d(X,Z), h(Z,Y).
Description for class b	class-b(X,Y) :- e(X,Y), f(X,X). class-b(X,Y) :- g(X), class-b(Y,X).	class-b(X,Y) :- e(X,Y), k(X,X). class-b(X,Y) :- g(X),class-b(Y,X).

by a set (ensemble) of models learned by stochastic application of the FOIL algorithm are more accurate than those produced by a single, deterministic application of the FOIL algorithm.

Previous work in learning multiple models consisting of rules has been done by [Gams89] and [Kononenko92]. The limitation of Gams' approach is that it does not retain the multiple models for classification. Instead, after learning several models of the data, a single model is constructed whose rules are chosen from the rules of the previously learned models. Kononenko & Kovacic ([Kononenko92]) have also done work in learning multiple models but in their work, each class in each model is described using just a single rule. Realizing that many classes cannot be accurately described using just a single rule, they advocate the use of a rule-selection metric that rewards rules that cover (i.e. match) training examples not covered by previously learned models for that class. But then this method does not reward learning more than one description for each disjunct (subclass) in the class.

There are two main empirical results of this paper. The first is the development of an approximation to the posterior probability of a rule-set model and the empirical demonstration that weighting the classifications of models by their posterior probabilities leads to more accurate classifications than those obtained by the model learned from a single, deterministic application of FOIL. The second result is that learning and using multiple models is especially helpful in data sets where there are many, apparently equally good rules to learn. To demonstrate these results, we adapt a relational learning algorithm (HYDRA, [Ali-Pazzani93]) to learn multiple models, yielding HYDRA-MM. HYDRA-MM learns multiple rule-set models, therefore it also learns more than one class description for each class. Because HYDRA is a *relational* learning algorithm, it can learn the more expressive relational rules as well as attribute-value rules.

20.2   Theoretical development

For classification, Bayesian probability theory advocates making the classification whose expected accuracy is maximal, with the expectation being taken over all possible models in the model space of the learning algorithm (this process is called Bayes averaging). In practice, we only compute this expectation over a small set of highly probable models. Using the notation in [Buntine90], let c be a class, $\mathcal{H}$ denote the model space, x be a test example, $\vec{x}$ denote the training examples and $\vec{c}$ denote the class labels of the training examples. Then, we should assign the test example x to the class c with the highest posterior probability:

$$E(p(c|x, \mathcal{H})) = \sum_{h \in \mathcal{H}} p(c|x, h)p(h|\vec{x}, \vec{c}) \tag{20.1}$$

Here, $p(h|\vec{x}, \vec{c})$ denotes the posterior probability of a model h given the training examples and their class labels. The posterior probability of the class given the test example x and the model h ($p(c|x, h)$) can also be thought of as the degree to which h endorses class c for example x. Equation 20.1 is only true when $\mathcal{H}$ is a finite or enumerably infinite set. If not, one also needs to integrate over all continuous variables ([Buntine90], pg. 99) to obtain the expectation $E(p(c|x, \mathcal{H}))$. In this paper, we will assume that one can get good approximations for the posterior without including the extensions for continuous variables. The good experimental results in section 20.5 suggest that significant accuracy improvements can be obtained despite this assumption.

Now we review how to compute the posterior probability $p(h|\vec{x}, \vec{c})$ of a decision tree. This form will be used in section 20.2.1 to develop the approximation for the posterior probability of a rule-set model. Applying Bayes rule to the expression for posterior probability yields:

$$p(h|\vec{x}, \vec{c}) = \frac{p(\vec{x}, \vec{c}|h)p(h)}{p(\vec{x}, \vec{c})} \propto p(\vec{x}, \vec{c}|h)p(h) \tag{20.2}$$

The $p(\vec{x}, \vec{c})$ term can be ignored because in this paper we are only interested in the *relative* posterior probabilities of the models. The term $p(h)$ refers to the prior probability of the model. The uniform distribution is used if there is no *a priori* reason to favor one model over another. We use the uniform distribution in this paper.

The term $p(\vec{x}, \vec{c}|h)$ is called the likelihood of the evidence (i.e. the training examples $\vec{x}$ together with their class labels $\vec{c}$) given the model. It measures how likely h is to produce the evidence $(\vec{x}, \vec{c})$. For classification, it is used in calculating how likely a model is to generate a set of training examples with their associated class labels.

In order to apply this theory, a way of calculating the likelihood function, $p(\vec{x}, \vec{c}|h)$, from data is needed. In order to do this, Buntine assumes that it is useful to partition the example space into V subsets where each subset should be modeled with its own set of parameters. That is, the assumption is that a good model can be built for each subset. (This is equivalent to assuming that confounding examples of one subset with those from another will cause modeling difficulties.) For classification, we will model each subset with C parameters where C represents the number of classes[2]. The i-th parameter of a subset will denote the probability of generating[3] an example of class i from that subset.

In order to derive a useful form for $p(\vec{x}, \vec{c}|h)$, we note that by assuming that the examples in the training set are independently generated, we can write:

$$p(\vec{x}, \vec{c}|h) = \prod_{i=1}^{N} p(x_i, c_i|h)$$

where N denotes the number of training examples. Furthermore, if we let $n_{j,k}$ denote the number of training examples of class j from the k-th subset (one of the V subsets), and we let $\phi_{j,k}$ denote the probability of generating a single example of class j in the k-th

---

[2]Note that each subset may correspond to an entire class or to some subclass of a class.

[3]Equivalently, the i-th parameter will represent the probability of finding an example of class i in the subset.

210   Kamal Ali and Michael Pazzani

subset, we can write:

$$p(\vec{x}, \vec{c}|h) = \prod_{k=1}^{V} \prod_{j=1}^{C} \phi_{j,k}^{n_{j,k}}$$

(20.3)

One can then show ([Buntine90]) that the contribution to the posterior from the k-th subset can be modeled by:

$$\frac{B_C(n_{1,k} + \alpha, \ldots, n_{C,k} + \alpha)}{B_C(\alpha, \ldots, \alpha)}$$

(20.4)

where B_C is the C-dimensional beta function and α is a parameter[4] which denotes the "weight" in number of examples that should be associated with the prior estimate $(1/C)$ of $\phi_{j,k}$. Putting equations 20.3 and 20.4 together, we get the likelihood:

$$p(\vec{x}, \vec{c}|h) = \prod_{k=1}^{V} \frac{B_C(n_{1,k} + \alpha, \ldots, n_{C,k} + \alpha)}{B_C(\alpha, \ldots, \alpha)}$$

(20.5)

Note that the postulated subclasses appear as leaves for decision trees. Finally, we get an expression for the posterior probability, $p(h|\vec{x}, \vec{c})$ that can be computed from data:

$$p(h|\vec{x}, \vec{c}) \propto p(h) \times \prod_{k=1}^{V} \frac{B_C(n_{1,k} + \alpha, \ldots, n_{C,k} + \alpha)}{B_C(\alpha, \ldots, \alpha)}$$

(20.6)

To summarize the preceding sections: in order to decide which class maximizes

$$E(p(c|x, \mathcal{H})) = \sum_{h \in \mathcal{H}} p(c|x, h)p(h|\vec{x}, \vec{c})$$

one can use equation 20.6 for the term $p(h|\vec{x}, \vec{c})$. For the endorsement term $(p(c|x, h))$, it is preferable to use a Laplace estimate ([Kruskal78], p. 256; $\frac{n_{i,j} + \alpha}{n_{.,j} + C\alpha}$) rather than the maximum likelihood estimate $(\frac{n_{i,j}}{n_{.,j}})$ because the maximum likelihood estimate assigns a value of 1 to a leaf covering just one example of a class and no examples of other classes. For such a leaf, it is not very likely that the true estimate of $p(c|x, h)$ is 1.

20.2.1 *Posterior probability of a rule-set model*

The learning algorithm presented in this paper learns rule-set models. A single rule-set model is a set of class descriptions, one per class. Each class description, in turn, is a set of rules. This section uses the method of deriving the posterior probability of a decision tree to develop a similar method for a rule-set model.

The derivation of the expression for the posterior probability of a tree only depends on assuming that the example space can be partitioned into V disjoint subsets. For trees, if we assume that each leaf of a learned tree models one of those subsets, we can estimate statistics of that subset from the numbers of training examples covered by that leaf. However, the difference between the tree and the rule-set model is that any example in the example space can only belong to one leaf whereas it is possible that rules may be learned in a way such that an example is covered by more than one rule of a class, or by more than one rule of different classes. To avoid this problem *within* a class description,

---

[4]Because we will use a single value of α for all the V subsets, α does not need j, k subscript.

we will assume that some ordering is imposed on the rules within each class description. Furthermore, we will add a default rule to the end of each class description. For the i-th class description, the default rule is $Class_i(...) \leftarrow true$. Thus we have the situation where each class description forms a partitioning of the example space. Because each class forms its own partitioning of the example space, the situation where a test example matches rules of more than one class is also not a problem. For decision trees, the posterior is a function of the way in which the example space is partitioned by the leaves of the decision tree. Analogously, for rule-set models, the posterior of each class description is a function of the way in which the example space is partitioned by the rules of that class description. To compute the posterior of the rule-set model then, we just take the geometric average of the posteriors of the class descriptions.[5] One way of interpreting this approach is that each class description provides an estimate of the posterior using a partitioning of the training data and that we are just forming the average of those estimates.

Let C denote the number of classes in the data, M denote a learned rule-set model, R_i denote the set of rules (including the default rule) for class i after some ordering has been imposed on those rules, $n_{1,ij}$ denote the number of training examples of class i covered by rule j of class i and $n_{2,ij}$ denote the number of training examples of classes other than class i covered by rule j of class i. Then, the posterior probability of a rule-set model is the geometric average of the estimates provided by each class description:

$$p(M|\vec{x}, \vec{c}) \propto p(M) \times \left(\prod_{i=1}^{k} \prod_{ij \in R_i} \frac{B(n_{1,ij} + \alpha, n_{2,ij} + \alpha)}{B(\alpha, \alpha)} \right)^{1/C} \qquad (20.7)$$

Note that we are assuming that if a rule of a class covers a training example of some other class, it does not matter which other class that example belongs to - the loss function for every other class is the same.

In order to use a rule-set model for classification using equation 20.1, one other term must be estimated from data. This is the $p(c|x, h)$ term - the degree to which model h endorses class c for test example x. To compute the endorsement for some class i, we will only consider the rules in the i-th class description of model h. Next, we need to impose an ordering on the rules of the i-th class description. To do this, we define $r_{ij}(x)$ to be a random variable that takes on value **true** if example x satisfied the conditions of rule j (of class i) and takes on value **false** otherwise. The rules of the i-th class description are then ranked in order of most accurate first where the accuracy of a rule is defined as $p(i|r_{ij}(x) = true)$. The training data is used to estimate the accuracy. Let $n_{1,ij}$ denote the number of training examples of class i covered by the j-th rule of class i, and let $n_{2,ij}$ denote the number of training examples of other classes covered by the rule. Then, we use the Laplace estimate of $p(i|r_{ij}(x) = true)$ from the training data:

$$p(i|r_{ij}(x) = true) \approx \frac{n_{1,ij} + 1}{n_{1,ij} + n_{2,ij} + 2}$$

Finally, the endorsement $p(i|x, h)$ is assigned the accuracy of the first (examined in order of most accurate first) rule in the i-th class description of h that is satisfied by the test example x. Because of the default rule and the ordering imposed on rules of a class

---

[5]The geometric average of numbers $x_1 \ldots x_n$ is $(\prod_i x_i)^{1/n}$.

description, it is guaranteed that exactly one rule of each class description will match each test example. Such an ordering is needed firstly to produce a *partition* of the training data as required by the framework of [Buntine90] and secondly to avoid the issue of how to combine evidence from rules (called the "rule overlap problem," [4]).

The foregoing paragraphs illustrate how to compute the posterior probability of a rule-set model and how to estimate it from data. They also explain how to estimate the degree of endorsement. Now we will consider how to *search* for models with high posterior probabilities.

20.2.2 *Bayesian statistics in learning*

As Buntine ([Buntine90]) has observed, if our goal is to find the most probable models, posterior probability should be used as a rule-selection (or tree-selection) metric. During greedy hill-climbing for example, one should prefer adding the decision node or condition to the model which maximizes the posterior probability of the model. As we will be concerned with a learning algorithm that learns a rule by successively specializing the rule (i.e. by greedily adding conditions to the body of the rule), we will add the condition $\mathcal{L}$ which maximizes

$$gain = p(h + \mathcal{L}|\vec{x}, \vec{c}) - p(h|\vec{x}, \vec{c}) \tag{20.8}$$

where $h + \mathcal{L}$ denotes the model with the addition of the condition $\mathcal{L}$ to the rule currently being learned. The difference in posterior probability is referred to as the "gain" and this gain metric is called the Bayes gain metric.

This general principle is instantiated for rule-learning as follows. During learning for class i, there are only two "virtual" classes: training examples of class i are called "positive training examples" and examples of other classes are called "negative training examples." The addition of the condition $\mathcal{L}$ to the rule separates out these two virtual classes into two subsets: T: the examples that match the new rule and, F: those that do not. By observing that this situation is analogous to a binary test at an internal node of a decision tree, we obtain the following for the contribution to the posterior probability by $\mathcal{L}$:

$$gain = p(\mathcal{L}) \times \frac{B(p + \alpha_1, n + \alpha_2)}{B(\alpha_1, \alpha_2)} \times \frac{B(P - p + \alpha_1, N - n + \alpha_2)}{B(\alpha_1, \alpha_2)} \tag{20.9}$$

where p and n denote the numbers of positive and negative training examples covered by the rule and P and N denote the numbers of training examples left uncovered by previous rules. We can disregard the prior probability $(p(\mathcal{L}))$ of the condition $\mathcal{L}$ because we are assuming the uniform prior distribution for all rule-set models. We make this assumption for parsimony rather than because the theory will not permit other kinds of prior distributions.

Note however that equation 20.9 is symmetric: that is, it will assign the same credit to a condition that excludes as many examples as to a condition that includes those same examples. This poses no problem for decision trees, but when learning rules, we are only interested in modeling the T subset: the subset of examples that are included by the rule. Therefore, we modify the definition of gain in equation 20.8 to be 0 if the ratio of positive examples to negative examples decreases as a result of addition of $\mathcal{L}$.

Now we consider how to test the theory presented in this section using the learning algorithm HYDRA ([Ali-Pazzani93]) that learns relational rules for noisy, multiclass tasks.

20.3   HYDRA: Learning a single rule set model

HYDRA uses a separate and conquer control strategy based on FOIL ([Quinlan90]). FOIL only learns on data sets consisting of two classes, one of which must be identified as the "positive" class. FOIL only learns a description for the positive class. In the "separate and conquer strategy" a rule is learned and then the training examples of the positive class which are covered by that rule are taken out of the training set. Subsequent rules are learned using all the negative examples and the remaining positive examples. Learning terminates when no positive examples are left in the training set. To use this algorithm, the user does not need to specify the number of rules to learn or the length of the rules to be learned. The pseudo-code for FOIL is presented in table 20.2. We adapt FOIL to learn for multiclass tasks (tasks with more than two classes) simply by calling FOIL once for each class in the data. Although FOIL uses the information gain metric ([Quinlan90]) to decide which rule condition to add to the rule next, we will discuss FOIL in the context of using the Bayes gain metric.

FOIL begins to learn a rule for $class\text{-}a(X,Y)$, for example, by starting with the rule that has the empty body (the body of this rule is satisfied by any example, positive or negative):

$$class\text{-}a(X,Y) \leftarrow$$

Then, it ranks each possible condition that it might add to the body of the rule by calculating how much the posterior probability of the model would be increased if the empty body were to be replaced by that condition. FOIL adds the rule condition which yields the largest gain. FOIL continues to add rule conditions until the body of the rule is not true for any negative example. Then, it removes positive examples covered by the rule and learns subsequent rules for the remaining training examples. The process terminates when no positive examples are left in the training set.

HYDRA differs from FOIL in learning a description for each class in the data. Another major difference between HYDRA and FOIL is that HYDRA attaches a reliability measure to each rule. For the experiments in this paper, the training data Laplace estimate of the "accuracy" of the rule (subsection 20.2.1) is used as the rule reliability measure. After learning has finished, the rules within each class description are ordered from most to least accurate.

Classification in HYDRA, which learns a single class description per class, proceeds as follows. If a test example satisfies (non-default) rules of just one class then that class is assigned to the test example. If the test example satisfies (non-default) rules from more than one class, HYDRA chooses the class corresponding to the satisfied rule with the highest reliability. Ali & Pazzani ([Ali-Pazzani93]) show that this modification to FOIL makes the system much more accurate in noisy domains. If the test example does not satisfy any non-default rules of any class, it is assigned to the most frequently occurring class (the Laplace accuracy of the default rule associated with the most frequent class will be higher than the associated accuracies of the other default rules).

The Laplace accuracy of a rule covering p positive and n negative training examples is $(p+1)/(p+n+2)$. The advantage of using this rather than the maximum likelihood estimate $(p/(p+n))$ is that it does not assign an accuracy of 1 to a rule covering just a small number of positive examples and no negative examples. The reliability of such rules is quite

TABLE 20.2. Pseudocode for FOIL: the separate and conquer strategy.

```
FOIL(Positive-Examples,Negative-Examples,Metric):
 Let POS be Positive-Examples.
 Until POS is empty do:
    Let NEG be Negative-Examples.
    Set NewClause to the clause with the empty body
    Until NEG is empty do:
          Conquer: (add a condition to clause body)
          Choose a condition L using Metric
          Conjoin L to body of NewClause.
          Remove from NEG examples that dont satisfy NewClause.
    Separate: Remove from POS all positive examples that
            satisfy NewClause.
```

different from that of a rule covering say a hundred positive and no negative examples. Yet, both would have the same maximum likelihood accuracy estimate. Additionally, among rules covering no negative examples, the Laplace estimate rises monotonically with increasing coverage of positive training examples.

20.4   Learning multiple models with HYDRA

Stochastic (randomized) search is used by HYDRA-MM to find multiple models. Ideally, one would want to find the models which have the highest posterior probability but in practice we use stochastic greedy search with respect to posterior probability to try to find probable models. If there are local maxima in the posterior probability space, we may not find the globally most probable model but we will find models that occupy local maxima.

During the process of deciding which rule condition to add to the body of a rule being learned, HYDRA-MM stores all candidates whose gain is within some factor β of the gain of the best candidate. HYDRA-MM then chooses a rule condition stochastically from this set; the probability of a rule condition being chosen is equal to its contribution to posterior probability. Thus, rule conditions that add more to posterior probability have a correspondingly greater probability of being chosen.

Classification in HYDRA-MM: To compute the expectation of a class given a test example and the set of learned rule models, we use equation 20.1, using the posterior probabilities of rule-set models and their degrees of endorsement. The test example is assigned to the class with the highest expectation. Another way of viewing equation 20.1 is that for a given class c, it collects endorsement evidence from each rule-set model. Endorsements of rule-set models with higher posterior probabilities are given correspondingly greater weight.

20.5   Experimental Results

The main goal of these experiments is to verify that learning multiple rule-set models in a way consistent with Bayesian probability theory leads to improvements in classification

TABLE 20.3. Accuracies obtained by learning multiple models. The numbers in the 3rd-to-last column give the significance level (SIG) at which 11 class descriptions (CDs) accuracy differs from that of 1 deterministic description according to the paired 2-tailed t-test (sign test for DNA). NS - not significant, NA - t-test not applicable. KRK 160,20 denotes learning from 160 training examples with 20% artificial class noise on the King-Rook-King task. β is 0.8.

Domain	Deter-ministic	1 CD	2 CDs	5 CDs	11 CDs	SIG.	ties	# Train eg.s
Promoters	65.1	75.0	77.2	85.9	84.8	98	24.0%	105
Lymphography	79.3	80.5	80.5	82.0	82.8	98	21.1%	99
Cancer	71.2	71.5	71.1	71.1	71.1	NS	9.3%	181
KRKP	94.6	94.3	94.8	94.8	94.9	NS	18.7%	200
Document	98.3	97.4	98.4	99.0	99.5	NA	25.2%	220
Students	86.1	85.4	86.9	88.9	90.4	99	7.3%	100
KRK 160,20	92.3	91.6	91.8	91.9	92.0	NS	7.2%	160
KRK 320,20	95.5	94.9	95.3	95.6	95.5	NS	5.8%	320

accuracy when compared to the accuracy of a single rule-set model learned by deterministic hill-climbing using the same training set. Another goal is to try and understand why learning multiple models leads to large gains in accuracy in some domains (e.g. DNA promoters in table 20.3) but more modest gains in other domains. Our hypothesis is that multiple class descriptions help most in domains where there are many, equally good rules to learn. We measure this by the percentage of attempts at adding a rule condition during learning in which more than one candidate condition equaled the highest posterior probability. That is, the criterion tries to measure how often more than one candidate offers exactly the same increase in posterior probability (a gain tie). In such a situation, the greedy deterministic learner is at a disadvantage because it cannot explore both search paths: it must arbitrarily choose one rule condition.

Table 20.3 presents results on the relational problems of learning the concept of "illegality" in the King Rook King chess data set ([Muggleton89]), deciding whether a part of an optically-scanned document contains the date of the document ([Esposito93]) and predicting whether a person is required to make payments on their student loan ([Pazzani-Brunk91]). We also tested HYDRA-MM on 4 attribute-value problems. The accuracies presented in the table are averages of thirty independent trials (except for the Document domain for which 10-fold cross-validation was used). Table 20.3 indicates that learning multiple descriptions improves accuracy by a statistically significantly margin on three of the domains and increases accuracy by a numerically significant margin on the Document domain.[6] No significant loss was incurred on any domain. The table indicates that using multiple models is most helpful on domains in which there are many gain ties.

Recursive class descriptions: In the document task ([Esposito93]) the problem is to determine whether an example representing a document or a part of a document is a "date-block". A "date-block" is that part of a document which contains the date (other parts being the body, the signature-block etc). The goal is automatic classification (using optical character recognition and a learned, relational rule-base) of documents into types of documents such as letters, orders, etc. For this domain, HYDRA and HYDRA-MM

---

[6]The t-test cannot be used for this domain as it assumes that the trials are independent which they certainly are not when doing 10-fold cross validation.

learned recursive rules such as:

$$\text{date-block}(X) \leftarrow \text{height-small}(X), \text{to-the-right}(X,Y), \text{date-block}(Y).$$

During learning, the extensional definition of *date-block* is used. That is, the recursive call to *date-block* is satisfied if the value bound to variable Y is a member of the set (the extensional definition) of positive examples for *date-block*. During classification of test examples, to determine if the recursive call is satisfied, we use the set of rules (the intensional description) learned for *date-block*. One issue that arises when recursive descriptions are mixed with multiple models is whether in order to check if the recursive call succeeds one should check all models of *date-block* or just the current model. That is, if the rule being matched against a test example comes from the i-th model, then to determine if a recursive call succeeds, should one check just rules of the i-th model or those of all models? We chose the former option in order to maintain the independence of the models. We also have to ensure that the matching process of a rule set (containing recursive rules) to an example terminates. Currently, we employ an absolute limit on the number of recursive calls to ensure termination. Table 20.3 shows that using multiple recursive models in the Document domain leads to greater accuracy than using a single recursive model.

20.6 Conclusion

There are four contributions of the work presented here. Firstly, we have presented a form for the posterior probability of a rule-set model in accordance with Bayesian probability theory. Secondly, we have provided some empirical evidence that that form leads to improvements in classification accuracy when compared to the accuracy obtained by the single description learned by deterministic greedy search over the same training set. Stochastic hill-climbing was used to learn more than one model from the same training set. Thirdly, we have characterized an experimental property (gain ties) which indicates when learning of multiple models is useful. Finally, we have demonstrated the learning of multiple models containing recursive rules. The resulting classifications were more accurate than those made by the single recursive description learned through deterministic hill-climbing.

20.7 REFERENCES

[Ali-Pazzani93] Ali K. and Pazzani M. (1993). HYDRA: A Noise-tolerant Relational Concept Learning Algorithm. In *Proceedings of the Thirteenth International Joint Conference on Artificial Intelligence*. Chambery, France: Morgan Kaufmann.

[1] Bergadano F., Giordana A. (1988) A Knowledge Intensive Approach to Concept Induction. In *Proceedings of the Fifth International Conference on Machine Learning.*, Ann Arbor, MA: Morgan Kaufmann.

[Berger85] Berger J. O. (1985). *Statistical Decision Theory and Bayesian Analysis*. Springer-Verlag, New York.

[Buntine90] Buntine W. (1990). *A Theory of Learning Classification Rules*. Doctoral dissertation. School of Computing Science, University of Technology, Sydney, Australia.

[2] De Raedt L. and Bruynooghe M. (1988). On Interactive concept-learning and assimilation. In D. Sleeman (Ed.), *Proceeings of the Third European Working Session on Learning.* (pp. 167-176). Pitman.

[Esposito93] Esposito F., Malerba D. and Semeraro G. (1992). Classification in Noisy Environments Using a Distance Measure Between Structural Symbolic Descriptions. *IEEE Transactions on Pattern Analysis and Machine Intelligence*, 14, 3.

[Gams89] New Measurements Highlight the Importance of Redundant Knowledge. In *European Working Session on Learning (4th : 1989 : Montpeiller, France)*. Pitman.

[Kononenko92] Kononenko I. and Kovacic M. (1992). Learning as Optimization: Stochastic Generation of Multiple Knowledge. In *Machine Learning: Proceedings of the Ninth International Workshop*. Aberdeen, Scotland. Morgan Kaufmann.

[Kruskal78] Kruskal W.H. and Tanur J.M. (1978). *International encyclopedia of statistics*. New York, NY: Free Press.

[Kwok90] Kwok S. and Carter C. (1990). Multiple decision trees. *Uncertainty in Artificial Intelligence, 4*, 327-335.

[Muggleton89] Muggleton S., Bain M., Hayes-Michie J. and Michie D. (1989). An experimental comparison of human and machine-learning formalisms. In *Proceedings of the Sixth International Workshop on Machine Learning*. Ithaca, NY. Morgan Kaufmann.

[3] Muggleton S. and Feng C. (1990). Efficient induction of logic programs. In *Proceedings of the First Conference on Algorithmic Learning Theory*. Tokyo. Ohmsha Press.

[Pazzani-Brunk91] Pazzani M. and Brunk C. (1991). Detecting and correcting errors in rule-based expert systems: an integration of empirical and explanation-based learning. *Knowledge Acquisition, 3*, 157-173.

[Pazzani-Kibler91] Pazzani M. and Kibler D. (1991). The utility of knowledge in inductive learning. *Machine Learning, 9, 1*, 57-94.

[Quinlan90] Quinlan R. (1990). Learning logical definitions from relations. *Machine Learning, 5, 3*.

[4] Segal R. and Etzioni O. (1994). "Learning Decision Lists Using Homogoneous Rules" in *Proceedings of the Twelfth National Conference on Artificial Intelligence*, Seattle, WA: AAAI Press.

[Smyth92] Smyth P. and Goodman R. (1992). Rule Induction Using Information Theory. In G. Piatetsky-Shapiro (ed.) *Knowledge Discovery in Databases*, Menlo Park, CA: AAAI Press, MIT Press.

21
Picking the Best Expert from a Sequence

Ruth Bergman[†] and Ronald L. Rivest[††]

Artificial Intelligence Laboratory[†]
Massachusetts Institute of Technology
Cambridge, MA 02139

Laboratory for Computer Science[††]
Massachusetts Institute of Technology
Cambridge, MA 02139

ABSTRACT We examine the problem of finding a good expert from a sequence of experts. Each expert has an "error rate"; we wish to find an expert with a low error rate. However, each expert's error rate is unknown and can only be estimated by a sequence of experimental trials. Moreover, the distribution of error rates is also unknown. Given a bound on the total number of trials, there is thus a tradeoff between the number of experts examined and the accuracy of estimating their error rates.

We present a new expert-finding algorithm and prove an upper bound on the expected error rate of the expert found. A second approach, based on the sequential ratio test, gives another expert-finding algorithm that is not provably better but which performs better in our empirical studies.

21.1   Introduction

Suppose you are looking for an expert, such as a stock broker. You have limited resources and would like to efficiently find an expert who has a low error rate. There are two issues to face. First, when you meet a candidate expert you are not told his error rate, but can only find this out experimentally. Second, you do not know a priori how low an error rate to aim for. We give here an algorithm to find a good expert given limited resources, and show that the algorithm is efficient in the sense that it finds an expert that is almost as good as the expert you could find if each expert's error rate was stamped on his forehead (given the same resources).

If each expert's error rate were stamped on his forehead then finding a good expert would be easy. Simply examine the experts one at a time and keep the one with the lowest error rate. If you may examine at most n experts you will find the best of these n experts, whose expected error rate we denote by b_n. You cannot do any better than this without examining more experts.

Since experts do not typically come marked with their error rates, you must test each expert to estimate their error rates. We assume that we can generate or access a sequence of independent experimental trials for each expert.

If the number of available experts is finite, you may retain all of them while you test

---

[1] *Learning from Data: AI and Statistics V.* Edited by D. Fisher and H.-J. Lenz. ©1996 Springer-Verlag.

[2] Supported by ARO grant N00014-89-J-1988, NSF grant 9217041-ASC (funded in part by the DARPA HPCC program), NSF grant CCR-9310888, and the Siemens Corporation.

them. In this case the interesting issues are determining which expert to test next (if you cannot test all the experts simultaneously), and determining the best expert given their test results. These issues have been studied in reinforcement learning literature and several interesting algorithms have been developed [Watkins 89, Sutton 90, Sutton 91, Kaelbling 90].

Here we are interested in the case where we may test only one expert at a time. The problems in this case are: (1) what is the error rate of a "good" expert, and (2) how long do we need to test an expert until we are convinced that he is good or bad?

First consider the case that we have a predetermined threshold such that an error rate below this threshold makes the expert "good" (acceptable). This is a well-studied statistical problem. There are numerous statistical tests available to determine if an expert is good; we use the ratio test which is the most powerful among them. The ratio test is presented in section 21.3.1.

However, in our problem formulation we have no prior knowledge of the error rate distribution. We thus do not have an error-rate threshold to define a good expert, and so cannot use the ratio test. The algorithm in section 21.3.2 overcomes this limitation by setting lower and lower thresholds as it encounters better experts. Section 21.3 contains the main result of this paper: our algorithm finds an expert whose error rate is close to the error rate of the best expert you can expect to find given the same resources.

Section 21.4 presents a similar expert-finding algorithm that uses the sequential ratio test [Wald 47] rather than the ratio test. [Wald 47] shows empirically that the sequential ratio test is twice as efficient as the ratio test when the test objects are normally distributed. While the theoretical bound we give for the sequential-ratio expert-finding algorithm is weaker than the bound for the ratio-test expert-finding algorithm, empirical results with specific distributions in section 21.5 indicate that the former algorithm performs better in practice.

21.2   An AI Application: Learning World Models

Consider the problem of learning a world model where rules describe causal relationships of the environment. A rule has the form

$$\text{precondition} \rightarrow \text{action} \rightarrow \text{postcondition}$$

with the meaning that if the preconditions are true in the current state and the action is taken, then the postcondition will be true in the next state. These are predictive rules as in [Drescher 89], as opposed to the prescriptive rules in reinforcement learning [Watkins 89, Holland 85] or operators in Soar [Laird et al. 87].

An algorithm to learn rules uses triples of previous state, S, action, A, and current state to learn. It may isolate a postcondition, P, in the current state, and generate preconditions that explain the postcondition from the previous state and action. For any precondition PC that is true in state S, the rule $PC \rightarrow A \rightarrow P$ has some probability p of predicting incorrectly. To learn a world model, the algorithm must find the rules with low probability of prediction error, and discard rules with high probability of prediction error.

The problem of finding a good rule to describe the environment is thus an expert-finding problem. It fits into the model discussed here since (1) each rule has an unknown error rate, (2) the distribution of rules' error rates is unknown and depends both on the environment and the learning algorithm, and (3) the learning algorithm can generate arbitrarily many rules.

21.3   Finding Good Experts from an Unknown Distribution

First, let us reformulate the expert-finding problem as a problem of finding low error-rate coins from an infinite sequence $c_1, c_2, \ldots$ of coins, where coin c_i has probability r_i of "failure" (tails) and probability $1 - r_i$ of "success" (heads). The r_i's are determined by independent draws from the interval $[0, 1]$, according to some unknown distribution. We want to find a "good" coin, i.e. a coin with small probability r_i of failure (error). We are not given the r_i's, but must estimate them using coin flips (trials).

The main result of this section is:

Theorem 1 *There is an algorithm (algorithm **FindExpert**) such that when the error rates of drawn coins are unknown quantities drawn from an unknown distribution, after t trials, with probability at least $1 - 1/t$, we expect to find a coin whose probability of error is at most $b_{t/\ln^2 t} + O(\frac{1}{\sqrt{\ln t}})$.*

This theorem states that after t trials, we expect the algorithm to find an expert that is almost as good as the best expert in a set of $t/\ln^2 t$ randomly drawn experts (who would have error rate $b_{t/\ln^2 t}$). We note that our result depends in a natural manner on the unknown distribution.

Recall that in t trials if the experts' error rates are known we can find the best of t experts' error rates (b_t). Compared to this, our algorithm must examine fewer experts because it must spend time estimating their error rates. For some distributions (such as for fair coins) $b_{t/\ln^2 t}$ and b_t are equal, while for other distribution they can be quite far apart.

The rest of this section gives the ratio test and our algorithm for finding a good expert.

21.3.1   The Ratio Test

Since we do not know the error rates of the coins when we draw them, we must estimate them by flipping the coins. If we knew that "good" coins have error rate at most p_1, we could use standard statistical tests to determine if a coin's error rate is above or below this threshold. Because it is difficult to test coins that are very close to a threshold, we instead use the ratio test, which tests one hypothesis against another. In this case the hypotheses are that the coin has error rate at most p_0, versus that the coin has error rate at least p_1, where p_0 is a fixed value less than p_1.

The Problem   Given a coin with unknown rate of failure p.

Test if $p \leq p_0$ vs. $p \geq p_1$. Accept if $p \leq p_0$. Reject if $p \geq p_1$.

Requirements   The probability of rejecting a coin does not exceed α if $p \leq p_0$, and the probability of accepting a coin does not exceed β if $p \geq p_1$. [3]

---

[3] We choose the ratio test since it has the most power, i.e., for a given α, i.e. it gives the least β (probability of accepting when the hypothesis H_0 is wrong (see [Rice 88].)

The Test Let m be the number of samples, and f_m be the number of failures in m samples. The ratio test is

$$\begin{aligned} reject \quad & \text{if } f_m \geq (p_0 + \sqrt{\tfrac{\ln 1/\alpha}{2m}})m \\ accept \quad & otherwise \end{aligned}$$

21.3.2   An Algorithm for Finding a Good Expert

We know how to test if a coin is good given a threshold defining a good error rate, but when we do not know the error-rate distribution we can not estimate the lowest error rate b_t that we can expect to achieve in t trials. The following algorithm overcomes this handicap by finding better and better coins and successively lowering the threshold for later coins.

The algorithm for finding a good coin is the following.

Algorithm 1 FindExpert
Input: *t, an upper bound on the number of trials (coin flips) allowed.*

Let BestCoin = Draw a coin.
Flip BestCoin $\ln^3 t$ times to find $\hat{p}$.
Set $p_1 = \hat{p}$.
Repeat until all t trials are used
$\quad$ *Let $p_0 = p_1 - \epsilon(p_1)$, where $\epsilon(p_1) = \sqrt{4/\ln(t)}$.*
$\quad$ *Let Coin = Draw a coin.*
$\quad$ *Test Coin using the ratio test:*
$\quad\quad$ *Flip Coin $m = \ln^2 t$ times.*
$\quad\quad$ *Accept if $f_m < (p_1 - \epsilon(p_1)/2)m$.*
$\quad$ *If the ratio test accepted then*
$\quad\quad$ *Set BestCoin = Coin.*
$\quad\quad$ *Flip BestCoin an additional $\ln^3 t$ times to find an improved $\hat{p}$.*
$\quad\quad$ *Set $p_1 = \hat{p}$.*
Output BestCoin.

The proof that **FindExpert** satisfies the statement of Theorem 1 is too lengthy for this paper. The following is a high level summary of the proof.
Description of the proof: Since the error-rate distribution is unknown, we do not have any estimate of b_t, so the algorithm uses better and better estimates. It starts with a random coin and a good estimate of its error rate. It prepares a test to determine if a new coin is better than the current coin (with high probability). Upon finding such a coin it prepares a stricter test to find a better coin, and so on. We show that *the time to test each coin is short,* and thus *we see many coins.* Since *we almost always keep the better coin* we can *find a coin whose error rate is at most the expected best error rate of the coins that the algorithm saw (plus a small correction).*

21.4   A Faster (?) Test for Experts

A disadvantage of the ratio test in the previous section is that the length of each test is fixed. This length is chosen so as to guarantee (with high probability) a good determination as to whether the tested coin has error rate at least ϵ better than the current best coin. For coins that are much better or much worse, it may be possible to make this determination with many fewer trials.

The sequential ratio test given by [Wald 47] solves precisely this problem. After each coin toss it assesses whether it is sufficiently sure that the tested coin is better or worse than the current best coin. If not, the test continues. The sequential ratio test thus uses a *variable* number of flips to test a coin. One can hope that for same probability of erroneous acceptances and rejections, the sequential ratio test will use fewer coin flips than the ratio test. Although the worst case sample size is larger for the sequential ratio test, [Wald 47] shows that in experiments with normally distributed error rates the sequential test is on average twice as efficient as the ratio test. Section 21.5 gives our experimental results comparing expert-finding algorithms based on the ratio test and on the sequential ratio test.

The rest of this section gives the sequential ratio test and the corresponding expert-finding algorithm.

21.4.1 The Sequential Ratio Test

This section describes the sequential ratio test due to [Wald 47].

The Problem Given a coin with unknown failure rate p.

Test if $p \leq p_0$ vs. $p \geq p_1$. Accept if $p \leq p_0$. Reject if $p \geq p_1$.

Requirements The probability of rejecting a coin does not exceed α if $p \leq p_0$, and the probability of accepting a coin does not exceed β if $p \geq p_1$.

The Test Let m be the number of samples, and f_m be the number of failures in m samples.

Reject if

$$f_m \geq \frac{\log \frac{1-\beta}{\alpha}}{\log \frac{p_1}{p_0} - \log \frac{1-p_1}{1-p_0}} + m \frac{\log \frac{1-p_0}{1-p_1}}{\log \frac{p_1}{p_0} - \log \frac{1-p_1}{1-p_0}}.$$

Accept if

$$f_m \leq \frac{\log \frac{\beta}{1-\alpha}}{\log \frac{p_1}{p_0} - \log \frac{1-p_1}{1-p_0}} + m \frac{\log \frac{1-p_0}{1-p_1}}{\log \frac{p_1}{p_0} - \log \frac{1-p_1}{1-p_0}}.$$

Otherwise, draw another sample.

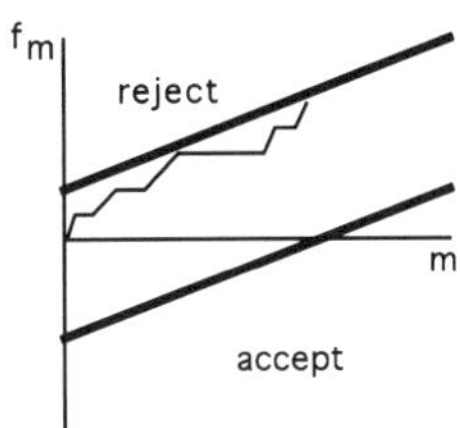

The sequential ratio test defines two lines with different intercepts and the same slope. Above the upper line is a reject region. Below the lower line is the accept region. The test generates a random walk starting at the origin which terminates when it reaches one of the two lines.

21.4.2 Finding a Good Expert Using the Sequential Ratio Test

The algorithm for finding a good coin using the sequential ratio test is as follows.

Algorithm 2 SeqFindExpert*:*
Input: *t, an upper bound on the number of trials allowed.*

Let BestCoin = Draw a coin.
Flip BestCoin $\ln^3 t$ *times to find* $\hat{p}$*.*
Set $p_1 = \hat{p}$*.*
Repeat until all t trials are used:

> *Let* $p_0 = p_1 - \epsilon(p_1)$*, where* $\epsilon(p_1) = \sqrt{\frac{4p_1(1-p_1)}{\log t}}$*.*
> *Let Coin = Draw a coin.*
> *Test Coin using the sequential ratio test*
>   *with parameters* p_0*,* p_1*, and* $\alpha = \beta = 1/t^2$*.*
> *If the sequential test accepts then*
>   *Set BestCoin = Coin.*
>   *Flip BestCoin* $\log^3 t$ *more times to find an improved* $\hat{p}$*.*
>   *Set* $p_1 = \hat{p}$*.*

Output BestCoin.

Because the worst case number of coin flips for the sequential ratio test is larger than the (fixed) number of coin flips for the ratio test, the bound we now prove for **SeqFindExpert** ratio test is not as strong as the bound shown above for **FindExpert**.

Theorem 2 *There is an algorithm (***SeqFindExpert***) such that when the coins are drawn according to an unknown error-rate distribution, after t trials, with probability at least* $1-1/t$*, we expect to find a coin whose probability of error is at most* $b_{t/\log^3 t} + O(\frac{1}{\sqrt{\log t}})$*.*

Theorem 2 shows that algorithm **SeqFindExpert**, which uses the sequential ratio test to find a low error-rate coin from coins drawn according to an unknown distribution, does almost as well as we can do if coins were labeled with their error rates, but see only $t/\log^3 t$ coins. The proof of Theorem 2 is similar to the proof of Theorem 1. The bound in Theorem 2 is not as tight as the bound for the **FindExpert**. In practice, however, **SeqFindExpert** often performs better because the test lengths are much shorter than the worst case test length used to prove Theorem 2.

For some distributions, such as the uniform distribution, the coins tested are typically *much* worse than the current best. (After seeing a few coins the algorithm already has a fairly good coin and most coins are much worse.) Thus, the sequential ratio tests will be short. When the error rates are uniformly distributed we expect that the algorithm **SeqFindExpert** will see more coins and find a better coin than **FindExert**. This argument is confirmed by our empirical results below. Our results also show the superiority of **SeqFindExpert** when the error rates are drawn from a (truncated) normal distribution.

21.5   Empirical Comparison of **FindExpert** and **SeqFindExpert**

	Coins Tested	Test Length	Best Estimated Error Rate	Best Actual Error Rate
FindExpert	7.6	49	.1085	.1088
SeqFindExpert	21	44	.0986	.0979

(a) Uniform distribution; limit of $t = 1000$ trials.

	Coins Tested	Test Length Test Length	Best Estimated Error Rate	Best Actual Error Rate
FindExpert	66	100	.0185	.0187
SeqFindExpert	230	33	.01	.0101

(b) Uniform distribution; limit of $t = 10000$ trials.

TABLE 21.1. Empirical Comparison of **FindExpert** and **SeqFindExpert** with the uniform distribution. The numbers in the tables are averaged over 1000 runs.

To compare the performance of **FindExpert** and **SeqFindExpert** we ran experiments for uniform and normally distributed error rates. (The normal distribution was truncated to lie within the interval $[0, 1]$.) Table 21.1 gives results for both algorithms on the uniform distribution. All results reported are an average over 1000 repeated executions of the algorithm. Table 21.1(a) contains the average of 1000 runs each with trial limit $t = 1000$. Table 21.1(a) shows that the **SeqFindExpert** algorithm had shorter average test lengths and therefore tested more experts. **SeqFindExpert** was able to find experts with lower actual error rate (.0979 on the average compared with .1088 for **FindExpert**). The table contains both the average actual error rate of the best experts that the algorithm found and the average error rate from experiments for the same experts. Table 21.1(b) shows that given more time ($t = 10000$ trials) to find a good expert **SeqFindExpert** performs significantly better than **FindExpert**. The average test length is much shorter and the resulting best error rate is .0101 compared with .0187.

Experiments with the normal distribution used a normal with mean 0.5 and variance 0.3. These results are reported in table 21.2. Note that for this distribution most coins have error rate close to .5. Table 21.2(a) reports the average of 1000 executions with trial limit 1000. It is interesting that the **SeqFindExpert** both tested more experts and had a longer average test length. The long average test is due to a few very long tests (to compare close experts), but most tests are very short. As expected, the average error probabilities of the best coin is lower for the **SeqFindExpert** algorithm. Table 21.2(b) shows that with a longer limit of 10000 trials the **SeqFindExpert** algorithm performs much better than **FindExpert**, giving an average best error rate of .2741 compared with .3352.

The experimental results in this section show that **SeqFindExpert** performs better than **FindExpert** for two distributions with different characteristics. The experimental results agree with the theoretical analysis in that some sequential tests are quite long (longer than the ratio tests), but the experiments also show that on the average the sequential test lengths are short especially when the trial limit is large. The average test

	Coins Tested	Test Length	Best Estimated Error Rate	Best Actual Error Rate
FindExpert	13	49	.4361	.4395
SeqFindExpert	29	61	.4144	.4204

(a) Normal distribution; limit of $t = 1000$ trials.

	Coins Tested	Test Length	Best Estimated Error Rate	Best Actual Error Rate
FindExpert	85	100	.3292	.3352
SeqFindExpert	470	31	.2670	.2741

(b) Normal distribution; limit of $t = 10000$ trials.

TABLE 21.2. Empirical Comparison of **FindExpert** and **SeqFindExpert** with the Normal Distribution (mean 0.5, variance 0.3) truncated at 0 and 1. The numbers in the tables are averaged over 1000 runs.

length is short when the time limit is large because the best expert is already much better than the average population.

21.6   Conclusions

This paper presents two algorithms to find a low error expert from a sequence of experts with unknown error-rate distribution, a problem that arises in many areas, such as the given example of learning a world model consisting of good rules. The two algorithms **FindExpert** and **SeqFindExpert** are nearly identical, but use the ratio test and sequential ratio test respectively to determine if an expert is good.

Theorem 1 shows that **FindExpert** finds an expert which is the best expert of $t/\ln^2 t$, given trial limit t. This result is strong in the sense that it shows only a factor of $\ln^2 t$ loss from testing over the best expert we could find in t trials if we knew the exact error rate of each expert. Theorem 2 gives a weaker bound for **SeqFindExpert**. Empirical results in section 21.5, on the other hand, indicate that **SeqFindExpert** performs better than **FindExpert** in practice (at least for the uniform and normal distributions).

The obvious open question from this work is to prove that **SeqFindExpert** expects to find a lower error-rate expert for general or specific distributions than **FindExpert**.

Acknowledgments: Ruth Bergman and Ronald L. Rivest are supported by ARO grant N00014-89-J-1988, NSF grant 9217041-ASC (funded in part by the DARPA HPCC program), NSF grant CCR-9310888, and the Siemens Corporation.

21.7   REFERENCES

[Drescher 89]   Drescher, G. L. (1989), Made-Up Minds: A Constructivist Approach to Artificial Intelligence, PhD thesis, MIT.

[Holland 85]   Holland, J. H. (1985), Properties of the bucket brigade algorithm, *in* 'First International Conference on Genetic Algorithms and Their Applications', Pittsburg, PA, pp. 1–7.

[Kaelbling 90]   Kaelbling, L. P. (1990), Learning in Embedded Systems, Technical Report TR-90-04, Teleos Research.

[Laird et al. 87] Laird, J. E., Newell, A. and Rosenbloom, P. S. (1978), 'SOAR: An Architecture for General Intelligence', *Artificial Intelligence* **33**, 1–64.

[Rice 88]   Rice, J. A. (1988), *Mathematical Statistics and Data Analysis*, Wadsworth & Brooks/Cole, Pacific Grove, CA.

[Sutton 90]   Sutton, R. S. (1990), First Results with DYNA, an Integrated Architecture for Learning, Planning, and Reacting, *in* 'Proceedings, AAAI-90', Cambridge, Massachusetts.

[Sutton 91]   Sutton, R. S. (1991), Reinforcement Learning Architectures for Animats, *in* 'First International Conference on Simulation of Adaptive Behavior', The MIT Press, Cambridge, MA.

[Wald 47]   Wald, A. (1947), *Sequential Analysis*, John Wiley & Sons, Inc., Chapman & Hall, LTD., London.

[Watkins 89]   Watkins, C. (1989), Learning from Delayed Rewards, PhD thesis, King's College.

22
Hierarchical Clustering of Composite Objects with a Variable Number of Components

A. Ketterlin, P. Gançarski and J.J. Korczak

LSIIT, URA 1871, CNRS
7, rue René Descartes
Université Louis Pasteur
F-67084 Strasbourg Cedex
France

ABSTRACT This paper examines the problem of clustering a sequence of objects that cannot be described with a predefined list of attributes (or variables). In many applications, a fixed list of attributes cannot be determined without substantial pre-processing. An extension of the traditional propositional formalism is thus proposed, which allows objects to be represented as a set of components, i.e. there is no mapping between attributes and values. The algorithm used for clustering is briefly illustrated, and mechanisms to handle sets are described. Some empirical evaluations are also provided to assess the validity of the approach.

22.1   Introduction

The basic process of clustering consists of finding coherent groupings of objects in a data set. Each grouping must exhibit *sufficient* internal cohesiveness, while maintaining enough distance to other members of the partition. This definition of clustering importantly constrains many techniques and algorithms devoted to the automated discovery of clusters [KR93].

The area of statistics concerned with this problem, namely the field of cluster analysis, essentially deals with numeric and/or categorical data, where objects are represented as feature vectors [DLPT82]. This representation is equivalent, in essence, to propositional logic (or an attribute-value formalism). Derived clusters are usually represented as centroids in the instance-space. Centroids are used for classification of new data with the help of a distance measure, or with some implicit probability distribution.

On the other hand, machine learning research on conceptual clustering [MS83], conducted during the last decade, focuses on the representation of instances and concepts; the goal is to derive some effective knowledge, in the form of *generalizations* (or concepts), about the domain under study. *Concept formation* is defined as incremental conceptual clustering [FPL91]: incrementality is the ability to maintain a conceptual structure that is updated after each observation.

Many machine learning clustering algorithms still use propositional logic to describe the data, but recent work enriches the language used to describe concepts discovered through clustering. Some systems use a extended form of propositional logic, where con-

---

cepts are defined by logical formulae [MS83]. Others use *probabilistic concepts* [Fis87], where a concept is defined by a probability distribution on each dimension. Other work extends clustering techniques to higher level languages: KBG [Bis92] deals with first-order logic, and KLUSTER [KM94] uses a KL-ONE-*like* language *that mitigates computational complexity, while still retaining consider representational power.

However, there are domains where numeric data must be considered. In such domains, logic-based formalisms are difficult to apply. In addition to our concern with numeric data, this paper describes an extension of attribute-value formalisms that allows objects to be represented as somewhat *structured*: in this framework an object is built up from any number of *components*. Earlier work on related extensions to propositional formalisms has been termed *structured concept formation* (see[TL91]), since objects exhibit several levels of structural detail.

Section 22.2 briefly reviews a simple clustering algorithm that we extend to cluster over objects represented as sets. This extension is described in Section 22.3. Section 22.4 gives three examples of our clustering algorithm in action.

22.2 Concept Formation

The underlying concept formation algorithm used throughout this paper is COBWEB.[2]. This algorithm forms a hierarchy of clusters from a sequence of objects. Each object is an attribute-value list. Derived clusters (or *concepts*) are represented in a probabilistic form. Attributes are *typed*: the algorithm is able to handle nominal variables as well as numeric ones. In the case of numeric attributes, an underlying normal distribution is assumed, whose parameters are estimated. Therefore, for each numeric attribute, a cluster stores the estimated mean and variance with respect to the covered objects. The variances help define a global predictivity score for one cluster, named Π, which is an average of the inverses of the standard deviations over all attributes.

Each object is incorporated into the cluster hierarchy in turn. When incorporating a new object, a search is performed by hill-climbing through a space of cluster hierarchies. The object is sorted down through the current hierarchy. At each level several operators are tentatively applied to the current partition, some of them having restructuring properties. The best of these operators is definitely applied, and the process restarts one level deeper (see [Fis87, GLF89] for a complete description of the algorithm).

When several distinct partitions are generated, a heuristic called *category utility* is used to select between them. *Category utility* evaluates the global *quality* of a single partition. This evaluation is based on the individual predictivity of each cluster involved. The partition $(C, \{C_1, \ldots, C_K\})$ is evaluated by:

$$\frac{1}{K} \sum_{k=1}^{K} P(C_k)[\Pi(C_k) - \Pi(C)]$$

where each $\Pi(C_k)$ measures the individual predictivity of C_k. This expression is an average of the predictivity-gain when stepping from the cluster C to one of its sub-clusters.

---

[2]In this paper, we consider COBWEBto be the algorithm described in [Fis87], with an extension to deal with numeric attributes developed in CLASSIT(see [GLF89]), but without any other extension from CLASSIT.

The individual predictivity of a cluster is defined as:

$$\Pi(C_k) = \frac{1}{I} \sum_{i=1}^{I} \Pi(A_i, C_k)$$

where $\Pi(A_i, C_k)$ quantifies how the attribute A_i is predictive in C_k (i.e. how precisely values of A_i can be predicted for members of C_k).

In this framework, a learning problem is defined on a set of attributes. Observations must be represented by a value for all (or some) of the attributes. Concepts (or clusters – both terms are used interchangeably in this context) must represent some distribution of values for each attribute of the learning problem. Attributes may be of different types. The definition of a type of attribute must thus include a value-space, a way to represent a distribution of such values, and a predictivity measure for such a distribution. As originally described in [Fis87], COBWEB provides the definitions for nominal (categorical) attributes. CLASSIT [GLF89] provides the definitions for numeric (continuous) attributes. The LA-BYRINTH system [TL91] further extends COBWEB to deal with 'composite' objects by allowing 'structured' attributes. The next section introduces a new type of attribute, called a 'set' attribute, defines the representation of values and distributions for this attribute type, and gives a predictivity evaluation measure for distributions over the values of a set attribute.

22.3   Clustering Sets

22.3.1   Representation of Objects

The aim of this paper is to describe a conceptual clustering system that allows objects to be represented as sets of *components* (i.e. sub-objects). The representation formalism is inspired by propositional logic. In that framework each object is a list of attribute-value pairs. Our system allows a value to be a set of objects (instead of being a single numeric or categorical quantity); all objects of a set are described with the same set of attributes. Members of such a set value are less abstract objects, each of which describes some aspect of the more global (abstract) object. To illustrate this representation scheme consider an example domain of *quadruped mammals* that is found in the literature: each object consists of several sub-objects (cylinders), each of which is described along several, identical numeric attributes. An object of that domain may be written as:

```
Dog1 = [ cyls={[r=13.9,l=25.1], [r=6.5,l=10.0], [r=4.7,l=6.3]} ]
```

In this case the object `Dog1` is described by only one attribute, named `cyls`, whose value is a set containing three sub-objects. Each of these sub-objects is described with the same two numeric attributes (`r` and `l`).[3].

---

[3]In this paper only single attribute domains will be considered. The mechanisms described here apply to one attribute, whose values are sets of objects, but with no loss of generality, the method can be extended to domains described by several attributes, each of them having sets as values

22.3.2   Clustering Strategy

Section 22.2 briefly explained how to build a cluster hierarchy. We noted that this problem can be cast as a problem of quantifying how predictive a cluster is. In the case of the domain described above, the question is: "How can one quantify the predictivity of a set of animals?". The answer to this problem lies in the fact that a separate cluster hierarchy can be built for each level of abstraction. In the animal domain it means that two cluster hierarchies are maintained: one for the cylinders (the component-cluster hierarchy) and one for the animals (the composite-cluster hierarchy). The results of incorporating the cylinders forming an animal are used during the incorporation of the animal. That is, the same process is repeated at each level of structural abstraction. This is a *component-first* strategy, where parts are clustered before the whole [TL91]. The algorithm may be summarized as follows:

1. for each component

   1a. incorporate the component in the component-cluster hierarchy

   1b. update the description of the composite (replace the component by a reference to the cluster it reached)

2. incorporate the composite object into the composite-cluster hierarchy

Each 'incorporate' operation is a call to COBWEB. The first call operates at a lower level (i.e. in the component space) than the last one, which operates in the composite space.

It follows from this sketch that the integration of a composite object (e.g. an animal) is done with each component (e.g. a cylinder) replaced by a reference to a component-cluster. What is thus needed is a way to quantify the predictivity of a set of sets of component-cluster references. This problem divides itself into two distinct phases, explained in the next two sections.

22.3.3   Representation of Clusters

The first step is to find an adequate representation for clusters covering objects whose description includes set-valued attributes. Since members of values (i.e. sub-objects or components) are hierarchically clustered, they can be replaced by their most specific covering cluster: the initial instance space is replaced by a cluster hierarchy (which is a partially ordered space). A conceptual (or intensional) representation of a set of composite objects must rely on a conceptual 'vocabulary' for the components. This language is provided by the component-cluster hierarchy: clusters of components will be used to characterize clusters of composite objects.

The second step is to select an adequate subset of composite clusters. An intuitive way to characterize a set of sets is by way of their mutual intersection (or overlap). However, instead of having simple members, one may apply the same line of reasoning to complex (structured objects). Generalizing a set of sets means finding a set of clusters describing all the sets of components, i.e. for each set under consideration, there exists a mapping from that set to the generalization. Hierarchically-organized clusters of components clearly help to find such a description.

The conceptual representation of a set of composite objects will thus be a set of component-clusters: these clusters are called *central-clusters*, and must have the following properties:

1. Each central cluster must cover at least one member of each value. This means that each central cluster must characterize the intersection between all the sets under consideration.

2. All the members of the sets must be covered by central clusters. This means that central clusters must cover all the elements of all the sets, thus avoiding 'marginal' intersections between only parts of the sets (that would leave some other components uncovered).

3. Central clusters have to be as specific as possible. This means that the set of central clusters is the most 'precise' conceptual characterization of the intersection between all the sets.

It is important to note that in cases where only one composite object is covered, the set of central clusters used to represent the composite cluster is the set of component-clusters covering the components of the object.

22.3.4 *Predictivity Evaluation*

The second step is to evaluate the predictivity of the set of central clusters. Since each central cluster can be found in the component-cluster hierarchy, it is labeled with its individual predictivity (named Π). A straightforward way to compute predictivity for a set of clusters is to average their individual predictivity score. This was the approach taken in our implementation, even though other methods of combination could be explored. However, since all central clusters do not cover the same proportion of objects, the contribution of each central cluster is weighted by the proportion of components it covers. The predictivity of a set of L central clusters $\{\gamma_1, \ldots, \gamma_L\}$ is thus:

$$\sum_{l=1}^{L} \frac{n_o(\gamma_l)}{N_o} \Pi(\gamma_l)$$

where $n_o(\gamma_l)$ is the number of members covered by γ_l, and $N_o = \sum_{l=1}^{L} n_o(\gamma_l)$ is the total number of components. Given the definition of central clusters, $n_o(\gamma_l)$ is guaranteed to be at least equal to the number of covered objects. The weight associated with a central cluster will be called its *coverage*. The overall predictivity is thus a tradeoff between the predictive ability and the coverage of central clusters.

22.3.5 *Incrementality*

The original COBWEB algorithm, which worked with a purely propositional formalism, addressed the problem of concept formation, which is defined as incremental conceptual clustering. Incremental ability is crucial in most applications. It is thus important to check that incrementality is preserved when extending the knowledge representation language.

The problem may be stated as: "Having a composite-cluster description and a new composite object to incorporate, how can one compute the new description of the composite-cluster?" The process can be divided into two phases:

- generalize any central cluster that does not cover at least one of the components of the new object

- if any component of the incoming object remains uncovered by a central cluster, generalize the "nearest" central cluster

This is only a sketch of the procedure. Importantly, any generalization must be performed carefully. A generalization is the replacement of one central-cluster by its parent in the component-cluster hierarchy. This process is equivalent to the application of a "climb-generalization-tree" operator [Mic83], with the property that the generalization tree is itself built and maintained by the system. Nevertheless, this simple procedure ensures that the definition of central clusters is maintained. The two phases correspond to the first two conditions in the definition of central clusters. It has to be noted that the addition of an object to a composite cluster may only generalize the central clusters describing it. This does not imply that the predictivity of that composite cluster decreases, since a loss in predictivity may be compensated by an increase in coverage.

22.3.6 *Complexity*

This section investigates the computational cost of the set-clustering process, and focuses particularly on the integration of a new composite object into an existing composite-cluster.

The problem of structured concept formation has already been addressed by the LABYRINTH system [TL91][4]. The difference between LABYRINTH and the system described in this paper is that LABYRINTH is designed to find a binding between the components and a predefined set of attributes. Once the binding is determined, the task is reduced to composite-object clustering. But this binding process has high computational cost: $O(d!)$ if an exhaustive search is performed (d being the number of components per object, which is fixed in LABYRINTH), $O(d^2)$ or $O(d^3)$ if heuristic solutions are used (see [TL91]). This binding is computed each time an object is added to a cluster. Note that this cost may be penalizing when d is high (see the next section for examples of such situations).

In contrast, our system solves the mapping by finding a mutual intersection between all of the sets. Hence, the mapping process is replaced by the search for central clusters. Let us consider a composite cluster, C, covering N objects, $S_1, \ldots, S_N$. Let $n_i = |S_i|$. It is easy to see that C will be represented by at most $\omega = \max_i\{n_i\}$ central clusters. Hence, the integration of the next composite object in C may lead to at most ω generalizations during the first step of the updating process. The second step may apply only $|S_{N+1}|$ times. The whole updating process is thus linear in the average number of components per composite object.

This notable decrease in complexity may be explained by the fact that instead of considering all possible bindings between the components and a predefined set of attributes, the set-clustering mechanism uses the component-cluster hierarchy to solve the partial-matching problem. Moreover, the binding searched for by LABYRINTH appears as an epiphenomenon: a global structure appears by way of the representation of composite clusters in terms of central clusters, rather than searching for a fixed number of components.

---

[4]LABYRINTH's formalism also includes relational information between components, which are not discussed here.

22.4 Empirical Results

22.4.1 The Simplified Quadruped Mammals Domain

This domain has already been used in the literature on structured concept formation to demonstrate the abilities of the CLASSIT algorithm [GLF89]. It has been simplified here for explanatory purposes. In this domain objects represent some perceptual sketch of an animal: each object is composed of three cylinders (instead of eight in the original domain) representing the head, the torso and one leg of the animal. Each cylinder is described by two (instead of 9 in the original domain) numeric attributes (the radius and length). This domain illustrates a special case of set clustering. Since all the objects have the same number of components (three here), the problem is equivalent to clustering objects wih no knowledge of the binding between attributes and values.

Since these objects are artificially generated from four available models (cat, dog, horse or giraffe),[5] the goal is to discover these classes from the unlabeled data. In our experiment, 20 objects were randomly generated (five from each model), each one being made of three cylinders. The system constructed a cluster hierarchy of animals: it thus built two hierarchies (one for the cylinders and one for the animals). Results are sketched on Figure 1. Note that each terminal class of Figure 1 is, in fact, further developed.

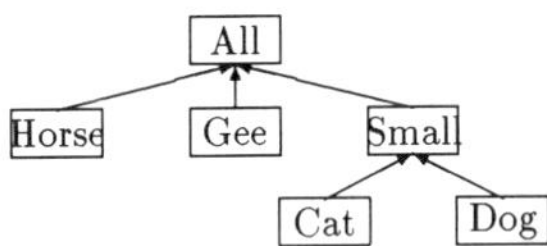

FIGURE 1. Results in the quadruped mammals domain.

The system found the four classes of animals even though all of them are not at the first level. More important is the fact (not shown in Figure 1) that the conceptual representation of each of these classes included three central clusters, corresponding to the three cylinders. The cluster labelled 'Small' covers all small animals (dogs and cats), and its conceptual representation includes only two cylinder clusters. The interpretation of such a conceptual representation is that a small animal is an animal made of two cylinders of one type (described by the first central cluster) and one cylinder of another type. Such a representation could not have been obtained if a fixed structure (i.e., a predetermined number of attributes) had been used.

22.4.2 Object Recognition Domains

Character Recognition

The second experiment involves a simplified form of pattern recognition. The basic idea is to consider a pattern as a set of pixels. To compensate for the loss of information on the spatial arrangement of pixels, the representation of each individual pixel is based on several convolutions.

---

[5]The instance-generator is available at the "UCI Repository of Machine Learning Databases and Domain Theories" at Irvine, at `ftp.ics.uci.edu`, as `/pub/machine-learning-databases/quadrapeds`

Data are drawn from a set of four alphabetic characters (shown on Figure 2), each of them printed in four directions. Each pattern was convolved with the Laplacian of a Gaussian at several different scales. Each pixel shown in black in the original pattern is considered as a component of that pattern. Each component is described by the values of the convolutions at its position. These values can be seen as local characteristics of the pixel in the original pattern. In fact, as shown in [MH80], convolution by the Laplacian of a Gaussian can be used as an edge detector (a value near zero, meaning that the pixel is on an edge in the original image). In the experiments reported here, three convolutions were used with the σ parameter respectively equal to 5/2, 3 and 7/2 (see [MH80] for details about the meaning of this parameter). The convolutions were directly applied to the iconic images.

Note also that the four sets representing one character (one set for each 'direction') contain exactly the same objects. Thus, objects from the same class have the same number of components, which is not true between classes. Data are shown on the left side of Figure 2. The system was thus given 16 composite-objects with a total of 1488 components. Results are shown on the right side of Figure 2.

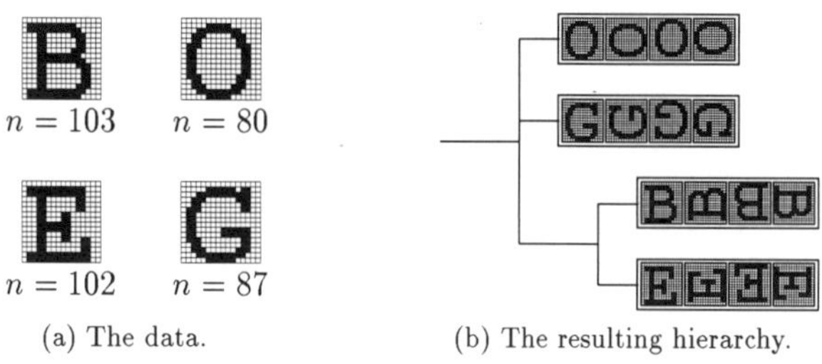

<table>
<tr><td align="center">

B   **O**

$n = 103$ $n = 80$

E   **G**

$n = 102$ $n = 87$

(a) The data.

</td><td align="center">

(b) The resulting hierarchy.

</td></tr>
</table>

FIGURE 2. The character-recognition data-set.

As was expected, the system discovered one cluster for each class of characters. As noted earlier, a fixed structure (i.e. a predefined set of attributes) cannot model such a problem.

Rotated Polygons

The third experiment was similar to the second, but with more realistic rotation angles. The design was as follows: each pattern was computed from an underlying polygon, to which a rotation was applied, whose angle was randomly drawn. The rotated figure was then discretized, leading to a grey-levels icon. The icon was then treated the same way as the alphabetic characters in the previous experiment. Each non-white pixel was considered a component, and local characteristics (i.e. values of the convolution with the Laplacian of a Gaussian) were used to describe it. Parameters of the convolutions were set to the same values as in the previous experiment.

Three 'models' were used to generate the data. Each model was randomly rotated by three different angles, and each time expressed as a set of pixels. Figure 3 shows the data with the angle of rotation and the number of pixels. This domain is an illustration of the most complex representational case in which objects do not have the same number of components, even if they are from the same class.

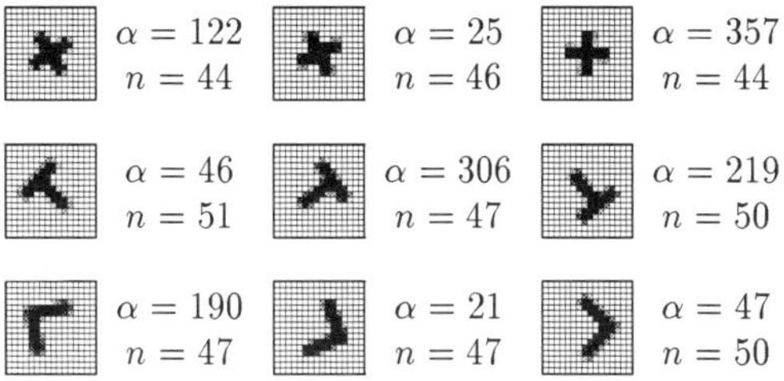

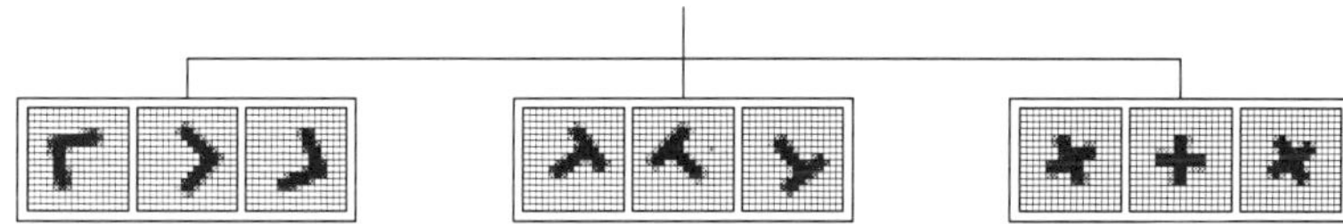

FIGURE 3. Nine rotated polygons.

The result of the clustering is shown on Figure 4. Results are as good as in the naïve case, and suggest that the set clustering algorithm is able to discover rotational-invariant clusters of patterns.

FIGURE 4. Results in the pattern recognition domain.

22.5   Conclusion

The set clustering mechanism described in this paper allows concept formation to take place in domains where a structure can be extracted. Particularly, it resolves the task of concept formation from composite objects. In the case where each set is restricted to one member, and several sets are used to describe each object, the problem reduces to composite-object clustering. In the general case, the overlap between sets is used to form generalizations. Experimental results illustrate the performance of the algorithm and demonstrate abilities that can not be attained by purely propositional algorithms.

The experiments described in the previous section raise some questions about the nature of the discovered clusters, particularly in the case of the pattern-recognition problems. The representation of a pattern (as a set of pixels) is somewhat unusual, and the clusters that are derived are expressed in terms of pixel-classes. It is thus impossible to ask the system to 'illustrate' what a cross is, for instance. But a new cross will be recognized as such. More precisely, a new cross will be recognized as being similar to previous ones. It seems that this is a case of a knowledge structure with no explicit, internal representation of known clusters. The component-cluster hierarchy may be said to reside at a sub-symbolic level (assuming that patterns are at the symbol level).

The work presented in this paper is part of a project on learning and image processing. Earlier work on this project established the adequacy on concept formation techniques for image segmentation [KBK94]. The next step is to study the abilities of the same algorithm on an object recognition problem. Preliminary results are described in this paper. The important point is that the same algorithm is used for both tasks (image segmentation and object recognition), and that this algorithm is totally unsupervised. Future work will likely include the integration of both phases (i.e. object recognition in conjunction with segmentation).

22.6 REFERENCES

[Bis92]   G. Bisson. Conceptual clustering in a first order logic representation. In *Proceedings of the Tenth European Conference on Artificial Intelligence*, pages 458–462. J. Wiley and Sons, 1992.

[DLPT82]   E. Diday, J. Lemaire, J. Pouget, and F. Testu. *Eléments d'Analyse de Données*. Dunod, Paris, 1982.

[Fis87]   D. H. Fisher. Knowledge acquisition via incremental conceptual clustering. *Machine Learning*, 2:139–172, 1987.

[FPL91]   D. H. Fisher, M. J. Pazzani, and P. Langley, editors. *Concept Formation: Knowledge and Experience in Unsupervised Learning*. Morgan Kaufmann, 1991.

[GLF89]   J. H. Gennari, P. Langley, and D. H. Fisher. Models of incremental concept formation. *Artificial Intelligence*, 40:11–61, 1989.

[KBK94]   J. J. Korczak, D. Blamont, and A. Ketterlin. Thematic image segmentation by a concept formation algorithm. In *Proceeding of the European Symposium on Satellite Remote Sensing*, 1994.

[KM94]   J-Ü Kietz and K. Morik. A polynomial approach to the constructive induction of structural knowledge. *Machine Learning*, 14:193–217, 1994.

[KR93]   J. J. Korczak and M. Rymarczyk. Application of classical clustering methods to digital image analysis. Technical report, Université Louis Pasteur, Strasbourg, France, 1993.

[MH80]   D. Marr and E. Hildreth. Theory of edge detection. *Proceedings of the Royal Society of London*, B. 207:187–217, 1980.

[Mic83]   R. S. Michalski. A theory and methodology of inductive learning. In R. S. Michalski, J. G. Carbonell, and T. M. Mitchell, editors, *Machine Learning: An Artificial Intelligence Approach*. Morgan Kaufmann, 1983.

[MS83]   R. S. Michalski and R. E. Stepp. Learning from observation: Conceptual clustering. In R. S. Michalski, J. G. Carbonell, and T. M. Mitchell, editors, *Machine Learning: An Artificial Intelligence Approach*. Morgan Kaufmann, 1983.

[TL91]   K. Thompson and P. Langley. Concept formation in structured domains. In Fisher et al. [FPL91].

23
Searching for Dependencies in Bayesian Classifiers

Michael J. Pazzani

Department of Information and Computer Science
University of California, Irvine
Irvine, CA 92717
pazzani@ics.uci.edu
Phone: (714) 824-5888
Fax: (714) 824-4056

ABSTRACT Naive Bayesian classifiers which make independence assumptions perform remarkably well on some data sets but poorly on others. We explore ways to improve the Bayesian classifier by searching for dependencies among attributes. We propose and evaluate two algorithms for detecting dependencies among attributes and show that the backward sequential elimination and joining algorithm provides the most improvement over the naive Bayesian classifier. The domains on which the most improvement occurs are those domains on which the naive Bayesian classifier is significantly less accurate than a decision tree learner. This suggests that the attributes used in some common databases are not independent conditioned on the class and that the violations of the independence assumption that affect the accuracy of the classifier can be detected from training data.

23.1   Introduction

The Bayesian classifier (Duda & Hart, 1973) is a probabilistic method for classification. It can be used to determine the probability that an example j belongs to class C_i given values of attributes of an example represented as a set of n nominally-valued attribute-value pairs of the form $A_1 = V_{1_j}$:

$$P(C_i | A_1 = V_{1_j} \& \ldots A_n = V_{n_j})$$

If the attributes are independent, this probability is proportional to Equation 23.1

$$P(C_i) \prod_k P(A_k = V_{k_j} | C_i) \tag{23.1}$$

Equation 23.1 is well suited for learning from data, since the probabilities $\hat{P}(C_i)$ and $\hat{P}(A_k = V_{k_j} | C_i)$ may be estimated from the training data. To determine the most likely class of a test example, the probability of each class is computed with Equation 1. A classifier created in this manner is sometimes called a simple (Langley, 1993) or naive (Kononenko, 1990) Bayesian classifier. One important evaluation metric for machine learning methods is the predictive accuracy on unseen examples. This is measured by randomly selecting a subset of the examples in a database to use as training examples and reserving the remainder to be used as test examples. In the case of the simple Bayesian classifier, the training examples are used to estimate probabilities and Equation 23.1 is then used

---

Domain	N	Naive Bayes	ID3
credit	250	.840	.798
horse-colic	200	.810	.783
iris	100	.931	.912
pima-diabetes	500	.755	.723
wine	125	.980	.934
wisc-cancer	500	.973	.952

TABLE 23.1. Accuracy of the naive Bayesian classifier and a simple decision tree learner.

Domain	N	Naive Bayes	ID3
glass	150	.417	.786
krkp	300	.843	.961
mushroom	800	.940	.996
voting	300	.904	.930

TABLE 23.2. Accuracy of the naive Bayesian classifier and decision tree learner on different problems from the UCI archive.

to predict the class of highest estimated probability of each example in the test set. The predictive accuracy is the proportion of agreements between predicted and actual classes.

On many problems, the accuracy of the naive Bayesian classifier is equal to or greater than that of more sophisticated machine learning algorithms. For example, Table 23.1 compares the naive Bayesian classifier to ID3, a simple decision tree algorithm (Quinlan, 1986) on several problems from the UCI archive of machine learning databases (Murphy & Aha, 1995). On each problem, both algorithms were run 24 times on the same training sets and tested on the same disjoint test sets consisting of all examples not in the training set. The number of examples in the training set is given in the second column, the mean accuracy of the Bayesian classifier in the third, and the mean accuracy of the decision tree learner in the fourth column. On each problem, the Bayesian classifier is significantly more accurate at the .05 level using a paired two-tailed t-test. However, on other problems, the reverse pattern is observed (see Table 23.2). On each problem, the Bayesian classifier is significantly less accurate than the decision tree at the .05 level using a paired two-tailed t-test and in one case (glass) the difference in accuracy is greater than 30%.

One possible explanation for the poor performance of the Bayesian classifier on this last set of problems is that the major assumption of the classifier, that the attributes are independent within each class does not hold. In this paper, we address this issue by searching for dependencies among pairs of attributes. For example, if there are three independent attributes, the product of Equation 23.1 would be as follows:

$$\hat{P}(A_1 = V_{1_j}|C_i)\hat{P}(A_2 = V_{2_j}|C_i)\hat{P}(A_3 = V_{3_j}|C_i)\hat{P}(C_i) \qquad (23.2)$$

However, if it is known that A_1 and A_3 are not conditionally independent on class membership[2], and A_2 is not relevant in this problem, then Equation 23.3 should be used:

$$\hat{P}(A_1 = V_{1_j}\&A_3 = V_{3_j}|C_i)\hat{P}(C_i) \qquad (23.3)$$

---

[2]In this paper, we will use "conditionally independent" to mean independent conditioned on class membership.

We say that the two attributes A_1 and A_3 are *joined* if Equation 23.3 is used instead of Equation 23.2. A more accurate classifier can result from joining the correct groups of attributes. Joining A_1 and A_3 and eliminating A_2 would be useful if the function to be learned were $A_1 \oplus A_3$. For example, consider classifying an example in which A_1, A_2 and A_3 were equal to 1. In this case, $\hat{P}(A_1 = 1|Class = True)$, $\hat{P}(A_2 = 1|Class = True)$, $\hat{P}(A_3 = 1|Class = True)$, $\hat{P}(A_1 = 1|Class = False)$, $\hat{P}(A_2 = 1|Class = False)$, $\hat{P}(A_3 = 1|Class = False)$, $\hat{P}(Class = True)$, and $\hat{P}(Class = False)$ and would all be close to 0.5 if estimated from from randomly selected examples, and the classifier that assumed independence (i.e., using Equation 23.2) would not be accurate at predicting whether an example is positive or negative. However, if Equation 23.3 were used, the classifier would be accurate since $\hat{P}(A_1 = 1 \& A_3 = 1|Class = True)$ would be 1 and $\hat{P}(A_1 = 1 \& A_3 = 1|Class = False)$ would be 0.

Joining is an operation that creates a new compound attribute that replaces the original two attributes in the classifier. The possible values of the new attribute are all possible combinations of the the the values of the original attributes. For example, if there are two attributes **height** (with values **tall** and **short**) and **weight** (with values **heavy** and **light**), joining these two original attributes will create a new attribute called **weight_height** (with values **tall_heavy**, **tall_light**, **short_heavy** and **short_light**). In order to allow joining on continuously-valued attributes, we discretize them into k equal intervals. We have used a value of 5 for k on all domain and have not experimented with other possible values. Note that joining two attributes differs from conjoining them as is commonly used in some induction systems (e.g., Schlimmer, 1987; Ragavan & Rendell, 1993) because the compound attribute formed by joining is not a Boolean attribute.

Formally, two attributes A and B are independent within each class C_i if for all values V_j of A and V_k of B, $P(A = V_j \& B = V_k|C_i) = P(A = V_j|C_i)P(B = V_k|C_i)$. In practice, when these probabilities are estimated from training data, these quantities will rarely be equal. One approach to deal with this problem is to assume that attributes are conditionally independent unless there is a significant difference according to some statistical criteria (Kononenko, 1991). Here, we consider an alternative approach that does not directly address the question of conditional independence. Instead, we ask, for the purposes of maximizing predictive accuracy, whether it is better to join two attributes. If conditionally independent attributes are joined and the probabilities of joined attributes are estimated from training data, a less accurate classifier may result because the probability estimates of the joined attributes are less reliable than the estimates of individual attributes. This occurs because joined attributes have more values and as a consequence, on average there are fewer examples for each value when compared to the same attributes unjoined.

In this paper, we explore two alternative methods of joining attributes while learning Bayesian classifiers: Forward Sequential Selection and Joining (FSSJ) and Backward Sequential Elimination and Joining (BSEJ). Both methods are related to approaches that have been used in other learning methods for selecting relevant features (Kittler, 1986; John, Kohavi, Pfleger, 1994).

23.2   Selecting and Joining Attributes

Both of the algorithms for joining attributes that we explore have a similar control structure. Each algorithm maintains a set of attributes (which is a subset of the attributes used to describe the examples) to be used by the classifier and a description of which attributes have been joined. Each algorithm estimates the accuracy of a Bayesian classifier by leave-one-out cross-validation on the training data using the attributes joined as indicated. Leave-one-out evaluation is used because it allows a single Bayesian classifier to be constructed on the entire training set. To classify a training example, the contribution of that example to the probability estimates is subtracted out (Langley, 1993). Each algorithm considers a set of possible operations (such as joining two attributes) and selects the operation that most improves the accuracy of the classifier as measured by leave-one-out cross validation. The two algorithms differ in how they create an initial Bayesian classifier and the operators they use to improve upon the classifier. Both algorithms use what John, Kohavi and Pfleger (1994) call the wrapper model because attribute joining operates without knowledge of how the Bayesian classifier operates. Instead, the attribute joining algorithms are only concerned with the accuracy of the resulting classifiers, as determined by leave-one-out cross-validation on the training data.

23.2.1 Forward Sequential Selection and Joining

The forward sequential selection and joining (FSSJ) algorithm initializes the set of attributes to be used by the Bayesian classifier to the empty set. A Bayesian classifier with no attributes simply classifies all examples to the most frequent class that occurs in the training data. Next, two operators are used to generate new classifiers:

a. Consider adding each attribute not used by the current classifier as a new attribute conditionally independent of all other attributes used in the classifier.

b. Consider joining each attribute not used by the current classifier with each attribute currently used by the classifier.

At each step in the classifier, every addition and every joining of an unused attribute with a used attribute is considered and evaluated using leave-one-out on the training data. If no change makes an improvement, the current classifier is returned. Otherwise, the change that makes the most improvement is retained and the process of modifying the classifier is repeated. Note that by repeated applications of the second operator, more than two attributes may be joined. However, the original attributes cannot be joined separately with other attributes.

The forward selection and joining algorithm differs from forward sequential selection (Kittler, 1986; John, Kohavi and Pfleger, 1994; Moore and Lee, 1994; Caruana and Freitag, 1994) in that forward sequential selection and joining has the ability to join attributes, while forward sequential selection can only select subsets of attributes. Selecting a subset of the original attributes is useful in some classifiers such as nearest neighbor that have problems with irrelevant attributes. Irrelevant attributes do not tend to be a major problem for Bayesian classifiers since, for an irrelevant attribute A, $\hat{P}(A = V|C_i) = \hat{P}(A = V|C_j) = \hat{P}(A = V)$. However, when there are dependencies among attributes, it might be better to delete one attribute than assume that two attributes are conditionally independent. Langley and Sage (1994) have used forward sequential selection with some promising results. In Section 23.3, we will present experimental evidence showing that there is additional benefit in joining attributes.

In the worst case, $O(A^3)$ Bayesian classifiers are constructed and evaluated (where A is the number of attributes) using FSSJ, since at most A steps (either adding or joining) occur, and the worst step requires considering joining O(A) unused attributes to O(A) used attributes.

23.2.2   Backward Sequential Elimination and Joining

The backward sequential elimination and joining (BSEJ) algorithm initially creates a Bayesian classifier treating all attributes as conditionally independent. It uses two operators for considering new hypotheses:

a. Consider replacing each pair of attributes used by the classifier with a new attribute that joins the pair of attributes.

b. Consider deleting each attribute used by the classifier.

Like the FSSJ algorithm, the BSEJ algorithm considers all one step modifications of the algorithm, evaluates these using leave-one-out cross validation on the training data, and permanently makes the change with the greatest improvement. If no change results in an improvement, the current classifier is returned. Otherwise, changes to the modified classifier are considered. By repeated application of the joining operator, several attributes may be joined. Like FSSJ, in the worst case, $O(A^3)$ Bayesian classifiers are evaluated using BSEJ.

23.3   Experimental Results

We ran experiments comparing the algorithms for selecting and joining attributes to the naive Bayesian classifier. These experiments were run on the same domains and in the same manner as the previous experiments. For FSSJ and BSEJ, we also recorded the number of attributes selected for use by the classifier and the number of joins that were made. The results are shown in Table 3.

The first four domains are domains on which a decision tree learner is substantially more accurate than the naive Bayesian classifier. Forward sequential selection and joining of attributes significantly ($p < .05$ using a paired, two-tailed t-test) increases the accuracy on three of the domains (as indicated by a + following the accuracy) but significantly decreases the accuracy on one (as indicated by a − following the accuracy) when compared to the naive Bayesian classifier. The last six domains are ones in which the naive Bayesian classifier is more accurate than a decision tree learner. Forward sequential selection and joining of attributes has no significant effect on four of these domains, and significantly decreases accuracy on two.

Backward sequential elimination and joining of attributes improves the accuracy of the Bayesian classifier on all four problems when the Bayesian classifier is less accurate than the decision tree, and has no significant effect on the six problems on which the Bayesian classifier is more accurate than the decision tree. In addition, on the glass problem, there is a large amount of improvement. On this problem, both FSSJ and BSEJ create classifiers that use approximately three of the nine available attributes. By comparing the total number of attributes (second column) to the number of attributes used after BSEJ (eight

Domain	Atts	Naive Acc	FSSJ Acc	FSSJ Atts	FSSJ Joins	BSEJ Acc	BSEJ Atts	BSEJ Joins
glass	9	.417	.765+	3.0	2.0	.768+	3.2	1.8
krkp	36	.843	.793–	4.4	1.4	.931+	33.5	6.0
mushroom	22	.940	.984+	2.3	1.0	.992+	21.0	3.9
voting	16	.904	.949+	2.7	.5	.920+	13.1	2.6
credit	15	.840	.836	5.0	2.7	.833	14.3	4.5
horse-colic	21	.810	.809	5.5	2.3	.802	20.1	6.5
iris	4	.931	.942	2.1	.5	.938	3.8	1.0
pima	8	.755	.740	3.7	2.0	.749	7.7	4.0
wine	13	.980	.954–	3.3	1.9	.975	12.8	0.8
wisc-cancer	9	.973	.959–	3.5	1.7	.971	8.6	0.8

TABLE 23.3. A comparison of the attribute joining algorithms on ten databases from the UCI Archive.

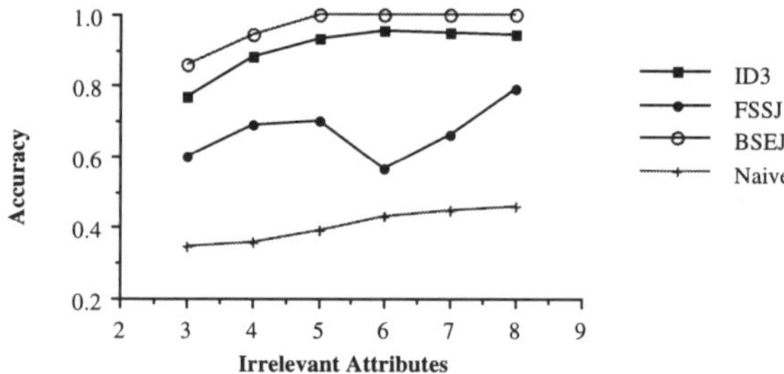

FIGURE 1. The accuracy of four learning algorithms at learning exclusive-or with irrelevant attributes as a function of the number of irrelevant attributes when training on 50% of the data and testing on the remaining 50%.

column), one can see that few attributes are eliminated by BSEJ. In contrast, FSSJ eliminates many more attributes than BSEJ resulting a much simpler classifier. On most problems, this simplicity makes little difference in accuracy. However, on some problems (e.g., krkp– Rook King Pawn chess end games), using fewer attributes and making fewer joins results in a classifier that is less accurate than BSEJ.

The naive Bayesian classifier is limited in expressiveness in that it can only learn linearly separable functions (Langley & Sage, 1994; Rachlin, Kasif, Salzberg & Aha, 1994). Therefore, even with many training examples, it does not approach 100% accuracy on some problems. By joining attributes, BSEJ and FSSJ have the potential to overcome this limitation. To demonstrate how joining attributes allows a Bayesian classifier to learn non-linearly separable functions, we consider learning *exclusive-or* of two attributes, a function that is very difficult for a naive Bayesian classifier. We tested *exclusive-or* functions with 3, 4, 5, 6, 7, and 8 irrelevant attributes included in the example description. For each function, we ran 20 trials in which 50% of the examples were randomly selected for training and the remaining 50% were used to test the accuracy. On each training set, we tested ID3, the naive Bayesian classifier, the Bayesian classifier with Backward Sequential Elimination and Joining, and the Bayesian classifier with Forward Sequential Selection and Joining. Figure 2 shows the mean accuracy of the four algorithms, plotted as a function of the number of irrelevant attributes. Paired t-tests at the .001 level indicate

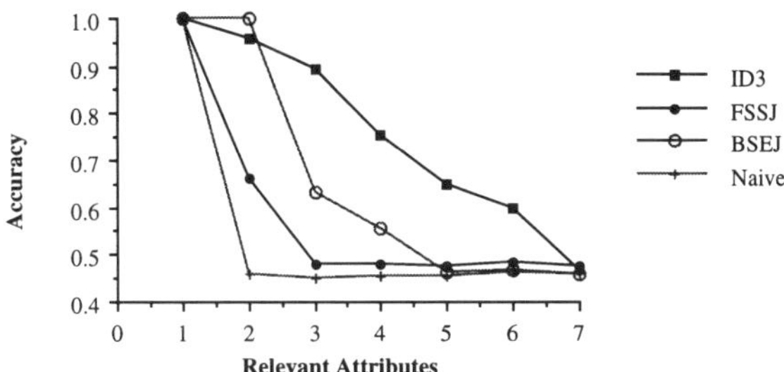

FIGURE 2. The accuracy of four learning algorithms at learning parity functions as a function of the number of relevant attributes. For each function, there were a total of 10 attributes and 1024 total examples. The algorithm is trained on 50% of the data and tested on the remaining 50%.

that BSEJ is more accurate than the other algorithms on this problem. This particular problem illustrates a potential weakness of the FSSJ algorithm which does particularly poorly on this problem. In order for FSSJ to consider joining two attributes, it must first construct a classifier in which one of the attributes is used independently. However, on exclusive-or, the Bayesian classifier with just one relevant attribute would not perform better than simply guessing the most frequent class and the correct relevant attributes are rarely joined. In contrast, BSEJ starts with the naive Bayesian classifier of all attributes and considers all pairs of joins. Since it always considers joining the two relevant attributes, it usually makes this join (especially on larger data sets) and consequently has a very high accuracy.

A potential weakness of the algorithm for joining attributes is that in order to join more than two attributes, it must first join two of the attributes and later join other attributes with the original two. This will not occur unless forming the first pair results in an increase in accuracy. This problem is common to all algorithms that use hill-climbing search. In a second experiment on artificial data, we evaluate the ability of the four algorithms to learn various parity functions. A parity function is true iff an odd number of the attributes have a value of true. We varied the number of relevant attributes from 1 to 7 and included irrelevant attributes so that each example had a total of 10 attributes. We ran 20 trials of each function, choosing 512 examples for training and 512 examples for testing. The results displayed in Figure 3 show that the Bayesian classifier with BSEJ is more accurate than the naive Bayesian classifier when there are fewer than 5 relevant attributes in the parity function.

When there are between 3 and 6 relevant features, ID3 is significantly more accurate than the Bayesian classifier with BSEJ. This shows that although BSEJ only considers joining pairs of attributes, it can improve the accuracy of the Bayesian classifier when there are more than two interacting attributes. However, there is still room for improvement since ID3 is more accurate. Although it would be possible to consider joining three (or more attributes), the computational complexity makes it impractical for most databases.

Domain	Naive Bayes	BSEJ	FSS
glass	.417	.768+	.737
krkp	.843	.931+	.739
mushroom	.940	.992+	.984
voting	.904	.920	.956+
credit	.840	.833	.830
horse-colic	.810	.802	.811
iris	.931	.938	.941
pima-diabetes	.755	.749	.741
wine	.980	.975+	.957
wisc-cancer	.973	.971+	.956
x-or	.453	.921+	.444
Parity	.360	.556+	.446

TABLE 23.4. Accuracy of two methods for improving the naive Bayesian classifier: BSEJ proposed in this paper, and FSS proposed by Langley & Sage (1994).

23.4   Related Work

Kononenko (1991) proposes a method for identifying that attributes are conditionally independent based upon a statistical test that determines the probability that two attributes are not independent. The algorithm joins two attributes if there is a greater than 0.5 probability that the attributes are not independent. Experimental results with this method were disappointing. On two domains, the modified Bayesian classifier had the same accuracy as the naive Bayesian classifier, and on the other two domains tested, the modified Bayesian classifier was one percent more accurate; but, it is not clear whether this difference is statistically significant.

Langley & Sage (1994) have used a forward sequential selection method for eliminating attributes from Bayesian classifiers. In Table 4, we compare BSEJ and the forward sequential selection (FSS) method used by Langley and Sage on the 10 domains from the UCI archive, plus one exclusive-or function (with 4 irrelevant attributes) and one parity function (with 4 relevant and 6 irrelevant features). We ran this experiment in the exact same manner as described in Section 1 and performed paired two-tailed t-tests comparing the accuracy of BSEJ and FSS. Table 4 indicates when one of the methods for improving upon the naive Bayesian classifier is significantly better than another at the .05 level by placing a "+" after the accuracy. The BSEJ method is more accurate than FSS on 7 problems, and the FSS is more accurate on one (voting). Although Langley and Sage advocate using FSS method for dealing with correlated attributes, the experimental results indicate that joining a pair of correlated attributes is more useful than ignoring one of a pair of correlated attributes.

Langley (1993) proposes the use of "recursive Bayesian classifiers" to address the non-linearly separable problem of Bayesian classifiers. Although the algorithm worked on an artificial problem, it did not provide a significant benefit on any naturally occurring databases.

Bayesian networks (Pearl, 1988) offer a more complex way of representing attribute dependencies. Although algorithms exist for learning Bayesian networks from data (Cooper & Herskovits, 1992), it has not yet been demonstrated that inducing Bayesian networks from training data results in more accurate classifiers than naive Bayesian classifiers.

23.5   Limitations and Future work

There are several limitations of the backward sequential elimination and joining method we have proposed. First, unlike the naive Bayesian classifier, it is not an incremental algorithm. Second, for efficiency reasons, it only considers joining pairs of attributes. Although it can join more than two attributes, this must be done in multiple steps. Finally, the algorithm requires symbolic data only, so numeric domains must be discretized, losing some information.

The Bayesian classifier is a natural choice for minimizing misclassification costs. The probability than an example belongs to each class is returned by the Bayesian classifier and this information, together with information on the cost of various errors can be used to choose the class with the least expected cost of error. Pazzani, Merz, Murphy, Ali, Hume, and Brunk (1994) evaluate several algorithms for reducing the misclassification cost of learning algorithms and show that the Bayesian classifier performs well at reducing misclassification costs when it is as accurate as other learners. Since we have improved upon the accuracy of the Bayesian classifier, it is possible that the BSEJ algorithm will provide additional benefits in reducing misclassification costs.

23.6   Conclusions

We have shown that, when learning Bayesian classifiers, searching for dependencies among attributes results in significant increases in accuracy. We proposed and evaluated two algorithms and shown that the backward sequential elimination and joining algorithm provides the most improvement. The domains on which the most improvement occurs are those domains on which the naive Bayesian classifier is significantly less accurate than a decision tree learner. This suggests that the attributes used in some common databases are not conditionally independent and that the violations of the conditional independence assumption that affect the accuracy of the classifier can be detected from training data.

Acknowledgements: The research reported here was supported in part by NSF grant IRI-9310413 and ARPA grant F49620-92-J-0430 monitored by AFOSR. I'd like to thank Pedro Domingos, Dennis Kibler, Pat Langley and Kamal Ali for advice on Bayesian classifiers, Cullen Schaffer for advice on cross-validation and David Schulenburg and Clifford Brunk for comments on an earlier draft of this paper.

23.7   References

Almuallim, H., and Dietterich, T. G. (1991). Learning with many irrelevant features. In *Ninth National Conference on Artificial Intelligence*, 547-552. MIT Press.

Caruana, R., and Freitag, D. (1994). Greedy attribute selection. In Cohen, W., and Hirsh, H., eds., *Machine Learning: Proceedings of the Eleventh International Conference*. Morgan Kaufmann

Cooper, G., and Herskovits, E. (1992). A Bayesian method for the induction of probabilistic networks from data. *Machine Learning*, 9, 309-347.

Danyluk, A., and Provost, F. (1993). Small disjuncts in action: Learning to diagnose errors in the telephone network local loop. *Machine Learning Conference*, pp 81-88.

Duda, R., and Hart, P. (1973). *Pattern classification and scene analysis.* New York: John Wiley & Sons.

John, G. Kohavi, R., and Pfleger, K. (1994). Irrelevant Features and the subset selection problem *Proceedings of the Eleventh International Conference on Machine Learning.* New Brunswick, NJ.

Kittler, J. (1986). Feature selection and extraction. In Young & Fu, (eds.), *Handbook of pattern recognition and image processing.* New York: Academic Press.

Kononenko, I. (1990). Comparison of inductive and naive Bayesian learning approaches to automatic knowledge acquisition. In B. Wielinga (Eds..), *Current trends in knowledge acquisition.* Amsterdam: IOS Press.

Kononenko, I. (1991). Semi-naive Bayesian classifier. *Proceedings of the Sixth European Working Session on Learning.* (pp. 206-219). Porto, Portugal: Pittman.

Langley, P. (1993). Induction of recursive Bayesian classifiers. *Proceedings of the 1993 European Conference on Machine Learning.* (pp. 153-164). Vienna: Springer-Verlag.

Langley, P., and Sage, S. (1994). Induction of selective Bayesian classifiers. *Proceedings of the Tenth Conference on Uncertainty in Artificial Intelligence.* Seattle, WA

Moore, A. W., and Lee, M. S. (1994). Efficient algorithms for minimizing cross validation error. In Cohen, W. W., and Hirsh, H., eds., Machine Learning: *Proceedings of the Eleventh International Conference.* Morgan Kaufmann.

Murphy, P. M., and Aha, D. W. (1995). UCI Repository of machine learning databases. Irvine: University of California, Department of Information & Computer Science. Machine-readable data repository `ftp://ics.uci.edu:/pub/machine-learning-databases`.

Pazzani, M., Merz, C., Murphy, P., Ali, K., Hume, T., and Brunk, C. (1994). Reducing Misclassification Costs. *Proceedings of the Eleventh International Conference on Machine Learning.* New Brunswick, NJ.

Pearl, J. (1988). *Probabilistic reasoning in intelligent systems: Networks of plausible inference.* San Mateo, CA: Morgan Kaufmann.

Quinlan, J.R. (1986). Induction of decision trees. *Machine Learning, 1*, 81-106.

Rachlin, Kasif, Salzberg, and Aha, (1994). Towards a better understanding of memory-based reasoning systems. *Proceedings of the Eleventh International Conference on Machine Learning.* New Brunswick, NJ.

Ragavan, H., and Rendell, L. (1993). Lookahead feature construction for learning hard concepts. *Machine Learning: Proceedings of the Tenth International Conference.* Morgan Kaufmann

Schlimmer, J. (1987). Incremental adjustment of representations for learning. *Machine Learning: Proceedings of the Fourth International Workshop.* Morgan Kaufmann

Schaffer, C. (1994). A conservation law of generalization performance. *Proceedings of the Eleventh International Conference on Machine Learning.* New Brunswick, NJ.

Part V
General Learning Issues

24
Statistical Analysis of Complex Systems in Biomedicine

Harry B. Burke, M.D., Ph.D.

New York Medical College
Department of Medicine
Valhalla, NY 10595
914-347-2428 (voice)
914-347-2419 (fax)
burke@unr.edu

ABSTRACT
The future explanatory power in biomedicine will be at the molecular-genetic level of analysis (rather than the epidemiologic-demographic or anatomic-cellular levels). This is the level of complex systems. Complex systems are characterized by nonlinearity and complex interactions. It is difficult for traditional statistical methods to capture complex systems because traditional methods attempt to find the model that best fits the statistician's understanding of the phenomenon; complex systems are difficult to understand and therefore difficult to fit with a simple model. Artificial neural networks are nonparametric regression models. They can capture any phenomena, to any degree of accuracy (depending on the adequacy of the data and the power of the predictors), without prior knowledge of the phenomena. Further, artificial neural networks can be represented, not only as formulae, but also as graphical models. Graphical models can increase analytic power and flexibility. Artificial neural networks are a powerful method for capturing complex phenomena, but their use requires a paradigm shift, from exploratory analysis of the data to exploratory analysis of the model.

24.1 Introduction

In the past, most biomedical phenomena were analyzed at the demographic-epidemiologic or anatomic-cellular levels. Since phenomena at these levels is largely linear or nearly linear, traditional statistical models were very helpful. One result of these analyses is that, today, most biomedical variables are linear or nearly linear variables. But the future will not be like the past. The future explanatory power in biomedicine is at the molecular-genetic level of analysis. This level is characterized by complex systems, i.e., nonmonotonicity and complex interactions. Complex systems are difficult for traditional statistical models to capture because traditional methods require a priori information about the variables in order to represent the variables in the model. Thus, the traditional statistician must "explore" the data, and must explicitly model what is discovered. But exploration and explicit modeling is not always practical at the molecular-genetic level, where there can be twenty or more variables, and where the variables may interact in three-way and higher combinations.

There is evidence that cancer is a complex system and that future prognostic factors will be nonmonotonic and exhibit complex interactions. Cancer is primarily a genetic disease (Fearon [Fearon90]; Fishel et al. [Fisher93]; Leach et al. [Leach93]) and a complex system. Cancer genes do not act in isolation; oncogenes, suppressor genes, and genetic mutations cause cancer through the complex interaction of the genes and their products (Papadopoulos et al. [Papadopoulos94]; Steel [Steel93]). A cascade of genetic abnormalities is required to produce a cancer (Knudson [Knudson85]; Fearon and Vogelstein [Fearon90]). Thus, it cannot be assumed (1) that a gene or its product will be monotonic or that it will have an independent prognostic value before it is combined with other genes and/or their products, (2) that gene interactions are binary, or (3) that there will only be a few simple genetic interactions. Furthermore, it will probably not be possible to specify in advance of the analysis which complex genetic interactions exist. The need to capture nonmonotonicity and complex interactions exists because the prognostic value of the genetic changes and their products can depend on their nonmonotonic characteristics and interactions (Fearon and Vogelstein [Fearon90]).

24.2   Artificial Neural Networks

Artificial neural networks are a class of nonlinear regression and discrimination statistical methods, and they are of proven value in many areas of medicine (Westenskow et al. [Westenskow92]; Tourassi et al [Tourassi93]; Leong and Jabri [Leong92]; Palombo [Palombo92]; Gabor and Seyal [Gabor92]; Goldberg et al. [Goldberg92]; O'Leary et al. [O'Leary92]; Dawson et al. [Dawson91]; Wu et al. [Wu93]; Astin and Wilding [Astin92]; Weinstein et al. [Weinstein92]). In medical research, the most commonly used artificial neural networks are *feed-forward* networks of simple computing units that use backpropagation training. A feed-forward network is usually composed of three interconnected layers of nodes (computing unit): an input layer, a hidden layer, and an output layer. In our case, each input node corresponds to a patient variable. All nodes after the input layer sum the inputs to them and use a transfer function (also known as an activation function) when they send the information to the adjacent layer nodes. The transfer function is usually a sigmoid function such as the logistic. The connections between the nodes have adjustable weights that specify the extent to which the output of one node will be reflected in the activity of the adjacent layer nodes. These weights, along with the connections among the nodes, determine the output of the network.

Backpropagation consists of fitting the parameters (weights) of the model by a criterion function, usually square error or maximum likelihood, using a gradient optimization method. In feed-forward networks using backpropagation, the error between actual and expected network output units is propagated back from the output through the connections between nodes, in order to adjust the connection weights in the direction of minimum error.

The mathematical representation of a feed-forward network as described above can be viewed as a series of regression equations within a regression equation, where there can be as many regression equations as is necessary to fit the phenomenon. Thus,

$$h_j = f(w_{j1}^h x_1 + w_{j2}^h x_2 + \cdots + w_{jn}^h x_n) \tag{24.1}$$

$$o_k = g(w_{k1}^0 h_1 + w_{k2}^0 h_2 + \cdots + w_{kn}^0 h_n) \tag{24.2}$$

equation (24.1) specifies the output of each of the hidden nodes j, f is a nonlinear transfer function, w_{ji}^h is the weight from input unit i to hidden node j, and x_i is the value of an input variable. Equation (24.2) specifies the output or prediction, o_k, of the network, where g is a nonlinear transfer function, w_{kj}^0 is the weight of the connection from hidden unit j to output node k, and h_j is the hidden node output. It should be noted that equation (24.2), without equation (24.1) input, is equivalent to logistic regression, where g is the logistic function, w is the beta coefficient, and h is the x covariate. Artificial neural networks with sufficient hidden units can approximate any continuous function to any degree of accuracy (Hornik et al. [Hornik89]; Leshno et al. [Leshno93]).

24.3   Clinical Example

We have compared the prognostic accuracy of the TNM staging system (Beahrs et al, 1992) and an artificial neural network according to five year cancer-specific survival.

Data: We have used three data sets (each of approximately 5,000 cases) in these analyses: two from the American College of Surgeons, the Patient Care Evaluation (PCE) breast cancer and colorectal cancer data sets; the National Cancer Institute's Surveillance, Epidemiology, and End Results (SEER) breast cancer data set; and the Mayo Clinic prostate cancer data set. The variables in the PCE, SEER, and Mayo data sets are either binary or monotonic. The factors were selected in the past for collection because they were significant in a generalized linear model, e.g., logistic regression. There is no predictive model that can improve upon a generalized linear model when the predictor variables meet the assumptions of the model and there are no interactions.

Accuracy: There are three components to predictive accuracy: the quality of the data, the predictive power of the prognostic factors, and the prognostic method's ability to capture the power of the prognostic factors. This work focuses on the third component. Comparative accuracy is assessed by the area under the receiver operating characteristic (ROC) curve (Hanley and McNeil [Hanley82]). The receiver operating characteristic area varies from zero to one. When the prognostic score is unrelated to survival, the score is .5, indicating chance accuracy. The farther the score is from .5 the stronger the prediction model. Specifically, the TNM staging system's predictive accuracy is determined by comparing (using the area under the ROC curve) its prediction for each individual patient, where the prediction is the fraction of all the patients in that stage who survive, to each patient's true outcome.

Model: The artificial neural network results reported in this paper are based on back-propagation training which uses the maximum likelihood criterion function and the gradient descent optimization method and the "NevProp" software implementation for training. Significant differences in the receiver operating characteristic areas between the TNM staging system and the artificial neural network are tested following Hanley and McNeil [Hanley82]. The training data set (approximately 3,500 cases) is divided into training (approximately 2,000 cases) and stop-training (approximately 1,500 cases) subsets. Training is stopped when accuracy starts to decline on the stop-training data subset. All analyses employ the same training and testing (validation) data sets, and all results are based on the one time use of the testing data sets.

Results: A comparison of the accuracy of the TNM staging system and the artificial neural network (Table 24.1) using the PCE breast cancer data set, which examines breast cancer-specific five-year survival accuracy for only the TNM variables, demonstrates that

TABLE 24.1. Comparison of TNM staging system and artificial neural networks (all comparisons are for five year, cancer-specific survival).

DATA SETS	TNM	ANN
PCE 1983 Breast Cancer - TNM v	.720	.770
PCE 1983 Breast Cancer - 54 v	.720	.784
SEER 1977 Breast Cancer - TNM v, 10 yr	.692	.730
PCE 1983 Colorectal Cancer - TNM v	.737	.815
PCE 1983 Colorectal Cancer - 87 v	.737	.869
Mayo Clinic Prostate Cancer	.563	.811

the artificial neural network's predictions are significantly more accurate (TNM .720 vs. ANN .770, $p < .001$). Adding 51 commonly collected variables to the TNM variables further increases the accuracy of the artificial neural network (.784). Extending these results to the SEER breast cancer data set, with a breast cancer-specific ten year survival endpoint; using only the TNM variables, the artificial neural network's prognostic accuracy is significantly greater than the TNM staging system (TNM .692 vs. ANN .730, $p < .01$).

Using the PCE colorectal data sets, the predictive accuracy of the two methods can be compared. For only the TNM variables, the artificial neural network's prognostic accuracy is significantly greater than the TNM stage model in predicting colorectal-specific five year survival (TNM .737 vs. ANN .815, $p < .001$). Adding 84 commonly collected factors to the TNM variables further increases the accuracy of the artificial neural network (.869).

The Mayo Clinic data set demonstrates that, for prostate cancers represented in its corpus, the TNM staging system has a low prognostic accuracy (.563), and that the artificial neural network, with other commonly collected variables, is significantly more accurate (TNM .563 vs ANN .811, $p < .001$).

24.4   The Future of Artificial Neural Networks

To demonstrate the power of the artificial neural network to capture unanticipated non-monotonicities and complex interactions, a constructed nonmonotonic variable is added to the 54 PCE breast cancer variables. The constructed nonmonotonic variable consists of two normal distributions centered at zero, one having a standard deviation of 1 for patients who are alive at five years, the other having a standard deviation of 10 for patients who are dead by five years. If the artificial neural network cannot capture nonmonotonicity without a priori specification of the phenomena, then its accuracy should remain at .770 with the TNM variables and .784 with the 54 variables, on the test set. The artificial neural network does capture the predictive power of the nonmonotonic factor, and its accuracy increases to .948 with the TNM variables and to .961 with the 54 variables, on the test set (Table 24.2)

A constructed complex three-way interaction is added to the 54 PCE breast cancer variables. The artificial neural network captures the informative three-way interaction, from among the 29,260 possible three-way interactions, its accuracy increases from .784 to .942 on the test data set. It is the case that anticipated nonmonotonicity has, with varying degrees of success, been modeled by classical prediction models. Although it is

TABLE 24.2. PCE 1983 Breast Cancer Data: 5 year Survival Prediction Accuracy, nonmonotonic variable added or three-way interaction added.

| PREDICTION MODEL | nonmonotonic variable | | three way |
	TNM variables accuracy*	55 variable accuracy*	57 variable accuracy*
pTNM Stages	.720	.720	.720
Stepwise Logistic Regression	.762	.776	.776
Backpropagation ANN	.948	.961	.942

* The area under the curve of the receiver operating characteristic.

computationally intensive, classical prediction models can test for a predictive three-way interaction among 29,260 possibilities, but it is not clear how they would discover four-way and higher interactions of nonlinear variables. It can be concluded that artificial neural networks are powerful models; they can capture the explanatory power inherent in complex systems.

At the present time, using variables selected by traditional statistical methods, it is not required that an artificial neural network be more accurate than traditional statistical methods in order for it to be an appropriate statistical method for cancer prediction. Artificial neural networks can be recommended for cancer prediction because: (1) they are as accurate as the best traditional statistical methods (results not presented), (2) they are able to capture complex phenomena (e.g., nonmonotonicity and complex interactions) without a priori knowledge, and (3) they are a general regression method, therefore, if the phenomenon is not complex, so that accuracy can be maintained using a simpler model, artificial neural networks can be reduced to simpler models resulting in simpler representations.

There are several possible objections to artificial neural networks, including: (1) they require an analysis of the model. They capture phenomena without requiring prior exploration of the data, but they require exploration of the model. More will be said regarding model analysis later in the paper. (2) Some believe that artificial neural networks are overparameterized because they can have a large number of weights. Overfitting can be prevented by keeping the weights small, thereby reducing the effective number of degrees of freedom. This can be accomplished by penalizing large weights, or stopping the itera-tive fitting algorithm before the weights have grown to their full size. It is often the case that, when one of these methods is used, predictive accuracy is better than it would be if we used a smaller model and fit the data without restriction. When a method is used that reduces the weights that are not being increased by the input variables, the weights to the hidden layer shrink, and when there are only linear relationships present, as the hidden layer weights approach zero the neural network approximates a generalized linear model.

(3) It is thought that artificial neural networks are less "transparent" (the importance of the variables is less obvious), than traditional statistical models. This view of transparency fundamentally misunderstands the situation. Artificial neural networks are as complex as is necessary to capture the phenomenon. Generally, if the phenomenon is complex, the model must be complex. If the phenomenon is simple enough to be captured by

simple models, then artificial neural networks can be reduced to a simple model, and the importance of covariates is easily observed. For example, if the phenomenon is linear, then a two layer (no hidden layer) artificial neural network with linear transfer functions, is mathematically identical to linear regression, and the weights of the artificial neural network are identical to the beta coefficients of the linear regression model. Therefore, model transparency (i.e., ease of variable interpretation) is properly understood as a function of complexity and accuracy. For a simple phenomena, a properly chosen simple model is easily interpretable. For a complex phenomenon (e.g., complex interactions) and a properly chosen model, increases in model complexity result in increases in accuracy if overfitting is avoided. Increases in model complexity reduce the transparency of both traditional statistical models and artificial neural network statistical models.

24.5 Domain Knowledge and Model Knowledge

Domain knowledge is information regarding the phenomena acquired by examining the data, and model knowledge is information regarding the phenomena acquired by examining the empirically derived model. Both are required for understanding phenomena, but their relative importance in the overall analysis can differ. In traditional statistics domain knowledge is the dominant approach. The statistician talks to the researcher, who suggests where the statistician can explore the data. The statistician then examines the data, finds the best fitting model, and performs inferential calculations to determine variable significance and importance. But this is not the only possible approach to understanding phenomena. Selection of the best (most accurate predictions) model can be based on either knowing the relationships between the predictors and the relationship of the predictors to the phenomenon, and selecting the model that best captures these relationships, or, if the relationships are not known a priori, selecting a model that is capable of capturing any relationships. This latter approach, selecting the model that can capture any phenomena, is very different from the traditional approach. It requires that the model be explored rather than the data. The relationships are captured in the model, and one decomposes the model in order to discover the phenomena.

It should be noted that, in terms of models, there is no difference between prediction and classification. Prediction and classification differ in the questions being asked, i.e., the character of the data and the type of outcome. Thus robotic control, from a model-theoretic perspective, does not differ from cancer-outcome prediction: vision is classification and movement is prediction.

There are some aspects of the analysis of a phenomena that are domain specific, and some that are model specific. For example, the number of hidden layer nodes (i.e., sub-regression equations) is domain specific. The number of hidden layer units cannot be determined, a priori, by any analytic method because the number of units depends on the complexity of the phenomena; a simple phenomena requires no hidden units, a complex may require five or ten hidden units.

An example of model analysis is the determination of whether the phenomenon exhibits nonmonotonicities or complex interactions. The approach is to compare the results of a two-layer neural network with those of a three-layer neural network. If the two-layer is significantly less accurate than the three-layer neural network, then there are nonmonotonic relationships, interactions, or both. If there is no difference between the two models, then the model can be simplified. If there is a difference, then a complex (three-layer) model must be used to capture the nonlinearities and interactions.

For simple phenomena, e.g., phenomena that do not require the use of a hidden layer in an artificial neural network, artificial neural networks are as transparent as other statistical models. For complex models sensitivity analysis can determine the contribution of input variables in the artificial neural network prediction (Intrator [Intrator93]). But sensitivity analysis is not adequate because complex relationships, represented by complex mathematical equations, are not easily understood. To understand these complex relationships visual models are needed. Buntine [Buntine94] points out "Graphical operations manipulate the underlying structure of a problem unhindered by the fine detail of the connecting functional and distributional equations. This structuring process is important in the same way that a high-level programming language leads to higher productivity over assembly language."(p 160) The ability to represent and manipulate artificial neural networks, in terms of graphical models, provides power and flexibility in model analysis.

24.6 REFERENCES

[Astin92] Astin ML, & Wilding P. (1992) Application of neural networks to the interpretation of laboratory data in cancer diagnosis. Clin Chem 1992;38:34-38.

[Beahrs92] Beahrs OH, Henson DE, Hutter RVP, & Kennedy BJ. (1992) Manual for staging of cancer, 4th ed. Philadelphia: JB Lippincott, 1992.

[Buntine94] Buntine WL. (1994) Operations for learning with graphical models. J Art Intell Res 1994;2:159-225.

[Dawson91] Dawson AE, Austin RE, & Weinberg DS. (1991) Nuclear grading of breast carcinoma by image analysis. J Clin Pahol 1991;95(Suppl):S29-S37.

[Fearon90] Fearon ER, & Vogelstein B. (1990) A genetic model for colorectal cancer. Cell 1990;61:759-67.

[Fisher93] Fishel R, Lescoe MK, & Rao MRS et al. (1993) The human mutator gene homolog MSH2 and its association with hereditary nonpolyposis colon cancer. Cell 1993;75:1027-36.

[Gabor92] Gabor AJ, & Seyal M. (1992) Automated interictal EEG spike detection using artificial neural networks. Electroencephalogr Clin Neurophysiol 1992;83:271-80.

[Goldberg92] Goldberg V, Manduca A, & Ewert DL. (1992) Improvement in specificity of ultrasonography for diagnosis of breast tumors by means of artificial intelligence. Med Phys 1992;19:1275-81.

[Hanley82] Hanley JA, & McNeil BJ. (1982) The meaning of the use of the area under the receiver operating characteristic (ROC) curve. Radiology 1982; 143:29-36.

[Hornik89] Hornik K, Stinchcombe M, & White H. Multilayer feedforward networks are universal approximators. Neural Networks 1989;2:359-66.

258   Harry B. Burke

[Intrator93] Intrator O, Intrator N. (1993) Neural networks for interpretation of nonlinear models. Proceedings of the Statistical Computing Section, American Statistical Society, San Francisco CA; 1993:244-9.

[Knudson85] Knudson AG Jr. (1985) Hereditary cancer, oncogenes, and antioncogenes. Cancer Res 1985; 45: 1437-43.

[Leach93] Leach FS, Nicolaides NC, & Papadopoulos N et al. (1993) Mutations of a mutS homolog in hereditary nonpolyposis colorectal cancer. Cell 1993;76:1215-25.

[Leong92] Leong PH, & Jabri MA. (1992) MATIC - an intracardiac tachycardia classification system. PACE 1992;15:1317-31.

[Leshno93] Leshno M, Lin VY, Pinkus A, & Schocken S. (1993) Multilayer feedforward networks with a nonpolynomial activation function can approximate any function. Neural Networks 1993;6:861-67.

[O'Leary92] O'Leary TJ, Mikel UV, & Becker RL. (1992) Computer-assisted image interpretation: use of a neural network to differentiate tubular carcinoma from sclerosing adenosis. Modern Pathol 1992;5:402-5.

[Palombo92] Palombo SR. (1992) Connectivity and condensation in dreaming. J Am Psychoanal Assoc 1992;40:1139-59.

[Papadopoulos94] Papadopoulos N, Nicolaides NC, Wei Y, & et al. Mutation of a mutL homolog in hereditary colon cancer. Science 1994; 263:1625-29.

[Steel93] Steel M. (1993) Cancer genes: complexes and complexities. Lancet 1993;342:754-5.

[Tourassi93] Tourassi GD, Floyd CE, Sostman HD, & Coleman RE. (1993) Acute pulmonary embolism: artificial neural network approach for diagnosis. Radiology 1993;189:555-58.

[Weinstein92] Weinstein JN, Kohn KW, Grever MR, Viswanadham VN, Rubenstein LV, & Monks AP. (1992) Neural computing in cancer drug development: predicting mechanism of action. Science 1992;258:447-51.

[Westenskow92] Westenskow DR, Orr JA, & Simon FH. (1992) Intelligent alarms reduce anesthesiologist's response time to critical faults. Anesthesiology 1992;77:1074-9.

[Wu93] Wu Y, Giger ML, Doi K, Vyborny CJ, Schmidt RA, & Metz CE. (1993) Artificial neural networks in mammography: application to decision making in the diagnosis of breast cancer. Radiology 1993;187:81-87.

25
Learning in Hybrid Noise Environments Using Statistical Queries

Scott E. Decatur

Aiken Computation Laboratory
Division of Applied Sciences
Harvard University
Cambridge, MA 02138

ABSTRACT We consider formal models of learning from noisy data. Specifically, we focus on learning in the *probability approximately correct* model as defined by Valiant. Two of the most widely studied models of noise in this setting have been *classification noise* and *malicious errors*. However, a more realistic model combining the two types of noise has not been formalized. We define a learning environment based on a natural combination of these two noise models. We first show that hypothesis testing is possible in this model. We next describe a simple technique for learning in this model, and then describe a more powerful technique based on statistical query learning. We show that the noise tolerance of this improved technique is roughly optimal with respect to the desired learning accuracy and that it provides a smooth tradeoff between the tolerable amounts of the two types of noise. Finally, we show that statistical query simulation yields learning algorithms for other combinations of noise models, thus demonstrating that statistical query specification truly captures the generic fault tolerance of a learning algorithm.

25.1 Introduction

An important goal of research in machine learning is to determine which tasks can be automated, and for those which can, to determine their information and computation requirements. One way to answer these questions is through the development and investigation of formal models of machine learning which capture the task of learning under plausible assumptions.

In this work, we consider the formal model of learning from examples called "probably approximately correct" (PAC) learning as defined by Valiant [Val84]. In this setting, a learner attempts to approximate an unknown target concept simply by viewing positive and negative examples of the concept. An adversary chooses, from some specified function class, a hidden $\{0,1\}$-valued target function defined over some specified domain of examples and chooses a probability distribution over this domain. The goal of the learner is to output in both polynomial time and with high probability, an hypothesis which is "close" to the target function with respect to the distribution of examples. The learner gains information about the target function and distribution by interacting with an example oracle. At each request by the learner, this oracle draws an example randomly according to the hidden distribution, labels it according to the hidden target function, and returns the labelled example to the learner. A class of functions $\mathcal{F}$ is said to be PAC learnable if

---

there exists an algorithm which works for every target function in $\mathcal{F}$.

Whereas previous models required the learner to exactly determine the hidden concept but allowed the learner to use unbounded time, the PAC model requires the learner to work in efficient time, yet only requires the hypothesis returned by the learner to be "close" to the target concept. These differences found in the PAC model seem to better reflect the requirements of learning in the real world. The PAC model has been widely adopted and there has been extensive research in providing algorithms and showing hardness results for this model (see [Ang92] for a survey).

However, one criticism of the PAC model is that the data used for learning is assumed to be noise free. In order to combat this deficiency, variations of PAC learning have been introduced which formalize the types of noise that might occur in a real training environment. Two of the most widely studied models of noise in computational learning theory have been *classification noise* [AL88] and *malicious errors* [Val85]. The classification noise model allows for common random occurrences (approaching 50% the time) of mislabelled examples, while the malicious error model allows for rare occurrences of adversarial corruption of the entire labelled example.

Although many algorithms have been constructed to learn in the presence of each type of noise separately, a more realistic model combining classification noise and malicious errors has not been formalized. In this paper, we define a learning environment based on a natural combination of these two noise types called *classification and malicious* (CAM) error. In this hybrid model, occasionally labelled examples are entirely corrupted, while the remaining examples have some fixed probability of being randomly misclassified. Learning in this noise model is at least as difficult as learning in each separate noise model.

We first show that one can perform hypothesis testing in the presence of CAM error, *i.e.* use only noisy data to detect which hypothesis from a set of hypotheses has the smallest error on *noise-free* data. We then show how one can use existing techniques to construct algorithms for learning in this noise model. Specifically, we show how to take an algorithm which tolerates classification noise and add to it the ability to tolerate malicious errors. We then characterize the limits of this technique in terms of the amount of noise tolerance and note that optimal noise tolerance cannot be achieved by these methods.

We next show a different technique for generating CAM-tolerant learning algorithms which tolerate strictly more noise. This technique is based on simulating *statistical query* (SQ) algorithms. In the SQ model of learning [Kea93], the learner may no longer view labelled examples, but instead may ask for estimates of the values of various statistics based on the distribution of labelled examples. Most classes which have PAC algorithms also have SQ algorithms and these SQ algorithms can be easily derived from their corresponding PAC algorithms [Kea93]. Thus, we show that all of these classes are learnable in the presence of simultaneous classification noise and malicious error.

Our technique provides a smooth tradeoff between the amount of tolerable classification noise and malicious error. We also show that for *any* SQ algorithm, the classification noise tolerance achieved is optimal while the malicious error tolerance is within a logarithmic factor of optimal with respect to the desired accuracy of learning.

Finally, we describe how the definition of a hybrid noise environment can be extended to combine various other types of noise. Learnability in these new hybrid noise models may also be achieved through the simulation of statistical query algorithms. The generality of the usefulness of statistical query specification demonstrates its ability to capture the fault tolerance intrinsic to a learning problem.

25.2   Learning Models

In Valiant's PAC model of learning from labelled examples [Val84], an adversary selects both the hidden target $\{0,1\}$-valued function f from a known class of functions $\mathcal{F}$ and the hidden probability distribution D which is defined over the domain of f. The domain of f constitutes the set of possible examples. This set is often the Boolean hypercube $\{0,1\}^n$, in which case n is the common length of all examples. The learner is given access to an example oracle $EX(f, D)$ which when polled by the learner returns $\langle x, f(x) \rangle$, an example x drawn randomly according to D and its correct labelling with respect to f. The learner is also given accuracy parameter $\varepsilon \in (0,1)$ and confidence parameter $\delta \in (0,1)$. In time polynomial in n, $1/\varepsilon$, and $1/\delta$, the learner must output an hypothesis h which, with probability at least $1 - \delta$, has the following property: given an example x drawn randomly according to D, the probability that $f(x) \neq h(x)$ is at most ε. If h has this property, we say that it is ε-close to f on D.

We next describe two variants of PAC learning which model noise in the learning process. These learning models, in addition to the statistical query model described below, all differ from the standard PAC model in which oracle they use to interact with f and D. Yet, all models still require the learner to output, with probability at least $1 - \delta$, an hypothesis h which is ε-close to f on D, *i.e.* with respect to noise-free labelled examples.

Angluin and Laird [AL88] introduced the model of PAC learning with *classification noise* in which the learner has access to a noisy example oracle $EX_{\text{CN}}^{\eta}(f, D)$. When a labelled example is requested from this oracle, an example is chosen according to distribution D, and returned. With probability $1 - \eta$, the correct labelling of the example according f is returned, while with probability η, the incorrect classification is returned. The learner is given η_b, an upper bound on the noise rate, such that $0 \leq \eta \leq \eta_b < 1/2$. The running time of a learning algorithm is allowed to be polynomial in $\frac{1}{1/2 - \eta_b}$ in addition to the usual parameters. We say that a class is learnable with classification noise if it is learnable for some constant classification noise rate $\eta > 0$.

The PAC model with *malicious errors* was introduced by Valiant [Val85] and studied further by Kearns and Li [KL88]. In this model, the learner has access to an example oracle $EX_{\text{MAL}}^{\beta}(f, D)$. When a labelled example is requested from this oracle, with probability $1 - \beta$, an example is chosen according to distribution D, correctly labelled according to the hidden target concept f, and returned to the learner. However, with probability β, a malicious adversary selects any example it chooses and labels it either positive or negative. Kearns and Li [KL88] showed that for "distinct" classes (virtually all interesting classes are distinct), it is impossible to tolerate a malicious error rate of $\beta \geq \frac{\varepsilon}{1+\varepsilon}$.

The statistical query (SQ) model [Kea93], a further variant of the noise-free PAC model, has been a useful tool in the construction of *noise-tolerant* PAC algorithms. In the SQ model, the PAC example oracle $EX(f, D)$ is replaced by a statistics oracle $STAT(f, D)$. The learner interacts with $STAT(f, D)$ by asking it queries of the form $[\chi, \tau]$ where χ is a $\{0,1\}$-valued function on *labelled* examples and $\tau \in (0,1)$ is the *tolerance* of the query. The query is a request for the value P_χ, the probability that $\chi(x, l) = 1$ where $\langle x, l \rangle$ is a labelled example drawn randomly from $EX(f, D)$. Thus, P_χ is the probability of drawing a labelled example that has "property" χ. The statistics oracle returns an approximation

$\hat{P}_\chi$ such that $|\hat{P}_\chi - P_\chi| \leq \tau$.

The *query space* Q of an SQ algorithm is the set of all possible queries χ which the algorithm asks the statistics oracle on all possible runs. The *tolerance* of an SQ algorithm is the lower bound on the tolerances of the queries the algorithm makes to the oracle. A class is said to be SQ learnable if: (1) there exists an SQ algorithm which makes only a polynomial number of queries, (2) there exists a polynomial bound on the time required to evaluate every χ used by the algorithm, and (3) there exists a polynomial bound on the inverse of the tolerance of the algorithm.

Kearns [Kea93] has shown that if a class has an SQ algorithm, then it is learnable with any amount of classification noise $\eta < 1/2$. Decatur [Dec93] has shown that if a class has an SQ algorithm with tolerance τ, then it is learnable with malicious error $\beta = \Theta(\tau)$.[2] Both results are based on the simulation of SQ algorithms in the respective noise models. Despite the noise in the examples being used, the simulations are able to effectively reproduce a noise-free statistics oracle with high probability.

25.3   Learning with Classification Noise *and* Malicious Errors

We formally define *classification and malicious* error (CAM) as a PAC variant using an example oracle with the following behavior:

$$EX_{\text{CAM}}^{\eta,\beta}(f,D) = \begin{cases} EX(f,D) & 1-\eta-\beta \\ EX(\neg f, D) & \eta \\ \text{Adversary} & \beta \end{cases}$$

The learner is told η_b such that $\eta \leq \eta_b < \frac{1}{2}$ and given $EX_{\text{CAM}}^{\eta,\beta}(f,D)$. In time polynomial in $1/\varepsilon$, $1/\delta$, n, and $\frac{1}{1/2-\eta_b}$, the learner must output an hypothesis h which with probability at least $1-\delta$ is ε-close to f on D.

25.3.1   Hypothesis Testing using a CAM Example Oracle

In this section, we show how to use the $EX_{\text{CAM}}^{\eta,\beta}$ oracle to perform so called "hypothesis testing" in order to determine which hypotheses of a given set have relatively small error with respect to the *noise-free* example oracle. We first prove a theorem which states an upper bound on the number of labelled examples from a CAM oracle sufficient to rank two hypotheses with different error rates.

Theorem 1 *Let h_1 and h_2 be two hypotheses with error rates ε_1 and ε_2 such that $\varepsilon_2 - \varepsilon_1 = \gamma > 0$. If $\beta \leq \gamma(1/2 - \eta)/2$, then a sample S of size $O\left(\frac{(\varepsilon_1+\gamma)\log(1/\delta)}{\gamma^2(1/2-\eta)^2}\right)$ drawn from $EX_{\text{CAM}}^{\eta,\beta}(f,D)$ is sufficient to guarantee, with probability at least $1-\delta$, that h_1 disagrees with fewer labelled examples than h_2 in the sample S.*

---

[2]We often use $O, \Omega, \Theta, o, \omega$ when characterizing the asymptotic behavior of functions of variables which approach 0 as opposed to the standard usage in which variables approach ∞. For instance, since τ approaches 0, we use $O(\tau)$ to denote a function g for which there exists constants k and τ_0 such that for all $\tau \leq \tau_0$, $g(\tau) \leq k\tau$. We also make use of "soft" order notation when we are not concerned with lower order logarithmic factors. Specifically, when $b > 1$, we define $\tilde{O}(b)$ to mean $O(b\log^c b)$ for some constant $c \geq 0$. When $b < 1$, we define $\tilde{O}(b)$ to mean $O(b\log^c(1/b))$ for some constant $c \geq 0$. We define $\tilde{\Omega}$ similarly for some constant $c \leq 0$ and analogously define $\tilde{\Theta}, \tilde{o}, \tilde{\omega}$. When discussing asymptotics related to η, we consider $\eta \to 1/2$, or correspondingly $(1/2-\eta)^{-1} \to \infty$.

Proof: By generalizing a technique of Laird [Lai88], Aslam and Decatur [AD94] effectively show that two important parameters determine the number of labelled examples sufficient to perform hypothesis testing by *any* type of example oracle. These parameters are (1) t — the probability of drawing a labelled example on which the two hypotheses disagree; and (2) α — the conditional probability, given the hypotheses disagree with each other, that the hypothesis with smaller error disagrees with the label. They show that it is sufficient to draw a sample S of $O(\log(1/\delta) \cdot t^{-1}(1 - 2\alpha)^{-2})$ labelled examples to ensure with probability at least $1 - \delta$ that h_2 disagrees with more labelled examples in S than h_1 does. We simply derive the values of t and α for $EX_{\mathrm{CAM}}^{\eta,\beta}$.

Let d be the probability of drawing an example x according to D in which $h_1(x) = h_2(x) \neq f(x)$. Let t_1 be the probability of drawing a labelled example $\langle x, l \rangle$ from $EX_{\mathrm{CAM}}^{\eta,\beta}(f, D)$ in which $h_1(x) \neq h_2(x) = l$. Similarly, let t_2 be the probability of drawing a labelled example $\langle x, l \rangle$ from $EX_{\mathrm{CAM}}^{\eta,\beta}(f, D)$ in which $h_2(x) \neq h_1(x) = l$.[3] We then have

$$
\begin{aligned}
t_1 &= \beta + (\varepsilon_1 - d)(1 - \eta - \beta) + (\varepsilon_2 - d)\eta \\
t_2 &= (\varepsilon_2 - d)(1 - \eta - \beta) + (\varepsilon_1 - d)\eta
\end{aligned}
$$

and

$$
t = t_1 + t_2 = \beta + (\varepsilon_1 + \varepsilon_2 - 2d)(1 - \beta).
$$

The value of α is simply t_1/t, and therefore $t^{-1}(1-2\alpha)^{-2} = t(t-2t_1)^{-2}$. By our assumption that $\beta \leq \gamma(1 - 2\eta)/4$, we have

$$
\begin{aligned}
t^{-1}(1 - 2\alpha)^{-2} &= \frac{\beta + (\varepsilon_1 + \varepsilon_2 - 2d)(1 - \beta)}{[(\varepsilon_2 - \varepsilon_1)(1 - \beta - 2\eta) - \beta]^2} \\
&\leq \frac{\gamma(1 - 2\eta)/4 + (\varepsilon_1 + \varepsilon_2)}{[\gamma(1 - 2\eta)/2]^2} \\
&= \frac{\gamma(1 - 2\eta)/4 + (\gamma + 2\varepsilon_1)}{\gamma^2(1/2 - \eta)^2}. \\
&= O\left(\frac{\varepsilon_1 + \gamma}{\gamma^2(1/2 - \eta)^2}\right).
\end{aligned}
$$

$\square$

A standard application of a result such as Theorem 1 is that it allows one to select, with high probability, an hypothesis with error at most ε from a set of hypotheses containing at least one whose error is at most $\varepsilon/2$. Specifically, consider a set of N hypothesis in which one of these hypotheses, say h_1, is guaranteed to have error rate at most $\varepsilon/2$. By selecting the hypothesis from this set which has the smallest empirical error on a sufficiently large sample, we can be confident that the error rate of this hypothesis is no more than ε. For this procedure to work, it is enough for h_1 to have smaller empirical error than any hypothesis with true error more than ε. Thus, we allocate $\delta/(N - 1)$ probability of failure to each comparison of h_1 with the other hypotheses and note that for each hypothesis of error more than ε, the corresponding gap is $\gamma \geq \varepsilon/2$.

---

[3] Probabilities t_1, t_2, and d are with respect to the draw of the example and any randomization in h_1 or h_2. Since t_1 and t_2 actually depend on a dynamically playing adversary, we consider the worst possible case in which, at every opportunity, the adversary returns a labelled example $\langle x, l \rangle$ such that $l = h_2(x) \neq h_1(x)$.

Corollary 2 *Let $h_1, \ldots, h_N$ be hypotheses with error rates $\varepsilon_1 \leq \cdots \leq \varepsilon_N$ such that $\varepsilon_1 \leq \varepsilon/2$. If $\beta \leq \varepsilon(1/2 - \eta)/4$, then a sample S of size $O\left(\frac{\log(N/\delta)}{\varepsilon(1/2-\eta)^2}\right)$ drawn from $EX_{\mathrm{CAM}}^{\eta,\beta}(f, D)$ is sufficient to guarantee, with probability at least $1 - \delta$, that the hypothesis with the fewest disagreements on S has error no more than ε.*

25.3.2 Simple Strategies for CAM Learning

Applying Corollary 2 is feasible when one already has a relatively small set of candidate hypotheses. Otherwise, one is confronted with the CAM learning problem of distinguishing between all possible functions. We next examine how existing tools for both classification noise and malicious error learning may be combined to achieve CAM learning. Specifically, we consider the strategy of starting with an algorithm which tolerates one type of noise, and transforming it so that it additionally tolerates the second type of noise.

Given an algorithm which tolerates malicious errors, no transformation is known which adds the ability to tolerate classification noise. This problem is illustrated by the class of parity functions. The class of parity functions has a malicious error tolerant algorithm (yielded by the technique of "multiple-runs" [KL88] discussed below, on the noise-free algorithm for learning parity functions [HSW92]), yet it is not known how to learn parity functions even in the presence of classification noise alone.

Conversely, such a transformation does exist when starting with an algorithm which tolerates classification noise. Kearns and Li [KL88] describe a technique that takes a PAC algorithm with sample complexity m which does not tolerate malicious errors, and by running it many times, converts it into one which tolerates a malicious error rate $\beta = \Theta(\frac{\log m}{m})$. We can take an algorithm which tolerates classification noise, and use this technique to create one which tolerates CAM error. The amount of CAM error which can be tolerated using these techniques may be upper bounded by using lower bounds on the sample complexity of classification noise learning. Simon [Sim93] has shown that any algorithm for function class $\mathcal{F}$ which tolerates classification noise must have sample complexity at least $m = \Omega(\frac{\mathrm{VCDim}(\mathcal{F})}{\varepsilon(1/2-\eta)^2})$.[4] In fact, most known classification noise algorithms have sample complexity at least $m = \Omega(\frac{\mathrm{VCDim}(\mathcal{F})}{\varepsilon^2(1/2-\eta)^2})$. Thus, these techniques yield algorithms which tolerate CAM with any $\eta < 1/2$, but β at most $\tilde{O}(\varepsilon(1/2-\eta)^2/\mathrm{VCDim}(\mathcal{F}))$. Specifically, the dependence of β on ε and η is at best $\tilde{O}(\varepsilon(1/2-\eta)^2)$ and usually $\tilde{O}(\varepsilon^2(1/2-\eta)^2)$. We next show a different technique for deriving algorithms which tolerate more CAM error.

25.3.3 Simulating Statistical Queries for Improved CAM Learning

We show how to efficiently simulate an SQ algorithm in the PAC model in the presence of CAM error. The strategy is to perform a simulation similar to that used for classification noise alone, but to do so with sharper tolerances in order to also tolerate the malicious errors.

Theorem 3 *Given an SQ learning algorithm for $\mathcal{F}$ with minimum tolerance τ, one can construct a PAC learning algorithm for $\mathcal{F}$ which can tolerate CAM for all $\eta < 1/2$ and $\beta \leq \beta_* = \Theta(\tau(1/2 - \eta))$.*

---

[4]The VC-dimension [VC71], or VCDim, of a class $\mathcal{F}$ is a combinatorial measure which is commonly used to characterize the required sample complexity for learning $\mathcal{F}$. Note that $\mathrm{VCDim}(\mathcal{F}) \geq 1$.

Note: This theorem allows tolerance of as much classification noise in the CAM model as was possible in the model of classification noise alone. In the case of malicious error, the amount of error tolerable depends on the tolerance of the SQ algorithm (as was the case in the model of malicious error alone) and the classification noise rate. This theorem provides a smooth tradeoff of classification noise against malicious errors: as η approaches 0, the amount of malicious error tolerable approaches the level tolerable in the model of malicious error alone; similarly, as β approaches 0, η may approach $1/2$.

Proof: The simulation of the SQ algorithm draws a sample of labelled examples from the CAM example oracle, uses that sample to estimate all of the queries in the SQ algorithm, and outputs the hypothesis returned by the SQ algorithm. If the simulation of the *STAT* oracle is correct (*i.e.* within proper tolerances) for all queries made, then by the correctness of the SQ algorithm, the output hypothesis is ε-good. Therefore, the simulation algorithm must, with probability $1 - \delta$, answer all queries accurately.

The strategy to simulate queries is two-fold: bound the fraction of examples on which the malicious adversary plays and ensure that the remaining examples *very accurately* estimate quantities used to estimate P_χ in the presence of classification noise. The additional accuracy is required to accommodate error introduced by the malicious error examples. The quantities we use to estimate P_χ are based on the *classification noise alone* simulation of SQ algorithms of Aslam and Decatur [AD94].

For each query $[\chi, \tau]$, we wish to estimate to within $\pm\tau$ the value P_χ, the probability that a labelled example drawn from $EX(f, D)$ satisfies χ. Let P_χ^r be the probability that a labelled example drawn from $EX_{\text{CAM}}^{\eta,\beta}(f, D)$ satisfies χ given that the labelled example was not chosen by the malicious adversary. Let $P_{\bar{\chi}}$ be the probability that a labelled example drawn from $EX(f, D)$ satisfies χ if we negate the example's label. Define $P_{\bar{\chi}}^r$ similarly. These definitions yield the following equations:

$$P_\chi^r = \frac{(1 - \beta - \eta)P_\chi + \eta P_{\bar{\chi}}}{1 - \beta} \tag{25.1}$$

$$P_{\bar{\chi}}^r = \frac{(1 - \beta - \eta)P_{\bar{\chi}} + \eta P_\chi}{1 - \beta} \tag{25.2}$$

By solving Equation 25.2 for $P_{\bar{\chi}}$ and substituting into Equation 25.1, we get the following expression for P_χ:

$$P_\chi = \frac{(1 - \beta - \eta)P_\chi^r - \eta P_{\bar{\chi}}^r}{1 - \beta - 2\eta}$$

Thus, in order to estimate P_χ, it is enough to have estimates of P_χ^r, $P_{\bar{\chi}}^r$, η and β. We make use of the following lemma which states how accurately these quantities must be estimated or "guessed." We omit the proof of this lemma which is based on simple algebra.

Lemma 4 *Provided $\beta \leq \beta_* = \tau(1/2 - \eta)/16$, in order to estimate P_χ to within $\pm\tau$, it is sufficient to estimate P_χ^r and $P_{\bar{\chi}}^r$ each to within $\pm\tau(1/2 - \eta)/8$, to guess an $\hat{\eta}$ such that $|\eta - \hat{\eta}| \leq \tau(1/2 - \eta)/8$, and to use any $\hat{\beta} \leq \beta_*$.*

We simulate the SQ algorithm multiple times, each with a different "guess" for η. Each simulation attempts to achieve error no more than $\varepsilon/2$. On any of the runs in which the noise rate is guessed accurately enough, the hypothesis output is $\varepsilon/2$-good with high probability. A set of $O(\frac{1}{\tau(1/2-\eta_b)})$ guesses, each spaced $\tau(1/2-\eta_b)/4$ apart, would assure that for at least one guess $\hat{\eta}$, we have $|\eta - \hat{\eta}| \le \tau(1/2-\eta)/8$. Yet, by using unequally spaced noise rate guesses, it is possible to construct a set of $O(\frac{1}{\tau}\log\frac{1}{1/2-\eta_b})$ guesses with this same property [AD94]. We compare the hypotheses generated by all of the different runs and output the one with smallest empirical error on a random sample. Corollary 2 can be used to determine the sample size sufficient to ensure with high probability that this hypothesis is ε-good. Note that we allocate $\delta/2$ probability of failure to the hypothesis testing and $\delta/2$ probability of failure to the determination of the estimates within all of the runs.

We next determine the sample size sufficient to ensure, with probability $1-\delta/2$, that every P^r_χ and $P^r_{\bar{\chi}}$ is estimated accurately. If we define $\tau' = \tau(1/2-\eta)/8$ and $\theta = \tau'-\beta$, then assuming $\beta \le \beta_* = \tau(1/2-\eta)/16$, we have $\theta = \Theta(\tau(1/2-\eta))$. Let S be a sample of size m (to be determined below) labelled examples drawn from $EX^{\eta,\beta}_{\text{CAM}}$, let S_1 be the subset of S chosen by the malicious adversary and let $S_2 = S \setminus S_1$, *i.e.* examples chosen according to the correct distribution but possibly mislabelled. Let S^χ be the subset of S which satisfies χ. Define S^χ_1 and S^χ_2 similarly. Let our estimate for P^r_χ be $|S^\chi|/|S|$. We show that if both Conditions 1 and 2 (given below) hold, then for all $\chi \in \mathcal{Q}$, $\left|P^r_\chi - |S^\chi|/|S|\right| \le \tau'$. That is, the fraction of examples in the sample from $EX^{\eta,\beta}_{\text{CAM}}$ which satisfy χ is a good estimate for P^r_χ.

Condition 1: $\frac{|S_1|}{|S|} \le \beta + \frac{\theta}{2}$

Condition 2: $\forall \chi \in \mathcal{Q}, \left|\frac{|S^\chi_2|}{|S_2|} - P^r_\chi\right| \le \frac{\theta}{2}$

Condition 1 states that the malicious adversary played on no more than a $\beta + \frac{\theta}{2}$ fraction of the m examples. Condition 2 states that for all queries, the fraction of non-adversarially labelled examples satisfying χ is within $\frac{\theta}{2}$ of the expected fraction P^r_χ. In order to show these two conditions jointly imply accurate estimates, we simply verify that they imply $\forall \chi \in \mathcal{Q}, |S_\chi|/|S| \le P^r_\chi + \tau'$ and $|S_\chi|/|S| \ge P^r_\chi - \tau'$:

$$\frac{|S^\chi|}{|S|} = \frac{|S^\chi_1|}{|S|} + \frac{|S^\chi_2|}{|S|} \le \frac{|S_1|}{|S|} + \frac{|S^\chi_2|}{|S_2|}$$
$$\le \left(\beta + \tfrac{\theta}{2}\right) + \left(P^r_\chi + \tfrac{\theta}{2}\right)$$
$$= P^r_\chi + \tau',$$

$$\frac{|S^\chi|}{|S|} \ge \frac{|S_2|}{|S|} \cdot \frac{|S^\chi_2|}{|S_2|} \ge \left(1 - \beta - \tfrac{\theta}{2}\right)\left(P^r_\chi - \tfrac{\theta}{2}\right)$$
$$= P^r_\chi - \beta P^r_\chi - \tfrac{\theta}{2}\left[P^r_\chi + 1 - \beta - \tfrac{\theta}{2}\right]$$
$$\ge P^r_\chi - \beta - \tfrac{\theta}{2} \cdot 2$$
$$= P^r_\chi - \tau'.$$

Condition 1 requires the empirical fraction of a set of independent trials of a Bernoulli random variable to be no more than $\theta/2$ larger than its expected fraction, while Condition 2 requires similar empirical fractions to be within $\pm\theta/2$ of their expected values. Using standard Chernoff bounds [AV79] and uniform convergence results [VC71], to ensure that (with probability $1-\delta/2$) both conditions hold, it is sufficient to use a sample of size $m = O(\frac{1}{\theta^2}\log\frac{|\mathcal{Q}|}{\delta})$ if the query space $\mathcal{Q}$ is finite or a sample of size $m = O(\frac{q}{\theta^2}\log\frac{1}{\theta}+\frac{1}{\theta^2}\log\frac{1}{\delta})$

if $\mathcal{Q}$ is infinite with finite VC-dimension q. Alternatively, one could use a separate sample for each of the N queries made by all runs of the algorithm, thus using a total of $m = O(\frac{N}{\theta^2} \log \frac{N}{\delta})$ examples. (See [Dec95] for a more complete discussion of sample sizes.) $\square$

25.3.4 Optimality of CAM Tolerance

Theorem 3 states that *any* class which is SQ learnable is PAC learnable with CAM for all $\eta < \frac{1}{2}$ and $\beta = O(\tau(1/2 - \eta)) = \tau/\Omega(\frac{1}{1/2-\eta})$. Aslam and Decatur [AD93] have shown that for any class learnable by SQ, the dependence of τ on ε is $\Omega(\varepsilon/\log(1/\varepsilon))$. Therefore, the dependence of β on η and ε need *never* be worse than $\Omega(\frac{\varepsilon}{\log(1/\varepsilon)}(1/2 - \eta))$.

The classification noise tolerance of Theorem 3 matches the information theoretic limit of $1/2$ and is therefore optimal. We demonstrate the optimality of the dependence of β on η and ε in the following two theorems. Together, these theorems show the error tolerance of Theorem 3 to be within a $\log(1/\varepsilon)$ factor of optimal.

Theorem 5 *No class is PAC learnable with CAM $\forall \eta < 1/2$ and $\beta = g_1(\tau)/o(\frac{1}{1/2-\eta})$ for any function $g_1(\cdot)$.*

Proof: By contradiction, if $\beta = g_1(\tau)/o(\frac{1}{1/2-\eta})$, then for any fixed τ (and therefore fixed $g_1(\tau)$) there exists an $\eta < 1/2$ for which $\beta \geq g_1(\tau)/[\frac{g_1(\tau)}{1/2-\eta}] = 1/2 - \eta$, in which case $\eta + \beta \geq 1/2$. But no class is learnable under such conditions since the malicious adversary is then able to make the CAM oracle simulate a standard classification noise oracle with noise rate $1/2$, making every label look like a random bit. $\square$

Theorem 6 *No distinct class is PAC learnable with CAM of $\beta = \omega(\varepsilon) \cdot g_2(1/2 - \eta)$ for any function $g_2(\cdot)$.*

Proof: For fixed $\eta < \frac{1}{2}$, $\beta = \omega(\varepsilon) \cdot g_2(1/2 - \eta) = \omega(\varepsilon)$. However, as stated above, Kearns and Li [KL88] have shown that no distinct class is learnable with $\beta \geq \frac{\varepsilon}{1+\varepsilon} = \Theta(\varepsilon)$. $\square$

For many commonly studied classes (such as conjunctions, decision lists, and k-DNF) $\tau = \Theta(\varepsilon/\mathrm{VCDim}(\mathcal{F}))$, in which case the malicious error tolerable is $\beta = \Theta(\varepsilon(1/2-\eta)/\mathrm{VCDim}(\mathcal{F}))$. This error tolerance from SQ simulation is strictly greater than that achievable by the techniques of Section 25.3.2, and has optimal dependence on both ε and η.

25.4 Other Hybrid Noise Models

In Section 25.3 we defined a hybrid noise environment in which an example oracle exhibited both classification noise and malicious errors. We may instead define hybrid noise environments based on other types of noise models. Two other such "noise" models are *distribution shift* and *distribution restricted* learning [Dec93]. In both of these models, the examples are labelled correctly, but the distribution of examples is somehow modified. We show that any combination of these noise models, along with classification noise and/or malicious errors, define a hybrid noise environment in which we can efficiently simulate any statistical query algorithm. Once again, the statistical query algorithm quantifies how much error we can tolerate in estimating a query. To each individual type of noise, we may allocate a part of that fault tolerance. Thus, the tolerance of the SQ algorithm quantifies the generic fault tolerance of the algorithm. Additionally, statistical queries have been used to learn in a different hybrid noise environment which incorporates both attribute noise and classification noise [DG95].

25.4.1 Individual Noise Models

In the model of distribution shift, the learning algorithm is given access to an example oracle which draws examples according to a distribution which differs from the distribution on which it will be tested, yet examples are always labelled correctly. The only restriction on the distribution used for training is that it be "close" to the distribution used for testing. A class is learnable with distribution shift σ if there exists an algorithm which can tolerate training on any distribution which is within distance σ from the testing distribution. The distance between two distributions P and Q over the domain X is measured as follows:

$$d(P, Q) = \max_{A \subseteq X} |P(A) - Q(A)|.$$

Note that this definition of distance is equivalent to $d(P,Q) = \frac{1}{2} \sum_{x \in X} |P(x) - Q(x)|$. If $\mathcal{F}$ is PAC learnable, then it is learnable with distribution shift σ if and only if $\sigma = O(\varepsilon)$ [Dec93].

In distribution restricted learning, a learner is promised that the distribution of examples belongs to a specified class of distributions $\mathcal{D}$. We can use a distribution restricted SQ algorithm to learn on distributions outside of the promised class of distributions $\mathcal{D}$. The larger class of distributions on which learning is possible is determined by the tolerance τ of the SQ algorithm. Specifically, the larger class of distributions is composed of all distributions within distance $O(\tau)$ of some distribution in $\mathcal{D}$ [Dec93].

25.4.2 The Hybrid Noise Model

The model which combines all four of these individual noise models is described by the following learning environment. There is a testing distribution D_1 which is close to some distribution $D \in \mathcal{D}$ (i.e. $d(D, D_1) \leq \gamma$) and there is a training distribution D_2 which is close to D_1 (i.e. $d(D_1, D_2) \leq \sigma$). The example oracle used for training is as follows:

$$EX_{\mathrm{HYB}}^{\eta,\beta,\gamma,\sigma}(f, D_1) = \begin{cases} EX(f, D_2) & 1 - \eta - \beta \\ EX(\neg f, D_2) & \eta \\ \text{Adversary} & \beta \end{cases}$$

An SQ algorithm for the class $\mathcal{F}$ on distributions $\mathcal{D}$ can be converted (by simulating statistical queries) into a PAC algorithm for learning $\mathcal{F}$ in this environment. The simulation is valid for all $\eta < 1/2$, and $\beta + \gamma + \sigma = O(\tau(1/2 - \eta))$. One again, there is a smooth tradeoff between all of the different types of noise and the amount of noise tolerable can be shown to be optimal in terms of its dependence on τ. The simulations and proofs of optimality are similar to those show earlier and details are given in the full paper.

Acknowledgements

Thanks to Robert Schapire whose questions prompted this research and to Javed Aslam and Leslie Valiant for helpful comments on earlier drafts of this paper. This research was supported by an NDSEG Doctoral Fellowship and by NSF grant CCR-92-00884. Author's current address is: Laboratory for Computer Science, Massachusetts Institute of Technology, 545 Technology Square – Room 313, Cambridge, MA 02139, *sed@theory.lcs.mit.edu*.

25.5 REFERENCES

[AD93]   Javed Aslam and Scott Decatur. General bounds on statistical query learning and PAC learning with noise via hypothesis boosting. In *Proceedings of the 34th Annual Symposium on Foundations of Computer Science*, pages 282–291, November 1993.

[AD94]   Javed Aslam and Scott Decatur. Improved noise-tolerant learning and generalized statistical queries. Technical Report TR-17-94, Harvard University, July 1994.

[AL88]   Dana Angluin and Philip Laird. Learning from noisy examples. *Machine Learning*, 2(4):343–370, 1988.

[Ang92]   Dana Angluin. Computational learning theory: Survey and selected bibliography. In *Proceedings of the 24th Annual ACM Symposium on the Theory of Computing*, 1992.

[AV79]   Dana Angluin and Leslie G. Valiant. Fast probabilistic algorithms for Hamiltonian circuits and matchings. *Journal of Computer and System Sciences*, 18(2):155–193, April 1979.

[Dec93]   Scott Decatur. Statistical queries and faulty PAC oracles. In *Proceedings of the Sixth Annual ACM Workshop on Computational Learning Theory*, pages 262–268. ACM Press, July 1993.

[Dec95]   Scott Decatur. *Efficient Learning from Faulty Data*. PhD thesis, Harvard University, 1995.

[DG95]   Scott Decatur and Rosario Gennaro. On learning from noisy and incomplete examples. In *Proceedings of the Eighth Annual ACM Workshop on Computational Learning Theory*. ACM Press, July 1995.

[HSW92]   David Helmbold, Robert Sloan, and Manfred K. Warmuth. Learning integer lattices. *SIAM Journal on Computing*, 21(2):240–266, 1992.

[Kea93]   Michael Kearns. Efficient noise-tolerant learning from statistical queries. In *Proceedings of the 25th Annual ACM Symposium on the Theory of Computing*, pages 392–401, San Diego, 1993.

[KL88]   Michael Kearns and Ming Li. Learning in the presence of malicious errors. In *Proceedings of the 20th Annual ACM Symposium on Theory of Computing*, Chicago, Illinois, May 1988.

[Lai88]   Philip D. Laird. *Learning from Good and Bad Data*. Kluwer international series in engineering and computer science. Kluwer Academic Publishers, Boston, 1988.

[Sim93]   Hans Ulrich Simon. General bounds on the number of examples needed for learning probabilistic concepts. In *Proceedings of the Sixth Annual ACM Workshop on Computational Learning Theory*, pages 402–411. ACM Press, 1993.

[Val84]   Leslie Valiant. A theory of the learnable. *Communications of the ACM*, 27(11):1134–1142, November 1984.

[Val85]   Leslie Valiant. Learning disjunctions of conjunctions. In *Proceedings of the Ninth International Joint Conference on Artificial Intelligence*, 1985.

[VC71]   V.N. Vapnik and A.Ya. Chervonenkis. On the uniform convergence of relative frequencies of events to their probabilities. *Theor. Probability Appl.*, 16(2):264–280, 1971.

26

On the Statistical Comparison of Inductive Learning Methods

A. Feelders[t] and W. Verkooijen[tt]

University of Twente[t]
Department of Computer Science
P.O.Box 217, 7500 AE Enschede
The Netherlands
feelders@cs.utwente.nl

Tilburg University[tt]
Department of Economics
P.O. Box 90153, 5000 LE Tilburg
The Netherlands
W.J.H.Verkooijen@kub.nl

ABSTRACT Experimental comparisons between statistical and machine learning methods appear with increasing frequency in the literature. However, there does not seem to be a consensus on how such a comparison is performed in a methodologically sound way. Especially the effect of testing multiple hypotheses on the probability of producing a "false alarm" is often ignored.

We transfer multiple comparison procedures from the statistical literature to the type of study discussed in this paper. These testing procedures take the number of tests performed into account, thereby controlling the probability of generating "false alarms". The multiple comparison procedures selected are illustrated on well-known regression and classification data sets.

26.1 Introduction

Recent interactions between the statistical and artificial intelligence communities (see e.g. [Han93, CO94]), have led to many studies that compare the performance of empirical statistical and machine learning methods on real-life data sets; examples are the StatLog project ([MST94]), the Santa Fe Time Series Competition ([WG94]), and comparisons reported in numerous journal articles (e.g., [KWR93, RABCK93, WHR90, TAF91, TK92, FG93]).

We observe that there is no consensus in the research community on how such a comparative study is performed in a methodologically sound way.

The ranking of k preselected methods is usually performed by training (estimating in statistical terminology) the methods on a single data set, and estimating their respective mean prediction errors (MPE) from a hold-out sample. The methods are subsequently ranked according to their estimated MPEs. Some studies use, in our view appropriately, statistical significance testing in order to make this ranking. However, the effect that comparing multiple methods has on the probability of generating a "false alarm" (claiming

---

[1]*Learning from Data: AI and Statistics V.* Edited by D. Fisher and H.-J. Lenz. ©1996 Springer-Verlag.

that one method is better than another, when in fact it is not) is to our knowledge ignored in the literature.

The statistical analysis of comparative studies, method-ranking in particular, is addressed in this paper. Specifically, we address *methodological* issues of studies in which the performance of several regression and classification methods are compared on real-life data sets.

26.2   Ranking Methods by Significance Testing

The ranking of methods by simply ordering them by the point estimates of their prediction errors should be extended by statistical significance testing. Appropriate tests are those concerned with the difference between means (regression) and proportions (classification). The standard t-test for testing the difference between two sample means $\overline{Y}_1$ and $\overline{Y}_2$ which come from *independent* normal distributed populations, leads to the following confidence interval for the difference

$$\theta_1 - \theta_2 \in [(\overline{Y}_1 - \overline{Y}_2) \pm t_{(\alpha/2,\nu)}\, \hat{\sigma}_{\text{diff}}] \tag{26.1}$$

where $\hat{\sigma}_{\text{diff}}$ equals $\sqrt{\hat{\sigma}^2_{\overline{Y}_1} + \hat{\sigma}^2_{\overline{Y}_2}}$. Here, α denotes the probability of making a Type I error, i.e. of claiming that the population means are different, when in fact they are not. In the standard comparative experiment, however, the MPEs are all estimated from the *same* test sample, which makes them highly correlated. Therefore, a *paired sample t-test* should be used instead. The dependence within the pairs only changes the standard error of the difference $\hat{\sigma}_{\text{diff}}$, which now becomes

$$\hat{\sigma}_{\text{diff}} = \sqrt{\hat{\sigma}^2_{\overline{Y}_1} + \hat{\sigma}^2_{\overline{Y}_2} - 2\operatorname{cov}(\overline{Y}_1, \overline{Y}_2)} \tag{26.2}$$

When the variables are positively correlated, the covariance will have a positive value and thus the variance and standard error of a difference between means will be *less* for matched than for unmatched samples. Consequently, the confidence intervals become smaller (given the same α value), which results in more powerful tests. In conclusion, neglecting the dependence between the samples generally results in tests that are too conservative.

Often the estimated MPEs of more than two, say k, methods are being compared. The first idea that comes to mind is to test each possible difference by a paired t-test with probability of Type I error of size α. The problem with this approach is that the probability of making at least one Type I error over the whole family of t-tests (one test per pair of methods being compared) exceeds α by an amount that increases with the number of tests made. For J *statistically independent* tests the probability of making at least one Type I error, better known as the *familywise error* rate (FWE), is $1-(1-\alpha)^J$. When J is large, say 20, this can be a large probability; for $\alpha = 0.05$ there is a probability of 0.64 for one or more Type I errors. This means that the probability that we incorrectly claim to have found one or more significant differences is 64%. Such an incorrect claim is called a "false alarm". When the tests are statistically dependent of each other, such as pairwise difference tests, the FWE becomes even larger. Thus, when enough pairwise tests are performed one will with high probability find one or more "significant" differences, i.e. false alarms. This

Observations	Functions						Total
	f_1	f_2	$\ldots$	f_i	$\ldots$	f_k	
1	Y_{11}	Y_{12}	$\ldots$	Y_{1i}	$\ldots$	Y_{1k}	$Y_{1.}$
2	Y_{21}	Y_{22}	$\ldots$	Y_{2i}	$\ldots$	Y_{2k}	$Y_{2.}$
$\vdots$	$\vdots$	$\vdots$	$\vdots$	$\vdots$	$\vdots$	$\vdots$	$\vdots$
j	Y_{j1}	Y_{j2}	$\ldots$	Y_{ji}	$\ldots$	Y_{jk}	$Y_{j.}$
$\vdots$	$\vdots$	$\vdots$	$\vdots$	$\vdots$	$\vdots$	$\vdots$	$\vdots$
n	Y_{n1}	Y_{n2}	$\ldots$	Y_{ni}	$\ldots$	Y_{nk}	$Y_{n.}$
Total	$Y_{.1}$	$Y_{.2}$	$\ldots$	$Y_{.i}$	$\ldots$	$Y_{.k}$	
Means	$\overline{Y}_{.1}$	$\overline{Y}_{.2}$	$\ldots$	$\overline{Y}_{.i}$	$\ldots$	$\overline{Y}_{.k}$	

TABLE 26.1. One-way repeated measures lay-out.

problem is known as the *multiplicity effect* or *selection effect*. Statistical procedures have been designed to take into account and properly control for the multiplicity effect; they are called *multiple comparison procedures*.

A crude approach to deal with the multiplicity effect is the Bonferroni method [2], which rejects the pairwise null hypothesis $\theta_i - \theta_{i'} = 0$ when the p-value is less than α/J, where α is the preset FWE level and J is the number of tests. This method neglects the dependency between the pairwise difference tests, and it further assumes normality of the data.

There are many alternative tests, ranging from slight adjustments of the Bonferroni method to very sophisticated techniques. The existing comparison procedures can roughly be categorised as analytical [HT87] or resampling based [WY93]. The former approach requires certain distributional assumptions of the underlying statistical model and typically uses table lookup to make a probability statement. The latter approach generates empirical distributions of the relevant statistics by resampling from the data set at hand, thereby removing the risk of making faulty statements due to unsatisfied assumptions. Evidently, the resampling approach involves much more computation than the analytical approach. In this paper we restrict ourselves to the analytical approach.

The characteristics of a particular experimental design often prescribe adjustments to general tests for differences or make special purpose tests necessary. The experimental design that captures the subject of this study is the *one-way repeated measures design*, which is displayed in Table 26.1. In such designs, blocks consisting of a random sample of, say, n experimental units drawn from a large population constitute the random factor. Each unit is measured under k different conditions. The conditions of measurements are fixed in advance, and constitute the treatment factor. In the terminology of this study, experimental units correspond to the observations from the test set, and the treatment factor corresponds with the regression or classification model type.

---

[2]This method orginally due to R.A. Fisher ([Fis35]) is popularly known as the Bonferroni method since it uses the Bonferroni inequality (which says that the probability of a union of events is less than the sum of the individual event probabilities).

26.3 Pairwise Comparisons for Regression

The general setting of this section is Table 1 (the one-way repeated measures design with k different prediction models that predict the observations from the *same* random test set of size n). The deviation of the predicted value from the true value is assumed to be measured as squared error, but any other error measure could be inserted equally well (e.g. absolute error). When the observations are not randomly drawn from a population, but result from a (highly) autocorrelated time series, the subsequent approach seems not to be justified. Diebold and Mariano [DM94] discuss the comparison of predictive accuracy of two time series models; they leave the multiple comparison problem for further research.

Let $\mathbf{Y}_j = (Y_{j1}, Y_{j2}, \ldots, Y_{jk})$ denote the vector of prediction errors for the jth observation $(1 \leq j \leq n)$. The following model is assumed:

$$\mathbf{Y}_j = \mathbf{M}_j + \mathbf{E}_j \quad (1 \leq j \leq n) \tag{26.3}$$

where all the $\mathbf{M}_j = (M_{j1}, M_{j2}, \ldots, M_{jk})$ and $\mathbf{E}_j = (E_{j1}, E_{j2}, \ldots, E_{jk})$ are distributed independently of each other as k-variate normal vectors, the former with mean vector $\boldsymbol{\theta} = (\theta_1, \theta_2, \ldots, \theta_k)$ (the vector of model effects) and variance-covariance matrix $\boldsymbol{\Sigma}_0$, and the latter with mean vector $\mathbf{0}$ and variance-covariance matrix $\sigma^2 \mathbf{I}$. Thus the $\mathbf{Y}_j$'s are independent and identically distributed (i.i.d.) $N(\boldsymbol{\theta}, \boldsymbol{\Sigma})$ random vectors where $\boldsymbol{\Sigma} = \boldsymbol{\Sigma}_0 + \sigma^2 \mathbf{I}$.

Exact procedures for making pairwise comparisons among the θ_i's can be constructed if we impose special restrictions on the form of $\boldsymbol{\Sigma}$. The least restrictive of such models is the *spherical model*, which assumes that all pairwise differences of the sample means of the regression models have the same variance (for more details see [HT87, CH90, WBM91, Hay88]). In practice, however, this assumption will rarely be satisfied [HT87, Hay88].

Therefore Hochberg and Tamhane [HT87, page 215] propose a test, in case one is unsure about the sphericity assumption. They propose the following approximate $100(1 - \alpha)\%$ simultaneous confidence intervals for the pairwise differences $\theta_i - \theta_{i'}$:

$$\theta_i - \theta_{i'} \in \left[\overline{Y}_{.i} - \overline{Y}_{.i'} \pm |M|_{k^*,n-1}^{(\alpha)} \sqrt{\frac{S_{ii} + S_{i'i'} - 2S_{ii'}}{n}} \right] \quad (1 \leq i < i' \leq k) \tag{26.4}$$

where $|M|_{k^*,n-1}^{(\alpha)}$ is the upper α point of the Studentized maximum modulus distribution (see [HT87, Table 6]) with parameter $k^* = k(k-1)/2$ and degrees of freedom $n - 1$; and where $S_{ii'}$, is the estimated (co)variance between $Y_{.i}$ and $Y_{.i'}$:

$$S_{ii'} = \frac{\sum_{j=1}^n (Y_{ji} - \overline{Y}_{.i})(Y_{ji'} - \overline{Y}_{.i'})}{n - 1} \quad (1 \leq i, i' \leq k) \tag{26.5}$$

The Studentized maximum modulus distribution is defined as the distribution of

$$\max_{1 \leq i \leq k} |T_i|$$

where $|T_i|$ is the modulus of the k-variate t-distribution with ν degrees of freedom and common correlation of zero.

We use an empirical analysis of the *Boston housing data*[3] to illustrate this procedure. We compare the performance of four regression models, designated f_1 through f_4. The first model f_1 is a linear model trained with OLS; models f_2 through f_4 are feed-forward neural networks with two, four, and six hidden units respectively. The data set is split into two parts: the first part (400) is used to estimate the parameters of the model; the second part (106) is used to measure the model's performance. The observed average squared prediction error $\overline{Y}_{.i}$ for method f_i ($i = 1, 2, 3, 4$) are $\overline{Y}_{.1} = 6.38\text{e-}3$; $\overline{Y}_{.2} = 3.49\text{e-}3$; $\overline{Y}_{.3} = 2.94\text{e-}3$; and $\overline{Y}_{.4} = 3.00\text{e-}3$.

	f_1	f_2	f_3	f_4
f_1	1.68e-04	3.76e-05	3.46e-05	4.02e-05
f_2	–	3.01e-05	1.54e-05	1.27e-05
f_3	–	–	2.24e-05	2.67e-05
f_4	–	–	–	4.15e-05

TABLE 26.2. The $S_{ii'}$ matrix for the Boston housing case.

	θ_2	θ_3	θ_4
θ_1	[5.9e-4 , 5.2e-3]	[1.1e-3 , 5.7e-3]	[1.1e-3 , 5.8e-3]
θ_2	–	[-4.3e-4 , 1.5e-3]	[-9.2e-4 , 1.9e-3]
θ_3	–	–	[-7.2e-4 , 6.2e-3]

TABLE 26.3. All pairwise confidence intervals.

Suppose that it is of interest to make all pairwise comparisons among the four methods, using a Type I familywise error rate $\alpha = 0.10$. In Table 26.2 the $S_{ii'}$ values are displayed in the cells, calculated according to (26.5). Obviously, we have $|M|_{6,105}^{(0.1)}$, which equals 2.135 ([HT87, Table 6]), and we construct the confidence intervals for the pairwise differences $\theta_i - \theta_{i'}$ according to (26.4). Table 26.3 shows that the linear model performs significantly worse than the three neural network models: the confidence intervals cover positive values only. Among the neural network models no significant differences can be observed; there is no statistical evidence for preferring neural network models with more than two hidden units.

26.4   Pairwise Comparisons for Classification

In this section we discuss significance testing for the comparison of two or more *classification* methods. Again we notice that it is not appropriate to use a standard test based on the assumption of independent samples. Instead, we use, as suggested by Ripley ([Rip93]), McNemar's test ([MM77]), when only two classification methods are compared. This test is normally used to test for differences between proportions in paired sample designs. The comparisons performed in this section should not be considered as serious evaluations of the methods involved; they are purely illustrative.

---

[3]This data set can be obtained by anonymous ftp from `lib.stat.cmu.edu` with user `statlib`

	I_{nn}	C_{nn}	Total
I_{lda}	61	23	84
C_{lda}	32	268	300
Total	93	291	384

TABLE 26.4. Incorrect and Correct classifications of lda and nn

26.4.1   Testing a single hypothesis

A hypothesis test involving the application of linear discriminant analysis and a feed-forward neural network to the *diabetes* data set[4] illustrates the use of McNemar's test. The 768 observations in this dataset were divided in a training set and a test set of 384 observations each.

Table 26.4 summarizes the result of using the linear discriminant function (lda) and a neural network (nn) estimated on the training set to classify the observations in the test set. The cells (C_{lda}, C_{nn}) and (I_{lda}, I_{nn}) contain the number of cases classified correctly and incorrectly by both the linear discriminant function and the neural network respectively. Since we want to test

$$H_0 : \text{MPE}_{lda} = \text{MPE}_{nn} \qquad \text{against} \qquad H_a : \text{MPE}_{lda} \neq \text{MPE}_{nn},$$

only the cells (I_{lda}, C_{nn}) and (C_{lda}, I_{nn}) of this table are of interest. Here MPE is defined as the proportion of misclassifications made on the population of interest. When an observation falling in the (I_{lda}, C_{nn}) cell of this table is defined as a success, then the number of successes is binomially distributed with $n = (I_{lda}, C_{nn}) + (C_{lda}, I_{nn}) = 55$ and $\pi = 0.5$, under the null hypothesis. Observing 23 successes out of 55 trials (with $\pi = 0.5$) has a p-value 0.28; this p-value was calculated using an exact binomial test. According to any conventional significance level, we should conclude that H_0 cannot be rejected. Whether or not conventional significance levels are appropriate in this kind of hypothesis test is debatable. It is not clear that a Type I error is the most severe error one can make.

26.4.2   Testing Multiple Hypotheses

We will now consider the case where $k > 2$ classification functions are compared. In this comparison we use the same training and test set as in the above example. We performed a comparative study, including linear discriminant analysis (f_1), quadratic discriminant analysis (f_2), a classification tree (f_3), and two feed-forward neural networks (f_4 and f_5 respectively) which differ only in the value of the weight decay parameter used. All pairwise comparisons are performed, which amounts to a total of $k^* = 10$ pairwise comparisons. Table 26.1 presents the general lay-out of a study that compares k classification functions. In this matrix Y_{ji} has the value zero if f_i classifies observation j correctly, and one otherwise.

For our comparative study we have: $n = 384$, $k = 5$, $\overline{Y}_{.1} = 0.219$, $\overline{Y}_{.2} = 0.232$, $\overline{Y}_{.3} = 0.31$, $\overline{Y}_{.4} = 0.211$, $\overline{Y}_{.5} = 0.229$.

---

[4]This data set can be obtained by anonymous ftp from `ics.uci.edu` in the directory `pub/machine-learning-databases`

	θ_2	θ_3	θ_4	θ_5
θ_1	[−0.071,0.045]	[−0.149,−0.033]	[−0.05,0.066]	[−0.068,0.048]
θ_2	–	[−0.136,−0.02]	[−0.037,0.079]	[−0.055,0.061]
θ_3	–	–	[0.041,0.157]	[0.023,0.139]
θ_4	–	–	–	[−0.076,0.04]

TABLE 26.5. 95% confidence intervals for all pairwise differences.

The pooled variance of any pairwise difference $\overline{Y}_{.i} - \overline{Y}_{.i'}$ for this design can be written as ([MM77], p. 180)

$$\hat{\sigma}^2_{\text{diff}} = \frac{2\left(k\Sigma^n_{j=1}Y_{j.} - \Sigma^n_{j=1}Y^2_{j.}\right)}{n^2 k(k-1)}$$

We can now construct $100(1-\alpha)\%$ simultaneous confidence intervals for all pairwise differences $\theta_i - \theta_{i'}$ as follows

$$\theta_i - \theta_{i'} \in \left[\overline{Y}_{.i} - \overline{Y}_{.i'} \pm Z^\nu_{k^\bullet:1-\alpha/2}\hat{\sigma}_{\text{diff}}\right] \quad (1 \leq i < i' \leq k)$$

where θ_i denotes the population proportion of incorrect classifications of f_i, and $\nu = n-1$ denotes degrees of freedom. The distribution of Z is based on the Student t distribution, adjusted for the number of comparisons k^* involved ([Dun61]). Tables for this statistic can be found in ([MM77],[Dun61]). As one would expect, the value of $Z^\nu_{k^\bullet:1-\alpha/2}$ increases with the number of comparisons k^*, leading to wider confidence intervals.

Table 26.5 provides 95% confidence intervals for all pairwise differences. We used $Z^\infty_{10:0.975} = 2.81$ and $\hat{\sigma}_{\text{diff}} = 0.0205$, leading to confidence intervals that are $2 \times (2.81 \times 0.0205) = 0.116$ wide. If the interval of $\theta_i - \theta_{i'}$ contains zero, then there is no significant evidence that classification functions f_i and $f_{i'}$ differ in their true prediction error. From Table 26.5 we conclude that f_3 (the classification tree) performs significantly worse than *all* other functions; among these other functions no significant difference has been found.

26.5   Conclusion and future research

In this paper we proposed statistical procedures to perform studies which compare the predictive accuracy of several regression or classification functions. These procedures explicitly take into account the fact that multiple comparisons are made, and control for the probability of giving a "false alarm". We have attempted to show the relevance of multiple comparison procedures to a type of study that we encountered frequently in the recent Artificial Intelligence (AI) and Machine Learning literature.

Although the general difficulties induced by the multiplicity effect and by the dependency among observations are easy to grasp, finding "the right" testing procedure is much more difficult. The literature on the subject is somewhat ambiguous, and requires a high level of statistical knowledge, which AI-researchers do not always possess.

We do not claim that the methods presented here are the best for the given purpose: they are *examples* of tests that can be used for the comparison of the prediction accuracy of different functions. We hope that in future research more attention will be given to this subject, and perhaps more appropriate methods will be found.

One interesting approach may be to expand the empirical analyses by incorporating re-sampling based multiple comparison techniques. This will enhance the practical relevance for the machine learning engineer, since the assumptions made by analytical approaches may not be satisfied in practical situations. With the computer power currently available, the computational effort required by resampling techniques does not seem to be an obstacle anymore.

26.6 REFERENCES

[CH90]   M.J. Crowder and D.J. Hand. *Analysis of repeated measures*. Chapman & Hall, London, 1990.

[CO94]   P. Cheeseman and R.W. Oldford, editors. *Selecting models from data: AI and statistics IV*. Lecture notes in statistics nr. 89. Springer-Verlag, New York, 1994.

[DM94]   F. Diebold and R. Mariano. Comparing predictive accuracy. Technical report, University of Pennsylvania, 1994.

[Dun61]   Olive Jean Dunn. Multiple comparisons among means. *Journal of the American Statistical Association*, 56:52–64, 1961.

[Fis35]   R. A. Fisher. *The design of experiments*. Olivier & Boyd, Edinburgh, 1935.

[FG93]   D. Fletcher and E. Goss. Forecasting with neural networks: an application using bankruptcy data. *Information & Management*, 24:159–167, 1993.

[Han93]   D. J. Hand, editor. *Artificial intelligence frontiers in statistics: AI and statistics III*. Chapman & Hall, London, 1993.

[Hay88]   W. Hays. *Statistics*. Holt, Rinehart and Winston, Inc, Fort Worth, 1988.

[HT87]   Y. Hochberg and A. Tamhane. *Multiple comparison procedures*. Wiley & Sons, New York, 1987.

[KWR93]   J. Kim, H. Weistroffer, and R. Redmond. Expert systems for bond rating: a comparative analysis of statistical, rule-based and neural network systems. *Expert Systems*, 10:167–171, 1993.

[MM77]   L. Marascuilo and M. McSweeney. *Nonparametric and distribution-free methods for the social sciences*. Brooks/Cole Publishing Company, Monterey, CA, 1977.

[MST94]   D. Michie, D.J. Spiegelhalter, and C.C. Taylor, editors. *Machine learning, neural and statistical classification*. Ellis Horwood, New York, 1994.

[Pre95]   Lutz Prechelt. A quantitative study of neural network learning algorithm evaluation practices. In *Proc. 4th Intl. Conf. on Artificial Neural Networks*, Cambridge, UK, June 26-28, 1995.

[RABCK93] A. Refenes, M. Azema-Barac, L. Chen, and S. Karoussos. Currency exchange rate prediction and neural network design strategies. *Neural Computing & Applications*, 1:46–58, 1993.

[Rip93] B.D. Ripley. Flexible non-linear approaches to classification. In V. Cherkassky, J.H. Friedman, and H. Wechsler, editors, *From Statistics to Neural Networks: Theory and Pattern Recognition Applications*. Springer-Verlag, 1993.

[TAF91] Z. Tang, C. de Almeida, and P. Fishwick. Time series forecasing using neural networks vs. box-jenkins methodology. *Simulation*, 57:303–310, 1991.

[TK92] Kar Yan Tam and Melody Y. Kiang. Managerial applications of neural networks: the case of bank failure predictions. *Management science*, 38(7):926–947, 1992.

[WBM91] B. Winer, D. Brown, and K. Michels. *Statistical principles in experimental design*. McGraw-Hill, New York, 1991.

[WG94] A. Weigend and N. Gershenfield. *Time series prediction: forecasting the future and understanding the past*. Addison-Wesley, Reading, 1994.

[WHR90] A. Weigend, A. Huberman, and D. Rumelhart. Predicting the future: a connectionist approach. *International Journal of Neural Systems*, 1(3):193–209, 1990.

[WY93] P.H. Westfall and S.S. Young. *Resampling-Based Multiple Testing*. John Wiley & Sons, New York, 1993.

27
Dynamical Selection of Learning Algorithms

Christopher J. Merz

Information and Computer Science Department
University of California, Irvine
Irvine, CA 92717
cmerz@ics.uci.edu
(714)824-3491

ABSTRACT Determining the conditions for which a given learning algorithm is appropriate is an open problem in machine learning. Methods for selecting a learning algorithm for a given domain have met with limited success. This paper proposes a new approach to predicting a given example's class by locating it in the "example space" and then choosing the best learner(s) in that region of the example space to make predictions. The regions of the example space are defined by the prediction patterns of the learners being used. The learner(s) chosen for prediction are selected according to their past performance in that region. This dynamic approach to learning algorithm selection is compared to other methods for selecting from multiple learning algorithms. The approach is then extended to weight rather than select the algorithms according to their past performance in a given region. Both approaches are further evaluated on a set of ten domains and compared to several other meta-learning strategies.

27.1 Introduction

Determining the conditions for which a given learning algorithm is appropriate is an open problem in machine learning. Methods for selecting a learning algorithm for a given domain (e.g. [Aha92, Breiman84]) or for a portion of the domain ([Brodley93, Brodley94]) have met with limited success. This paper proposes a new approach that dynamically selects a learning algorithm for each example by locating it in the "example space" and then choosing the best learner(s) for prediction in that part of the example space. The regions of the example space are formed by the observed prediction patterns of the learners being used. The learner(s) chosen for prediction are selected according to their past performance in that region which is defined by the "cross-validation history."

This paper introduces DS, a method for the dynamic selection of a learning algorithm(s). We call it "dynamic" because the learning algorithm(s) used to classify a novel example depends on that example. Preliminary experimentation motivated DW, an extension to DS that dynamically *weights* the learners' predictions according to their regional accuracy.

Further experimentation compares DS and DW to a collection of other meta-learning strategies such as cross validation ([Breiman84]) and various forms of stacking ([Wolpert92]). In this phase of experimentation, the meta-learners have six constituent learners that are heterogeneous in their search and representation methods (e.g. a rule learner, CN2 [Clark89]; a decision tree learner, C4.5 [Quinlan93]; an oblique decision tree learner, OC1 [Murthy93]; an instance-based learner, PEBLS [Cost93]; a k-nearest neighbor learner, k-

---

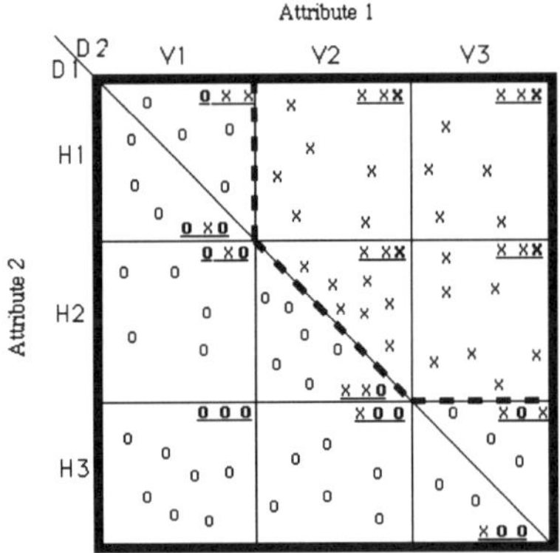

FIGURE 1. Dividing up the example space by response patterns.

nn [Cost93], and a naive Bayesian learner, BAYES [Cost93]). DS and DW are consistently among the top performers for a collection of ten domains.

The paper begins by giving motivation for the DS algorithm (Section 27.2). Section 27.3 describes the method for building a "cross-validation history" for a collection of learning algorithms that is then used in the DS algorithm for making predictions on novel examples. Section 27.4 briefly describes the DS algorithm and reports experimental results comparing DS and other methods for selecting a learning algorithm(s). An extension to DS, DW, is briefly described in Section 27.5 along with further experiments that compare DW with its predecessor DS and a variety of other meta-learning methods. A discussion of related work is given in Section 27.6.

27.2   Motivation

The main goal of this research is to determine the strengths and weaknesses of a collection of learning algorithms within a given domain. The approach taken is to characterize the example space in terms of the performance of each algorithm. Given this information, novel examples can be placed in the example space and the strongest learner(s) for that region can be referred to for a prediction.

To characterize the example space, we simply use the patterns of predictions from the constituent learners to define regions in the example space. For example, Figure 1 shows how the models built by three decision tree learning algorithms can be used to create these regions. Model V is made up of two vertical splits on Attribute 1 with three "leaves"; V1, V2, and V3. Model H has two horizontal splits on Attribute 2 with three leaves; H1, H2, and H3. Model D has one split which is an oblique split on both attributes forming two leaves; D1 and D2. The training examples are plotted according to their respective values for Attribute 1 and 2, and the symbol used to plot each example is its class (i.e., "X" or "O"). The leaves of the three models divide the training examples up into twelve areas. Each area has an underlined triplet showing the prediction of V, H, and D, respectively.

These triplets form seven distinct "regions" characterizing the example space (i.e., OXX, XXX, OXO, XXO, OOO, XOO, and XOX).

Although Figure 1 is a contrived example, it does show the potential utility of having this characterization of the example space. The bold dashed line highlights the learned boundaries we would like to use to partition the examples. For regions near the left side of the top leg of the boundary, we would like to rely on model V's predictions more. Model D's prediction would be the preferred for the diagonal leg of the boundary, and Model H would be the preferred predictor for regions just below the right leg of the boundary. Regions further from the boundary may have more than one "reliable" predictor where there is more agreement between the models.

Next, a way of determining which learner(s) are superior within a given region is needed. Since most learners are capable of attaining high accuracies on training data (i.e., decision trees can continue to partition until class purity within each partition is reached), the strengths of the learners being used may not appear all that different for the examples of a training partition. To obtain a more robust evaluation of how well each learner does on each training example, several cross-validation runs were done on the training data. As each training example became a "test" example in these cross-validation runs, the accuracy of each learner was recorded to accumulate a history of how well each learner classifies each example.

Returning to the example in Figure 1, this approach would show that the training examples which fell into the OXX region tended to be more accurately classified by learner V than the other two learners. Thus, we would defer to model V for test examples which produced an OXX prediction pattern. The next two sections elaborate on how this "cross-validation history" is built and how it is used.

27.3   Generating a Cross-Validation History

A common method for deciding which classification algorithm to use on a given domain is cross validation ([Breiman84]) where all the examples for a domain are divided randomly into v (approximately) equal partitions or folds. Each algorithm being evaluated is then trained on $v - 1$ partitions and tested on the remaining partition v times. Normally, the algorithm with the best average cross-validation accuracy is chosen as the "best" classifier for all examples of that domain. Instead, we rely on cross validation to build a **cross-validation history** to be used by DS for dynamically deciding which algorithm(s) to use for classification as a function of each (test) example.

This history is generated from the examples in the training partition. Given a set of training examples we form the history by building two m by n matrices where m is the number of training examples and n is the number of learning algorithms. The first is the **response matrix** which contains the predictions of the models built from the training data on the training data (i.e., the (i,j)th cell contains the prediction of model j for training example i).

The second matrix, the **performance matrix**, contains the number of times each model was correct when that training example appeared as a test example in a cross-validation run on the training partition. That is, k v-fold cross validations are run for each model on the training partition so the (i,j)th cell in the performance matrix is the number of times (out of k) that example i was correctly predicted by model j when it

Given: **R:**   An m X n response matrix where
                         m = the number of examples and
                         n = the number of models
            P: An m X n performance matrix
            EX: A test example
Return: **C:**   the predicted class of **EX**

1. Get row vector of responses, $\mathbf{R}_{EX}$, by polling each of the n models for **EX**.

2. Find set of "closest" row(s) to $\mathbf{R}_{EX}$ in **R**.

3. Use corresponding rows in **P** to compute local accuracy for each model.

4. Let models tied with highest local accuracy vote to choose **C**.

FIGURE 2. The DS classification algorithm.

appeared as a test example in a cross-validation run.

Using the models built from the training data we can get a row vector of model responses for a single test example. This row vector can be compared to the row vectors for the training examples in the response matrix. The set of row vectors with the most similar response patterns then represent a region defined from the perspective of the models built.[2] The corresponding rows in the performance matrix are then used to determine the accuracy of each of the models for the region defined by the examples with similar response patterns.

27.4   The DS Algorithm

The algorithm for dynamically selecting models, DS, for a given test example is described in Figure 2. The first step is to let all of the learned models classify the test example to form a row-vector of responses, $\mathbf{R}_{EX}$. In step 2, the "closest" rows are all the rows tied with the highest level of agreement with $\mathbf{R}_{EX}$. The local accuracy computed in step 3 is calculated by averaging the performances only on the rows selected in step 2. The algorithm with the highest local accuracy is then selected in step 4. If a tie occurs, the algorithms having the highest local accuracy vote with equal weighting.

Initial experiments with DS used 24 variations of the OC1 program [Murthy93] as different learning algorithms against which to do dynamic selection. The different entropy measures (6), pruning options (off/on with 20 percent of training data for pruning), and types of splits (univariate/multivariate) combined for 24 unique configurations of OC1. All other parameters had the default settings. In forming the history, 3-fold cross validation was done ten times.

Four domains with all numeric attributes (Glass, Iris, Breast Cancer Wisconsin, and Liver Disorders) were chosen from the UCI repository for evaluation because OC1 is

---

[2]We use the number of matched responses to measure similarity.

TABLE 27.1. Glass domain results.

Method	Number of Examples		
	50	100	150
SAM	**54.01**	65.25	69.57
CVM	55.65	63.23	**67.12**
DS	58.90	65.25	70.48
Best Individual Model	58.62 (#9)	64.35 (#12)	68.09 (#10)

limited to datasets with numeric features. The Glass dataset consists of 214 examples with nine attributes describing forensic information of a glass sample. The class attribute specifies ten possible glass types. The Iris dataset contains 150 examples, each with four numeric attributes describing an Iris flower's sepal and petal measurements, and a class attribute specifying the three types of Iris flowers. In the Breast Cancer dataset of 470 examples, there were nine attributes describing a tumor, and the class attribute specifying whether it was malignant or benign. For the Liver Disorders dataset, there are five blood test attributes and one alcohol consumption attribute and a class attribute specifying whether a disorder existed.

DS was compared to two other methods for selecting learning algorithms. The Select-All Majority (SAM) method weights each algorithm's prediction equally and returns the most frequent prediction as its prediction. The cross-validation majority (CVM) method returns the prediction of the algorithm having the highest cross-validation accuracy in the performance table, and, in the case of ties, returns the majority vote of the tied algorithms. Tables 27.1 through 27.4 report the test accuracies for each methods for each domain where 25 percent of the examples were held back for testing and various numbers of training examples were used. Each cell entry is an accuracy averaged over 30 runs. Bold cell entries in the SAM and CVM rows indicate a statistically significant difference between that method and DS.[3] Also included in the tables is a row reporting the best individual model's performance. The number in parentheses next to these entries is the corresponding configuration of OC1 which had that performance and is meant to track which configuration is best as the number of examples changes. The identification of the best individual model stems from a post-experiment analysis of the test accuracies attained by each OC1 variant.

The Glass domain results given in Table 27.1 show significantly better performance for DS over SAM with fewer examples and for CVM for larger example sets. DS has the best performance overall for all example set sizes. The best individual model changes for each training set size, thus no one configuration is doing well across the learning curve. The performance of CVM is not as good as the best individual model because that best individual configuration is not identified by CVM as the unique best model based on cross-validation performance. In fact, the individual best model may not be selected as one of the best models by CVM at all. Therefore, CVM may consider the predictions of the individual best model along with the predictions of other, non-optimal models.

Table 27.2 shows that DS does as well as or better than the other selection methods in the Iris domain. This difference is most prevalent with fewer examples. The best individual

---

[3]All references to statistical significance are using a paired two-tailed t-test at the .05 confidence level.

TABLE 27.2. Iris domain results.

Method	Number of Examples		
	25	50	100
SAM	88.22	94.66	95.33
CVM	89.55	94.11	95.66
DS	89.55	95.00	95.33
Best Individual Model	90.66 (#11)	95.77 (#2)	96.45 (#2)

TABLE 27.3. Breast Cancer domain results.

Method	Number of Examples		
	100	200	300
SAM	**93.65**	**94.59**	**95.50**
CVM	91.95	93.38	94.16
DS	92.61	93.79	94.81
Best Individual Model	92.37 (#17)	94.08 (#17)	94.51 (#18)

model configuration changes as the training size increases and consistently does the best. However, CVM fails once again to recognize the best individual learning algorithm.

In Table 27.3, we see that for the Breast Cancer dataset DS consistently outperforms CVM. However, the SAM method maintains a slight (1 percent or less) but significant edge over DS. One possible explanation for this is that the majority of OC1 configurations are well suited for this domain giving SAM a consistent edge.

Performance in the Liver Disorders domain is similar to the Breast Cancer domain. DS does better than CVM and significantly worse than SAM. The performance of SAM on the last two domains led to further experimentation discussed in the next section.

A couple more comments are in order for these results. First, one can argue that the CVM method could do better with a greater number of folds and/or runs. However, in at least two of the domains, even if CVM were to choose the best model configuration, DS would consistently do as well or better. Second, for two of the domains evaluated (Glass and Iris), DS has better relative performance when fewer examples are available.

27.5   Extending Dynamic Selection to Dynamic Weighting

In our preliminary experimentation, the limited success of DS and the greater success of SAM motivated an extension of DS to consider a broader range of the constituent learners

TABLE 27.4. Liver Cancer domain results.

Method	Number of Examples	
	100	200
SAM	**67.02**	**70.25**
CVM	64.42	67.37
DS	64.65	68.62
Best Individual Model	65.19	68.75

TABLE 27.5. Meta-learning results

Domain	M-Bayes	M-C4.5	M-CN2	M-PEBLS	SAM	CVM	DS	DW	BI
wave	77.30	71.22	70.89	69.46	**77.88**	72.17	74.97	75.03	72.85 **3**
iris	93.75	93.98	94.19	93.76	95.66	94.18	95.66	**95.98**	94.70 **5**
breast	**95.91**	95.60	95.32	94.01	**95.90**	94.50	94.26	94.33	94.99 **4**
glass	64.70	64.31	65.16	63.00	**72.39**	63.39	70.92	72.00	66.52 **2**
liver	68.86	66.45	64.29	58.82	**69.87**	64.52	66.26	66.94	66.71 **2**
credit	**86.45**	86.15	86.20	78.38	85.09	84.68	84.84	85.26	84.86 **1**
voting	**95.35**	94.68	94.72	92.36	**95.35**	94.95	94.22	94.26	95.01 **1**
krk	92.48	92.27	92.38	89.83	**93.38**	91.90	92.58	93.13	92.32 **1**
heart	79.83	79.36	79.78	72.08	**80.12**	79.54	79.80	80.01	80.16 **3**
tumor	38.9	39.68	42.71	33.86	44.30	**44.86**	39.68	40.34	44.91 **4**

when making its predictions like SAM does, while still retaining its dynamic nature. To do this, we decided to weight each constituent's vote according to its regional accuracy. Recall that this information is available in the performance matrix (see step 3 in Figure 2. This is a simple extension to DS, which can be thought of as a degenerate weighting strategy where each of the **best** regional learners is weighted with a one and the others are weighted with a zero.

To evaluate DW, as well further evaluate DS, experiments were done using a more heterogeneous collection of constituent learners, six additional domains, and several new methods for classification using a suite of learners. The new constituent learners were: CN2 [Clark89], a rule learner; C4.5 [Quinlan93], a decision tree learner; PEBLS [Cost93], an instance-based learner; k-nn [Cost93], a nearest neighbor learner, and Bayes [Cost93], a naive Bayesian learner. OC1 was also still used, but only with its default configuration, and only on the all-numeric domains.

The new domains were also chosen from the UCI data repository. The wave domain consisted of 300 examples with 40 numeric attributes. The credit database has 6 numeric attributes and 9 symbolic attributes with three classes and 690 examples. The congressional voting domain has 16 symbolic attributes and 435 examples with two classes. In the king-rook-king (krk) domain there were 35 symbolic features, 830 examples and two classes. Finally, the primary tumor domain contains 339 examples, 17 symbolic attributes and 22 classes.

In addition to DW, DS, SAM, and CVM several new methods for learning at the meta-level were evaluated. These methods were *stacking* [Wolpert92] extensions to some established learning strategies: M-Bayes, M-CN2, M-C4.5, and M-PEBLS. Stacking is described in more detail in the next section.

Table 27.5 shows performance results of the meta-learners on the ten domains. Bolded accuracies denote the meta-learner with the best accuracy for a given domain. The last column is the accuracy of the best individual (BI) constituent learner. The bold number next to it is the constituent with the best overall accuracy for that domain (i.e., 1 = C4.5, 2 = CN2, 3 = PEBLS, 4 = Bayes, 5 = OC1, 6 = k-nn).

The most striking result is that SAM has the leading accuracy on all but three domains where it has the second highest accuracy. None of these "wins" are statistically significant over all the other methods, but it was significantly better than all but the closest competitor in the wave, liver, and tumor domains, and significantly better than all but the closest two competitors in the iris, breast, glass, and king-rook-king domains.

We factored out SAM to see how well the other meta-learners did and found that DW does best on four of the domains (i.e., heart, krk, glass, and iris). This difference was statistically significant for the latter two domains. M-Bayes does best on five of the domains (i.e., wave, breast, liver, credit, and voting) and this difference was statistically significant on the wave domain. DW was also a strict improvement on DS, and both methods consistently outperform CVM, M-C4.5, M-CN2, and M-PEBLS except in a few cases.

Relating these performances to the best individual constituent, we see that CVM and M-PEBLS consistently do worse. All the other meta-learners demonstrate a capacity to exceed the best individual constituent. It should also be noted that each constituent was the best individual at least once, and no one constituent dominated.

To summarize these results, we see that it is fairly common to exceed the performance of the best individual constituent with three of the meta-learning strategies (i.e., SAM, DW, and M-Bayes). We also see that the most effective way to do this is by taking the "simple majority" of the constituents (i.e., SAM). However at times it does pay to go against the simple majority and use an alternative meta-learning strategy. This situation must occur when one or more of the constituents has found a niche in the example space and the meta-learner is capable of exploiting it.

27.6   Related Work

The approach taken here is most akin to the approach of stacked generalization as described by [Wolpert92] in which learning algorithms at the meta level learn how to classify novel examples by observing the prediction patterns of the constituent learners. These prediction patterns are generated using v-fold cross validation where each training example is converted to a "meta-learning example" by using as its attributes the predictions of the constituent learners for this example when it was in the test fold of the cross-validation run. This is different than DS/DW where the attributes of the meta-learning examples are the predictions of the constituents built on *all* the training data. Also, in selecting/weighting algorithm(s) for voting, DS/DW has the extra information of the cross-validation history which can be thought of as constructed attributes. Thus, DS/DW can be viewed as a form of stacked generalization augmented by constructive induction.

In cross validation ([Breiman84]), the appropriateness of each learning algorithm is judged by the corresponding average cross-validation accuracy and the algorithm with the best accuracy is chosen for that domain. There are a couple of limitations to this approach. One is that a single learning algorithm is selected as the "appropriate" one for an entire domain. An alternative we explore here is to select a learning algorithm(s) for a given example. Another potential drawback to the cross-validation approach is that several algorithms may have the same cross-validation accuracy but may give different classifications for many examples. Which one is really the "best" if they are not classifying the same examples correctly? Ideally, we would like to use these patterns of predictions to identify regions in the example space where each learning algorithm is more (or less) accurate. This research addresses both of these limitations by considering the actual prediction patterns of a set of learning algorithms rather than a summary of those predictions (i.e., accuracy) to dynamically select the algorithm(s) with the best performance history for similarly classified training examples.

Another approach for selecting the best algorithm for a given domain is given by [Aha92] where rules are derived which characterize the conditions under which various learning algorithms do well based on certain high-level properties of a domain (e.g., number of examples, number of attributes, number of classes, types of attributes, class and attribute noise). However, this method relies on having a good characterization of the domain, which is often difficult to obtain. DS/DW emphasizes the selection of a learning algorithm at the example level rather than at the domain level since it avoids the need for characterization rules by instead considering each learning algorithm's past performance on "similar" examples from that domain.

[Brodley93] proposes a knowledge-based approach to building hybrid decision structures by using heuristics to select the best algorithm at a given stage of learning. These heuristics are created from practical knowledge about how to detect when a generalization is a good fit or when to switch to a different search bias during the course of learning. Conditions in these rules include checks for certain pathologies in the learning process such as how often features are tested. Also, they use information about the domain such as the ratio of the number of examples to the number features. The development of these heuristics is a difficult ad hoc manual process based on the practitioner's past experience. DS/DW avoids this process by building a collection of models and dynamically choosing which one(s) to use based on past performance in the given region of the domain.

[Kwok90] selects several *good* models (i.e., rules/trees with high posterior probabilities) manually and average their predictions. This avoids the difficult problem of calculating posterior probabilities for rules [Buntine89]. The two main drawbacks to this approach are the manual derivation of the models, and the assumption that the majority prediction is always the best one to use. Our approach builds the models automatically and retains models which may not have high posterior probabilities, but still may be useful in certain regions of the example space.

27.7   Conclusion

Several methods for dynamically and statically selecting/weighting the predictions of a suite of learning algorithms were described and evaluated. Two of these approaches (i.e., DS and DW) are novel ways of dynamically weighting a collection of constituent learners for a given test example. Several of the dynamic approaches (i.e., DW, M-Bayes, and DS) consistently outperformed the static approach of cross validation for a collection of ten domains. However, the naive approach of "simple majority," SAM, tends to have the best performance. Further research exploring when it is appropriate to go "against" the simple majority is needed.

27.8   REFERENCES

[Aha92] Aha, D. W. (1992). Generalizing from case studies: A case study. *Proceedings of the Ninth International Machine Learning Conference* (pp. 1-10). San Mateo, CA: Morgan Kaufmann.

[Breiman84] Breiman, L., Friedman, J.H., Olshen, R.A. & Stone, C.J. (1984). *Classification and Regression Trees.* Belmont, CA: Wadsworth.

[Brodley93] Brodley, C. E. (1993). Addressing the Selective Superiority Problem: Automatic Algorithm/Model Class Selection. *Proceedings of the Tenth International Machine Learning Conference* (pp. 17-24). San Mateo, CA: Morgan Kaufmann.

[Brodley94] Brodley, C. E. (1994). Recursive Automatic Bias Selection for Classifier Construction. To appear in the special issue of Machine Learning on "Bias Evaluation and Selection." Kluwer Academic Publishers.

[Buntine89] Buntine, W. (1989). Decision tree induction systems: a bayesian analysis. *Uncertainty in Artificial Intelligence 3* (pp. 109-127). North-Holland, Amsterdam.

[Cheeseman88] Cheeseman, P., Kelly, J., Self, M., Stutz, J., Taylor, W., & Freeman, D. (1988). AutoClass: A Bayesian classification system. *Proceedings of the Fifth International Machine Learning Conference* (pp. 54–64). Ann Arbor, MI: Morgan Kaufmann.

[Clark89] Clark, P., & Niblett, T. (1989). The CN2 Induction Algorithm. *Machine Learning, 3*, 261–284.

[Cost93] Cost, S. & Salzberg, S. (1993). A Weighted Nearest Neighbor Algorithm for Learning with Symbolic Features. *Machine Learning, 10*, 57–78

[Kwok90] Kwok, S. W., & Carter, C. (1990). Multiple decision trees. *Uncertainty in Artificial Intelligence 4* (pp. 327-335). North-Holland, Amsterdam.

[Murthy93] Murthy, S., Kasif, S., Salzberg, S., & Beigel, R. (1993). OC1: Randomized induction of oblique decision trees. *In Proceedings of AAAI-93* (pp. 322-327). Washington DC: AAAI Press.

[Quinlan93] Quinlan, R. (1993). *C4.5 Programs for Machine Learning.* San Mateo, CA: Morgan Kaufmann.

[Rumelhart86] Rumelhart, D. E., Hinton, G. E., & Williams, R. J. (1986). Learning Interior Representation by Error Propagation. *Parallel Distributed Processing, 1* 318–362. Cambridge, MASS.: MIT Press.

[Wolpert92] Wolpert, D. H. (1992). Stacked Generalization. *Neural Networks, 5*, 241–259.

28
Learning Bayesian Networks Using Feature Selection

Gregory M. Provan[†] and Moninder Singh[††]

Institute for Decision Systems Research[†]
4984 El Camino Real, Los Altos, CA, 94022
provan@camis.stanford.edu

Dept. of Computer and Information Science[††]
University of Pennsylvania
Philadelphia, PA 19104-6389
msingh@gradient.cis.upenn.edu

ABSTRACT This paper introduces a novel enhancement for learning Bayesian networks with a bias for small, high-predictive-accuracy networks. The new approach selects a subset of features that maximizes predictive accuracy prior to the network learning phase. We examine explicitly the effects of two aspects of the algorithm, feature selection and node ordering. Our approach generates networks that are computationally simpler to evaluate and display predictive accuracy comparable to that of Bayesian networks which model all attributes.

28.1 Introduction

Bayesian networks are being increasingly recognized as an important representation for probabilistic reasoning. For many domains, the need to specify the probability distributions for a Bayesian network is considerable, and learning these probabilities from data using an algorithm like K2 [Cooper92] could alleviate such specification difficulties.

We describe an extension to the Bayesian network learning approaches introduced in K2. Our goal is to construct networks that are simpler to evaluate but still have high predictive accuracy relative to networks that model all features. Rather than use all database features (or attributes) for constructing the network, we select a subset of features that maximize the predictive accuracy of the network. Then the learning process uses only the selected features as nodes in learning the Bayesian network. We examine explicitly the effects of two aspects of the algorithm: (a) feature selection, and (b) node ordering.

Our experimental results verify that this approach generates networks that are computationally simpler to evaluate and display predictive accuracy comparable to the predictive accuracy of Bayesian networks that model all features. Our results, similar to those observed by other studies of feature selection in learning [Caruana94, John94, Langley94a, Langley94b], demonstrate that feature selection provides comparable predictive accuracy using smaller networks. For example, by selecting as few as 18% of the features for the

---

[1] *Learning from Data: AI and Statistics V.* Edited by D. Fisher and H.-J. Lenz. ©1996 Springer-Verlag.
[2] This work was supported by NSF grants #IRI92-10030 and #IRI91-20330, and NLM grant #BLR 3 RO1 LMO5217-02S1.

gene-splice domain, we obtained a predictive accuracy of 94.9% (as opposed to 96.2% with all features). However, the reduction of features is not always as significant (e.g. 69% for the letter recognition domain), which implies that feature selection should be used advisedly, and its effect is domain- and data-dependent.

The remainder of the paper is organized as follows. Section 28.2 introduces the Bayesian network learning algorithm which we modify, the K2 algorithm. Section 28.3 describes our new learning approach. Section 28.4 outlines the experimental design, and Section 28.5 summarizes the experimental results. Section 28.6 compares and contrasts our approach with other related work. Finally, we summarize our contributions in Section 28.7.

28.2   Bayesian Network Learning

We now define the learning algorithm used in K2.[3] Assume that we have a database D of m cases, where each case contains an instantiation for each of a set Z of n discrete features. B_S denotes a belief network structure representing the features in Z. In a belief network, a node represents a feature, and the absence of an arc between two nodes denotes the independence of the two nodes given the remaining network structure. The posterior probability of a network given the data, $P(B_S|D)$, is proportional to the joint probability, so networks can be ranked according to their joint probabilities. The single most likely network is given by $B_{S_{max}} = \mathrm{argmax}_{B_S}[P(B_S|D)]$.

K2 constructs a Bayesian network from a set of features as follows. K2 selects the network (out of a set of possible networks exponential in the number of network nodes) which maximizes the network's posterior probability, $P(B_S, D)$. K2 requires an ordering on the features from which the network will be constructed. Given an ordering $n_1, n_2, ..., n_m$ of the m features, K2 takes each successive feature in the ordering, adds it as a node n_i in the network, and creates parents for n_i in a greedy fashion: rather than evaluate all subsets of network nodes $n_1, n_2, ..., n_{i-1}$ as parent nodes, K2 selects as a parent node the the *single* node in $\{n_1, n_2, ..., n_{i-1}\}$ which most increases the posterior probability of the network structure. New parent nodes are added incrementally to n_i as long as doing so increases the posterior probability of the network given the data.

Our new approach proceeds in two phases. The first phase computes a subset $\Delta \subseteq Z$ of features that generates the network B_ζ with highest predictive accuracy, where B_ζ denotes the network formed from the subset $\zeta \subseteq Z$ of features. The second phase computes the network (from the set of features Δ) which maximizes the predictive accuracy over the test data.

The learning algorithm that we use, called CB, is a modified version of K2 [Singh95]. Whereas K2 assumes a node ordering, CB uses conditional independence (CI) tests to generate a "good" node ordering, and then uses the K2 algorithm to generate the Bayesian network from the database D using this node ordering. The CB algorithm starts by using CI tests of order 0 and keeps constructing networks for increasing orders of CI tests as long as the predictive accuracy of the generated network keeps increasing. Since CB uses the K2 algorithm to generate the Bayesian network from a particular ordering, CB is correct in the same sense that K2 is [Singh95]. Singh and Valtorta show the importance

---

[3]We use the nomenclature used in the papers on K2 by Herskovits and Cooper [Herskovits90, Cooper92].

of deriving a good node ordering [Singh95], given the $n!$ possible node orderings on n features.

28.3   Feature Selection Algorithm

We implemented the Feature Selection Algorithm using the CB algorithm in both the attribute selection as well as the network construction phase. We call this approach K2-AS, since it uses the basic K2 algorithm allied with *Attribute Selection* in the attribute selection phase. The algorithm we use is what has been described as a *wrapper model* [John94], in that "the feature subset selection algorithms conducts a search for a good subset using the induction algorithm itself as part of the evaluation function" [John94, page 124].

Our learning approach consists of two main steps, attribute selection and network construction. In the attribute selection phase, we choose the set of nodes from which the final network is constructed. In the network construction phase, we construct the network from the subset of attributes selected in the previous phase. Finally, we test the predictive accuracy of the network.

The algorithm used for the attribute selection phase is a forward selection algorithm, in that it starts with an empty set of features and adds features using a greedy search. This forward selection is just like K2. We now describe the different phases of the algorithm:

- **attribute selection phase:** In this phase, K2-AS chooses the set of attributes Δ, $\Delta \subseteq Z$ (Z is the set of all attributes) from which the final network is constructed. The algorithm starts with the initial assumption that Δ consists of only the class variable $class_{var}$. It then adds incrementally that attribute (from $Z - \Delta$) whose addition results in the maximum increase in the predictive accuracy of the network constructed from the resulting set of nodes. When there is no single attribute whose addition increases predictive accuracy, the algorithm stops adding attributes. We define $\Phi(\Delta)$ to be the predictive accuracy of the set Δ of attributes, and $\mathcal{G}(M)$ to be the network constructed from the set M of nodes. This phase can be described as follows:

$$\Delta \leftarrow \{class_{var}\}$$
$$\Phi_{old} \leftarrow \Phi(class_{var})$$
$$\text{NotDone} \leftarrow \text{True}$$
$$\text{while NotDone do}$$
$$\qquad \forall\, x \in Z - \Delta, \text{ let } B_{S_x} \leftarrow \mathcal{G}(\Delta \cup \{x\})$$
$$\qquad \Phi_{new} \leftarrow \max_x \Phi(B_{S_x})$$
$$\qquad z = \arg\max_x \{\Phi_x\}$$
$$\qquad \text{if } \Phi_{new} > \Phi_{old} \text{ then}$$
$$\qquad\quad \Phi_{old} \leftarrow \Phi_{new}$$
$$\qquad\quad \Delta \leftarrow \Delta \cup \{z\}$$
$$\qquad \text{else NotDone} \leftarrow \text{false}$$
$$\text{end \{while\};}$$

- **network construction phase:** K2-AS uses the final set of attributes Δ selected in the attribute selection phase to construct a network using training data. Once again, the CB algorithm is used to construct the final Bayesian network.

- **network evaluation phase:** In order to test the quality of the network, we test the network for its predictive accuracy on the test data.

28.4   Experimental Design

	Features	Nodes Selected	Cases (TOTAL)	Cases in *training* set	Cases in *evaluation* set	Cases in *test* set
Gene-splice	60	11	3175	1000	1000	1175
Soybean	35	9	630	215	215	200
Chess	36	5	3196	500	500	2196
Letter	16	11	20000	8000	8000	4000
Voting	16	2	435	100	100	235

TABLE 28.1. Summary of databases and learning approaches. The first two columns summarize the total number of features and the number of features selected by K2-AS (on average). The last four columns summarize information about how we split the databases.

We divide the database into three parts. The first two parts are used for the attribute selection phase: the first part, the *training* data, is used for learning the network (from the current subset of nodes and the node under consideration); the second part, the *evaluation* data, is used to test it for predictive accuracy (to decide whether to add the new node to the set of selected nodes). Once we have finished with the selection of the subset of attributes to be network nodes, we use the first part of the database to learn the network using the CB algorithm (the network construction phase). The third part of the database, the *test* data, is then used for determining the predictive accuracy of the network derived from the network construction phase. We performed inference on the networks using the Lauritzen-Spiegelhalter inference algorithm as implemented in the HUGIN [Andersen89] system.

The K2-AS approach trades off the time required to construct a network from the full feature set (as done in K2) with precomputing a feature subset and subsequently constructing a network with this feature subset. The attribute selection phase in K2-AS adds a modest amount of computational expense to the network induction. At each iteration of the attribute selection phase, K2-AS constructs a network (for each of the nodes not in the current set of selected nodes) by adding each node to the current set of selected nodes and then perform inference using the constructed network. At each stage, since a very small subset of the features is used, the network generated is very small and so the network construction as well as the inference phase is very fast (a matter of seconds for the databases considered).

Note that the attribute selection phase depends on selecting a relatively small number of features to create Bayesian networks that are efficient to perform inference on. Our results confirm the success of this approach, in that less than a dozen features were selected in each of domains studied, as shown in the first two columns of Table 28.1.

28.5 Results

We have performed a set of experiments to compare the networks generated by our approach with those created by CB. We tested this method on four databases acquired from the University of California, Irvine Repository of Machine Learning databases [Murphy92], namely Michalski's Soybean database, Slate's Letter Recognition database, the Gene-Splicing database due to Towell, Noordewier, and Shavlik,[4], Schlimmer's voting database and Shapiro's Chess Endgame database.

	K2-AS	CB	C4.5
Chess	94.10	94.63	98.40
Gene-splice	94.89	96.21	93.67
Soybean	88.03	89.45	93.10
Letter recognition	82.35	81.54	79.72
Voting	95.15	94.83	92.21

TABLE 28.2. Comparison of predictive accuracy for the basic CB approach and for K2-AS. The predictive accuracies show are averaged over 20 trials.

Table 28.1 summarizes the five databases used in terms of number of cases and features, and compares the number of features with the number of nodes selected. The K2-AS approach always selects fewer features to be nodes in the network. The ratio of nodes selected (excluding the node for the class variable) to total database features ranges from 15% for the gene-splice domain to 69% for the letter recognition domain.

Table 28.2 compares the predictive accuracy obtained for the basic CB approach and for the K2-AS approach. The predictive accuracies were obtained by averaging over 20 trials. For four of the five databases tested, the difference between K2-AS and CB was *not* statistically significant. It was only in the case of the Gene-splice database that K2-AS was significantly poorer. However, note that even though, in the Gene-splice case, the predictive accuracy was almost 2% less in the case of K2-AS, the network which had 2% lower predictive accuracy was one sixth the size of the full network created by the CB approach: i.e., of the 60 features, 11 were deemed to be important to the predictive accuracy, resulting in a network one-sixth the size of the network generated using all 60 features. Note also that the time required for performing inference on the networks induced using K2-AS was substantially less than than required for the networks induced using CB. This was particularly true for the network for the chess domain, which is very densely connected.

For the sake of comparison, Table 28.2 also depicts the average predictive accuracy obtained on the various databases by using a decision tree algorithm, C4.5. For this purpose, we used the IND (version 2.1) software. Note, however, that this paper does not attempt to achieve the accuracy found on the five databases studied using other approaches, such as decision trees. Instead, the purpose of this paper is *not* to identify the Bayesian network with the highest predictive accuracy, but to identify a parsimonious

---

[4]We used the database used to recognize genes in DNA Sequences (which we call gene-splicing database), as created by Towell, Noordewier, and Shavlik.

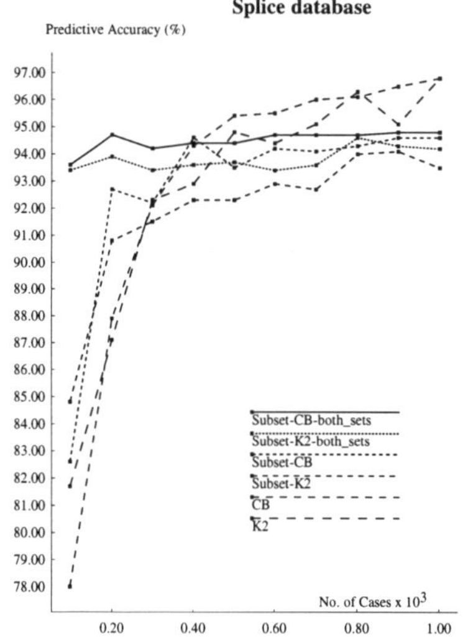

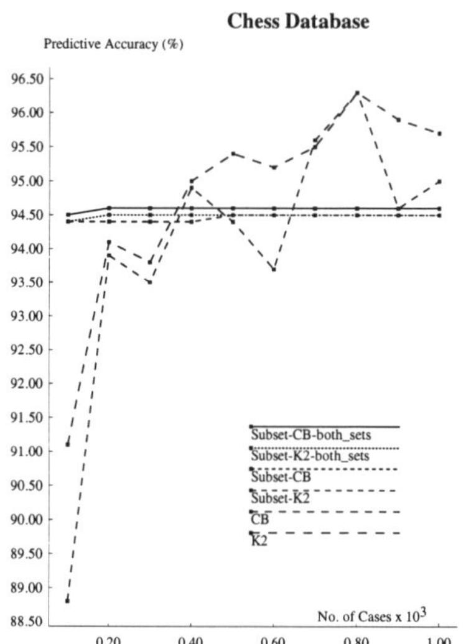

FIGURE 1. Induction rates for Gene-splice and Chess databases

model with *good* predictive accuracy. It is possible to compute multiple models and average over them [Madigan93] to obtain the best predictive accuracy, and we hope to take this approach in future work. In addition, we restrict our attention to Bayesian networks. To fairly compare the best possible predictive accuracy of other approaches to the predictive accuracy obtained using reduced-attribute Bayesian networks is not useful; rather, the averaged-model approach should provide a predictive accuracy for such comparison.

Another advantage of K2-AS over CB is that since it selects a much smaller set of features, the induction rate for the Bayesian networks learnt by K2-AS from the smaller set is much faster than that of the networks learnt by CB from the entire set of features. Figure 1 shows the induction rates for the Splice and Chess databases. These curves were generated from a single trial, unlike the predictive accuracies which were averaged over multiple trials. In these graphs CB derived an ordering from the data and K2 was given the " pseudo-optimal" node ordering selected by CB which resulted in a network with the best predictive accuracy. We denote the feature selection algorithms using the "subset" prefix.[5]

Since the number of attributes used by the subset-selection algorithm for these two domains is small relative to the total number of attributes (approximately 15%), few cases are needed to learn the networks, and the learning rate curve is almost flat. In addition, the predictive accuracy of the reduced-feature networks is within 2% of the networks with all attributes. Hence, in these domains K2-AS significantly reduced network size with little loss of predictive accuracy.

Figure 2 shows the induction curves for the Soybean database for a single trial. The

---

[5]The curves with suffix "both-sets" refer to predictive accuracies obtained by using the first two parts of the database for prior probabilities, as opposed to using just the first part.

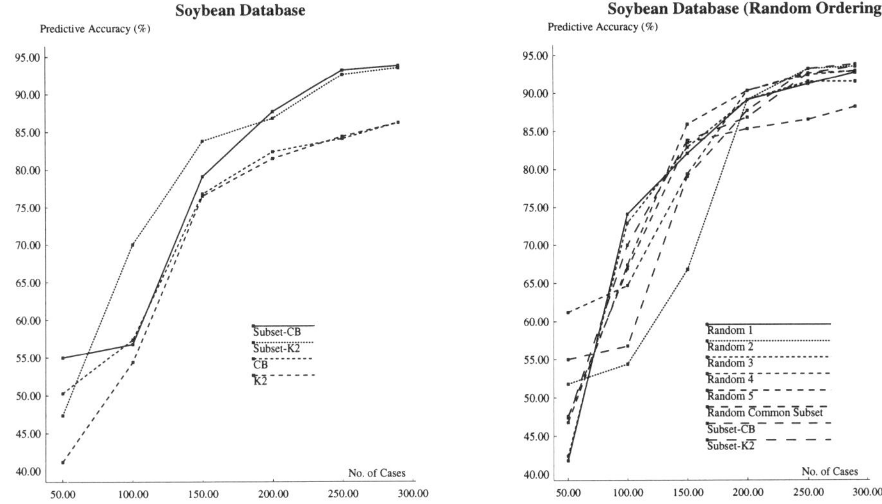

FIGURE 2. Induction rates for Soybean databases

number of attributes used by the subset-selection algorithm for this domain is 26%. Since the reduced-attribute network is closer in size to the full-attribute network, the learning rates for subset-based and full-attribute-set approaches are comparable.

The right-hand-side graph of Figure 2 shows the learning rates for the soybean database for a variety of node orderings: five random orderings, the average of these five orderings, the ordering computed by CB (subset-CB), and the pseudo-optimal ordering for K2 (subset-K2). This figure shows that although there is some variability in predictive accuracy and learning rate for random orderings, the final predictive accuracy is relatively insensitive to the ordering chosen: the predictive accuracy results for all but one of the random orders are within 2% of the predictive accuracy of the pseudo-optimal ordering.

	Nodes	Nodes Chosen	Nodes Always Chosen
Gene-splice	60	9	6
Soybean	35	11	6
Chess	36	5	4

TABLE 28.3. Nodes always selected (regardless of ordering) over multiple random orderings

We also compared the nodes selected by these eight orderings (random, CB-derived and pseudo-optimal) over a single trial, as shown in Table 28.3. Table 28.3 shows that each ordering contained an identical set of nodes; we call these the relevant nodes. For example, for the soybean database, 6 of the 11 features selected were common over all the runs using random orderings. In addition, of the networks nodes which are not always selected, there was a subset that was selected in most runs using random orderings; we call these weakly relevant nodes. This approach thus provides an empirical method for determining relevance of nodes in a Bayesian network.

28.6   Related Work

Feature selection has been widely used in statistics and pattern recognition, and its use within the computational learning community has become quite widespread within the last few years. In statistics, research on feature selection has focused primarily on selecting a subset of features within linear regression. Techniques developed include sequential backward selection [Marill63], branch&bound [Narendra77, Xu89], and search algorithms [Siedlecki88]. A 1993 meeting of the Society of AI and Statistics was dedicated to papers on "Selecting Models from Data"[Cheeseman94], and contains a large number of papers on feature selection. This statistical approach to subset selection shares many principles with other statistical notions of information minimality, like MDL. For example, Dawid discusses the close relation between subset selection and the MDL principle in [Dawid92].

The computer vision community has extensively studied feature selection [Dejviver82], and has formed a sub-community within computational vision called pattern recognition. The mainstream vision community typically does not use statistical feature selection to identify relevant features, but makes assumptions about what the features should be present to represent, for example, a class of objects.

Feature selection has received considerable attention in the last few years within the computational learning community, using both filter-based and wrapper-based approaches [John94]. A filter model filters out less relevant features using an algorithm different from the induction algorithm used for the learning, and a wrapper model uses induction algorithm itself for feature selection.

Three filter-model approaches that have been taken are: the Relief algorithm [Kira92a, Kira92b] (this has been extended in [Kononenko94]), the FOCUS algorithm [Amuallim91], and an extended nearest-neighbor algorithm [Cardie93]. Wrapper-based approaches have been studied in [John94, Caruana94, Langley94b], among others.[6]

A growing consensus in this research is that the success of feature selection is strongly correlated to the data itself, as well as to the algorithm employed. Many domains studied (for example, many domains described in the University of California, Irvine database) have a relatively small number of features, namely the features have been pre-selected for their relevance. It is expected that in such domains feature selection may not make a significant impact. One exception is the study of cloud classification by Aha and Bankert [Aha94], in which a set of 204 attributes were significantly pruned, leading the greatly improved performance.[7] Better understanding of data sets and of domains may lead to a deeper understanding of the role of feature selection, and improved performance from feature selection algorithms.

---

[6]Langley [Langley94a] presents a thorough review of feature selection approaches studied within the Machine Learning literature.

[7]In this domain it is likely that the features were not pre-selected for relevance, as there was no *a priori* knowledge of relevance and irrelevance.

28.7   Conclusion

This paper introduces a feature-selection approach for learning Bayesian networks using a greedy search (i.e. K2-based) algorithm called CB. Selecting a subset of features prior to learning the networks significantly improves the inference efficiency of the resulting networks, and achieves a predictive accuracy comparable to networks learned using the full set of attributes.

We believe that the benefits of having a simpler network (which greatly reduces the inference time) can outweigh the slight reduction in predictive accuracy and the one-time cost incurred during network construction, especially in cases where the network generated using the entire set of features may be too large to even allow inference.

In addition, we have showed that the predictive accuracy of the learned Bayesian network is relatively insensitive to node ordering, and that this approach can identify the most relevant subsets of nodes in a Bayesian network.

Acknowledgements

The second author gratefully acknowledges the use of HUGIN provided by Hugin Expert A/S, Denmark for his PhD research. We also thank Wray Buntine and RIACS / NASA Ames Research Center for making IND Version 2.1 available to us.

28.8   REFERENCES

[Aha94] Aha, D.W. and R.L. Bankert (1994). Feature selection for case-based classification of cloud types. In *AAAI Workshop on Case-based Reasoning*, 106–112, Seattle, WA. AAAI Press.

[Amuallim91] Amuallim, H. and T.G. Dietterich (1991). Learning with Many Irrelevant Features. In *Proc. Conf. of the AAAI*, 547–552. AAAI Press.

[Andersen89] Andersen, S.K., K.G. Olesen, F.V. Jensen and F. Jensen (1989). "HUGIN— a Shell for Building Bayesian Belief Universes for Expert Systems". *Procs. IJCAI*, 1080–1085.

[Buntine92] Buntine, W.L. and T. Niblett (1992). A further comparison of splitting rules for decision-tree induction. *Machine Learning*, 7.

[Cardie93] Cardie, C. (1993). Using Decision Trees to Improve Case-based Learning. In *Proc. Machine Learning*, 25–32. Morgan Kaufmann.

[Caruana94] Caruana, R. and D. Freitag (1994). Greedy attribute selection. In W. Cohen and H. Hirsch, editors, *Proc. Machine Learning*, 28–36. Morgan Kaufmann.

[Cheeseman94] Cheeseman, P. and W. Oldford, editors. (1994). *Selecting Models from Data: AI and Statistics IV*. Springer-Verlag.

[Cooper92] Cooper, G.F. and E. Herskovits (1992). A Bayesian Method for the Induction of of Probabilistic Networks from Data. In *Machine Learning 9*, 54–62.

[Dawid92] Dawid, A.P. (1992). Prequential Analysis, Stochastic Complexity and Bayesian Inference. In J.M. Bernardo, J. Berger, A. Dawid, and A. Smith, editors, *Bayesian Statistics 4*, 109–125. Oxford Science Publications.

[Dejviver82] Dejviver, P. and J. Kittler (1982). *Pattern Recognition: A Statistical Approach*. Prentice-Hall.

[Herskovits90] Herskovits, E. and G.F. Cooper (1990). KUTATO: An Entropy-Driven System for Construction of Probabilistic Expert Systems from Databases. In *Proc. Conf. Uncertainty in Artificial Intelligence*, 54–62.

[John94] John, G., R. Kohavi, and K. Pfleger (1994). Irrelevant features and the subset selection problem. *Proc. Machine Learning*, 121–129. Morgan Kaufmann.

[Kira92a] Kira, K. and L. Rendell (1992). A practical approach to feature selection. In *Proc. Machine Learning*, 249–256, Aberdeen, Scotland. Morgan Kaufmann.

[Kira92b] Kira, K. and L. Rendell (1992). The Feature Selection Problem: Traditional Methods and a New Algorithm. In *Proc. AAAI*, 129–134. AAAI Press.

[Kononenko94] Kononenko, I. (1994). Estimating attributes: Analysis and extension of relief. In *Proc. European Conf. on Machine Learning*, 171–182.

[Langley94a] Langley, P. (1994). Selection of relevant features in machine learning. In R. Greiner, editor, *Proc. AAAI Fall Symposium on Relevance*. AAAI Press.

[Langley94b] Langley, P. and S. Sage (1994). Induction of selective bayesian classifiers. In *Proc. Conf. on Uncertainty in AI*, 399–406. Morgan Kaufmann.

[Madigan93] Madigan, D., A. Raftery, J. York, J. Bradshaw, and R. Almond (1993). Strategies for Graphical Model Selection. In *Proc. International Workshop on AI and Statistics*, 331–336.

[Marill63] Marill, T. and D. Green (1963). On the effectiveness of receptors in recognition systems. *IEEE Trans. on Information Theory*, 9:11–17.

[Murphy92] Murphy, P.M. and D.W. Aha (1992). UCI Repository of Machine Learning Databases. Dept. of Information and Computer Science, Univ. of California, Irvine.

[Narendra77] Narendra, M. and K. Fukunaga (1977). A branch and bound algorithm for feature subset selection. *IEEE Trans. on Computers*, C-26(9):917–922.

[Siedlecki88] Siedlecki, W. and J. Sklansky (1988). On automatic feature selection. *Itnl. J. of Pattern Recognition and Artificial Intelligence*, 2(2):197–220.

[Singh95] Singh, M. and M. Valtorta (1995). Construction of Bayesian Network Structures from Data: a Brief Survey and an Efficient Algorithm. *Int. Journal of Approximate Reasoning*, **12**, 111–131.

[Xu89] Xu, L., P. Yan, and T. Chang. (1989). Best-first strategy for feature selection. In *Proc. Ninth International Conf. on Pattern Recognition*, 706–708. IEEE Computer Society Press.

29
Data Representations in Learning

Geetha Srikantan and Sargur N. Srihari

CEDAR, Department of Computer Science
State University of New York at Buffalo
202 UB Commons, 520 Lee Entrance
Amherst, NY 14228-2567
USA

ABSTRACT

This paper examines the effect of varying the coarse-ness (or fine-ness) in a data representation upon
the learning or recognition accuracy achievable. This accuracy is quantified by the least probability
of error in recognition or the *Bayes error rate*, for a finite-class pattern recognition problem.
We examine variation in recognition accuracy as a function of resolution, by modeling the granularity
variation of the representation as a refinement of the underlying probability structure of the data.
Specifically, refining the data representation leads to improved bounds on the probability of error.
Indeed, this confirms the intuitive notion that more information can lead to improved decision-
making. This analysis may be extended to multiresolution methods where coarse-to-fine and fine-
to-coarse variations in representations are possible.

We also discuss a general method to examine the effects of image resolution on recognizer performance. Empirical results in a 840-class Japanese optical character recognition task are presented.
Considerable improvements in performance are observed as resolution increases from 40 to 200 *ppi*.
However, diminshed performance improvements are observed at resolutions higher than 200 *ppi*.
These results are useful in the design of optical character recognizers. We suggest that our results
may be relevant to human letter recognition studies, where such an objective evaluation of the task
is required.

29.1 Introduction

We examine the critical role of the data representation in learning and classification. In
particular, we analyze the coarse to fine variation in representation induced when objects
are digitized at different resolutions by the image capturing mechanism. Figure 1 contains
images of Japanese characters digitized at different resolutions - 40 to 400 pixels per
inch (*ppi*). As resolution is decreased the shape and structure of individual characters is
progressively less well-defined. Discriminability of patterns at reduced granularity of their
representation is harder to establish as a consequence. Clearly, measurement of data is
critical and affects subsequent data engineering and analysis.

A formalization of the discriminability of patterns at varying resolutions is presented.
Improved bounds on the Bayesian error probability as a function of increased resolution are
proved via probabilistic refinement modeling. Effects of reduced resolution of performance
are demonstrated empirically in an application for Japanese optical character recognition,
involving over 800 character classes. Intrinsic information content of the data is measured

[1] *Learning from Data: AI and Statistics V.* Edited by D. Fisher and H.-J. Lenz. ©1996 Springer-Verlag.
[2] Geetha Srikantan's current address is Hughes Information Technology Systems, 1616 McCormick Dr.,
Upper Marlboro, MD 20774. Email: gsrikant@eos.hitc.com.

<u>Japanese Characters</u>

Pixels Per Inch

400	列	連	路	労	郎
300	列	連	路	労	郎
200	列	連	路	労	郎
100	列	連	路	労	郎
80					
66					
50					
40					

FIGURE 1. Two-Dimensional Images at Multiple Resolutions

via mutual information and shown to be a useful predictor of classifier performance on the data.

This study has a particular significance in the domain of optical character recognition (OCR) (an integral component of many document image understanding systems). By using coarse data representations at the OCR level much higher speeds of operation are possible, [15, 14, 12]. However, this is at the cost of accuracy in recognition. Trade-offs between speed and accuracy of the OCR component are to be established to aid in system design. Our empirical results provide design guidelines for practical pattern recognition problems.

We address the specific problem of recognition accuracy with respect to measurement dimensionality and resolution - both induced by image digitization parameters. This problem is complicated by several factors. In practical applications such as character recognition, there is some intrinsic overlap between character classes as the pattern shapes are often "non-orthogonal". For instance, character images of $\{O, Q\}$ and $\{I, l, 1, i\}$ appear visually similar, and are harder to distinguish between at lower resolutions. Additional error is introduced in measurements at lower resolution. As in all practical statistical applications, the effects of finite training set size is an additional complication. Hence, the general problem of relating the data resolution to classification accuracy is a hard one. Previous work in pattern recognition focused on the effect of general measurement/feature dimensionality on performance when finte samples or training sets are used. More details on previous work and how our results relate to them follow in Section 29.2.4.

In Section 29.2, the theoretical framework and results are discussed. Empirical work in the domain of Japanese character recognition is presented in Section 29.3. Concluding remarks follow in Section 29.4.

29.2 Probabilistic Refinement Models Varying Data Resolution

29.2.1 Bayesian Formalism

Given a potentially infinite set Ω of distinct objects, the goal in pattern recognition [6] is to find an effective method of classifying these objects into a finite set of classes, Θ, where $|\Theta| \ll |\Omega|$. In other words, we seek a mapping $\phi : \Omega \to \Theta$, such that given any arbitrary member of Ω, ϕ should automatically classify it to a single $\theta \in \Theta$. In the Bayesian formalism, the random variable X, taking values is the set Ω is used to estimate the value of random variable S, which takes values is Θ. That is, the observations are used to determine the class identity. We assume *a priori* knowledge of P the probability distribution of Θ and $W = W(x|s)$, the conditional probability matrix of x given s. The Bayes error rate - recognition error of the optimal Bayesian classifier [6] - is the best performance possible for a given problem. Bounds on the Bayes error rate are useful to know as they determine absolute limits in the performance of any pattern recognition system.

29.2.2 Probabilistic Refinement

We model the variation in granularity of the data as a refinement of the underlying probability structure [3]. Assume that each point x_i in the original space gets subdivided into i_k points, and that the probability measure of x_i is also subdivided among the new points in X'. This is called a refinement [3] P', of P. The refinement of the conditional probabilities W is expressed as W'.

29.2.3 Refinement vs. Bounds on Probability of Error

Information-theoretic terms are introduced in this section, followed by our extension of previous information-theoretic bounds on error probability under the refinement model.

The uncertainty in S given by another random variable X can be assessed by the conditional entropy or equivocation [4]. If X is a random variable (or vector, in our case) over the space V with a well-defined probability distribution $P(x)$, such that for each $x \in V$, $p(.|x)$ is well-defined, then the equivocation is defined as:

$$H(S|X) = -\sum_{x \in V} p(x) \sum_{\theta} p(\theta|x) \log p(\theta|x).$$

Mutual information, [4], $I(P,W)$ is a measure of dependency between random variables, S and X, defined as follows:

$$I(S;X) = I(P;W) \; = \; \sum_{\theta \in \Theta} \sum_{x \in X} P(\theta)W(x|\theta) log \frac{W(x|\theta)}{\sum_{\theta \in \Theta} P(\theta)W(x|\theta)} \tag{29.1}$$

The following relations hold between $I(P;W)$ and $H(S|X)$ [4], where $H(X)$ and $H(S)$ are the self-information of random variables X and S, respectively.

$$I(P;W) = I(S;X) = H(X) - H(X|S) = H(S) - H(S|X) \tag{29.2}$$

It can be shown that equivocation is a non-increasing function of probabilistic refinement and that the mutual information is a non-decreasing function of probabilistic refinement (for proofs, refer [3, 16]).

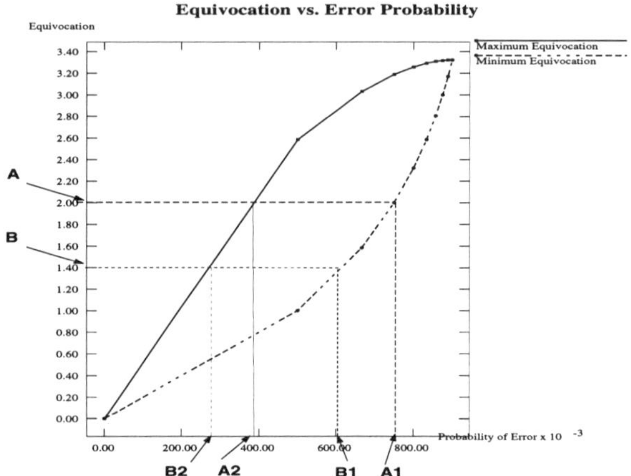

FIGURE 2. Resolution and Equivocation vs. Probability of Error

Recent bounds on the error probability in terms of the equivocation [7] are extended in this paper. As refinement improves information theoretic estimates [3], an important consequence is the improvement in the bounds on error probability. Theorem 29.2.1 is predicts lower bounds on error probability as the equivocation decreases. By relation 29.2 a similar result can be proved for the mutual information measure between measurements and class identities. That is, improved bounds on error probabilty are proved as a function of increasing values of mutual information estimate.

Theorem 29.2.1 *Bounds on probability of error decrease with increase in sampling rate.*

Proof:
Observe that equivocation decreases with refinement induced by increasing the sampling rate. As $\phi^*()$ and $\Phi()$ are monotonic and continuous functions and defined based on $H(S|X)$, both functions decrease as $H(S|X)$ decreases.

A pictorial proof of this theorem is shown in Figure 2. The relation between the equivocation and probability of error predicts all the achievable pairs of $(H(S|X), \Pi(S|X))$, as shown in Figure 2, for the case of $M = 10$. Equivocation is higher at the point A, which models the low sampling rate data representation, than at B which models a higher sampling rate representation. Correspondingly, the upper and lower bounds on the error probability assuming that equivocation is at A, are shown to be $A1$ and $A2$ respectively. Observe that the upper and lower bounds on the error probability for lower value of the equivocation at B, are reduced to $B1$ and $B2$ respectively. Hence, if $H_2 \leq H_1$, where H_1 is the equivocation at A and H_2 is the equivocation at B, then the possible (h, π) pairs possible can be bounded more tightly for H_2 than for H_1.

29.2.4 Discussion

In interpreting the results in Theorem 29.2.1, the lack of a $1 - 1$ map between the mutual information and error probability, must be noted. In other words, there exist probability distributions P_1 and P_2, such that $H_1 < H_2$, $I_1 > I_2$ and $P_{e_1} > P_{e_2}$, where (H_1,H_2), (I_1,I_2) and (P_{e_1},P_{e_2}) are the corresponding equivocation, mutual information and error probabilities, respectively, at the two distributions. That is, decreased equivocation measures do not necessarily imply reduced error probabilty nor does increased mutual information measures necessarily imply that the error probabilty of the distribution will be lower. As a result, in the Theorem 29.2.1 improved *bounds* on error probability are proved rather than improvement in the error probability itself. While these bounds on error probability are of great interest, they do require a priori knowledge of P the probability distribution on the finite set of pattern classes, and W the conditional probability distribution matrix. Such statistics are seldom known in practical problem domains.

The analysis presented above can be extended to multiresolution and hierarchical goal-directed approaches in artificial intelligence. Essentially low resolution implies higher uncertainty and increasing the resolution reduces the uncertainty. In multiresolution object recognition we see that the bounds on recognition error probability improve with increasing resolution of the data.

Previous Work

While there has been no previous systematic study of the problem of relating data resolution (as determined by image digitization rate) and recognition accuracy, certain related research issues have been addressed previously. A brief summary of related work is discussed below. The main distinction between our research and previous work is also mentioned.

- Image or signal "sampling"[3] Most of this research has been concerned with the reconstruction of the original signal given measurements at the sample points [10, 11]. Note, however that in pattern recognition the goal is to discriminate between patterns of distinct classes, rather than exact reconstruction of the original object. In our analysis, we do not require exact reconstruction of the original signal for the purpose of recognition.

- *Mean recognition accuracy* of statistical pattern recognizers [8] examines the effect of measurement dimensionality on recognition performance, with attention to finite training set size effects. Finite-class pattern recognition problems defined by the number of training elements of each class and the class-conditional probability P for each quantized cell of a measurement space was considered. The *mean recognition accuracy* is defined as the probability of correct recognition averaged over all sets P and over all sets of training elements of the assumed size. Examining mean recognition accuracy as a function of (a) the pattern measurement complexity (or features) and (b) the data set size, [9] (and references therein) show that peaking in performance for a finite measurement complexity is possible under certain conditions. Under other conditions monotonic improvement in performance with increase in measurement complexity is shown. Several assumptions such as independent and binary measurements, recoverability of smaller measurement sets from larger ones [18] have been used in the monotonicity arguments. This body of work examines

---

[3]"sampling" in Digital Signal Processing refers to the points in time or space, where a signal's value is measured or 'sampled'.

general measurement spaces without reference to a structured representation such as resolution induced representations that we have addressed.

- *Theoretical bounds on error probability* in pattern recognition and information theory are studied firstly to determine analytic bounds in recognition problems and secondly as a guideline for practical pattern recognition system design. We have extended some of these results to the problem at hand.

- Baird [2] discusses a model-based approach to evaluating the intrinsic error in recognition of machine-printed characters digitized at low resolutions. Wang and Pavlidis [19] have examined the trade-off between quantization and resolution in a character recognition approach. They demonstrate the importance of grayscale in improving recognition performance at low resolutions. In the following section we describe our computational model for examining resolution effects in a pattern recognition application. The methods we describe are quite general and can be applied to a wide range of pattern recognition problems - including handwritten and machine-printed character recognition, as well as other object recognition problems.

29.3   Empirical Results

In practical problems, we are restricted to finite sized training sets and these bounds are approached in approximate fashion only. In our experiments we determine error probabilty based on the performance of nearest neighbor classifiers. As these classifiers are non-parametric, empirical data directly influences performance evaluation.

A description of algorithms, experiments and empirical results in Japanese character recognition are presented in this section. We demonstrate that the performance of the classifiers improves with resolution, (that is, without peaking in performance for a fixed resolution). Also information theoretic measures are shown to be useful predictors of error probability or recognition accuracy. An estimate of the intrinsic error in recognition at low resolutions is achieved by these results.

29.3.1   Design Philosophy

Empirical evaluation of resolution effects on a pattern recognition task requires data at multiple resolutions. We simulate multi-resolution data for reasons described below. As each character image, at any resolution, can have an arbitrary dimension, direct comparison of any two images is not possible. We generate fixed-length feature vectors from each image to simplify the classification task. A K-nearest neighbor classifier was used in our experiments, as it is close to optimal in performance. In addition to evaluation of recognition performance from image feature vectors of the testing set, we also evaluated the information content in feature vectors of the training set at multiple resolutions.

Multi-resolution Simulation

For our application in optical character recognition, character images were extracted from document images digitized at a fixed resolution. This involved a combination of manual and automatic character extraction methods. Extraction and labeling of character images

Data Description			
# of Classes	Training	Testing	Comments
840	21131	41601	Machine-print Japanese alphabet

TABLE 29.1. Japanese Data Set

from large documents is a time-consuming and labor-intensive task. In order to avoid repeating this process of character image extraction at multiple resolutions, we have devised a method for data simulation at multiple digitization resolutions, via multirate filter theory [5]. Japanese character images captured at 400 *ppi* were modified to multiple resolutions between 40 and 400 *ppi* artificially using this method. For more details refer [16].

Japanese Character Dataset

The data used in the experiment is described in Table 29.1. Machine-printed Japanese data set images originally digitized at a resolution of 400 ppi are reduced to resolutions of $\{40, 50, 80, 100, 120, 160, 200, 240, 300, 320, 350, 360\}$ ppi. We used images of only 840 character classes out of approximately 3300 classes in the JIS Level I code. This choice of data subset was made, as the other classes did not contain sufficient data in this database.

Experiment Details

Directional features [17] are extracted from train and test data. Figure 3 shows directional features in a Japanese character image that were encoded into a vector of length 192. A K-nearest neighbor classifier, with $K = 3$, is used for classification of the test data into one of the 840 classes. That is, each training feature vector is stored with the class identity; each test feature vector is compared with all the stored data to find the K nearest neighbors using a Euclidean distance metric. The most common class among the K neighbors is selected as the class identity for the test image.

As explained in Section 29.2.2, mutual information, $I(P, W)$ between the measurements or features and class identity improvements predict improved bounds on error probability. To demonstrate this empirically, we estimate the value of mutual information between feature measurements and class identities of training images at each resolution and plot this as a function of resolution.

Results

The results are shown in Figures 4 and Figure 5. The recognition accuracy (which is the dual of the error probability) is graphed on the y-axis for each resolution (x-axis) in Figure 4. Recognition performance is seen to improve drastically as resolution increases from 40 to 200 it ppi. Thereafter, diminished returns are observed for increased resolution of the data.

In Figure 5 the intrinsic information content of features extracted from images at multiple resolutions is calculated via the mutual information 29.1 between the feature vectors and class identities is graphed at each resolution. It is observed that the mutual information and recognition accuracy both improve as a function of the resolution. Hence the

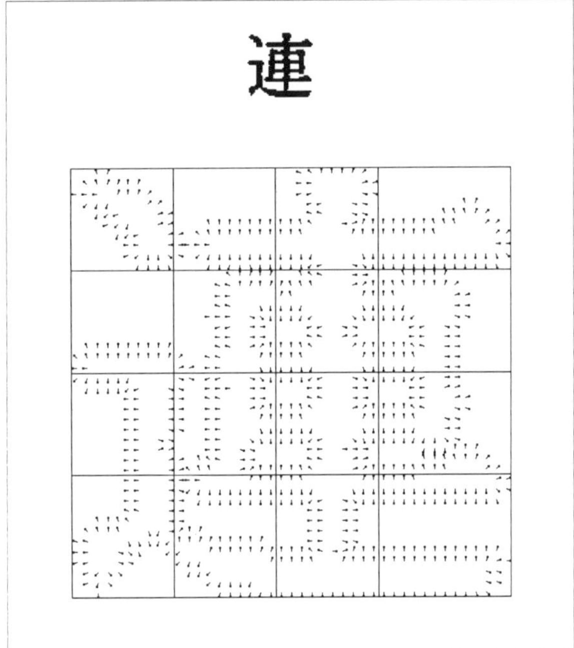

FIGURE 3. Directional Features used in Character Recognition

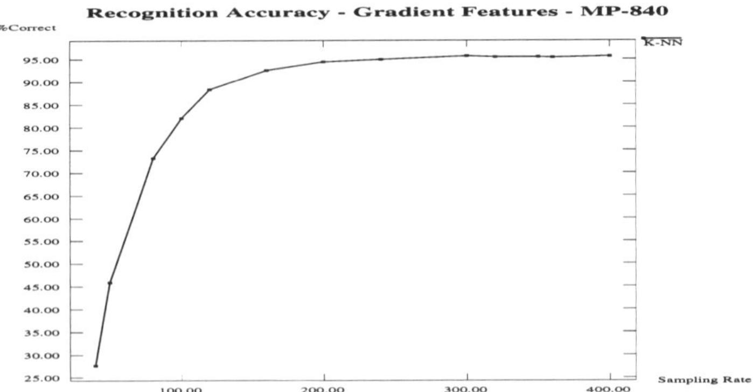

FIGURE 4. Recognition accuracy vs. Sampling rate - Gradient features: Machine-print Japanese Alphabets (840 classes)

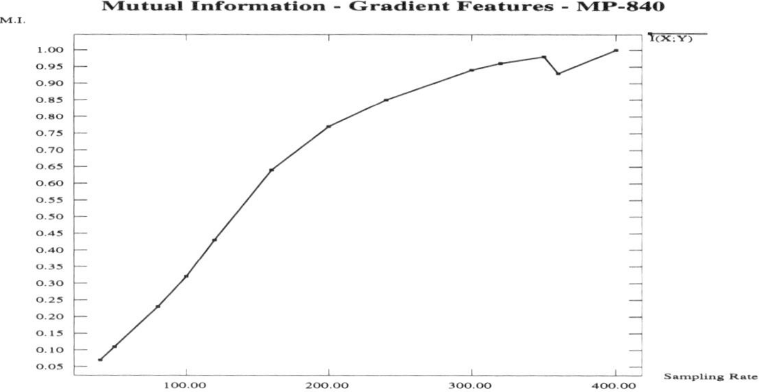

FIGURE 5. Information measure vs. Sampling rate - Machine-print Japanese Alphabets (840 classes)

empirical information estimate is a good predictor of the recognition accuracy or error probability in a practical pattern recognition problem also.

29.4   Summary

We have demonstrated the importance of data representation on recognizer performance both analytically and empirically. Extension of bounds on error probability under varying resolutions has been presented. A general methodology to examine the effect of varying resolution on classification performance has been devised and demonstrated. Experiments in Japanese optical character recognition demonstrate the effect of resolution on recognition in a practical problem.

While our empirical work is in the domain of optical character recognition, the theory and methods are easily extensible to other problems in object recognition. Our results establish that the point of diminishing returns is 300 *ppi* in the case of Japanese machine-prinetd character recognition. Our results contribute towards the design of practical pattern recognition systems where choices regarding the data resolution have to be made.

We suggest also that our results can be used as an objective or ideal observer in calibrating certain human visual tasks, such as letter recognition [1, 13]. Much work in human letter recognition is geared towards identifying the limits of human visual capability and in determining visual acuity and disorders by comparison with some standardized reference. We hope that our results will be useful in this domain also.

Further analysis of the relation between resolution and machine or human learning/recognition must take into account domain-specificity (as object frequencies vary over wide ranges across domains) and feature dependence of the problem.

29.5   REFERENCES

[1] K. R. Alexander, W. Xie, and D. J. Derlacki. Spatial-frequency characteristics of letter identification. *Journal of the Optical Society of America - A*, 11:2375–2382, 1994.

[2] H. S. Baird. Document image defect models and their uses. In *Proceedings of ICDAR, 1993*, 1993.

[3] R. E. Blahut. *Principles and Practice of Information Theory*. Addison-Wesley, 1990.

[4] T. M. Cover and J. A. Thomas. *Elements of Information Theory*. John Wiley, 1991.

[5] R. E. Crochiere and L. R. Rabiner. Interpolation and decimation of digital signals: A tutorial review. *Proceedings of the IEEE*, 69:300–331, 1981.

[6] R. O. Duda and P. E. Hart. *Pattern Classification and Scene Analysis*. John Wiley, 1973.

[7] M. Feder and N. Merhav. Relations between entropy and error probability. *IEEE Transactions on Information Theory*, 1994.

[8] A. K. Jain and B. Chandrasekaran. Dimensionality and sample size considerations in pattern recognition practice. *Handbook of Statistics - Classification, Pattern Recognition and Reduction of Dimensionality, Ed. P. R. Krishnaiah & L. N. Kanal*, 2:835–855, 1982.

[9] B. Kanal, L. & Chandrasekaran. On dimensionality and sample size in statistical pattern recognition. *Pattern Recognition*, 3:225–234, 1971.

[10] D. Lee, T. Pavlidis, and G. W. Wasilkowski. A note on the trade-off between sampling and quantization in signal processing. *Journal of Complexity*, 3:359–371, 1987.

[11] A. V. Oppenheim and R. W. Schaeffer. *Discrete-time Signal Processing*. Prentice-Hall, 1989.

[12] P. Palumbo, S. N. Srihari, J. Soh, R. Sridhar, and V. Demjanenko. Postal address block location in real time. *IEEE Computer*, pages 34–42, 1992.

[13] D. H. Parish and G. Sperling. Object spatial frequencies, retinal spatial frequencies, noise, and the efficiency of letter discrimination. *Vision Research*, 31:1399–1415, 1991.

[14] S. N. Srihari. High-performance reading machines. *Proceedings of the IEEE*, 80:1120–1132, 1992.

[15] S. N. Srihari and J. J. Hull. Character recognition. *Encyclopaedia of Artificial Intelligence*, 1, 1992.

[16] G. Srikantan. *Image Sampling Rate and Image Pattern Recognition*. Doctoral Dissertation, Department of Computer Science, SUNY at Buffalo, 1994.

[17] G. Srikantan and S. N. Srihari. A study relating image sampling rate and image pattern recognition. In *CVPR-94*. IEEE Press, 1994.

[18] W. G. Waller and A. K. Jain. On the monotonicity of the performance of bayesian classifiers. *IEEE Transactions on Information Theory*, 24:392–394, 1978.

[19] L. Wang and T. Pavlidis. Direct gray-scale extraction of features for character recognition. *IEEE Transactions on Pattern Analysis and Machine Intelligence*, 15:1053–1067, 1993.

Part VI

EDA: Tools and Methods

30
Rule Induction as Exploratory Data Analysis

Jason Catlett

AT&T Bell Laboratories
Room 2T-412
600 Mountain Avenue
Murray Hill, NJ 07974
USA

ABSTRACT This paper examines induction of decision rules for purposes of exploratory data analysis, and presents various tools and techniques for this. Decision tables provide a compact, consistent format that allows several fairly large classifiers to be examined on a page. Variation in rulesets arises naturally to a surprisingly large degree from small differences in sampling from a training set, and extreme differences can be provoked by altering parameters related to misclassification costs. We argue that this variation can be beneficial to human understanding by providing multiple views of the data. Unfortunately the goal of simplifying classifiers to improve their clarity conflicts with the requirement in some applications for accurate class probability estimates. These phenomena are explained and exhibited on several datasets, including an easily understood illustration using rulesets for a domain based on the card game Twenty-one. Techniques for displaying rulesets in the form of decision tables are also presented.

30.1 Introduction and motivation

Exploratory data analysis (EDA) [24] is "the manipulation, summarization, and display of data to make them more comprehensible to human minds, thus uncovering underlying structure in the data and detecting important departures from that structure [1]." Despite its venerable goal of producing comprehensible knowledge [19], and its increasingly statistical bent, the literature of Machine Learning (ML) contains only a few papers [3], [14] that consider software for induction as a tool to assist these shared goals. In real applications a long iterative process of experimentation and data analysis [2] is almost always necessary [8], [6]. This paper describes tools and techniques for the induction and presentation of sets of decision rules [7], to help that process.

The next section describes a thought experiment that will help explain and clarify the phenomena investigated here. Section 30.3 describes the Twenty-one domain used. Section 30.4 shows how diversity that helps understand the data can be provoked by specifying a variety of misclassification costs. Section 30.5 examines the large degree of diversity that arises naturally. Section 30.6 surveys related work.

---

[1] *Learning from Data: AI and Statistics V.* Edited by D. Fisher and H.-J. Lenz. ©1996 Springer-Verlag.

30.2 Thought experiment

This section demonstrates how two classifiers with identical accuracy or goodness of fit can give very different impressions of the data from which they were built. It also shows how somewhat rare classes can be overlooked.

Consider inducing a ruleset from examples of two classes, + and –, using a single attribute with values V1 and V2, and the following probability distribution: P(+ & V1) = 0.6, P(– & V1) = 0.0, P(+ & V2) = 0.2, and P(– & V2) = 0.2. Two rulesets that could reasonably be induced from this data are (1) classifier A, which always predicts +, with a probability estimate (PE) of 0.8, and (2) classifier B, which predicts + with a PE near 1.0 for instances with value V1, and gives a PE of 0.5 for each of the two classes for instances with value V2. Both classifiers have the same error rate, but they err in different ways. Classifier A has the advantage of being smaller; B delivers more accurate probability estimates. Unless the analyst examines the count information associated with specific rules, A gives a stronger impression than B that the positive class prevails.

The main point of this thought experiment is to show that due to the discrete choices that have to be made by any rule induction algorithm, a variety of very different candidate classifiers may be available, apparently similar or indistinguishable in error rate.

A secondary point concerns the probability estimates delivered by the classifiers. Some applications require not just a decision, but accurate probability estimates. Here classifier B is superior to A. The price paid is its greater complexity. We face a conflict between the need for accurate estimates and desire to keep the classifiers small and comprehensible.

Now suppose that we change part of the distribution very slightly so that P(+ & V2) = 0.22, and P(– & V2) = 0.18. Given enough data, most rule induction systems will produce only classifier A. This "tyranny of the majority" hides a great deal of information that is potentially useful in data analysis. The 50-50 borderline can be effectively changed in induction systems that allow specification of unequal misclassification costs, such as described by [11] or [17]. In the former paper, which deals only with the case of binary classes, a single number (there called the *loss ratio* (LR)) expresses the ratio of the cost of a false positive error to a false negative error. The default of LR = 1 is the normal case of equal misclassification costs.

In many ML applications one class is comparatively rare and of special interest: in manufacturing and medicine, periods during which a system is behaving abnormally; in information retrieval (IR), pieces of text relevant to some particular category; in marketing, customers who take certain actions (desirable or undesirable). The goal of induction during data analysis may be simply to explore the areas of the attribute space where the rare class is comparatively more common, even if it nowhere reaches a majority. We need ways to see those areas.

30.3 Description of illustrative domain

In order to illustrate the issues above on a domain of plausible complexity, but which is comprehensible without specialist knowledge, I constructed a classification task based on the popular casino game *Twenty-one*, also called *Blackjack*. [15]. Several departures from the real game are made for convenience. Cards are drawn randomly and with replacement from a deck and dealt. Initially two cards are dealt to the player and one card is dealt to the dealer, which the player can also see. Kings, queens and jacks count as 10 points; aces as 1 or 11 at the holder's choice; all other cards as their numeric value. With a total of less than 21 the player may choose either to *sit* or *hit* (to decline or request an additional card). If the player ever exceeds 21, the dealer wins. If the player sits with less than 21, a dealer with a total below 17 is required to take extra cards to attain this minimum; if the dealer exceeds 21 or reaches 17 with less than the player's total, the player wins.

To keep the number of classes to two, the following simplifications are made: with *blackjack* (21 with only two cards) the player wins even if the dealer has the same, but the dealer wins all other draws. We exclude complicated player options such as splitting, doubling, and insurance.

In this simulation the class we aim to predict is whether the player or dealer will win, given the following four attributes, each preceded by their abbreviated names: ptc: the total of the player's first two cards (counting aces as 1); pha: whether the player has at least one ace; duc: the dealer's "up" card; phs: which of the following two equiprobable strategies the player follows: to sit immediately, or to hit and continue doing so until the hand totals 17 or more. We do not attempt to predict whether the player will hit or sit, although some guidelines can be extracted from the rules as to which would be more advantageous in certain situations.

The training sets were of size 1,000 except where otherwise specified. Most of the experiments use the original C4.5, but where a loss ratio other than 1 is specified, a version modified as described in [11] was used. A single independent test set of 10,000 examples was used across all experiments; training sets were generated independently. In most cases figures for both the trees and rulesets built from them are given; where only one is given it refers to rulesets. The class p (player wins) is considered the positive class. The test set contained 6,480 negative examples, so the default error from a classifier that always gives the majority class is 35.2%.

30.4   Provoking diversity to aid EDA

To exhibit the benefit to exploratory data analysis of varying the LR, I have chosen three rulesets for Twenty-one built from the same training set using different values for the loss ratio. These rulesets have been automatically reformatted from C4.5's output and presented in the form of *decision tables*.

Table 30.4 is the simplest; this ruleset is delivered by very small values for LR, such as 1/32. Attribute names are listed in the rightmost four columns. The columns indicate (left to right)

- the class of the rule's conclusion (p for player, d for dealer),

- the rule's identifying number (which in this case happens to be 1, but in general the ordering is arbitrary),

- the percentage of examples that the rule *covers* (i.e. those examples whose attribute values satisfy all the rule's conditions),

- the error rate on those examples matching the rule (Rule one is "sure-bet" at 0.0% errors), and

Class	Rule #	% Coverage	% Error	ptc	pha	duc	phs
p	1	6.1	0.0	[11,–)	y		
d	2	85.9	33.4		n		
d	default	8.0	40.2				
Classifier Totals		100.0	32.0				

TABLE 30.1. A ruleset with low proportion classified positive

- rule conditions: a blank slot indicates the attribute did not appear in any condition of the rule (this is sometimes called a *don't care*); a numeric range indicates the attribute value must be in that range (this is actually two conditions except in the case of ranges open at one end); and a discrete value indicates the condition is a test of equality for that value.

Thus the first rule is exactly the blackjack rule described above: "if the player's total is 11 or more and the player has an ace, then the player will win." The next line, Rule 2, simply states that if the player has no ace, then the dealer will win (the class is again given in the subsequent line). This rule is wrong in about a third of the cases. The last line specifies *default class* as d. This *default rule* is used in the 8% of cases where neither of the above rules holds; it has a higher error rate. The last line specifies the collective error rate of the classifier on the entire test set; in this case it is 32.0%.

Table 30.4 contains two rulesets: the upper one produced by the unmodified C4.5 (with a LR of 1.0); the lower one using LR=4.0. The upper ruleset gives ten rules for class p; the figures to the right of the p give the total coverage of the rules for that class and the error rate (here on an independent test) for those examples covered. Where rules are not mutually exclusive, and example is counted under the first rule it satisfies. C4.5 groups all rules for a class together (with the possible exception of the default rule, as in the lower ruleset), so it effectively orders the classes: the rules for the second class are considered only if none of the rules for the first class is satisfied.

The main point of this section is that varying the LR parameter can yield useful, enlightening rules that may otherwise be missed. The clearest example is the rule for blackjack (Rule 1, the only p-class rule in Table 1): on many training sets it does not appear in the ruleset built by the standard C4.5; in its stead is a less accurate rule of slightly wider coverage with a lower threshold for the player's total count and a condition on the dealer's count (Rule 9 in Table 2 upper). (Its appearance becomes more frequent as training sets become larger.) Another example is the first rule in Table 2 lower, which predicts that if a player hits with high total when the dealer is showing a 9 or 10 (reducing his chances of exceeding 21), then the dealer is very likely to win. This rule is extremely obvious and intuitive to anyone who has played the game, yet no comparable rule appears in the original C4.5 ruleset. Because of the "masking" effect of higher-priority rules, the conjecture that the original ruleset may nonetheless be getting these cases right could not be dismissed at a glance, but this only highlights the point: altering the loss ratio can make some rules clearer, in both senses: easier to understand, and more accurate and clear-cut. This capability can be very helpful in exploratory data analysis.

As forewarned in the thought experiment, the simplicity of these rulesets is bought at the expense of less specific information on the class-probability value assigned to an unseen

Class	Rule #	% Coverage	% Error	ptc	pha	duc	phs
p	9	6.5	7.6	[9,−)	y	[5,−)	
p	41	1.2	50.0	[11,11]		[6,8]	
p	7	2.8	37.9	[7,−)	y	(−,4]	
p	5	1.3	76.9	[5, 6]	y	(−,9]	
p	62	2.8	37.9	(−,15]		[6,6]	s
p	43	1.2	75.0	[13,14]		[6,7]	
p	1	0.2	50.0	(−,2]			h
p	31	0.9	44.4	[10,10]		[5,8]	h
p	95	4.7	25.0	[20,−)		[2,−)	s
p	61	2.3	45.8	[16,19]		[4,5]	s
p consequent totals		23.9	32.8				
d	75	2.6	29.6	[12,19]		[8,8]	s
d	3	0.8	62.5	[3,3]			h
d	68	0.7	42.9	(−,5]	n	[7,−)	s
d	27	0.9	44.4	[9,9]		(−,8]	h
d	42	31.9	22.4	[12,−)			h
d	87	17.5	28.5	[6,19]	n	[9,−)	
d	64	10.0	37.3		n	(−,3]	
d	8	0.4	25.0	[7,8]	y	[5,9]	
d	10	1.1	18.2	(−,8]	y	[10,−)	
d consequent totals		65.9	27.5				
d	default	10.4	37.7				
Classifier Totals		100.0	29.8				

Class	Rule #	% Coverage	% Error	ptc	pha	duc	phs
d	28	2.7	14.3	[17,19]		[9,−)	h
d	22	1.8	16.7	[14,18]		[9,9]	
d	20	1.0	20.0	(−,9]	n	[9,9]	
d	3	0.4	25.0	[4,8]	y	[10,−)	h
d	9	0.8	12.5	[15,19]		(−,1]	s
d	6	0.7	28.6	(−,8]	n	(−,1]	
d consequent totals		7.4	17.4				
p	2	13.7	32.9		y		
p	10	4.9	26.0	[20,−)			s
p	29	7.8	62.5	[17,−)		[2,−)	s
p	12	37.1	61.8	(−,19]		[2,8]	
p	27	22.4	73.4	(−,16]			
p consequent totals		85.9	58.2				
d	default	6.7	13.0				
Classifier Totals		100.0	52.2				

TABLE 30.2. A ruleset built by the unmodified C4.5 (upper part) and a ruleset built to have a high proportion classified positive (lower part)

example: a more complex classifier can choose from a larger set of values. For example, Table 1 partitions the attribute space into three regions, each covered by one of the three rules (the blackjack rule, the rule that an aceless player will lose, and the default rule), whereas the other rulesets have many more rules for each class. Each rule is associated with a class-probability value, just as a decision tree gives a separate class-probability at each leaf. A large set of possible class-probability values will be important in applications requiring an accurate estimate of each example's class-probability, but not in those where the only question of interest is whether the value is above or below a certain threshold. In the latter case the clarity offered by the simpler model may make them attractive.

30.5   Natural diversity

This section shows that even the original C4.5 produces rulesets with similar accuracy but which classify a very different fraction of the examples as positive. In the boxplots [24] on the right-hand side of Figure 1, the black circles indicate the error rate for rules averaged over 50 training sets of four sizes from 1,000 to 30,000, with LR=1 (the original C4.5). The notches indicate the 95% confidence interval for the mean; we see from the fact that each successively larger training set has no overlap with the previous notched range that significantly more accurate classifiers are being obtained. The other horizontal lines indicate five percentiles, at 10, 25, 75 and 90. The hollow circles are datapoints above the 90th percentile. The variance is smaller for larger sets, because random subsets are more similar. Still, 1000 examples is already generous for Twenty-one given its comparatively small attribute space (two are binary and the other two have 11 and 19 values, so on average each possible example appears more than once).

The left-hand side of Figure 1 shows on the same scale the corresponding figures for the percentage of examples classified positive. The range is clearly much wider than error rate. A classifier that indicates 12% of unseen examples positive gives a very different impression to one that indicates 27% positive. Other domains I have examined also exhibit comparable variation.

To check that the variation was not due to variation in the percentage of positive examples in the training set, I ran the same experiment with stratified training sets controlled to contain exactly the number of positive examples, but keeping the generation of examples otherwise independent and random. This slightly decreased the ranges for trees, but slightly increased them for rules. The results were not substantially different on the stratified training sets, and same conclusion holds.

To check that these findings were not unique to the Twenty-one domain, I performed similar analyses on other datasets from the UCI Repository. The first, entitled "pima-indians-diabetes", has eight continuous attributes. Using a constant test set of 268 examples, for each of 50 trials, training sets of 400 examples were selected randomly and independently from the remaining 500 examples. The variance of percentage positive was also very large. One might expect that drawing a large majority of the examples from a smallish set would produce a far more homogeneous set of classifiers than would be obtained from sets where all the examples were generated independently. However, a quick test on the Twenty-one domain showed that this does not appear to be the case. The second domain, "euthyroid" from Quinlan's thyroid data [20], has seven continuous attributes, 21 binary and one other discrete attribute. With a fixed test set of 2,000

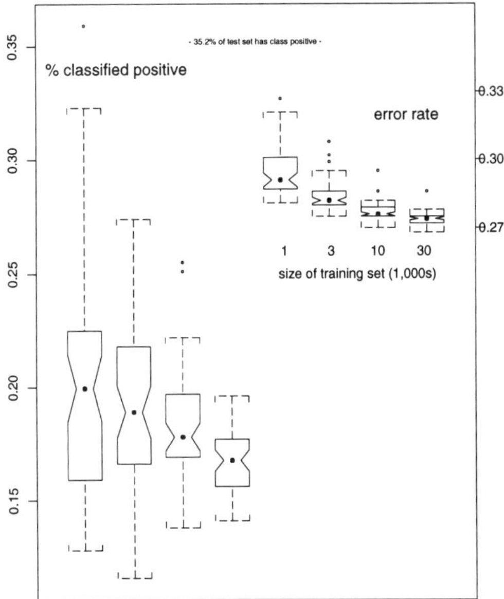

Figure 1: Boxplots of error rate and % positive for 1-30k egs

FIGURE 1. Range of percentage classified positive and error for Twenty-one

examples, 50 training sets of 1,500 examples were chosen from the remaining 5,218 examples. One class is very rare. In this domain rules exhibit much wider range of percentage positive than trees.

In summary, the classifiers produced from very similar training sets by the unmodified C4.5 exhibit considerable variance. This can be used to advantage in getting different views of the data.

30.6   Related work

Perhaps consistent with the fact that most decision-tree induction algorithms offer no way of requesting greater sensitivity to one class, Segal, Etzioni, & Riddle [21] find standard methods "biased towards producing *full-coverage classifiers*—functions that seek to map *any* example to the appropriate class. As a result, the algorithms can overlook patterns that occur in small subsets of the data. Yet, in our application, even minute patterns represent valuable information that we cannot afford to miss." They report improved accuracy from algorithms that perform exhaustive search of small concepts for the distinguished class.

Setting the LR to an extreme value is a useful way to get a classifier small enough to be understood. Another way is overpruning [5] , but the results are very different, because overpruning tries to minimize the error overall, whereas an extreme LR tries to minimize one type of error. They are different tools for different needs.

Quinlan [18] has pointed out that the MDL criterion used by C4.5 to choose among candidate rulesets is strangely indifferent to apparently clear-cut alternatives, and recently added to the code a preference bias towards classifiers with a percentage positive figure close to the percentage of positive examples in the training set. This gave an improvement in generalization performance. (Incidentally, the classifiers in Figure 1 almost all had a percentage positive below that of the minority positive class.) Those modifications do reduce some of the grosser variations, but they cannot remove the intrinsic problem described in Section 30.2: there may exist no deterministic classifier with a percentage positive similar to the training data, even with an enormous sample.

Breiman [4] proposes a method of "bagging" several diverse classifiers from the same data into a larger but more accurate one.

Michie [13], who has long been a strong advocate of "human friendly" induction, describes a method that has a similar affect to altering the LR. For each leaf in a decision tree, the ratio of the number of positive to negative training instances at that leaf is calculated. Leaves with a ratio higher than the user-specified threshold are left as positive; the others are deemed negative. This method is the basis of the modifications I used in [11] (without the pessimistic correction for small samples), before the conversion to decision rules is done.

Decision tables have been commonplace in decision analysis and systems analysis for decades [12], [9], [16], but have been practically ignored in the ML literature [23], [10], though they have appeared in some commercial software packages. Traditionally conditions are listed on the upper rows (versus the columns here), and a row is provided at the bottom for each class value (allowing multiple "actions," versus the assumption of mutually exclusive values here). One study reported that decision tables are harder to understand than decision trees [22], but I find that after a brief introductory explanation most people can work more quickly with tables than raw textual rules. For me the main benefit is allowing the eye to run up and down the columns of a single attribute, even across several rulesets. The decision tables for Twenty-one are fairly small, but it is not difficult to pack into a page some fifty rules across a dozen attributes.

30.7   Conclusions

This paper argued that rule induction can be a useful tool for exploratory data analysis if a variety of rulesets are examined. Variation arises naturally to a surprisingly large degree, and extreme differences can be provoked by altering the ratio of the cost of a false positive error to a false negative error.

Acknowledgements

I thank William Cohen, Ronny Kohavi, David Lewis and Matthew Partridge for their comments on various stages of this work, Doug Fisher for his editorial suggestions, Ross Quinlan for the code for C4.5, Alan Skea for assistance with Twenty-one, Kent Spackman for several references, and Kenji Yamanishi of NEC and colleagues at Bell Labs for discussions on MDL. Donald Michie converted me to the goal of machine learning as data analysis.

30.8   REFERENCES

[1] David F. Andrews. Data analysis, exploratory. In *International encyclopedia of statistics*, pages 97–106, NY, 1978. Free Press.

[2] George E. Box, William G. Hunter, and J. Stuart Hunter. *Statistics for Experimenters: An Introduction to Design, Data Analysis and Model Building.* Wiley, 1978.

[3] I. Bratko, I. Mozetic, and N. Lavrac. *KARDIO: a study in deep and qualitative knowledge for expert systems.* MIT Press, Cambridge, Massachusetts, 1989.

[4] Leo Breiman. Bagging predictors. In *Technical Report No. 421.* Department of Statistics, UC Berkeley, 1994.

[5] J. Catlett. Overpruning large decision trees. In *Proceedings of the Twelfth International Joint Conference on Artificial Intelligence*, volume 2, pages 764–769, Sydney, Australia, 1991. Morgan Kaufmann.

[6] Andrea P. Danyluk and Foster John Provost. Small disjuncts in action: learning to diagnose errors in the local loop of t. In *Proceedings ML-93*, pages 81–88, 1993.

[7] Thomas G. Dietterich and Ryszard S. Michalski. A comparative review of selected methods for learning from examples. In *Machine Learning: an AI approach*, Los Altos, CA, 1983. Morgan Kaufmann.

[8] Bob Evans and Doug Fisher. Process delay analysis using decision tree induction. *IEEE Expert*, 1993.

[9] C. Gane and T. Sarson. *Structured systems analysis: tools and techniques*. Prentice, 1979.

[10] Ron Kohavi. The power of decision tables. In *Proceedings ECML-95*, 1995.

[11] David D. Lewis and Jason Catlett. Heterogeneous uncertainty sampling for supervised learning. In *ML-94*, pages 148–156, 1994.

[12] J.R. Metzner and B.H. Barnes. *Decision table languages and systems*. Academic Press, 1977.

[13] D. Michie. Problems of computer-aided concept formation. In *Applications of Expert Systems*, volume 2, pages 310–333, Wokingham, UK, 1989. Addison-Wesley.

[14] D. Michie. Inducing knowledge from data: first principles. In *Keynote address at the Seventh International Conference on Machine Learning (did not appear in Proceedings)*, 1990.

[15] Albert H. Morehead and Geoffrey Mott-smith. *Hoyle's rules of games*. Signet, New York, 1959.

[16] Bernard M. E. Moret. Decision trees and diagrams. *ACM Computing Surveys*, 14(4):593–623, 1982.

[17] Michael J. Pazzani, Christopher Merz, Patrick M. Murphy, Kamal M. Ali, Timothy Hume, and Clifford Brunk. Reducing misclassification costs. In *Machine Learning: Proceedings of the Eleventh International Workshop*, New Brunswick, NJ, 1994. Morgan Kaufmann.

[18] J. R. Quinlan. The minimum description length principle and categorical theories. In *ML-94*, pages 233–241, 1994.

[19] J.R. Quinlan. Simplifying decision trees. *International Journal of Man-machine Studies*, 27:221–234, 1987.

[20] J.R. Quinlan, P.J. Compton, K.A. Horn, and L. Lazarus. Inductive knowledge acquisition: a case study. In *Applications of Expert Systems*, Glasgow, 1987. Turing Institute Press with Addison Wesley.

[21] Patricia Riddle, Richard Segal, and Oren Etzioni. Representation design and brute-force induction in a boeing manufacturing domain. *Applied Artificial Intelligence*, 8:125–147, 1994.

[22] Girish H. Subramanian, John Nosek, Sanakaran P. Raghunathan, and Santosh S. Kanikar. A comparison of the decision table and tree. *CACM*, 35(1):89–94, 1992.

[23] Atsushi Sugiura, Maximillian Riesenhuber, and Yoshiyuki Koseki. Comprensibility improvement of tabular knowledge bases. In *AAAI-93*, pages 716–721, 1993.

[24] John W. Tukey. *Exploratory data analysis*. Addison-Wesley, Reading, MA, 1977.

31

Non-Linear Dimensionality Reduction: A Comparative Performance Analysis

Olivier de Vel[†], Sofianto Li[†], and Danny Coomans[††]

Dept of Computer Science[†]
James Cook University
Townsville Q4811
Australia
olivier@cs.jcu.edu.au

Dept of Maths and Statistics[††]
James Cook University
Townsville Q4811
Australia
Danny.Coomans@jcu.edu.au

ABSTRACT We present an analysis of the comparative performance of non-linear dimensionality reduction methods such as Non-Linear Mapping, Non-Metric Multidimensional Scaling and the Kohonen Self-Organising Feature Map for which data sets of different dimensions are used. To obtain comparative measures of how well the mapping is performed, Procrustes analysis, the Spearman rank correlation coefficient and the scatter-plot diagram are used. Results indicate that, in low dimensions, Non-Linear Mapping has the best performance especially when measured in terms of the Spearman rank correlation coefficient. The output from the Kohonen Self-Organising Feature Map is easier to interpret than the output from the other methods as it often provides a superior qualitative visual output. Also, the Kohonen Self-Organising Feature Map may outperform the other methods in a high-dimensional setting.

31.1   Introduction

In many applications, dimensionality reduction is used to explore a data set to try to obtain some insight into the nature of the phenomenon that produced the data. We are often interested in understanding the structural relationships that exist in the feature space, such as clusters or data point density discontinuities. Measures that could be used to reveal such structural relationships include the inter-point distance, the "shape" of the data distribution etc... Such an understanding is very domain-dependent and may require a deep understanding of the structures, causality etc... that may exist between the features. In some cases, the application may impose a constraint on the serialisation or ordering of the topological mapping as in, for example, chronological ordering in a time series or regression analysis [CLN91].

In many applications, one important objective of dimensionality reduction is that it preserves as much as possible the structural relationships that exist in the data set when

---

performing the mapping from a high dimension to a lower one (usually two or three dimensions) while, at the same time, removing any redundancy in the data. For data characterised by low dimensions (i.e. where the number of features is small), the "important" or "interesting" relationships may be detected relatively easily by using pictorial methods such as histograms (one or two dimensions), scatter-plot diagrams, or even kinematic graphic techniques (such as small angle rigid body rotations in three dimensions). For a high-dimensional feature space, we have to resort to techniques that reduce or project the feature space into one, or more, lower two- or three-dimensional representations. Such techniques must be able to effectively handle the "curse of dimensionality" (due to the fact that high-dimensional space is sparse for small-to-medium sample sizes) and be able to ignore redundant and noisy features.

The most popular dimensionality reduction techniques are linear transformations, such as principal components analysis (PCA) [Fuk90]. However, PCA and many other similar approaches assume a linear constraint of input space and therefore would not perform satisfactorily for non-linear constraints of input space common in high-dimensional data. Furthermore, the covariance structure of data does not necessarily relate to the clustering of data points. Some limited work has been undertaken on the comparative performance of PCA as a dimensionality reduction technique (e.g. [BD91] and [MHP94]).

More general approaches to dimensionality reduction are those of non-linear methods which do not impose any input space constraints. Such techniques include Non-Linear Mapping (NLM) [Sam69], Multidimensional Scaling (MDS) [BL87] and various clustering algorithms [Fuk90]. Several neural networks techniques have also been introduced more recently as in, for example, Kohonen Self-Organising Feature Map [Koh88] and the Back Propagation algorithm [Sau89]. However, neither the comparative performances nor the quality of the results produced by these methods have been sufficiently investigated.

In this paper, we present three non-linear dimensionality reduction techniques: Non-metric Multidimensional Scaling, Non-Linear Mapping and the Kohonen Self-Organising Feature Map (SOFM), and provide a comparative analysis of the performance of each technique. We briefly give the theoretical background for each method and the techniques used for evaluating the comparative performance of each method.

31.2 Dimensionality Reduction Methods

31.2.1 Kohonen Self-Organising Feature Map

The Kohonen SOFM algorithm [Koh88] attempts to produce a distorted, but topographically-organised, projection of the input space where the similarity of the input vectors is converted into a proximity relationship in the projection. The map will attempt to preserve the "important" similarity relationships present in the high-dimensional input space while, at the same time, factoring out any redundancy latent in the input features [LGZ93].

31.2.2 Multidimensional Scaling

Multidimensional Scaling (MDS) is based on dissimilarity data which reflect the amount of dissimilarity between pairs of objects, events or concepts [Kru64]. Typically, the dissimilarity data are distances between all pairs of points in multidimensional space. Thus similar objects are close together and dissimilar objects are far apart. Consider a set of input patterns (also called *configuration*) $\{\mathbf{x}_i\}$ in n-dimensional space, with dissimilarity data which can be calculated from inter-point distances $\delta_{rs} = \| \mathbf{x}_r - \mathbf{x}_s \|$. The method attempts to find a configuration $\{\mathbf{y}_i\}$ in m-dimensional space, where $m < n$, with inter-point distances $d_{rs} = \| \mathbf{y}_r - \mathbf{y}_s \|$ such that $d_{rs} \approx \delta_{rs}$ for all r and s. In achieving $d_{rs} \approx \delta_{rs}$ $\forall\, r$ and s, a transformation from $\{\mathbf{x}_i\}$ to $\{\mathbf{y}_i\}$ is effected. MDS models with such dissimilarity data, together with linear transformations, are known as *metric Multidimensional Scaling* [BL87]. Rather than using distance magnitudes, it is possible to preserve the rank ordering of d_{rs} to be the same as that of δ_{rs}. In this case, the search for the $\{\mathbf{y}_i\}$ configuration should satisfy :

$$d_{rs} \approx f(\delta_{rs}) \tag{31.1}$$

where f is a monotonically increasing function satisfying:

$$\delta_{r_1 s_1} < \delta_{r_2 s_2} \quad \Longleftrightarrow \quad f(\delta_{r_1 s_1}) < f(\delta_{r_2 s_2}) \tag{31.2}$$

for some r_1, s_1 and r_2, s_2. This variant is referred to as *non-metric Multidimensional Scaling*.

Major operations on non-metric MDS are iterative calculations of the $\{\mathbf{y}_i\}$ configuration for which the monotonicity between d_{rs} and δ_{rs} must always hold. In order to ensure this condition, a measure of departure from monotonicity is defined so that adjustments on the $\{\mathbf{y}_i\}$ configuration can be made to improve the degree of monotonicity. This measure is called the *stress* value, S. The algorithm involves computing $\hat{d}_{rs}$ which minimises [Kru64]:

$$S^2 = \frac{\sum \sum_{r<s}(d_{rs} - \hat{d}_{rs})^2}{\sum \sum_{r<s} d_{rs}^2} \tag{31.3}$$

while holding the monotone constraint:

$$\hat{d}_{r_1 s_1} \leq \hat{d}_{r_2 s_2} \leq \ldots \leq \hat{d}_{r_m s_m} \tag{31.4}$$

Typically, $\hat{d}_{rs}$ are computed using a monotonic regression algorithm and the iterative calculation of the m-dimensional configuration $\{\mathbf{y}_i\}$ is achieved by using the gradient descent method.

31.2.3 Non-Linear Mapping

Non-Linear Mapping (NLM) is a similar concept to that of MDS and was introduced by Sammon in 1969 [Sam69]. Similar to metric MDS, NLM tries to preserve distances between points in the original data and the reduced dimensional data. The NLM method is also based on iterative calculations of the $\{\mathbf{y}_i\}$ configuration which minimises the error:

$$E = \frac{\sum \sum_{r<s} \frac{(\delta_{rs} - d_{rs})^2}{\delta_{rs}}}{\sum \sum_{r<s} \delta_{rs}} \tag{31.5}$$

As in the case of MDS, gradient descent is a typical minimisation procedure employed. A number of optimisation procedures are usually needed in all SOFM, non-metric MDS and NLM methods because the gradient descent procedure may become trapped in a local minimum. To avoid local minima, global optimisation procedures may have to be implemented by using, for example, simulated annealing and genetic algorithms.

31.3   Comparative Performance Analysis Techniques

Two methods of comparing the performance of each dimensionality reduction technique are presented: Procrustes analysis and the Spearman rank correlation coefficient. A third, empirical method, the *scatter-plot*, is used to plot inter-point distances in the original data space versus distances in the reduced dimension data space. This enables the data correlation to be visualised as a two-dimensional scatter-plot.

31.3.1   Procrustes Analysis

The Procrustes analysis method is aimed at measuring how well the shapes of two data configurations match one another. The "shape" of a configuration is commonly understood to refer to the geometrical attributes that remain invariant when the configuration is subject to a rigid body transformation (translation and rotation) and dilatation.

Let the input data configuration, viewed as a geometrical figure in $\Re^n$, consist of p labeled points and represented by a $p \times n$ matrix $\mathbf{X}$. Similarly, define the output data configuration as a $p \times m$ matrix $\mathbf{Y}$ in $\Re^m$, where $m \leq n$. It is assumed that the output configuration $\mathbf{Y}$ is in the $\mathbf{X}$ subspace i.e. point i in $\mathbf{X}$, $\mathbf{X}^{(i)}$, corresponds to point i in the $\mathbf{Y}$ configuration, $\mathbf{Y}^{(i)}$. In an attempt to match configurations $\mathbf{X}$ and $\mathbf{Y}$, $n - m$ columns of zeros are added to the $\mathbf{Y}$ configuration to obtain $m = n$. A typical measure of the degree of coincidence between the two configurations is the sum of the square of distances between corresponding points (M^2), i.e.

$$M^2 = \sum_{i=1}^{p} \|\mathbf{X}^{(i)} - \mathbf{Y}^{(i)}\|^2 \tag{31.6}$$

The two configurations are first translated, rotated and dilated to obtain the best possible fit. M^2 measures the "lack of fit", or *residual sum of squares*. Small values of M^2 result in a good match between two given shapes [Kar82]. Without presenting a detailed proof, the Procrustes Analysis procedure for evaluating the two configurations is summarised as follows [Kar82]:

1. *Translation*: Evaluate the *mean-centering* of $\mathbf{X}$ and $\mathbf{Y}$ by using

$$\mathbf{X}_j^{(i)} \leftarrow \mathbf{X}_j^{(i)} - \bar{\mathbf{X}}^{(i)} \tag{31.7}$$

$$\mathbf{Y}_j^{(i)} \leftarrow \mathbf{Y}_j^{(i)} - \bar{\mathbf{Y}}^{(i)} \tag{31.8}$$

   where $i = 1, \ldots, p$, $\quad j = 1, \ldots, n$ and $\bar{\mathbf{X}}^{(i)} = \frac{1}{n} \sum_{i=1}^{p} \mathbf{X}_j^{(i)}$ and $\bar{\mathbf{Y}}^{(i)} = \frac{1}{n} \sum_{i=1}^{p} \mathbf{Y}_j^{(i)}$.

2. *Rotation*: Rotate $\mathbf{Y}$ with respect to $\mathbf{X}$ by using $\mathbf{Y} = \mathbf{Y}\mathbf{Q}$ where $\mathbf{Q}$ is an orthogonal matrix given by $\mathbf{Q} = \mathbf{V}\mathbf{U}^T$ where $\mathbf{U}\mathbf{\Lambda}\mathbf{V}^T$ is the singular value decomposition of matrix $\mathbf{X}^T\mathbf{Y}$.

3. *Dilatation or Scaling*: Scale $\mathbf{Y}$ by a factor c where

$$c = \text{trace}(\mathbf{\Lambda})/\text{trace}(\mathbf{Y}\mathbf{Y}^T) \tag{31.9}$$

   At this stage, M^2 can be calculated directly from $c^2 + M_{min}^2 = 1$ (the residual sum square, M^2, obtained is minimum), provided both configurations have been standardised, that is, the sum of squared distances from the respective centroids is unity.

31.3.2 Spearman Rank Correlation Coefficient

The Spearman rank correlation coefficient is a method which measures how much the ranking of two groups of data agree with one another. Using the distance data calculated from scatter-plot diagrams (a graph of the inter-point distances in the original and reduced data spaces), the coefficient provides a measure of the correlation between the original data set and the reduced dimension data set.

The Spearman rank correlation coefficient, Γ, is defined as follows [Hay81]:

$$\Gamma = 1 - \frac{6\sum_{l=1}^{N} D_l^2}{N(N^2 - 1)} \tag{31.10}$$

where D_l is the scalar difference between each element of two ranked vectors l and N is the number of data. A strong correlation is indicated by a value close to 1, 0 meaning no correlation at all, and -1 meaning irrelevance.

31.4 Performance Results and Discussion

Experiments are undertaken on three data sets; (i) thyroid data, (ii) glass data, and (iii) fish data. The dimension of these data sets is 5, 10 and 12, respectively; the number of input pattern vectors is 100, 50 and 34, respectively, and the number of classes is 3, 4 and 3, respectively. The thyroid and glass data sets include outliers. All experiments produce two-dimensional data from the high dimensional input data, and the Euclidean distance is used as a standard distance measure in all techniques. All input data are transformed using one or more of the following: the Z-transform, the log transform, mean-centering and double-centering.

For the SOFM, the algorithm is run several times with different parameters (size of feature map, number of training iterations etc...), for each data set, to achieve optimal performance (minimal over-fitting and maximum output resolution). A typical map size was 10×10, and the number of iterations was 80,000. Figure 1 shows the results produced by the SOFM by repeatedly running the SOFM program on the thyroid data sets with different map sizes (i.e. different number of neurons). It can be seen that the performance is initially low when using a small map size. As the map size increases, the performance also increases until a saturation point is reached where upon performance declines. Similar results are observed with the glass and fish data sets. This phenomenon is due to *over-fitting* (over-parametrisation as a function of sample size). A relatively broad range of feature map sizes provides a satisfactory maximum performance. However, the range is seen to be sensitive to the type of data set used.

Scatter-plots are generated from the output produced by each reduction method and the comparative performance is evaluated by using Procrustes analysis and the Spearman rank correlation coefficient. More details of the experiments and optimisation procedure are given in [LdVC94] .

Results of the experiments are summarised in Figures 2 and Figure 3. For completeness, results for the linear PCA method are also included. As to be expected, the performance of the PCA techniques degrades rapidly as the dimensionality of the data increases.

NLM and, to a lesser extent, non-metric MDS appear to have an overall best performance in terms of the Spearman coefficient, whereas non-metric MDS has the best

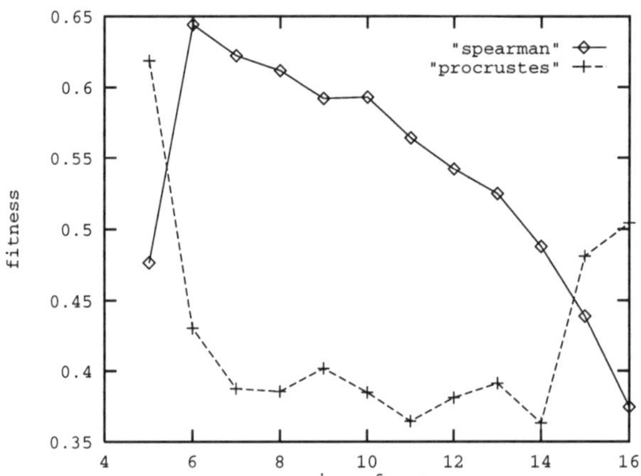

FIGURE 1. SOFM performance with different map sizes.

performance in terms of Procrustes analysis, particularly for low input dimensions. Non-metric MDS has a better performance in terms of reducing the residual sum of squares and preserving the shape of the data point distribution, whereas NLM is superior in terms of the rank correlation between the input and reduced data sets.

The SOFM algorithm generates output maps with better visualisation and interpretation capabilities; data points are more evenly distributed in the output space with the classes often clearly distinguished. However, in general, the performance of the SOFM algorithm is inferior to the other non-linear dimensionality reduction techniques in low dimensional input spaces. Some of the structure in the data point distribution is lost in the mapping process. Also, the SOFM algorithm is not always sensitive to outliers (particularly in the case of the thyroid data set), whereas outliers are easily identifiable when using NLM or MDS.

As the dimensionality of the input space increases, the relative performance of the SOFM algorithm improves significantly (particularly when using the Spearman rank correlation coefficient as the performance measure, where the value of the coefficient actually increases with dimensionality) and the SOFM algorithm may actually outperform both NLM and non-metric MDS when the dimension of input data is high. This suggests that the SOFM algorithm would seem to avoid the curse of dimensionality better than the other non-linear methods. Higher dimensionality may also improve the convergence rate of the SOFM algorithm, although this was not verified (see, however, [MSRK91]).

All dimensionality reduction methods, particularly the NLM and non-metric MDS, were shown to be sensitive to the presence of outliers. The presence of an outlier produced an unbalanced and distorted output map. This is as expected, given the sensitivity of the Euclidean metric to distance. Careful preprocessing (which was undertaken in the experiments reported here) must be implemented to remove all outliers and ensure robust dimensionality reduction.

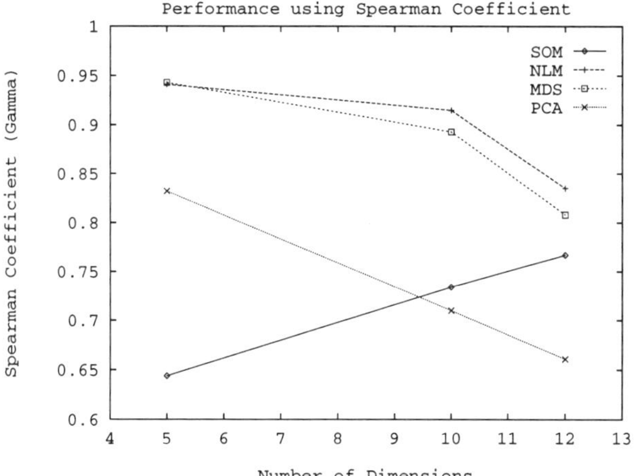

FIGURE 2. Comparative performance measured in terms of Spearman coefficient.

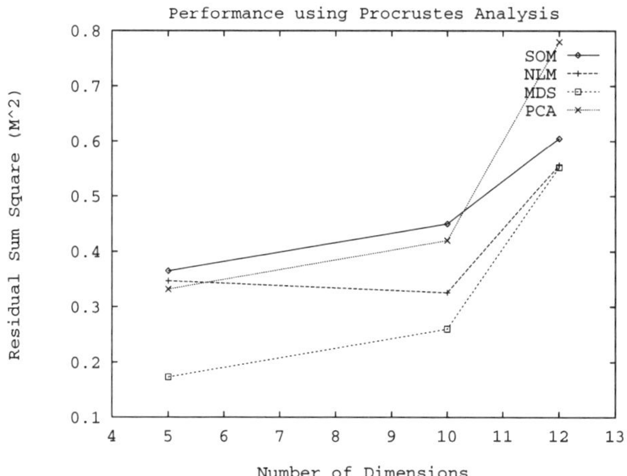

FIGURE 3. Comparative performance measured in terms of Procrustes analysis.

31.5 Conclusions

The use of Procrustes analysis and the Spearman rank correlation coefficient in comparing and contrasting non-linear dimensionality reduction methods was described. Both non-metric MDS and Non-Linear Mapping perform relatively well for low dimensional data. Non-metric MDS has a better performance in terms of minimising the residual sum of squares and preserving the shape of the data point distribution. NLM is superior to non-metric MDS in terms of the rank correlation between the input and reduced data sets. The performance of both non-metric MDS and NLM degrade quite rapidly as the dimensionality increases.

The Self-Organising Feature Map generates a more superior visual output than the other two methods in that it is relatively easy to interpret and the data points are well-clustered. However, in low dimensions, the performance of the SOFM algorithm is inferior to that of NLM and non-metric MDS. Unlike NLM and MDS, the performance of the SOFM algorithm increases with dimensionality (when performance is measured in terms of the Spearman rank correlation coefficient) and may outperform non-metric MDS and NLM for high-dimensional data. It would seem that the SOFM algorithm avoids the curse of dimensionality better than the other non-linear methods.

Further work using large scale simulations will need to undertaken to establish more rigourously the superiority of the SOFM in a high-dimensional context and its ability to ignore noisy and information-poor features.

31.6 REFERENCES

[BD91]   F. Blayo and P. Demartines. "Data analysis: how to compare Kohonen neural networks to other techniques". *International Joint Conference on Neural Networks*, 1:469–476, 1991.

[BL87]   I. Borg and J. Lingoes. *Multidimensional Similarity Structure Analysis.* Springer-Verlag, New York, 1987.

[CLN91]   V. Cherkassky and H. Lari-Najafi. "Constrained topological mapping for non-parametric regression analysis". *Neural Networks*, 4:27–40, 1991.

[Fuk90]   K. Fukunaga. *Introduction to Statistical Pattern Recognition.* Academic Press, Inc., New York, 1990.

[Hay81]   W.L. Hays. *Statistics.* CBS College Publishing, New York, 1981.

[Kar82]   M.J. Karson. *Multivariate Statistical Methods : An Introduction.* The Iowa State University Press, Ames, Iowa, 1982.

[Koh88]   T. Kohonen. *Self-organization and Associative Memory.* Springer-Verlag, Heidelberg, 1988.

[Kru64]   J.B. Kruskal. "Nonmetric multidimensional scaling: a numerical method". *Psychometrika*, 29:115–129, 1964.

[LdVC94] S. Li, O. de Vel, and D. Coomans. *Comparative performance analysis of non-linear dimensionality reduction methods*. Technical report, Department of Computer Science, James Cook University, Townsville, Australia, 1994.

[LGZ93]   X. Li, J. Gasteiger, and J. Zupan. "On the topology distortion in self-organizing feature maps". *Biological Cybernetics*, 70:189–198, 1993.

[MHP94]   F. Murtagh and M. Hernández-Pajares. *The Kohonen self-organizing map method: An assessment*. Technical report, Dept de Matemàtica Aplicada i Telemàtica, Universitat Politècnica de Catalunya, Barcelona, Spain, 1994.

[MSRK91] W.J. Melssen, J.R.M. Smits, G.H. Rolf, and G. Kateman. *Two-dimensional mapping of infrared spectra using a parallel implemented self-organising feature map*. Technical report, University of Nijmegen, 1991.

[Sam69]   J.W. Sammon. "A nonlinear mapping for data structure analysis". *IEEE Transactions on Computers*, 18:401–409, 1969.

[Sau89]   E. Saund. "Dimensionality-reduction using connectionist networks". *IEEE Transaction on Pattern Analysis and Machine Intelligence*, 11:304–314, 1989.

32
Omega-Stat: An Environment for Implementing Intelligent Modeling Strategies

E. James Harner and Hanga C. Galfalvy

Department of Statistics and Computer Science
P.O. Box 6330
West Virginia University
Morgantown, WV 26506 USA
Email: ejh@cs.wvu.edu

ABSTRACT Omega-Stat, a new data analysis and modeling paradigm, is built on Lisp-Stat—an object-oriented statistical programming environment. It contains extensible, reusable-component libraries for performing data management, multivariate analyses, modeling, and dynamic graphics. A point-and-click user interface allows instant access to all objects, including analysis and graphics objects. Modeling is done by adding new model objects, i.e., extended datasets, to a tree structure originally containing prototypes for linear, generalized linear, and nonlinear models. Knowledge, and methods for accessing this knowledge, are embedded within model objects and edge objects linking these models. This representation allows the modeling process to be studied by following the analysis trails of expert analysts. The objective is to provide an *expert consultant* that is accessible as part of man/machine interaction. Modeling strategies can then be built into Omega-Stat, by using prior knowledge and data-analytic heuristics, to guide the process of constructing the model tree and the iterative search for an "optimal" model.

32.1 Introduction

Omega-Stat is an integrated, user-friendly, extensible environment for data analysis, modeling, and dynamic graphics. It is based on Lisp-Stat, an object-oriented statistical programming environment developed by Tierney [Tierney90]. Early versions of Omega-Stat are described by Harner [Harner90, Harner91]. This paper presents the modeling component being developed within Omega-Stat. Statistical modeling, however, is completely integrated with the data management, dynamic graphics, and multivariate components of Omega-Stat.

The modeling environment builds on the ideas of many researchers, but those of Thisted [Thisted86] and Oldford and Peters [OldfordPeters86] are most relevant. Specifically, the modeling process generates a tree structure with nodes representing the data structures underlying the models. Different views of the model objects are possible, typically as reports or diagnostic plots. These view objects are specialized forms of regular Omega-Stat objects, e.g., dynamic projection plots.

A brief summary[2] of the fundamental statistical objects in Omega-Stat is presented first to provide a backdrop for the modeling system. The modeling environment, both the underlying objects and the strategies for analysis, is described next.

---

[2] A complete description is given in [Galfalvy94]

32.2   Basic Statistical Objects: Variables and Datasets

A *dataset*—an ordered collection of *variables* with value-lists indexed to the same set of sample *individuals*—is the principal data structure for performing statistical analyses in Omega-Stat. Most analyses begin by selecting variables from a dataset according to their role. These variables are the arguments to a constructor function which returns a child dataset with a pointer to its parent. For example, a single response variable, with role Y, and one or more explanatory variables, each with role X, are selected prior to choosing **Model** from the **Data Browser** menu (Figure 1). Analysis or graphical messages can also be sent to an individual variable. In this case, the constructor function returns a dataset, but the parent is a variable.

32.2.1   Variable and Dataset Metadata

Both dataset and variable objects contain *metadata* [Hand93], i.e., properties and descriptors about the variables and datasets themselves. A variable object has slots for its name, label, type (numeric or categoric), role (*Label, X, Y, Z*), and admissible values which, along with its actual values, define the variable. These attributes can be defined initially, specified later (except for the variable values), and changed at any time. Similar considerations apply to categoric variables, for which information on the order, actual and formal levels, level frequencies, and contrasts is saved. At any time, users can also have other numerical information computed and automatically saved within variable objects, e.g., equal count intervals used in constructing conditioning plots.

Metadata for datasets contain information on the nesting and crossing relationships among categoric variables, variable groupings, and state information (Section 32.2.2). As analysis proceeds, additional information is stored within datasets or their descendants. For example, the results of generalized singular value decompositions on groupings of numerical variables give information on collinearity, missing values, and multivariate outliers; on groupings of categorical variables, similar decompositions give information on dependency relationships and "cell outliers." The results of the former should be considered prior to and during regression analyses, whereas the latter is relevant to log-linear and analysis of variance models.

Metadata is used for data validation initially; subsequently, it is used to constrain the admissible analyses. Furthermore, this *a priori* specified (and computed) metadata is used along with diagnostics in assessing the efficacy of statistical models and in guiding statistical model selection. For example, models with count data as the dependent variable suggest examining the mean-variance relationship to determine if the Poisson, negative binomial, or some other probability model is appropriate.

32.2.2   Dataset Views and Controllers

Every statistical object currently created by a constructor function, e.g., a principal component, discriminant, or model instance, is an *extended dataset*, which has dependency relationships to its higher-level datasets. A dataset is said to be extended in the sense of Thisted [Thisted86]: i.e., it is a *data-analytic artifact* which has pointers to the original statistical and state variables, model-derived variable expressions, indicator variables, descriptors, and dependency views.

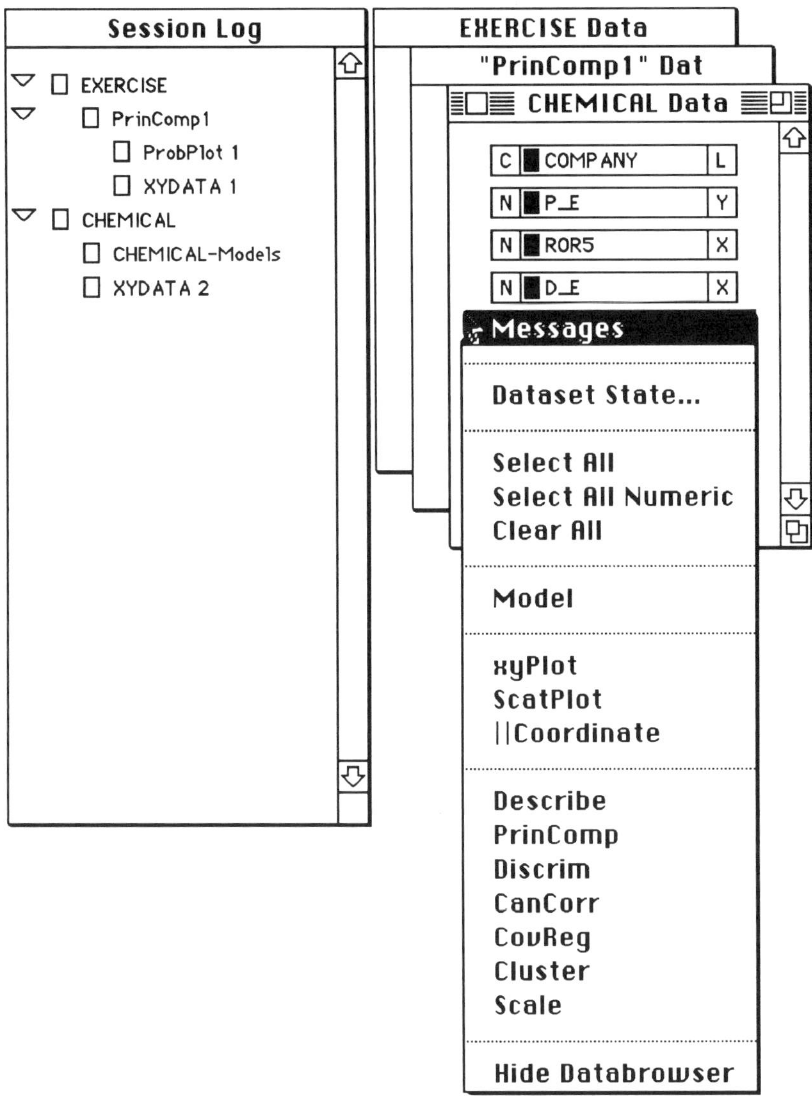

FIGURE 1. Session Log with dataset object subviews, data browsers, and a data browser menu.

Datasets, whether specified by the user or generated by a constructor function, are visualized iconically in the **Session Log** (Figure 1). Session-log-proto has a slot containing a pointer to a tree with the actual datasets as nodes and linkages representing parent-child relationships. This dependency tree is the basis for updating child dataset objects when a parent object is changed. Similar dependency relationships (and updating considerations) exist between dataset model objects and their views.

Within a child dataset, the underlying variables from a parent dataset are never replicated and derived variables are stored as expressions. This economy is necessary since numerous dependent datasets are created during the analysis process. For example, a principal component dataset has a slot containing pointers to the constructed principal variables (linear expressions with the right singular vectors as coefficients) and the variables underlying the model, another slot for the principal variable scalings (the singular values), but no slot for the canonical basis of the principal variables (the left singular vectors).

Statistical dataset objects are abstract data types with the following data access methods: *recognizers* for determining dataset types; *selectors* for extracting or modifying object slots, and *constructors* for creating dataset instances. Datasets, as abstract objects, exist independently of their views. Generally, datasets have multiple views: iconic, browser, attribute (slot), textual, and graphical.

The standard representation of a dataset is its iconic view. Dataset icons are actually subviews of the **Session Log**, which shows the hierarchical structure of related datasets and provides an analysis history mechanism (Figure 1). The controller for this subview is a popup menu for sending messages to the dataset object.

The **Data Browser** view has variable objects as subviews (Figure 1). The controller for variable icon subviews is a popup menu for sending messages to a variable, e.g., to visualize the variable values in a quantile plot. The messages depend on the type of the variable object, i.e., numeric or categoric. The controller for the **Data Browser** is a popup menu for sending messages to collections of variables. Depending on the message, variable roles need to be specified, e.g., as response (Y), explanatory (X), or conditioning (Z) variables. Dialog views of variables and datasets show their state information.

Datasets generally have one textual view, which summaries the numerical information inherent in the model represented by the dataset, and many graphical views. Graphical view constructors actually create a new dataset object which in turn is visualized. Many views are available or are being developed. These include various types of quantile, conditioning, and multivariate plots. For example, **xyPlot** provides "guided tours" [HurleyBuja90] of user-specified subspaces constrained by the roles of the variables.

Graphical views have two types of controllers (input devices): methods built into the view itself and object instances of controller prototypes. The latter controllers are reusable components which are available to any graphical view object. Although views and controllers have explicit links to each other and to their dataset object, datasets do not have explicit links to their views and controllers. In other words, statistical objects— even graphical objects—exist independently of their views. However, implicit links exist through the dependency trees of `session-log-proto`, since views (and child datasets) must be given the opportunity to reflect changes made to their underlying dataset object. Views and datasets can be updated automatically or by user intervention.

Datasets also contain state variables which determine how individual observations are represented in dataset views. These "variables" include point labels, symbols, colors, and states (invisible, highlighted, etc.) as values. State variables are defined in the top-level dataset and are "inherited" by child datasets. This forms the basis of the strong linking among dataset views, e.g., dynamic highlighting when the mouse is in brushing mode. Datasets can also contain user-defined logical (indicator) variables to flag outliers, identify groups, etc., and are specific to a dataset. A special logical variable defines a mask and is used to select, generally by highlighting points in a plot, a subset of the parent dataset for analysis.

Omega-Stat currently supports two general classes of datasets, those derived from `multivariate-proto` and those from `model-proto`, in addition to datasets underlying plot objects. The remainder of this paper focuses on dataset objects derived from `model-proto`.

32.3   Model Objects

All prototypes and model objects inherit from `model-proto`, which has no instances. `Model-proto` is a basic or foundation prototype; it contributes functionality and attributes that all models share. Currently, two prototypes inherit from `model-proto`: `linear-model-proto` and `nonlinear-model-proto` (Figure 2). Each of these is "specialized" by prototypes allowing generalized linear and nonlinear models, respectively.

32.3.1   The Model Dependency Tree

As described above, model objects are actually datasets which have report and graphical views. However, unlike multivariate dataset objects, only the foundation dataset object is recorded in the **Session Log**. All other "descendant" datasets are visualized in a **Model Browser** which gives an overall view of the model dependency tree and the inheritance links to prototypes (Figure 2). This approach was taken to eliminate complexity in the **Session Log**, since model dependency trees can grow rapidly and become deep. Furthermore, `session-log-proto` does not contain sufficient structure for developing intelligent modeling strategies.

The **Model Browser** is a powerful tool for communicating with model dataset objects, which are represented by iconic subviews. The controller for these model object subviews is a popup menu (Figure 2). It contains messages for producing summary and diagnostic views of the current model object and for creating new model objects with dependency links to the current object.

The modeling portion of Omega-Stat is an implementation of Thisted's paradigm for data analysis [Thistcd86], i.e., the model objects are extended datasets which are viewed as nodes in a dependency tree. Furthermore, edge subviews connecting the nodes form the basis of a semantic map. The edge object contains information on the heuristics used to generate the next model object (node). The controller for the edge subview is a popup menu which provides the mechanism for the data analyst to record modeling explanations.

Fitted models are *data-analytic artifacts* as discussed above. Depending on the parent prototype of a model, the extended dataset contains computed variables and expressions for fitted values, residuals, leverages, etc. These derived variables and variable expressions, together with slots for coefficients, etc., form the basis of methods used to assess the efficacy of the model. The resulting diagnostic and report views of the model datasets are invoked by popup menus.

32.3.2   Symbolic Representation of a Model

A model object also has slots containing the symbolic representation of the model. This representation, together with information on admissible values, is built interactively by

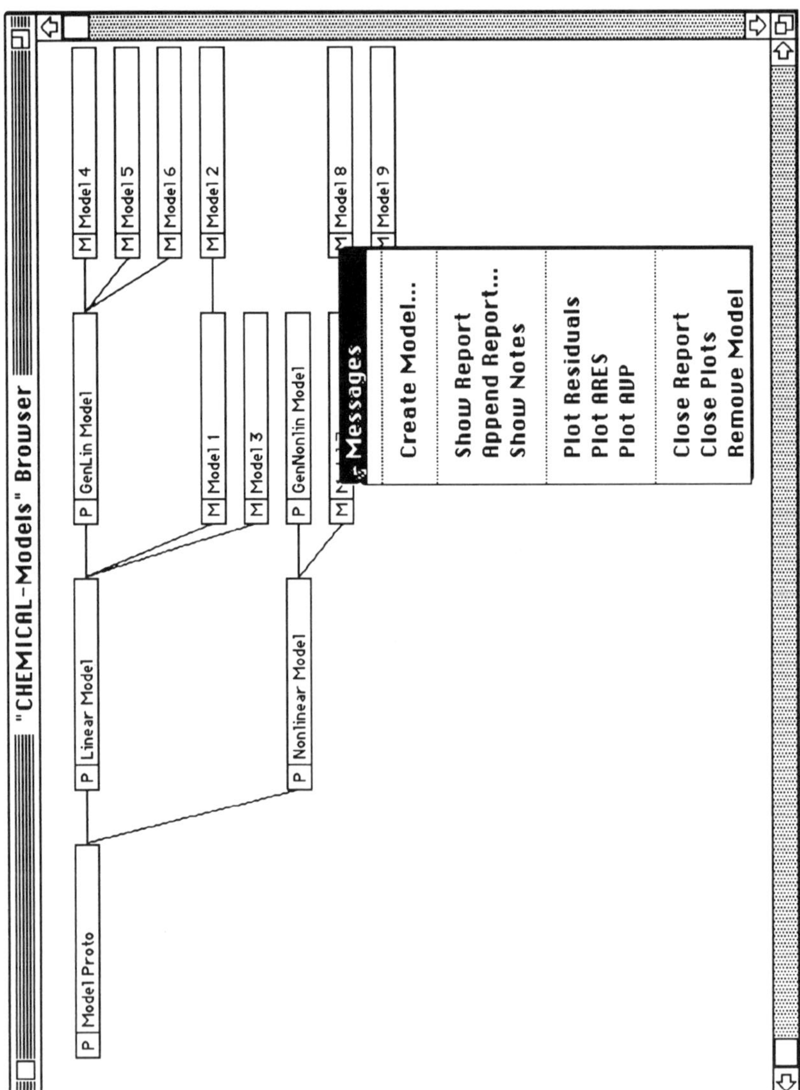

FIGURE 2. Model Browser with its model dependency tree and a model popup menu.

a point-and-click **Model Builder**. The **Model Builder** contains lists of possible variables (as determined by the parent dataset), algebraic symbols and functions, and model subexpressions.

The model is represented by a *mean function* and a *variance function*, together with information on scaling and parametrization when appropriate. Models containing *linear predictors* are fit to an ordered sequence of *terms*. These terms, comprising the linear part of the mean function, consist of variables, variable expressions, and interactions involving variables and variable expressions. Linear expressions can be expressed using a notational system with higher-level operators [McCullaghNelder89], but the symbolic representation is always reduced [CameronDixon92] to expressions containing two operators: '+' to separate terms and ':' to indicate interactions.

The estimable parameters are implicit and are not specified as part of the mean function. Terms containing parameters are either linearized by a Taylor series expansion or are solved conditionally, which then allows the model to be animated over the conditioning parameter.

Terms evaluate to vectors or matrices within the context of the dataset underlying the model object, e.g., terms containing categorical variables evaluate to matrices computed from columnwise products of contrast or indicator matrices using standard marginality constraints. The resulting vectors and matrices, one from each term, are joined to form a *model matrix* which is actually used in the fitting process.

The variables or variable expressions contained within at least one term of the reduced linear predictor become part of the dataset underlying the model object. Computed variables and variable expressions resulting from the analysis also become part of the dataset. Both the syntax and semantics of the mean function are based on the approach taken in S [ChambersHastie92]. A *scanner* and *parser* for the mean function, as well as a *constructor* for the model matrix, was developed initially by Hasebe [Hasebe94].

The variance function is a product of a (possibly known) dispersion parameter and a (possibly parametrized) function of both the the mean expression (represented by μ) and other variables not contained in the mean function. It is a numeric variable expression (and hence a variable object) and it evaluates to a vector which is then used as a weight variable in the (iterative reweighted) least squares algorithm. The link for generalized linear models is a user specified monotonic function of the mean expression. By default, it must satisfy the range constraint of the response variable.

The mean function for a nonlinear model is a parametrized expression involving the explanatory variables. Since it is a numeric variable expression, the mean function, given values for its parameters, evaluates to a vector. The variance function for a nonlinear model is identical to that of the (generalized) linear model described above.

Least squares, possibly weighted, or maximum likelihood is used as an optimizing criterion for all of the above models. However, other criteria, e.g., robust M-estimators, are also under development.

32.4 Modeling Strategies

Modeling entails *node exploration* and *node transformation* strategies. Messages selected from the controller of a model iconic subview either create another model (a node transformation) or produce summary and diagnostic views (node explorations).

32.4.1 Node Explorations

Node exploration strategies, which are directed at a specific model instance, consist of both low-level and intermediate-level methods. Low-level diagnostics use "primitives" already available in the extended model dataset, e.g., a residual expression, to produce views which are generally graphical. Intermediate-level explorations, such as collinearity diagnostics, build on the "primitives" to generate views, which are often both textual and graphical. Recall, however, that graphical views are themselves based on an extended datasets; thus, each specific model object not only has model dependents, but also dataset dependents associated with diagnostic graphical views. The dataset graphical views of a model can be visualized as subviews in its **Model Diagnostic Browser.**

Most of the diagnostics for model checking are based on methods owned by `model-proto`. These form a group of diagnostics which in their basic form apply directly to linear models, but specialized versions apply to generalized linear and nonlinear models. These include case-deletion diagnostics. Generalized singular value decompositions, with appropriate row and column weights, allow collinearity and other diagnostics to be made for linear models.

The inheritance structure of the model prototypes allow approximate diagnostic methods, based on linearizing higher-level models, to be replaced by more precise methods when available. This is the justification for having `generalized-linear-model-proto`, which contains linear models as a subset, inherit from `linear-model-proto` rather than the reverse. Each prototype in the inheritance chain has additional diagnostics; for example, checks on the link and variance function are added for generalized linear models.

The information gleaned from the model diagnostics are saved in slots in each of the model objects. This includes numerical or categorical information summarizing plots and other diagnostics. For example, assessments of trend using smoothing algorithms for added variable plots or explorations using animation in *ARES* plots indicate whether an omitted variable should be included, and in what way, as an explanatory variable [CookWeisberg94]. Currently, this diagnostic information is added to the model object by the analyst, but automated procedures are planned.

32.4.2 Node Transformations

Once diagnostics are "completed" on a model object, the modeler shifts attention to selecting the next model specification, i.e., finding a node transformation based on heuristics applied to the knowledge gained from exploring models in the dependency tree. Node-transformation strategies are termed high-level and also depend on protocols for evolving the dependency tree. For example, a model with minor changes relative to the currently selected model (e.g., deleted outliers) is created as a direct descendant of the current model, whereas a model with major changes (e.g., a different link function) is created as a descendant of a different model object or an appropriate prototype. The modeling process is helped by the flexibility of the symbolic representation of the model formula and the ease of changing state information. Furthermore, the analyst can move around the dependency tree to reexamine or perform more diagnostics on previously generated models.

Modeling is often done in small steps, but experts move purposely as they select the next transformation and narrow down the multitude of possible models to a manageable number. As has been mentioned previously, the modeler is encouraged to explain why a parent or prototype gave birth to an offspring. This process creates a semantic map which provides insight into the modeling process in general and an understanding of the generated models in particular.

32.5   Future Work

The current dataset model is limited in several areas. First, regular data (e.g., time series and spatial data), sparse data (e.g., species abundances), and other data structure are not implemented. Although these data types can be represented in datasets, this is often not efficient. More general data structures are possible within the Lisp-Stat object system, but the currently implemented statistical operations require a dataset (or variable) structure. As spatial and time series models and scientific visualization plots are added, other data structures will be developed.

Secondly, statistical database objects are not implemented (e.g., see [Michalewicz91]. Datasets are arranged hierarchically in **session-log-proto** and in **model-proto**, but more general dependency relationships and dataset constructor operators are needed. This added functionality will define the specifications for an object-oriented statistical database system, which is essential for large-scale research projects involving several related datasets.

Lastly, views are available for modifying state information for both variables and datasets. However, these views are limited to the dialog items available within Lisp-Stat. A spreadsheet view is needed to allow user-friendly viewing, editing, and modifying of variable and dataset objects. Since graphical views cannot scroll the data at reasonable speeds, alternative approaches are being investigated.

Model and plot objects are being generalized. For example, analyses using alternative objective functions are currently under development. This will include methods for fitting robust models. Other model classes, e.g., additive models, will soon be added to the modeling paradigm. Most Omega-Stat plot objects are being modified to allow animation on the basis of one or more conditioning variables. This capability is important for assessing model objects.

The search strategy for obtaining well-fitting models described thus far is informal, but it provides a method of developing man/machine strategies for modeling. After examining the semantic maps of expert (and perhaps non-expert) analysts' attempts at modeling targeted datasets, a rule-based system will be developed encompassing search strategies for reaching one or more acceptable models. This system should incorporate prior substantive knowledge of the problem domain, use metadata, interpret diagnostics (created by man or machine) of the models in the tree, represent this knowledge, and apply heuristics from this rule-based system. The objective is for the machine-resident "expert consultant" to suggest, but not enforce, the next node transformation. At each step, the search process is repeated until an "optimal" model is attained.

32.6   References

[CameronDixon92] Cameron, R. D. & Dixon A. H. (1992). *Symbolic Computing with LISP*. Englewood Cliffs, NJ: Prentice Hall.

[ChambersHastie92] Chambers, J. M. & Hastie, T. J. (1992). *Statistical Models in S*. Pacific Grove, CA: Wadsworth.

[CookWeisberg94] Cook, R. D. & Weisberg, S. (1994). *An Introduction to Regression Graphics*. New York, NY: John Wiley & Sons.

[Galfalvy94] Galvalvy, H.C. (1994). An object environment based on XLISP-STAT for multivariate analysis and dynamic graphics. Master's Project Report, Department of Statistics and Computer Science, West Virginia University.

[Hand93] Hand, D. J. (1993). Measurement scales as metadata. In D. J. Hand (Ed.), *Artificial Intelligence Frontiers in Statistics.* London: Chapman & Hall, 54–64.

[Harner90] Harner, E. J. (1990). Interactively developing models based on the exponential family. In K. Berk and L. Malone (Eds.), *Proceedings of the 21st Symposium on the Interface, 21*,110–115.

[Harner91] Harner, E. J. (1991). An exploratory data analysis and modeling system based on the exponential family. *Proceedings of the Statistical Computing Section of the Americal Statistical Association,* 70–78.

[Hasebe94] Hasebe, Y. (1994) A model parser and constructor for generalized linear models. Master's Thesis, Department of Statistics and Computer Science, West Virginia University.

[HurleyBuja90] Hurley, C. & Buja, A. (1990). Analyzing high-dimensional data with motion graphics. *SIAM J. Sci. Stat. Comp., 11,* No. 6, 1193–2111.

[Michalewicz91] Michalewicz, Z. (1991). *Statistical and Scientific Databases.* New York, NY: Ellis Horwood.

[McCullaghNelder89] McCullagh, P. & Nelder, J. A. (1989) *Generalized Linear Models, 2nd Edition.* London: Chapman & Hall.

[OldfordPeters86] Oldford, R. W. & Peters S. C. (1986). Implementation and study of statistical strategy. In W. A. Gale (Ed.), *Aritficial Intelligence and Statistics.* Reading, MA: Addison Wesley, 335–353.

[Thisted86] Thisted, R. A. (1986). Representing statistical knowledge for expert data analysis systems. In W. A. Gale (Ed.), *Aritficial Intelligence and Statistics.* Reading, MA: Addison Wesley, 267–284.

[Tierney90] Tierney, L. (1990). *LISP-STAT: An Object-Oriented Environment for Statistical Computing and Dynamic Graphics.* New York, NY: John Wiley & Sons.

33

Framework for a Generic Knowledge Discovery Toolkit

Pat Riddle, Roman Fresnedo, David Newman

Boeing Computer Services
Seattle, WA 98124-0346

33.1   Introduction

Industrial and commercial firms accumulate vast quantities of data in the course of their day-to-day business. The primary use of this data is to monitor business processes: inventory, maintenance actions, and so on. However this data contains much valuable information that, if accessible, would enhance the understanding of, and aid in improving the performance of, the processes being monitored.

Traditional statistical procedures provide some insight into this data, but they are often misused in non-expert hands. With the rapidly increasing quantity of data, it is no longer cost effective for trained statisticians to analyze all the data. The number of variables and observations in these datasets is often very large, and the number of candidate statistical models that might be considered is too large to permit manual systematic exploration. In this type of situation, a Knowledge Discovery (KD) tool is the most effective way to explore the data.

33.2   Generic Knowledge Discovery Toolkit

Our current goal is to develop a generic toolkit for KD. In our previous work we have developed a suite of tools and shown their potential on actual industrial data; these tools must now be enhanced to make them into generic tools. This suite of tools can be thought of as a collection of exploratory data analysis techniques; it uses nonparametric, multivariate methods to identify a large number of candidate models in the form of rules. The algorithms identify rules where the independent variables have a high probability of affecting the dependent variable. The effectiveness of these algorithms has been demonstrated at Boeing in several datasets containing massive numbers of variables and observations. We have successfully explored datasets with over 100,000 observations and datasets with over 140 variables.

Currently, the tools must be extensively hand-tuned by a data analysis expert (e.g., machine learning expert, statistician, etc.) to work with each dataset from each new domain. Hand-tuning refers to the initial data engineering, setting algorithm parameters, and analysis of the results. Data engineering refers to the transformation of the data from its initial form into an appropriate form for data analysis. The data analyst's expertise is

---

[1]*Learning from Data: AI and Statistics V.* Edited by D. Fisher and H.-J. Lenz. ©1996 Springer-Verlag.

currently a big factor in the successful use of this technology. We are currently automating this hand-tuning process. The reason we refer to it as a toolkit, is that the set of generic tools will be too complicated for a novice user to combine correctly. We envision a data analysis expert crafting a domain specific tool from the tools in the toolkit. This domain specific tool will be appropriate for novice use.

33.3   Knowledge Discovery Environment

The environment of our knowledge discovery work is slightly non-standard and worth a little elaboration. The results we produce (for instance rules predicting rejected parts) will not be used to create an expert system for determining whether a part should be rejected. Instead these rules will be examined by the process owner to learn more about their process and thereby achieve process improvement. Also our dependent variables typically have multiple causes, e.g. there is probably more then one cause responsible for rejected parts. We therefore typically expect multiple hypotheses, but there is no reason why these multiple hypothesis should partition the space (i.e., certain observations can be covered by multiple rules). These two factors were major reasons why we ventured away from a standard decision tree approach to our current approach.

Three terms common in machine learning are accuracy, positive coverage, and coverage and they are used throughout this paper. The accuracy is the (empirical) conditional probability of the rule's postcondition being true given that the rule's precondition is true. The positive coverage is the (empirical) conditional probability of the rule's precondition being true given that the rule's postcondition is true. The combination of these two measures gives the strength of the implication in each of the two directions. The rule's coverage is the probability that the rule's precondition is true.

The example used throughout this paper shows results from an actual dataset from the Boeing company. The variable names and values have been disguised for proprietary reasons but all the results are from this actual dataset. There were 6384 observations and 141 variables, which were ordinal (i.e., some discrete and some continuous) and categorical. The base rate of the dependent variable is 23.5%. The base rate is the proportion of the "value of interest" in the dependent variable. We transform the independent variable Y into a binary variable, called the target, by partitioning the range of Y into two sets: a set containing the value of interest, Target=True, and its complement. The rule is made up of a precondition and a postcondition (i.e., the target). The independent variables used in the precondition of the rule are also called predictor variables.

33.4   Boeing's Knowledge Discovery Framework

There are three main extensions to existing technology which are necessary to achieve this generic knowledge discovery capability. These research directions are:

1. to develop a statistically valid process for selecting and ordering the "good" rules produced by the algorithm and expressing the uncertainty associated with competing rules,

2. to develop a summarization and visualization methodology that allows process owners and data analysis experts to examine how the rules compete with and comple-

ment each other, and

3. to develop a methodology for deriving data representations which best capture the objectives of the process owners, and at the same time maximize the effectiveness of the rule extraction algorithms used for KD.

Figure 1 shows the interrelations between these three foci and the rule extraction algorithm. These three aspects of applying KD tools have received little attention from academic investigators. But more importantly, these foci are the aspects of this technology which must be resolved (at least partially) if KD is to become really useful for commercial applications.

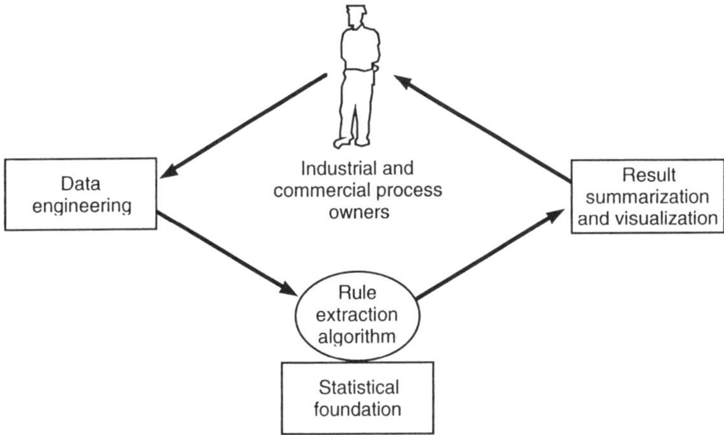

FIGURE 1. Boeing Knowledge Discovery Framework

The rule extraction algorithm on which we are currently focusing is the Brute/Beam-Brute algorithm. The dependent variable must be represented in categorical form, while the other variables can be either categorical or ordinal. Note that the dependent variable need only be categorical but is treated as binary once the value of interest is chosen. For instance if the dependent variable contains a value from the set duck, dog, cat and the value of interest is defined as "duck", the algorithm treats the variable as a "duck" "not duck" variable. The requirement that the dependent variable be categorical means that in cases where the dependent variable is ordinal it must be discretized. We have done this with datasets in the past by using quantiles.

Brute determines conjunctive rules which predict the target. It exhaustively explores all possible conjunctive rules over the independent variables up to the maximum depth, orders these rules using an evaluation function, and returns the top X rules where X is the size of the rule bucket (e.g., the top 200 rules ordered by the evaluation function). Brute was developed by Etzioni & Segal at University of Washington Computer Science Department in collaboration with Riddle at Boeing, based on their experiences using the IND algorithm [2] on Boeing datasets. IND simulates several decision tree algorithms including CART and C4.5. The difficulty in using decision tree algorithms in these domains led to the development of Brute, as discussed in [6]. This paper also contains a thorough description of the Brute algorithm and experiments comparing it to CART and C4.5. The major focus in developing Brute was to develop an algorithm which would discover a few good rules which may only cover a small number of observations, instead of focusing on covering the entire space of observations with a decision tree. This approach is based on the notion that we are dealing with domains which are fundamentally multiple causal domains.

In our use of Brute within Boeing, statistical measures (accuracy, positive coverage, and χ^2) derived by the algorithm during learning are an integral part of the resulting learned rule. This is a fundamental difference between our work and most of the historical work in the field of machine learning.

33.5 Statistical Foundation

Brute discovers collections of rules (models). We must determine how to select, rank, compare and validate these models. KD often generates very different rules which cover disjoint or nearly disjoint groups of observations in the dataset. Informally, we could say that rules can be competing, complementary or subsumed. Competing rules "explain" roughly similar sets of observations using different sets of predictor variables. Complementary rules "explain" roughly disjoint pieces of the space of observations. Subsumed rules cover nested subsets of the observation space, covering fewer and fewer points while becoming increasingly accurate. Widely varying complementary rules may indicate extreme heterogeneity in the dataset. The possibility of differing explanations for different parts of the dataset as complementary must be admitted.

Until we have a firmer probabilistic foundation and understanding of the search algorithms, the rules discovered by Brute will be considered descriptive statistics. Because of this, the only type of "bad rules" we are trying to prevent from being reported by our statistical methodology are those which are random. KD should not find rules when the target is statistically independent of the predictor variables.

	Target T	Target F
Precondition T	P_{11}	P_{12}
Precondition F	P_{21}	P_{22}

TABLE 33.1. 2x2 Table

Currently, Brute is run on all the data and use the χ^2 statistic as Brute's internal evaluation function to order the rules and keep the best ones in a very large bucket. The χ^2 statistic comes from the 2x2 table generated by the target being treated as binary and the rule's precondition being true or false. This is shown in Table 33.1. A 1000-fold bootstrap with the same data is used to determine the the 20 percentile of the accuracy of each rule. If the base rate is above this percentile, the rule is removed from consideration. It would be natural, for base rates closer to 0, to demand that the bootstrapped 20 percentile be an integer multiple of the base rate. Accuracy was chosen because it is the measure that users seem to care about the most. When accuracy is used as the evaluation function internal to Brute, however, it does not give good rule selections when the coverage of a rule is small. Any other simple statistics can be easily added to the bootstrap evaluator. The accuracy and positive coverage statistics given previously, can now be formally defined as $accuracy = P_{11}/(P_{11} + P_{12})$, $positive\ coverage = P_{11}/(P_{11} + P_{21})$

The other popular resampling technique, cross-validation, has been tried, but the maze of rules generated by Brute multiplies. For each cross-validation fold Brute would learn a new set of rules. Strong rules tend to appear in all subsamples almost unchanged, but weaker ones sometimes disappear or are greatly modified. In how many subsamples must a rule appear to be accepted? How much variation of its ranking and χ^2 across subsamples would be acceptable? The task of managing all this variation seems difficult.

As we stated, we are using χ^2 to order the rules and the bootstrapping on accuracy to define the cut-off point. χ^2 does a good job of ordering the rules, but cannot be used as a cut-off, since the size of the dataset has such a large impact on the results.

We did a number of experiments with the combination of χ^2 and bootstrap. One question which arises is whether the ordering imposed by χ^2 is close enough to the ordering given by the bootstrapped accuracy, so that rules left outside the χ^2 bucket would not found to be "good" rules by the bootstrap evaluator. The experiments show that the χ^2 and the bootstrapped accuracy correlate well enough that this will occur rarely. The bootstrap evaluator cannot be used as the internal evaluation function of Brute for computational reasons. The Brute process is very CPU-intensive, and the evaluation function must be modest in terms of its computational requirements. A further advantage of the bootstrap, not exercised yet, is that it allows the construction of simultaneous confidence intervals for all these rules (models) and their accuracies. Each rule is normally treated as an isolated model, but one must not forget that they are highly redundant and statistically dependent.

An attractive alternative to the bootstrap, which we are currently using, is based on permutation tests. A number of new datasets (usually 3) are created that have the same independent variables but where the target is randomly permuted. Then a search method (Brute or any other) is run on each of these datasets and choose our χ^2 cutoff value (or any other evaluation function) as the maximum χ^2 observed. A value as high as this could show up in a sample where the target is not related to the independent variables. This test is specific to a particular dataset. It takes into account the distributions of the variables, the size of the dataset, and the distribution of the target. Therefore it gives more precise answers, but must be recomputed for each new dataset.

We are exploring other evaluation functions: odds ratio[9]; Pearson's normalized χ^2[9], a Bayes factor[5] and the entropy statistic as used by CN2 [3]. Each evaluation function has its advantages and disadvantages and all of them have the problem of not having an absolute cutoff value.

The users' temptation to make inferences from these models will be great. We are exploring the application of graphical models [8] to be able to assess how much these models are supported by the data.

33.6   Result Summarization & Visualization

Model summarization and visualization tools are needed because the aggressive search strategies for rule extraction in our current algorithms produce too many rules (models) for detailed human review. After statistical screening and validation, the user is left with many rules that are either "competing", "complementary" or "subsumed". The set of rules produced are therefore naturally grouped into families. We currently use statistical hierarchical clustering techniques to automate the family creation process.

The intention is to help the process owners choose from among those models which pass the statistical screen of model selection. It is important to present as many of the valid models as possible to the process owners, as it is a priori unclear which of the models might make the most sense to them. Competing models will allow the user to work with several "working hypotheses" until further analysis sheds more light on the subject; they also show correlations between predictor variables that could suggest further statistical modeling. It is important to remember that when two rules are competing (e.g., aliasing for each other or collinear), they may only exhibit this behavior in the target (or positive) observations. This is very different than two variables which are directly aliasing for each other in all the observations. We use the term "aliasing" to specify that two variables are associated or correlated with each other. Complementary models may suggest actions with different cost/payback ratios, and thus allow the user to direct efforts where returns are higher or immediate.

33.6.1   Visualizing Multiple Rules

Our current approach to hierarchical clustering is to produce a matrix that has the rules as rows and 2x the observations as columns. For each rule we specify which observations satisfy its precondition of the rule and which observations satisfy both its precondition and postcondition (hence the doubling of observations). Once this matrix is created we do a hierarchical clustering with a euclidean distance metric. This metric is $d^2(R1, R2) = \sum_{obs}(R1(obs) - R2(obs))^2 + \sum_{obs}((R1 \wedge Target)(obs) - (R2 \wedge Target)(obs))^2$ The rules are considered close when their preconditions (R1 and R2) cover roughly the same points and roughly the same positives.

This allows us to see obvious sources of aliasing. An example is shown in Figure 2, only a portion of the hierarchical cluster tree for the 121 good rules produced by Brute is shown. Notice the rules on the branch marked X. These rules are clearly aliasing off the last conjunct.

33.6.2   Visualizing Single Rules

Single rules may be visualized in two ways. We have developed specific visualizations which present accuracy, positive coverage, and χ^2 for a rule to show how it varies as one variable moves a ordinal boundary. This is shown in Figure 3. We have also used standard statistical displays which were suggested by specific rules. This is demonstrated in Figure 4 which was suggested by the following rules: IF airline = X $\wedge$ component = 1 THEN target = T, and IF airline = X $\wedge$ component = 2 THEN target = T.

Further displays are currently under development. One interesting display shows a sequence of good rules with strictly increasing cumulative positive coverage. It is plausible, and has been our experience, that we can exhaust the set of excellent rules and yet leave a non-negligible proportion of positives unexplained. This last phenomena should be common in observational studies when some of the variables that explain positives are left unobserved.

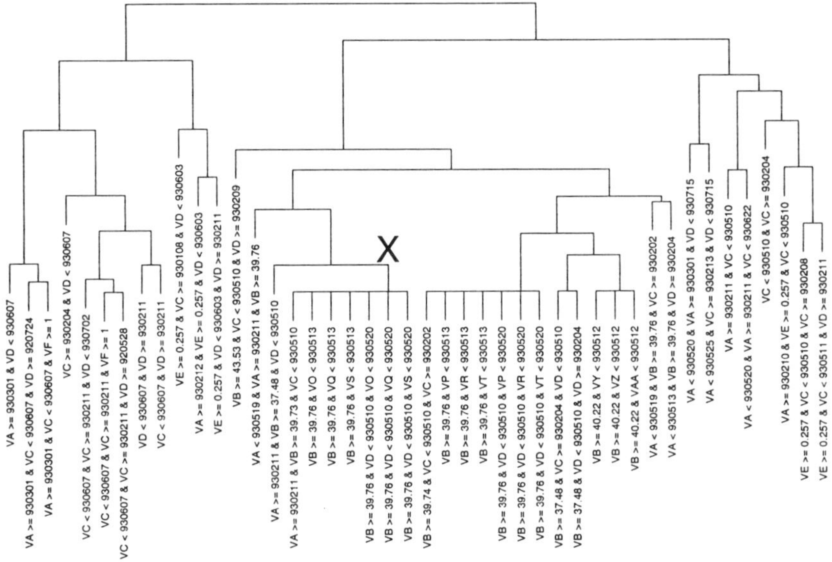

FIGURE 2. Rule Clustering

33.7 Data Engineering

It is easy to see that each of the different forms in which the data is represented will allow different types of patterns to be found in the data. The data representation is the result of formulating the problem by selecting the variables and specifying one as the single predicted variable. The success of the entire KD process is dependent upon the form in which the data is presented to the tool. The first step is data cleaning, where the data is made as error free and as consistent as possible. But even if clean data is presented in a bad form, no patterns may be found even if there exist patterns in the data. In principle there are many reformulations of the data (i.e., functions over the data) which could be computed.

We are creating a library of low level generic data transformations. A transformation is a function which translates one variable or combination of variables into another variable. Examples of these generic data transformations are discretizing and interpolating. We have become very proficient at reformatting the data by iterating between changes to the data representation and rule extraction algorithm runs. We are capturing this knowledge to embed it in a semi-automated procedure, a reformulation advisor, which a process owner can use to reformat his own data, so that the tool will perform well in that domain.

It is important to determine what effect the use of a lot of data transformations will have on the algorithm's statistical validation measures. This problem is another side of the overfitting problem mentioned above. Bear in mind our original description of a toolkit. We are developing generic tools which can be assembled together by a data analysis expert for a particular domain. A novice user then uses the tools within that domain to

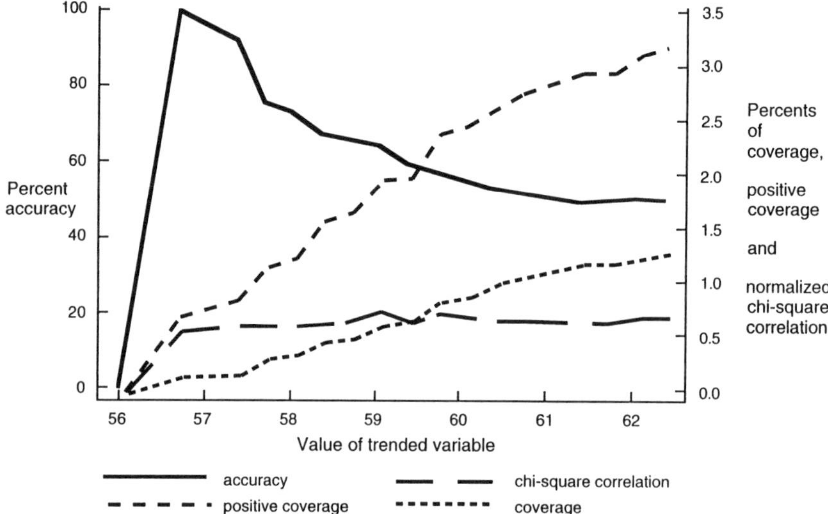

FIGURE 3. Trend Diagram

do his analysis. In this methodology a data analysis expert will be available to assure that overfitting does not occur.

One effect of variable-redefinition is Simpson's paradox [8] where the statistical measures of correlation can validate a model which is incorrect in the untransformed dataset. This paradox is caused by not making a natural distinction in the data, rather than making too many. So the use of data transformations to add more independent variables will ease the Simpson's paradox problem.

33.8 Related Work

At the framework level there is little related work to report. Most places are working on the discovery algorithms themselves (e.g., new Brutes) instead of total frameworks within which to place them. On the other hand each of the pieces of our framework has its own field of related work.

The Brute algorithm itself is most similar to ITRULE. The evaluation function they used is more tuned to finding high coverage rules instead of high predictive rules as is the Brute evaluation function. A comparison between Brute, ITRULE and other rule extraction systems can be seen in [6].

Our current visualization work is a small portion of the field of scientific visualization. We have not delved into the more advanced techniques in this field. Currently the more mundane aspects of this field have been sufficient for our framework.

Likewise the field of data engineering is vast in itself in the areas of representation reformulation and constructive induction. Since our data is in an attribute value repre-

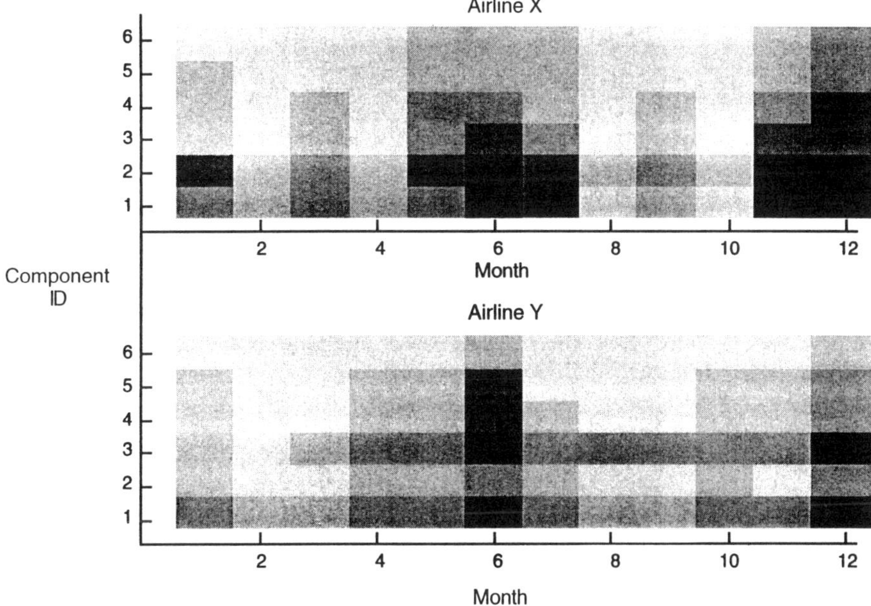

FIGURE 4. Image Plots

sentation, we have not had to use the more advanced techniques in this field either. But this is the portion of our framework which has received the least work and still remains a form of black art which requires the user to know his data well.

In the area of the statistical foundation of our rules, there is the whole history of statistical thought which is the vastest field yet. We have used numerous techniques form these fields most cited in the statistical foundation section itself. It is not easy to find in the statistical literature an analysis of the statistical validity of a collection of rules; graphical models [4] could provide a solution (see next section). We generally find ourselves retrofitting techniques developed for other purposes.

33.9 Future Directions

Even with the bootstrap pruning and the rule clustering, in some domains with many variables which alias each other, there will be too many good rules and the process owners will feel overwhelmed looking at them all. We are currently looking into the benefits of using a two pass system. For instance, upon examining the first set of rules which are produced, more data engineering can be done. First, variables which never appear in the rules can be removed. Notice this will only affect the efficiency of the algorithm not the results, unless it allows enough efficiency to allow the search to go to a lower depth (e.g., more conjuncts allowed per rule). Second, variables which obviously alias for one another can be combined. Remember we consider two variables to alias for one another if they can be substituted for each other in rules which predict the target. So the aliasing is seen through a projection on the target and may not extend to a total aliasing of the variables in all situations. When we discover aliased variables, all the duplicate variables but one can be removed, or actually combined. For instance if $V1 = A$ seems to be aliased by $V2 \geq 5$, then both these variables can be removed and a new variable added which is true when either $V1 = A$ or when $V2 \geq 5$ and false otherwise.

It is unclear whether in a two pass system both rule extraction passes would be performed by the Brute algorithm. We are also contemplating using a Graphical models algorithm in this second pass. This could be done by treating the Brute rules' entire preconditions as the random variables in the graphical model and see how these relate to each other. Alternatively the newly engineered variables from the first pass of Brute could be used as the random variables in the graphical model. The use of graphical models would give us more complete information about dependencies, but we are not certain it can be adapted for novice use.

33.10   REFERENCES

[1] L. Breiman, J.H. Friedman, R.A. Olshen, and C.J. Stone. *Classification and Regression Trees*. Wadsworth & Brooks, 1984.

[2] Wray Buntine and Rich Caruana. Introduction to IND and recursive partitioning. NASA Ames Research Center, Mail Stop 269-2 Moffet Field, CA 94035, September 1991.

[3] Peter Clark and Tim Niblett. The CN2 induction algorithm. *Machine Learning*, 3(4):261–284, March 1989.

[4] D. Madigan and A. E. Raftery. Model selection and accounting for model uncertainty in graphical models using Occam's window. *Journal of the American Statistical Association*, 89(4):1535–1546, 1994.

[5] R.E. Kass and A. E. Raftery. Bayes factors and model uncertainty. Technical Report 254, Department of Statistics, University of Washington, Seattle, Wash. 98195, U.S.A., 1991.

[6] P.J. Riddle, R. Segal, and O. Etzioni. Representation design and brute-force induction in a Boeing manufacturing domain. *Applied Artificial Intelligence*, 8(1):125–147, 1994.

[7] J.W. Tukey. An alphabet for statisticians' expert systems. In W. A. Gale, editor, *Artificial Intelligence and Statistics*. Addison-Wesley, Reading, Mass, 1986.

[8] J. Whittaker. *Graphical Models in Applied Multivariate Statistics*. John Wiley & Sons, Chichester, England and New York, 1990.

[9] Y.M.M. Bishop, S.E. Fienberg, and P.W. Holland. *Discrete Multivariate Analysis*. The MIT Press, 1975.

34
Control Representation in an EDA Assistant

Robert St. Amant and Paul R. Cohen

Computer Science Dept., LGRC
University of Massachusetts
Box 34610
Amherst, MA 01003-4610

ABSTRACT
To develop an initial understanding of complex data, one often begins with exploration. Exploratory data analysis (EDA) provides a set of statistical tools through which patterns in data may be extracted and examined in detail. The search space of EDA operations is enormous, too large to be explored directly in a data-driven manner. More abstract EDA procedures can be captured, however, by representations commonly used in AI planning systems. We describe an implemented planning representation for AIDE, an automated EDA assistant, with a focus on control issues.

34.1 Understanding Complex Data

Data analysis plays a central role in our attempts to understand the behavior of complex systems. Exploratory studies are one part of the descriptive process; they provide an informal prelude to experiments, in which questions and procedures can be refined. Exploration is a kind of detective work: to make a formal presentation of a case, the researcher must follow subtle and potentially conflicting clues to a set of possible conclusions. Exploratory results give rise to confirmatory studies in a cycle of successively more refined exploration and confirmation [Cohen95, Hand86].

Exploratory data analysis (EDA) offers a wide range of statistical tools for the early stages of analysis [Tukey77]. Simple exploratory procedures generate histograms of discrete and continuous variables, scatter plots and box plots for bivariate relationships, partitions of relationships that distinguish different modes of behavior, simplified functional relationships, and two-way tables such as contingency tables. From a consideration of these partial descriptions of data, a more complete picture emerges.

Viewed as search, exploration is a difficult problem. The flexibility of exploratory operators entails a large branching factor and an unbounded search space. If an exploratory analysis were to be driven purely by successive features discovered in data, the task would be impossible: Is a partitioning or a functional transformation appropriate? With what parameters? When should one stop? Though difficult and painstaking, exploration is nevertheless manageable in human hands. Several characteristics of exploration make this possible: relatively few general principles guide exploratory procedures; difficult problems are often decomposed into smaller or simpler parts; exploration is constructive, often relying on partial results and incremental improvement to reach solutions. We can draw a natural analogy between exploration and planning.

We have designed an Assistant for Intelligent Data Exploration, AIDE, to assist humans in the early stages of data analysis [StAmant94a, StAmant94b]. In AIDE, data-directed mechanisms extract simple observations and suggestive indications from the data. Scripted EDA operations then act in goal-directed fashion to generate simpler, deeper, and more extensive descriptions of the data. Control plans guide the EDA operations, relying on intermediate results for their decisions. The system is mixed-initiative, capable of autonomously pursuing high and low level goals while still allowing the user to guide or override its decisions.

AIDE is currently a prototype under development. Though capable of the analysis we present here, AIDE has not yet been extensively tested.

34.2   The Problem

In *Exploratory Data Analysis* [Tukey77, p.v], John Tukey describes EDA in this way:

> A basic problem about any body of data is to make it more easily and effectively handleable by minds—our minds, her mind, his mind. To this general end:
> - anything that makes a simpler description possible makes the description more easily handleable.
> - anything that looks below the previously described surface makes the description more effective.
> So we shall always be glad (a) to simplify description and (b) to describe one layer deeper.

Our design revolves around an understanding of the EDA process as the iterative application of four classes of operations: description, simplification, deepening, and extension.

Descriptive operations produce conventional summaries of data: single-valued statistics such as means and medians, or structures such as histograms and fitted lines. Such results can be directly used or interpreted by the user.

Simplification operations transform data to facilitate description. A log transform that straightens a skewed relationship is one example. Irregularities such as outliers are more easily detected in linear relationships than in nonlinear ones; a change in density can often enhance patterns in data; many statistical operations, even ones as simple as Pearson's correlation coefficient, rely on linearity. A straightening operation simplifies in that it enhances our observation, manipulation, and evaluation—in a word, our description—of the data.

Deepening operations increase the detail, accuracy, and precision of descriptions, as a microscope enhances observation by trading a global perspective for local detail. An example is the common practice of examining the residuals of a linear fit. The examination itself is carried out by simplification and description operations. Residual examination can bring to light structure not captured by the line, such as clustering, unequal variance in residuals, or local deviations from linearity.

With descriptive, simplifying, and deepening capabilities, EDA builds descriptions of single variables, bivariate relationships, tables, partitions and clusters. Local descriptions can have non-local, often widespread implications, however; these are examined by *extension* operations. When clustering is observed in two different relationships, for example, an extension operation prompts a comparison and an attempt to consolidate the descriptions of the relationships.

EDA operations give local detail about patterns in variables and relationships. If a dataset contains dozens or even hundreds of variables, however, it becomes difficult to determine where these operations should be applied. Modeling procedures serve to focus exploration. Such procedures include regression analysis, discriminant analysis, clustering, and causal modeling. In the case of regression, one explores to see whether there are nonlinear relationships between a regression variable and a regressor, or whether there are patterns in residuals [Daniel80]. In cluster analysis one searches for functional relationships that may distort cluster shapes [Chen74]. In causal modeling one searches for patterns that influence conditional independence [StAmant94b]. Modeling constrains the application of EDA operations and supplies context for interpretation of their results.

34.3   A Brief Example

Much of our research deals with the behavior of AI planners in demanding simulation environments. One such system is TRANSSIM, a transportation planner/simulator [Oates94]. In TRANSSIM, ships travel between ports carrying cargo along assigned routes. Bottlenecks and other occurrences change the environment in unexpected ways. The planner must react dynamically to these changes by rescheduling dock and ship assignments and rerouting cargo through different intermediate ports.

An early experiment examined relationships between resource costs. We collected measurements of cumulative port cost (P), ship cost (S), and dock cost (D), the number of ships (N), and trial duration (TD). We were particularly interested in the relationship between the resource costs P and S over the duration of a trial. While we set N at different values, we fixed per-day port, ship, and dock costs. In this example, we assume that a modeling procedure (here a regression or causal modeling procedure) requires examination of the relationship $\langle P, S \rangle$.

We begin with summary statistics for each variable. For S, cumulative ship cost, we find that the median is about 310K, the interquartile range 95K, and there is a slight skew toward lower values. More significantly, when we examine a histogram of S (Figure 1a), there are four distinct clusters in the data. Our preliminary partial description of S comprises the statistics and our observations about the clustering. Our observations of P are comparable.

When we turn to the relationship $\langle P, S \rangle$ (Figure 1b), we see a different pattern: the values fall into *five* clusters. The distinct separation in $\langle P, S \rangle$ values, as well as the observation that one of the modes in the histogram of S (Figure 1a) corresponds to a cluster twice as large as the others, leads us to return to our description of S. We establish an alternative possible description of S as containing five clusters, consistent with the clusters in $\langle P, S \rangle$.

Continuing with our analysis of $\langle P, S \rangle$, we see that values in the leftmost cluster (which we denote ps_a) can be fit by a straight line. In fact, this is true for all five clusters, though it is a different line in each case. Once we settle on the description of each cluster as a line, we can add the observation that the slopes of the lines decrease as the clusters move toward the right.

If we plot the central locations of the clusters ps_a through ps_e (i.e., the median coor-

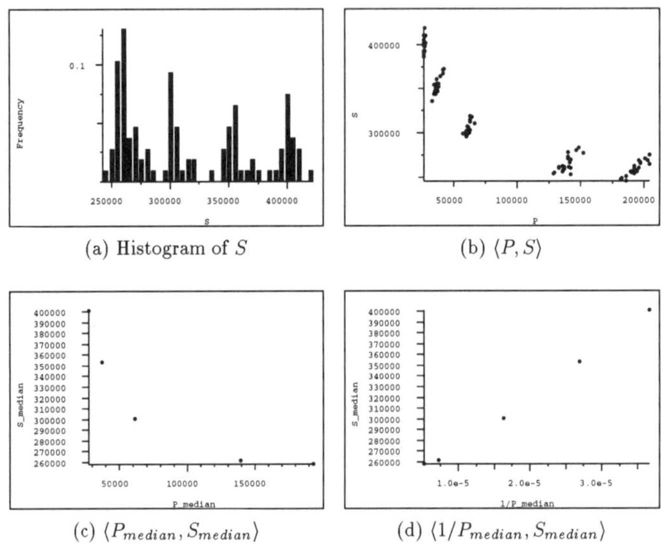

(a) Histogram of S (b) $\langle P, S \rangle$

(c) $\langle P_{median}, S_{median} \rangle$ (d) $\langle 1/P_{median}, S_{median} \rangle$

FIGURE 1. Exploring ship and port costs

dinates P_{med} and S_{med} for each cluster), as shown in Figure 1c, we see that these five summary points can be fit by a smooth curve. Further exploration shows that the curve is of the form $S_{med} = 1/P_{med}$. When we perform this transformation (Figure 1d) it straightens the curve, leaving no clear pattern in the residuals.

Extending the analysis, we find that the five discrete values of the variable N, the number of ships, correspond to the clusters found in the relationship $\langle P, S \rangle$. Because we have experimental control over the number of ships but not the cumulative costs, we count an observation about the former measurements as an explanation of the latter. In this case, N explains the clustering of resource costs.

This description, though brief, should give the flavor of the analysis. To summarize, we begin with initial descriptions of the data, such as the observation of gaps between adjacent values. From these we generate indications [Mosteller77], or suggestive characteristics: the data fall into clusters. Based on these indications we apply specific EDA procedures: we break the data down and analyze the clusters individually. These procedures may involve iterative refinement, as with the alternative descriptions of clusters in S. When possible, we simplify; instead of dealing with all data points, we work with a reduced dataset of just the median points of the clusters. We then follow two separate courses: we extend current results, here by considering other variables, and we deepen results, in this case by turning from a surface description of the clusters to an examination of the behavior of the data within each cluster. The result is a coherent, structured description of the data.

34.4   Data Manipulation

Datasets are the basic level of data representation. A dataset is an extension of the familiar relational table of attributes and values. Extensions capture implicit and explicit knowledge we have about the data. Datasets and attributes can be specialized as distinct types, the most common cases being relationships (or multivariate subsets of datasets) and variables (or univariate relationships). While there exist operations applicable only to variables, or to relationships, there is a large set of shared operations that make it desirable for all data to be handled in tabular form.

We manipulate a dataset in three distinct ways. The first kind of manipulation involves subdividing or combining elements of datasets. In order to construct fitted lines to each cluster in the TRANSSIM data, we partitioned the original data into five smaller datasets. We then examined each cluster independently, at a finer level of detail. We call an operation that breaks data into smaller but similar parts a *data decomposition*. A data decomposition generates a mapping from the elements of a dataset to membership in a set of new, derived datasets. Operations that bin data, exhaustively partition data, or nonexhaustively cluster data all decompose data in this way. The inverse operation, *data composition*, combines separate datasets. Data decompositions and compositions let us capture independent detail in subsets of data and recombine results.

The second kind of manipulation involves deriving new attributes for elements of the dataset. During the TRANSSIM analysis we performed a transformation of cluster medians to remove curvature. Operationally the procedure transformed a single attribute, P_{med}, into a new attribute, P', where $P' = 1/P$. We call such an operation an *attribute transformation*. A special type of transformation is attribute composition: consider the transformation from attributes port cost, ship cost, and dock cost to $S + P + D$, or total cost. Another type of transformation is attribute decomposition: to conclude that a given line is a good fit to a relationship, we need to generate the residuals by subtracting $\hat{P}$ from P—by decomposing the attribute into structure and residual.

Because dataset attributes may contain datasets as well, another useful transformation is the mapping of a statistic over the elements of an attribute. Our data decomposition of the $\langle P, S \rangle$ relationship generates five new datasets. More precisely, the result is a new dataset, PS_c, containing a single attribute "Clusters", whose value is these new datasets. We can now apply an attribute transformation to the "Clusters" attribute, to produce newa ttributes such as "x-median" and "y-median". We continue by fitting a line, a further attribute transformation. Functions for attribute transformation include arithmetic operations, exponentiation, and these higher level dataset operations.

The third kind of manipulation is *reduction*, which is the type of function performed by summary statistics such as means and medians. Correlations and partial correlations are similarly reductions of bivariate and multivariate relationships.

Designing a system around such general structures and operations is not simply an attempt at conciseness. Rather, the design is geared specifically toward the kinds of operations appropriate for EDA. Consider the example of generating a histogram for a discrete variable N in a dataset T. To build a histogram we divide a variable into bins and count the number of observations that fall into each bin. Using the operations describe above, we can implement a histogramming procedure as follows. First we apply a data decomposition to T. The result is a set of new datasets that form disjoint subsets of T, stored as an attribute "Data" in a new dataset, T_h. Our operations now work on T_h. An attribute transformation of "Data", based on the count statistic, gives the size of the bins. Another transformation using the mode statistic gives the value of N associated with each bin. These distinct values and their counts can then be displayed directly in histogram form.

This may seem an inordinate effort to produce such a basic structure, but consider a simple extension: the contingency table. To build a contingency table between N and C_{mem}, an attribute that records the membership of each data point in a specific cluster, we follow essentially the same procedure. This time the decomposition simultaneously bins N and C_{mem}. Decomposition operations are not limited to single variables; they can just as easily decompose two variables or an entire dataset. The result of the operation is a dataset, T_c, with a two-dimensional attribute containing the partitions, one for each unique value of N and C_{mem}. Again we transform by the count statistic. Calculating the N, C_{mem} values for each bin is equally straightforward. The desired contingency table data is the result. A simple variation of this procedure produces a box plot, by computing letter values rather than counts. More complex two-way tables can also be generated in a similar way, by calculating statistics for a third variable in each of the partitions.

Producing structures in this way has two abstract benefits: complex procedures can often be seen as natural extensions of existing procedures, and natural connections between conceptually similar structures become clear. While it is trivial to observe that a contingency table is a two-dimensional histogram, it is worth stressing that the two structures can be produced by very similar procedures, and that both procedures are different instances of a single canonical, divide-and-conquer EDA procedure. A data decomposition breaks a dataset into smaller parts; dataset-level attribute transformations compute a set of features of the reduced data; these features are explored both individually and in aggregate. The procedure is a simple, powerful example of generalization, as described in the conceptual clustering and constructive induction literature [Michalski83, Fisher86].

Planning offers a natural framework for representing EDA procedures. A planning representation can capture general procedures and intermediate goals, constructive derivation of results, and hierarchical problem decomposition—necessary elements for exploration. At the core of AIDE is a script-based planner with a library of partial plans, or scripts, which can be combined as required by the characteristics of the data.

```
(define-script variable-histogram
   :goal     (describe ?variable ?directives ?histogram)
   :vars     (variable directives (histogram :output))
   :features ((:content-type :categorical))
   :bindings ((histogram (partitions index count)))
   :grammar  (:SEQUENCE
              (script-partition variable 'identity
                                 :output histogram
                                 :names '(partition index))
              (script-transform histogram #'(lambda (partition)
                                              (script-reduce partition 'count))
                                 :key (attributes partitions)
                                 :name count)))
```

The scripting language is based on work in control planning for knowledge-based signal processing [Carver93]. In its simplest form, a script has a goal form with input and output

variables, a set of bindings, and a grammar of actions to execute. Grammar constructs
allow sequencing, iteration, mapping, and conditionalizing of actions. A script becomes
active when a goal is established at a higher level that matches its goal form; it succeeds
when its own grammar of actions completes. A script to generate a histogram, slightly
abbreviated, is shown above.

We have already mentioned variants of the histogramming procedure, which are implemented by similar scripts. Other scripts compute initial values for parameters of resistant
lines, transformations of skewed relationships, and so forth. Each script produces a set of
new or transformed structures, appropriately annotated and related to existing structures.
As scripts execute, a growing hierarchy of datasets, relationships, and their attributes is
generated.

34.5   Control Planning

The scripting representation lets AIDE search through a space of general procedures rather
than primitive data manipulations. Guiding the search are *indications,* or suggestive characteristics of the data. Indications involve evaluation of a statistic or descriptive structure
derived from the data. For example, evidence of clustering is an indication, as is the presence of outliers, and curvature in the residuals of a linear fit. Indications may be simple
computations, such as testing the skew of a distribution against a preset threshold. A more
complex indication involves constructing a nearest neighbors clustering for a relationship
and checking for outlying distances between adjacent clusters. In each case the presence
of an indication leads to the selection of scripts that can potentially explain its presence.

This representation restructures the search space of exploration into a more appropriate
form. Unfortunately, the problem of choosing an appropriate search operation still applies,
even at this higher level. The space remains too large to be searched with only data-driven
operations.

Control planning addresses the problem [Carver93]. Control plans are a simple extension of data manipulation scripts. A control plan grammar contains subgoals rather than
actions, and can implement more general procedures than those provided by scripts. Control plans guide the activation of data manipulation scripts and other control plans. The
control plan given below is a generalization of the histogramming script.

```
(define-control-plan* describe-cluster-behavior
  :goal        (select-control ?relationship ?directives ?result)
  :vars        (relationship directives (result :output))
  :features    ((:dataset-type relationship)
                (:cardinality 2))
  :indications ((clustering-indication *))
  :grammar     (:SEQUENCE cluster-subgoal
                          extract-components-subgoal
                          (:MAP (substructure*)
                                describe-substructure-subgoal)
                          compose-descriptions-subgoal
                          explore-composition-subgoal))
```

This control plan is used in the TRANSSIM analysis to describe the data in terms of the locations of the clusters, rather than the data points themselves. It decomposes a dataset into separate components, describes each component, and aggregates the results into a single structure, to be explored in turn. A lower level plan supplies descriptions in the form of the location of each cluster. While the behavior of this plan is similar to that of a script, it can be used more generally to explore the locations of the clusters, their sizes, their functional behaviors, and so forth. The entire process is an example of simplification, in which we work with an abstraction of the original data to facilitate a high level functional description.

The combination of control plans with data manipulation scripts further restructures the search space; focusing heuristics provide the final component. Actions that later turn out to be incorrect or irrelevant are inevitable in the analysis. This means that at points during the execution of a plan, an intermediate result can obviate some set of pending actions. When this happens, a focusing heuristic can abandon the current plan and select a new one based on the current state of knowledge.

More specifically, focusing allows context-sensitive information to influence the order in which subgoals are satisfied and the selection of appropriate parameter values. The most important role of focusing heuristics is to allow for user interaction in the analysis. We often find that information external to the data affects our decisions about the course of exploration. We may know that a relationship falls naturally into three clusters, rather than four, because of a known sampling bias, or that some variable value acts as a natural threshold between different behaviors in the data. Focusing heuristics take such user-supplied information into account where needed: in the decisions about which plans are applicable, and which input values are appropriate for their execution.

A broad view of the exploratory analysis, using the TRANSSIM example, is as follows. A modeling procedure generates a relationship $\langle P, S \rangle$ between port cost and ship cost. Exploration results are interpreted in the context of the model. The exploratory phase of the analysis begins with the generation of features and indications for the relationship: relatively high correlation, a hierarchical clustering with clear separation between individual groups, evidence of curvature, skew in both variables.

A focusing heuristic evaluates the relevant control plans, using information provided by the indications, and selects the plan describe-cluster-behavior. The rest are suspended, to be pursued when the first returns with more information. The selected control plan breaks down into three main subgoals: (a) the generation of clusters, (b) the exploration of each cluster, and (c) the exploration of the aggregated descriptive results of the clusters. Subgoal (a) can be satisfied by different plans: one that attends to clusters in the entire relationship, and others that detect clusters in the component variables. The bivariate clustering plan is selected, again by a focusing heuristic. This plan generates three candidate assignments of points to clusters; the heuristic threshold for the indication is not sensitive enough to make an unambiguous assignment. A single assignment is selected. As with plan selection, the remaining possibilities are suspended until later, if needed.

The selected cluster assignment gives a decomposition of the relationship into separate clusters. In control phase (b), each is explored in turn. A focusing heuristic, taking into account the context of the cluster description control plan, selects a data manipulation plan to determine the location of each cluster. Location is computed from the features of each cluster: from the mean, median, and other summary statistics. The values generated are returned and aggregated into a single relationship.

Subgoal (c) is then established to explore this new relationship. Based on indications for the reduced data, a new control plan is selected that will straighten the curved relationship and fit a line to the result. The linear fit is then explored in turn as a form of deepening.

Each decision point is controlled by a focusing heuristic. If a selected plan or variable value results in either failure or inadequate information, the focusing heuristic will be reactivated to make a different selection, to try to satisfy the goal in a different way. Focusing and refocusing give a needed flexibility to the exploration process. Sometimes a shallow exploration will be sufficient to show, for example, that there are no gross departures from linearity in a relationship; a focusing heuristic can evaluate indications in context and decide not to reactivate other pending plans. Sometimes a result can show that there are patterns deeper in the data; a focusing heuristic can direct attention toward the patterns in an opportunistic way.

34.6   Related Work

This work draws on a number of different sources. The clearest relationship is to early work in developing concepts of statistical strategy, or the formal descriptions of actions and decisions involved in applying statistical tools to a problem [Hand86]. The REX system, for example, implemented a strategy for linear regression [Gale86]. Comparable complex strategies were developed by others for collinearity analysis [Oldford86]. The goals of AIDE bear a resemblance to those of TESS [Lubinsky88], which supports analysis by accommodating user knowledge of context in a search good descriptions of data. AIDE extends this work by supplying an appropriate taxonomy for EDA operations and using this taxonomy to guide search at different levels of abstraction.

Central to AIDE is the opportunistic, incremental approach to discovery described by Tukey, Mosteller, and other advocates of EDA. There are some obvious difficulties: maintaining consistency in an incrementally growing set of descriptions can be difficult; in many cases local techniques can miss simple global patterns; local techniques can lead to a plethora of spurious results. Nevertheless many such problems can be alleviated in a system docs not always act autonomously, focuses on both data and on goals of the analysis, and pursues exploration paths with the aid of external knowledge and the context supplied by its own actions.

34.7   Acknowledgments

This research is supported by ARPA/Rome Laboratory under contracts F30602-91-C-0076 and F30602-93-C-0010. The U.S. Government is authorized to reproduce and distribute reprints for governmental purposes not withstanding any copyright notation hereon.

34.8   REFERENCES

[Chen74]   H. J. Chen, R. Gnanadesikan, and J. R. Kettenring. Statistical methods for grouping corporations. *Sankhya, Series B*, 36:1–28, 1974.

[Carver93]   Norman Carver and Victor Lesser. A planner for the control of problem solving systems. *IEEE Transactions on Systems, Man, and Cybernetics, special issue on Planning, Scheduling, and Control*, 23(6), November 1993.

[Cohen95]   Paul R. Cohen. *Empirical Methods in Artificial Intelligence*. MIT Press, 1995. In press.

[Daniel80]   Cuthbert Daniel and Fred S. Wood. *Fitting Equations to Data*. Wiley, 1980.

[Fisher86]   Douglas Fisher and Pat Langley. Conceptual clustering and its relation to numerical taxonomy. In William Gale, editor, *Artificial Intelligence and Statistics*. Addison-Wesley, 1986.

[Gale86]   W. A. Gale. REX review. In W. A. Gale, editor, *Artificial Intelligence and Statistics*. Addison-Wesley, 1986.

[Hand86]   D.J. Hand. Patterns in statistical strategy. In W.A. Gale, editor, *Artificial Intelligence and Statistics*, pages 355–387. Addison-Wesley, 1986.

[Lubinsky88]   David Lubinsky and Daryl Pregibon. Data analysis as search. *Journal of Econometrics*, 38:247–268, 1988.

[Michalski83]   R. S. Michalski and R. E. Stepp. Learning from observation: Conceptual clustering. In R. S. Michalski, J. G. Carbonell, and T. M. Mitchell, editors, *Machine Learning: An artificial intelligence approach*. Morgan Kaufmann, 1983.

[Mosteller77]   Frederick Mosteller and John W. Tukey. *Data Analysis and Regression*. Addison-Wesley, 1977.

[Oates94]   Tim Oates and Paul R. Cohen. Toward a plan steering agent: Experiments with schedule maintenance. In Kristian Hammond, editor, *Second International Conference on Artificial Intelligence Planning Systems*, pages 134–139. AAAI Press, 1994.

[Oldford86]   R. Wayne Oldford and Stephen C. Peters. Implementation and study of statistical strategy. In W.A. Gale, editor, *Artificial Intelligence and Statistics*, pages 335–349. Addison-Wesley, 1986.

[StAmant94a]   Robert St. Amant and Paul R. Cohen. A planning representation for automated exploratory data analysis. In D. H. Fisher and Wray Buntine, editors, *Knowledge-Based Artificial Intelligence Systems in Aerospace and Industry Proc. SPIE 2244*, 1994.

[StAmant94b]   Robert St. Amant and Paul R. Cohen. Toward the integration of exploration and modeling in a planning framework. In *Proceedings of the AAAI-94 Workshop in Knowledge Discovery in Databases*, 1994.

[Tukey77]   John W. Tukey. *Exploratory Data Analysis*. Addison-Wesley, 1977.

Part VII

Decision and Regression Tree Induction

35

A Further Comparison of Simplification Methods for Decision-Tree Induction

Donato Malerba, Floriana Esposito, and Giovanni Semeraro

Dipartimento di Informatica - Università degli Studi di Bari
via Orabona, 4 - 70126 Bari - Italy
{malerbad | esposito | semeraro} @ vm.csata.it

ABSTRACT   This paper presents an empirical investigation of eight well-known simplification methods for decision trees induced from training data. Twelve data sets are considered to compare both the accuracy and the complexity of simplified trees. The computation of optimally pruned trees is used in order to give a clear definition of bias of the methods towards overpruning and underpruning. The results indicate that the simplification strategies which exploit an independent pruning set do not perform better than the others. Furthermore, some methods show an evident bias towards either underpruning or overpruning.

35.1   Introduction.

Various heuristic methods have been proposed for the construction of a decision tree, among which the most widely known is the top-down approach. In top-down induction of decision trees (TDIDT) it is possible to identify three main issues [BFOS84]:

1. The definition of a decision process associated with the tree.

2. The determination of the tests in the nodes.

3. The determination of the leaves.

This paper is mainly concerned with the third one, which is deemed to be important because of its influence on the overfitting problem. In general, a tree *overfits* the training data when some of its branches degrade the classification performance on unseen cases, and, as a minor consequence, reduce the comprehensibility of the tree itself. This problem is particularly felt when several sources of noise and uncertainty affect the training data. In this case, indeed, the decision of stopping the growth of a tree when all examples in each leaf belong to the same class is too weak, since it does not prevent the generation of harmful branches. A typical approach adopted in many TDIDT systems grows a tree, T_{max}, to maximum depth, and then retrospectively removes those branches that seem superfluous with respect to predictive accuracy [Nib87]. The final effect should be that of improving the intelligibility of a decision tree without really affecting its predictive accuracy. Recently, Fisher [Fis92] and Schaffer [Sch93] have shown that overfitting avoidance by means of tree pruning is a form of bias (here intended as a set of factors that influence hypothesis selection) rather than a statistical improvement of the classifier. Nevertheless,

---

this conclusion should not imply the abandonment of pruning techniques, rather, it should stimulate the investigation of the biases of different methods so to apply the appropriate method, if any, to the problem at hand.

This paper investigates the behaviour of eight well-known simplification methods on twelve data sets taken from the UCI Machine Learning Repository [MA94]. In particular, we show how the computation of optimally pruned trees can be used to give a definition of bias of a method towards either underpruning or overpruning. This experimental approach generalizes an approach by Holte [Hol93], who found upperbounds on accuracy when decision tree depth was limited to one. The upperbound was computed by randomly splitting the dataset into a training and a test set, and then measuring the highest accuracy on the test set of all 1-level decision trees built on the training set. Holte did not consider the possibility of simplifying the trees and consequently the complexity of the induced trees. Conversely, we are interested in finding lower bounds on the complexity of the *most accurate* simplified trees, as well as upper bounds on the accuracy achievable by simplifying trees. The experimental design of Section 3 embodies a number of differences from previous studies on pruning methods; these differences allow us to draw some conclusions on the performance of the methods that are at variance with those already published in the literature or expected from the formulation of the methods themselves. A summary of the empirical results is reported in Section 4.

35.2   A brief survey of simplification methods.

The simplification methods considered in our study are briefly described in Table 1. Some of them start the simplification process from the root of T_{max} and proceed towards the leaves (top-down), while others move from the leaves to the root (bottom-up). Moreover, some methods simplify the tree in two steps: they first generate a set of subtrees of T_{max} by taking into account the complexity of the tree, then they select the best one in the set according to its estimated accuracy. In some cases, the estimate of the accuracy of a tree is computed on a *pruning set* which is independent of the set used for building the tree T_{max}. This means that the *training set* is actually split into two subsets: the *growing set* for building the tree T_{max} and the *pruning set* for simplifying it. Two of the three variants of the cost-complexity pruning method proposed by Breiman et al. require a pruning set (0SE and 1SE), while the other one (CV-0SE) uses cross-validation sets to estimate the predictive accuracy of the subtrees. In this latter case T_{max} is built on the whole training set and not on the smaller growing set. The version of the minimum error pruning (MEP) method, that we implemented for our experiments, is based on techniques proposed by Cestnik and Bratko [CB91], but differs in that we use a pruning set for choosing amongst trees pruned with different values of the parameter m.[2] Furthermore, our reduced error pruning (REP) algorithm is based on Quinlan's original formulation instead of that given by Mingers [Min89], so that we are guaranteed that it finds the smallest version of the most accurate subtree with respect to the pruning set. Details of the methods REP, MEP, 0SE, 1SE, and pessimistic error pruning (PEP) can be found in [EMS93].

---

[2] The parameter m determines the impact of the prior probability on the estimation of the error rate. In our experiments, we selected the following values of m: 0.5, 1, 2, 3, 4, 8, 12, 16, 32, 64, 128, 512 and 1024. For each value, MEP returns a subtree of T_{max}. The pruning set is used to select the best subtree.

Method	Steps	Pruning Set	Traversal Strategy	Operators	Characteristics
Reduced Error Pruning (REP) [Quinlan, 1987]	1	yes	bottom-up	pruning any branch	Finds the optimally pruned tree w.r.t. the pruning set.
Pessimistic Error Pruning (PEP) [Quinlan, 1987]	1	no	top-down	pruning any branch	Uses the continuity correction.
Minimum Error Pruning (MEP) [Cestnik & Bratko, 1991]	1-2	yes *last version*	bottom-up	pruning any branch	Based on the m-probability estimate of the error rate.
Critical Value Pruning (CVP) [Mingers, 1989]	2	yes	bottom-up	pruning branches of depth one	Selects the best tree on a subset of the space of all possible subtrees.
Cost-Complexity Pruning (0SE, 1SE and CV-0SE) [Breiman et al., 1984]	2	yes *0SE and 1SE*	all nodes considered together	pruning any branch	Can use cross-validation sets to estimate the accuracy.
Error-Based Pruning (EBP) [Quinlan, 1993]	1	no	bottom-up	pruning and grafting any branch	Estimates confidence intervals on the training set.

TABLE 35.1. Simplification methods considered in the experiments.

Finally, another characteristic of the methods is the type of *simplification* operators used to trim the tree, namely

- *Pruning branches of depth one.* In this case a branch of depth n can be pruned in n steps.

- *Pruning branches of any depth.*

- *Grafting a sub-branch* of a node t onto the place of t itself, thus removing only some of the nodes of the subtree rooted in t.

Currently, the only simplification method that combines grafting and pruning operators is the error based pruning (EBP) implemented in C4.5 [Qui93]. All pruning methods have been implemented as an extension of C4.5, so that we could use a well-tested system to generate the decision trees T_{max}'s.

35.3   Design of the experiment.

Major properties of the data sets considered in our experiments are summarized in Table 2. In particular, for each database we report the following information:
- The number of cases available in the data set.
- The number of classes of the training cases.
- The number of attributes used to describe each example.
- The number of attributes that are treated as real-valued.
- The number of non-numerical attributes with more than two values.
- The presence of null values in the description of an example.[3]

---

[3] Null values are treated according to the standard procedure implemented in C4.5.

database	No. cases	No. classes	No. attributes	Real-valued attributes	Multi-valued attributes	Null values	% Base error	Noise level	Uniform distrib.
Iris	150	3	4	4	0	no	66.67	low	yes
Glass	214	7	9	9	0	no	64.49	low	no
Led	1000	10	7	0	0	no	90	10%	yes
Hypo	3772	4	29	7	1	yes	7.7	no	no
P.-gene	106	2	57	0	57	no	50	no	yes
Hepat.	155	2	19	6	0	yes	20.65	no	no
Cleveland	303	2	14	5	5	yes	45.21	low	yes
Hungary	294	2	14	5	5	yes	36.05	low	no
Switzerland	123	2	14	5	5	yes	6.5	low	no
Long Beach	200	2	14	5	5	yes	25.5	low	no
Heart	920	2	14	5	5	yes	44.67	low	yes
Blocks	5473	5	10	10	0	no	10.2	low	no

TABLE 35.2. Major properties of the data sets considered in the experimentation.

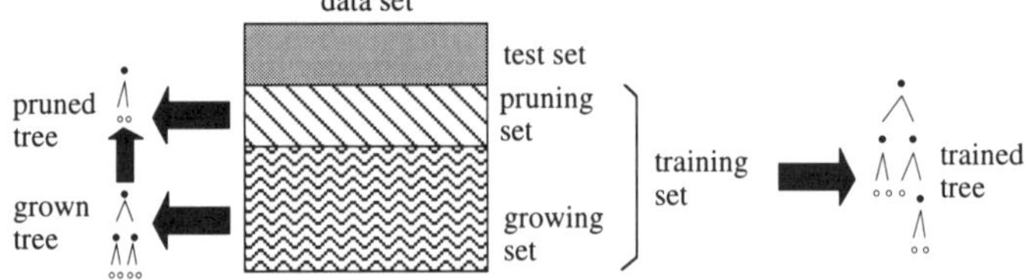

FIGURE 1. Partitioning of data available in each data set.

- The error rate obtained if the most frequent class is always predicted.
- The expected amount of noise in the database.
- The uniformity of distribution of training examples per class.

The data base Heart-disease is actually the join of four data sets with the same number of attributes but collected in four distinct places (Hungary, Switzerland, Cleveland and Long Beach). Of the 76 original attributes, only 14 have been considered, since they are the only ones deemed useful for the classification. Moreover, examples have been assigned to two distinct classes: no presence (value 0) and presence of heart diseases (values 1,2,3,4). For each data set considered, 25 experiments have been made by randomly partitioning the data set into three subsets: growing set (49%), pruning set (21%) and test set (30%) (see Figure 1). The union of the growing and pruning sets is called *training set*, and its size is just 70% of the whole data set. Therefore, the growing set contains the 70% of the cases of the training set, while the pruning set the remaining 30%. Both the growing and the training set are used to learn decision trees, that are called *grown tree* and *trained tree*, respectively. The former is used by those methods that need an independent pruning set in order to prune a decision tree, namely REP, MEP, CVP, and the cost-complexity pruning based on an independent test set and adopting the one standard error rule (1SE) or not (0SE) [BFOS84]. The latter is used by those methods that exploit the training set only, such as PEP, EBP, and the cost-complexity pruning based on cross-validation sets and not adopting the one standard error rule (CV-0SE).

In each experiment two statistics are recorded for pruned, grown and trained trees: the number of leaves (size) of the resultant tree and the error rate (e.r.) of the tree on the test set. As proposed by Buntine and Niblett [BN92], a two-tailed paired t-test is used to compare differences between pairs of either error or size averages. This is a major deviation from Mingers experimental design [Min89], where the ANOVA significance testing is preferred. Another major difference is that PEP, EBP, and CV-0SE have access to the whole training set rather than to the smaller growing set. Indeed, they simplify the trained trees and not the grown trees as the other ones. In a certain sense, there is a trade-off between the possibility of building a better tree T_{max} with all training cases and the lack of fresh cases for the pruning phase. Thus, *all* simplification methods can use the same set of examples, but REP, MEP, CVP, 0SE and 1SE will put aside some cases only for the pruning process. Finally, it is worthwhile to observe that Mingers investigated the possibility of interactions between four selection measures and pruning methods, while our study is limited to only one of those measures, namely the gain-ratio [Qui93].

In order to get an insight of the characteristics of some domains and pruning methods, we produced two decision trees for each experiment, called *optimally pruned grown-tree (OPGT)* and *optimally pruned trained-tree (OPTT)* respectively. The former is a grown tree that has been pruned by using the reduced error pruning on the *test set*. Thus, it is the best pruned tree we could produce from the grown tree because of the property of the reduced error pruning we mentioned in Section 2. Similarly, the OPTT is the best tree we could obtain by pruning some branches of the trained tree. OPGTs and OPTTs define an upper bound on the improvement in accuracy that pruning techniques can produce. Obviously, such an estimate is rather optimistic: the error rates of these optimal trees can even be lower than the corresponding Bayes optimal errors. However, optimally pruned trees are useful tools for investigating some properties of the data sets. For instance, by comparing the accuracy of the grown/trained trees with the accuracy of the corresponding OPGTs/OPTTs, it is possible to evaluate the maximum improvement produced by an ideal pruning algorithm. Moreover, the magnitude of differences in accuracy of OPGTs and OPTTs can help to understand if the ideal goal of those simplification methods that requires a pruning set is far from the ideal goal of the other methods. On the contrary, a comparison of the accuracy of the corresponding grown and pruned trees provides us with an indication of the initial advantage that some methods may have over the others.

The size of the optimally pruned trees can be exploited to establish a bias of the simplification methods towards either overpruning or underpruning. In this case, we should compare the size of an OPGT with that of the corresponding tree produced by those methods that do use an independent pruning set, while the size of an OPTT should be related to the result of the other methods. As a matter of fact, tree size is a coarse measure, since it may happen that a branch with p leaves is erroneously pruned while an equally-sized harmful branch is not. In this latter case, indeed, the size of the resultant tree is optimal even though there are both an overpruning and an underpruning problem. However, as we will show later, this does not prevent us from detecting the most serious of the two possible faults occurring in a simplification process. The real limitation of our empirical study concerns the comparison of the average size of OPTTs with that of trees produced by the EBP. This is due to the fact that REP, the method used to produce OPTTs, uses only a pruning operator, while EBP also adopts a grafting operator (see Table 1). Therefore, OPTTs are not optimal with respect to the space explored by EBP. Although

		Iris	Glass	Led	Hypo	P-gene	Hepat.	Clev.	Hung.	Switz.	L.B.	Heart	Block
GROWN	size	5.4	27.92	41.33	20.24	19	11.24	33.76	32.56	9.64	31.64	115.7	82.92
	e.r.	5.866	36.87	27.96	.622	27.12	22.81	30.07	25.5	13.84	33.27	24.72	3.65
OPGT	size	3.64	12.6	21.44	9.16	7.84	3.4	12.2	10.16	1.32	4.12	36.16	25.84
	e.r.	4.888	31.62	25.64	.399	18.50	15.83	22.99	16.91	4.622	23.6	18.77	2.609
TRAINED	size	6.84	36.32	44.68	27.24	25.6	16.76	50	47.88	11.76	43.48	164.6	111.9
	e.r.	5.598	35.38	27.48	.604	23.5	21.87	29.1	25.5	13.3	33.73	23.82	3.57
OPTT	size	4	15.08	22.04	9.36	10	4.36	16.36	9.64	1.16	4.92	45.96	30.44
	e.r.	4.442	28.31	25.31	.352	16.5	16.34	20.84	17.27	5.731	23.13	17.71	2.354

TABLE 35.3. Average size and error rate of the (optimally pruned) grown/trained trees.

we did not observe a frequent application of the grafting operator in our experiments, we plan to address this limitation in the future work.

35.4   Experimental results.

The average size and error rate of the (optimally pruned) grown/trained trees for each domain is reported in Table 3.

The ratio *(grown tree size/OPGT size)* ranges from 1.5 for the Iris data to 7.7 for the Long Beach data, while the ratio *(trained tree size/OPTT size)* is even greater than 10 for the Switzerland data. Such strong differences in size between some grown/pruned trees and their corresponding optimally pruned trees can be explained by looking at the base error column in Table 2. Indeed, for the "incriminated" data set, there are only two classes, one of which contains only 6.5% of cases. Since the learning system fails in finding an adequate hypothesis for those cases, the pruning method will generally tend to prune the tree up to the root.

Another point is the greater accuracy of trained trees: this means that those methods that require a pruning set are under a disadvantage with respect to the others. The only exception is represented by the database Hungary, in which the greater number of training instances does not increase the accuracy of the trees but does increase their complexity.

By restricting our attention to only two domains (Iris and Led[4]) also used in Mingers' empirical comparison, we could detect a certain discrepancy between the two figures on the average size of unpruned trees:[5] 6.9 compared to 5.4 in Iris, and 56.6 compared to 41.33 in Led. Such a difference may be attributed to the percentages of cases used to build the trees (60% in Mingers' study versus 49% in ours). Indeed, by looking at the average size of trained trees which are built on 70% of cases, this discrepancy is significantly reduced, at least for the Iris domain. The discrepancy for the Led data should be mainly attributed to the tree collapsing process, implemented in C4.5, that stops branching when the best split does not reduce the apparent error rate. However, such differences do not compromise the validity of our results, since simplification methods should be able to operate on any tree built by a TDIDT system.

---

[4]The Led domain is named Digit in Mingers' paper.

[5]Unfortunately, we do not know if the average size reported in Mingers' paper refers to trees build using the gain-ratio measure as in our experiments.

database	REP	MEP	CVP	0SE	1SE	PEP	CV-0SE	EBP	trained
Iris	5.689	6.222	5.866	5.778	7.645	5.332	5.6	5.065	5.598
Glass	*38.5*	38.19	36.87	38	*40.69*	35.31	*38.94*	35.88	35.38
Led	*28.15*	*28.24*	*27.99*	28.16	*29.52*	27.91	27.77	27.36	27.48
Hypo	**.515**	**.51**	.541	**.508**	.576	**.505**	**.496**	**.447**	.604
P. gene	23	24.37	25.62	24	25.62	23.38	23.5	21.75	23.5
Hepatitis	20.34	21.28	20.6	20.17	20.42	21.1	21.19	21.36	21.87
Cleveland	27.65	28.88	30.07	29.10	29.89	29.01	29.58	28.88	29.1
Hungary	**21.68**	**22.14**	27.41	**21.73**	**21.77**	**21.73**	**21.23**	**22.36**	25.5
Switzerland	**6.272**	12.65	**2.393**	**6.164**	**5.839**	**6.163**	**6.271**	**6.163**	13.3
Long Beach	**26.27**	**28.53**	**26.93**	**27.47**	**25.67**	**27.47**	**26.93**	**29.33**	33.73
Heart	23.54	23.67	24.13	23.81	24.29	**22.94**	23.35	**22.72**	23.82
Blocks	**3.22**	**3.34**	3.602	**3.19**	**3.35**	2.98	**3.17**	**3.05**	3.57

TABLE 35.4. Average error rates for different databases.

In Table 4 the average error rates concerning different simplification methods (*sophisticated* strategies) are reported together with the average error rate of unpruned trained trees (*naive* strategy). Both the sophisticated and the naive strategies have access to the same data, the training set, but the sophisticated one can either use some data for growing the tree and the rest for pruning it, or exploit all the data at once for building and pruning the decision tree. From a quick look at the table we can immediately conclude that the sophisticated strategies do generally worse than the naive one in the first three databases, namely Iris, Glass and Led. More precisely, sophisticated methods that exploit an independent pruning set yield trees that are more accurate than corresponding grown trees, which indicates that pruning is beneficial. Unfortunately, the grown tree is less accurate than the trained tree, so that the global effect is negative. The only two cases in which the sophisticated strategy seems to win are those in which the trained tree is pruned by means of the PEP and EBP methods, which do not use independent pruning sets. On the contrary, for the Hypo, Switzerland, Hungary, and Long Beach domains, almost all pruning methods perform well. In particular, for the Switzerland data, the worst result obtained with a sophisticated strategy is better than that obtained with the naive approach. This is not surprising, since we had already observed that the OPGT (OPTT) is much smaller than the corresponding grown (trained) tree, which means that in this domain techniques for simplifying decision trees are inherently beneficial. Finally, the 1SE shows the worst performance in five databases, while the EBP has the lowest error rate in five domains with a significant improvement in two of them.

In order to study the statistical significance of these results, we compared the error rates of the pruned trees with those of the corresponding trained trees. Thus, values reported in bold (italics) denote a significant improvement (worsening) in predictive accuracy at the confidence level of 0.10. It is easy to see that *tree simplification does not generally decrease the predictive accuracy.* The only exception is represented by the application of the 1SE rule with both an independent pruning set and cross-validation sets. Furthermore, no method is able to significantly reduce the error rate in six domains (Iris, Glass, Led, P-gene, Hepatitis and Cleveland), while all methods perform well when applied to the Long Beach database, and all but one improve the accuracy of the final tree in the Hungary and Switzerland domain. It is interesting to observe that PEP and EBP produce

database	REP	MEP	CVP	0SE	1SE	OPGT	PEP	CV-0SE	EBP	OPTT
Iris	3.4	4	*5.4*	3.44	**3**	3.64	3.76	4.48	*4.88*	4
Glass	11.04	*18.52*	*27.92*	14.12	**7.04**	12.6	*21.12*	18.16	*28.72*	15.08
Led	20.52	25.28	*36*	25.64	**12.32**	21.44	**18.32**	*30.24*	*30.6*	22.04
Hypo	**8.2**	*15.76*	*14*	8.64	**7.2**	9.16	9.32	9.68	*13.68*	9.36
P-gene	**6.4**	*10*	*13.36*	7.24	**4.36**	7.84	**8.44**	9.52	*15.28*	10
Hepatitis	2.64	*8.36*	3.6	2.56	**1.36**	3.4	5.44	**2.64**	*9.08*	4.36
Cleveland	10.92	15	*30.76*	9.88	**4.12**	12.2	*19*	**10.36**	*29.28*	16.36
Hungary	**7.32**	10.44	*21*	**5.24**	**2.28**	10.16	10.2	9.2	*17.12*	9.64
Switzerland	1.4	*8.92*	1.68	1.56	1	1.32	1.16	1.16	1.08	1.16
Long Beach	6.16	*17.16*	7.28	6.92	**1.4**	4.12	4.96	3.8	*10.88*	4.92
Heart	33.52	31.92	*58.2*	27.52	**7.2**	36.16	**21.92**	19.17	46.04	45.96
Blocks	24.68	*65.04*	*78.48*	28.12	**13.12**	25.84	*37.24*	**17.56**	*50.92*	30.44

TABLE 35.5. Average tree size for different databases.

significantly better trees for the same data sets, so that it is possible to postulate, on empirical grounds, the equivalence of the two methods that appear to be different in their formulation. By summarizing these results, we can conclude that there is no indication that methods exploiting an independent pruning set perform definitely better than the others. This claim is at a variance with that reported by Mingers [Min89], and should be attributed to the different design of the experiments.

The average size (number of leaves) of the simplified trees for different domains is reported in Table 5. As already pointed out, data concerning the REP, MEP, CVP, 0SE and 1SE should be compared to the average size of corresponding OPGTs, while the results of PEP, CV-0SE, and EBP should be related to those of OPTTs. At a glance we can say that REP is biased towards overpruning, since in nine domains it produces trees smaller than the optimal ones on the average. This result has a theoretical justification: REP takes the decision of removing a branch by considering data in the pruning set alone, without any care of the evidence provided by the cases used to build the tree. Therefore, when relatively few cases are reserved for pruning, this simple solution will lead to overpruning.[6] Another method clearly biased towards overpruning is 1SE. This is not surprising, since the one standard error rule was introduced to choose the simplest tree whose estimated accuracy is comparable to that selected by the 0SE. On the contrary, MEP, EBP and CVP underprune in almost all cases, and even worse, CVP does not prune at all in two domains, namely Iris and Glass. The reason for this odd result seems to be related to the use of the gain-ratio as selection measure. Indeed, when all the leaves in a tree are "pure", that is their examples belong to the same class, the gain-ratio computed in the deepest internal nodes is equal to the maximum value 1.0. Thus, the critical value should be greater than or equal to 1.0 in order to prune the nodes at the bottom of the tree,[7] but in this case all internal nodes should be pruned according to the same criterion. This means that in the first step critical value pruning may find only two trees, T_{max} and the

---

[6]Actually, in a separate study on this method we have also observed that, in most of the domains considered in this paper, the optimal size of the pruning set is 70% of the training set.

[7]Experiments on the CVP are made by setting a maximum critical value equal to 1.0 and a step equal to 0.01.

root tree, with the obvious consequence that the former one will be generally chosen in the second step. Obviously, this happens only in the case of pure leaves, as with the Iris and Glass domains. In fact, in all the other domains in which there is at least an impure leaf because of clashes (i.e., examples of distinct classes but with same attribute values) or because of tree collapse,[8] this issue is less evident. Also for the Switzerland data there is no underpruning problem, but in this case the reason is that the best tree is often the root tree. Results on the CVP reveal another discrepancy with Mingers' data, at least for the two common domains, Iris and Led. Unfortunately, in this case we have no explanation, and we can only hypothesize the adoption of different stopping criteria in Mingers' TDIDT system. The differences for the other two methods, namely REP and MEP, can be justified by the fact that our algorithms are different from his. Indeed, as already pointed out, the REP method we implemented is guaranteed to find the optimal tree with respect to the pruning set, while this seems not to be true for Mingers' formulation. Moreover, our MEP version is the latter proposed by Cestnik and Bratko [CB91], and not the original one proposed by Niblett and Bratko [NB86].

As done for the error rate, even for the tree size we tested the significance of the differences by means of two-tailed paired t-tests. This time a number in bold indicates a significant overpruning at a level of 0.10, while a number in italics represents a significant underpruning. Recall that the comparison involves OPGTs for those methods that operate on the pruning set, and OPTTs for the others. The tests confirm that MEP, CVP and EBP are biased towards underpruning, while 1SE prefer overpruning. It is worthwhile to note that the predictive accuracy is not necessarily increased when a simplification method produces trees which do not significantly differ in size from the correspondingly optimally pruned trees: in fact, pruning may help to simplify trees without improving their predictive accuracy. Moreover, by measuring the tree size we do not have detailed enough information to guarantee that only overpruning or underpruning occurred. Thus, we can observe a significant increase of error rate with respect to the optimal tree even when the size appears to be optimal (see the results of REP in the Glass domain). Finally, by ideally superimposing Tables 4 and 5, it is also possible to draw other interesting conclusions. For instance, in some databases, such as Hungary and Heart, overpruning produces better trees than underpruning. This latter result agrees with Holte's observation that even simple rules perform well on most commonly used data sets in the machine learning community [Hol93]. It is also a confirmation of the overfitting problem that affects TDIDT systems.

35.5 Conclusions.

In this paper, a new empirical study of eight well-known simplification methods for decision-tree induction has been presented. The two main differences with previous works performed by Quinlan [Qui87] and Mingers [Min89] are the diverse design of the experiments and the use of optimally pruned trees as yardsticks for comparing the accuracy and complexity of simplified and unsimplified trees. The main conclusions drawn in our

---

[8]Another default stopping rule implemented in C4.5 prevents from growing a tree when the number of cases per outcome is less than two. However, we invoked the system by setting the parameter $-m$ to 1, so forcing the system to consider also the cases of n-ary tests with one example per outcome.

study are the following:

- Putting aside some data for the pruning process is not generally the best strategy.

- In general, simplification methods do not significantly decrease the predictive accuracy of the final trees.

- MEP, CVP and EBP are biased towards underpruning, while 1SE tends to overprune.

Further work is needed to confirm and extend these results. In particular, artificial data sets with either attribute or class noise may provide us with meaningful information on the performance of different simplification methods.

35.6 REFERENCES

[BFOS84] L. Breiman, J. Friedman, R. Olshen, and C. Stone. *Classification and regression trees*. Wadsworth International, Belmont, CA, 1984.

[BN92] W. Buntine and T. Niblett. A further comparison of splitting rules for decision-tree induction. *Machine Learning*, 8:75–85, 1992.

[CB91] B. Cestnik and I. Bratko. On estimating probabilities in tree pruning. In *Proceedings of the EWSL-91*, pages 138–150, 1991.

[EMS93] F. Esposito, D. Malerba, and G. Semeraro. Decision tree pruning as a search in the state space. In P. Brazdil, editor, *Machine Learning: ECML-93*. Springer-Verlag, Berlin, 1993.

[Fis92] D. H. Fisher. Pessimistic and optimistic induction. Department of Computer Science, Vanderbilt University, 1992.

[Hol93] R. C. Holte. Very simple classification rules perform well on most commonly used datasets. *Machine Learning*, 11:63–90, 1993.

[MA94] P. M. Murphy and D. W. Aha. UCI repository of machine learning databases [machine-readable data repository]. Department of Information and Computer Science, University of California, Irvine, 1994.

[Min89] J. Mingers. An empirical comparison of selection measures for decision-tree induction. *Machine Learning*, 4:227–243, 1989.

[NB86] T. Niblett and I. Bratko. Learning decision rules in noisy domains. In *Proceedings of Expert Systems 86*, Cambridge, 1986. Cambridge University Press.

[Nib87] T. Niblett. Constructing decision trees in noisy domains. In I. Bratko and N. Lavrac, editors, *Progress in Machine Learning*. Sigma Press, Wilmslow, 1987.

[Qui87] J.R. Quinlan. Simplifying decision trees. *International Journal of Man-Machine Studies*, 27:221–234, 1987.

[Qui93] J. R. Quinlan. *C4.5: Programs for Machine Learning*. Morgan Kaufmann, San Mateo, 1993.

[Sch93] C. Schaffer. Overfitting avoidance as bias. *Machine Learning*, 10:153–178, 1993.

36
Robust Linear Discriminant Trees

George H. John

Computer Science Department
Stanford University

ABSTRACT We present a new method for the induction of classification trees with linear discriminants as the partitioning function at each internal node. This paper presents two main contributions: first, a novel objective function called *soft entropy* which is used to identify optimal coefficients for the linear discriminants, and second, a novel method for removing outliers called *iterative re-filtering* which boosts performance on many datasets. These two ideas are presented in the context of a single learning algorithm called DT-SEPIR, which is compared with the CART and OC1 algorithms.

36.1   Introduction

Recursive partitioning classifiers, or decision trees, are an important nonparametric function representation in statistics and machine learning (Friedman 1977, Breiman, Friedman, Olshen & Stone 1984, Quinlan 1986, Quinlan 1993). Their wide and successful use in fielded applications and their simple intuitive appeal make decision tree learning algorithms an important area of study. In this paper we address both of the main issues in decision tree induction: construction and pruning.

We follow the traditional recursive partitioning approach to tree building, but rather than partitioning the data on a single axis we instead employ a linear discriminant at each node to recursively split the data (Henrichon, Jr. & Fu 1969, Friedman 1977, Breiman et al. 1984). The issue of how to find a good discriminant naturally arises. We discuss problems in previous approaches and propose a new splitting criterion, *soft entropy*. The DT-SE algorithm uses recursive partitioning on this criterion to build a pure tree, one that correctly classifies each instance in the training set.

A problem with working on decision tree splitting criteria is that a decision tree is seldom used for prediction without first pruning the tree in an attempt to avoid overfitting and address the bias-variance tradeoff. Thus, results on the superiority of one splitting criterion over another should be in the context of the use of some regularization algorithm in order to mirror the results we expect to observe in practice.

To address the overfitting problem, we developed the DT-SEP algorithm which prunes the pure tree using a method similar to a stopping rule. Pruning involves removing an entire subtree and replacing it with a leaf. Implicit in this operation is the assumption that the input patterns which caused the subtree to be built are locally either unimportant or harmful. We suggest that these patterns might also be *globally* unimportant or harmful (they might have caused bad splits higher in the tree) and thus the induction algorithm should remove such patterns from the training set and build a new tree. This step of

---

[1] *Learning from Data: AI and Statistics V.* Edited by D. Fisher and H.-J. Lenz. ©1996 Springer-Verlag.

TABLE 36.1. Notation used in the rest of the paper.

Symbol	Meaning
$\vec{x} \in \mathcal{X}$	An instance. A vector of attribute values.
$y \in \mathcal{Y}$	The class or output value of an instance.
f	A target function mapping instances to output values. $f : \mathcal{X} \mapsto \mathcal{Y}$.
T	A training set. $T \subseteq \{(\vec{x}, y) \| \vec{x} \in \mathcal{X}, y = f(\vec{x}) + \epsilon\}$ for some *noise* term ϵ.
z	A training (labeled) instance. $z = (\vec{x}, y)$. $\mathcal{Z} = \mathcal{X} \times \mathcal{Y}$.
S	A weighted set of training instances. For each $z \in S$, $w_S(z) \in (0, 1]$ where w_S is the weighting function of S. $w_S(z) = 0$ for all $z \notin S$. Unweighted sets are a special case where $w_S(z) \in \{0, 1\}$.
$\theta \in \Theta$	A vector of parameters for a splitting function. In the case of linear splitting functions, θ is the vector of coefficients.
g_θ	A *soft* binary splitting function, mapping instances into $[0, 1]$. $g_\theta : \mathcal{X} \mapsto [0, 1]$. Given a weighted set S, define S/g_θ to be the tuple of weighted sets (S_0, S_1) gotten by applying g_θ to each $z_x \in S$ such that $w_{S_0}(z) = g_\theta(z_x)w_S(z)$ and $w_{S_0}(z) + w_{S_1}(z) = w_S(z)$. *Hard* splitting functions have the range $\{0, 1\}$.
I	A splitting criterion (impurity function) over binary splits, mapping two weighted sets to the real numbers $\mathbb{R}$.

filtering the pruned instances out of the training sample and rebuilding yields the DT-SEPIR (Iterative Re-filtering) algorithm. Though common in regression in the guise of *robust* (Huber 1977) or *resistant* (Hastie & Tibshirani 1990, Chapter 9) fitting, in the context of classification this appears to be novel.

In Section 36.2 we formalize the problem of finding a splitting function for an arbitrary splitting criterion as a function optimization problem. Section 36.3 discusses problems with splitting criteria used in previous work in decision trees, and presents the soft entropy criterion. Section 36.4 presents the pruning algorithm and discusses the iterative re-filtering method in more detail. Experimental results comparing DT with related algorithms are presented and discussed in Section 36.5. Section 36.6 discusses related work, and Section 36.7 gives directions for future work. Finally, Section 36.8 presents our conclusions.

36.2   Finding Splitting Functions

In this section we frame the problem of finding the best splitting function at each node in the tree as a function optimization problem. To define this correctly we require a bit of notation summarized in Table 36.1.

We view the problem of choosing a splitting function at a node of a decision tree as a function optimization problem. Given a weighted set S of training instances we desire to split, and some fixed parametric form of the splitting function g_θ, our goal is to find θ^* satisfying:

$$\theta^* = \operatorname*{argmin}_{\theta \in \Theta} I(S/g_\theta) \ . \tag{36.1}$$

In defining a problem as function optimization there are three important decisions: the domain in which the optimization should be performed, the objective function, and the function optimizer used. The domain, Θ, is defined by whatever functional form we set for g—all parameters remaining to be set constitute Θ. The optimization algorithm could be any method for unconstrained optimization, including methods using first derivative information if $\partial I/\partial\theta$, the derivative of the impurity measure with respect to the parameters, exists. Fortunately, by the chain rule $\partial I/\partial\theta$ can be separated into two terms: $\partial I/\partial g_\theta\ \partial g_\theta/\partial\theta$, so that we need only derive the derivatives for the splitting criterion with respect to the outputs of the splitting function (whatever it may be) and the derivative of the output of the splitting function with respect to its parameters θ (regardless of the impurity function being used).

The essential ingredient is the objective function: the splitting criterion I which we wish to minimize. All of the splitting criteria we are aware of—twoing (Breiman et al. 1984), entropy, Gini, delta (Morgan & Messenger 1973), gain-ratio (Quinlan 1986), C-sep (Fayyad & Irani 1992)—are defined on sets, but have straightforward extensions to weighted sets. The criteria are all just functions of the following counts: the total number of instances in S, the total number in S_0 and S_1, and the number of instances of each class in S_0 and S_1. For weighted sets, we define cardinality as $|S| = \sum_{z\in S} w_S(z)$. Simply by replacing sets with weighted sets and the standard cardinality measure by the new cardinality measure, we can define *soft* versions of all of our favorite splitting criteria.

As an example, consider the softened linear discriminant splitting function

$$g_\theta(\vec{x}) = 1/(1 + \exp(-\theta^T \vec{x})) \ . \tag{36.2}$$

Note that the range of g is $(0, 1)$, thus when using g_θ to partition a set S, all $z \in S$ are partially assigned to both S_0 and S_1.

The function optimization algorithm repeatedly evaluates $I(S/g_\theta)$ to get the value of the current θ, makes adjustments to θ (perhaps based on $\partial I/\partial\theta$) and evaluates the new θ. Eventually the optimization algorithm stops, outputting $\hat{\theta}$, its estimate of θ^* from Equation 36.1.

36.3   Building Trees with Soft Entropy

Although several approaches to building linear discriminant trees have used a soft splitting criterion (mean-squared error in logistic regression), and many other approaches have used a hard criterion based on entropy or Gini, we are aware of no approach combining the desirable properties of both. Below we discuss why we prefer a criterion based on entropy and why we prefer a soft criterion. We then give the details of the tree construction algorithm DT-SE.

36.3.1   Why Soft Entropy?

The left graph in Figure 1 shows several candidate hard splits of a small dataset. All of the splits shown have equivalent counts (number of instances of each class in each subset) and hence they are evaluated equally by a hard splitting criterion. Because they use sets instead of weighted sets, hard criteria possess limited granularity and are unable to distinguish between splits that are quite different. The unique best soft split is shown in Figure 1 on the right. Since the soft split assigns instances partially to both the left and right subset (in the case of Equation 36.2, based on the distance between the instance and

FIGURE 1. Left: all four splits have equivalent entropy (and Gini, delta, *etc.*). Right: the split found by minimizing soft entropy.

FIGURE 2. Left: The soft linear discriminant minimizing mean squared error fails to partition the data. Right: The soft linear discriminant minimizing soft entropy.

the splitting hyperplane), all hyperplanes on the left are assigned different values by the splitting criterion. Additionally, since soft criteria are differentiable they are amenable to the use of a wider variety of function optimizers.

Given that softness is a desirable property, much work has used the soft linear discriminant of Equation 36.2 combined with an error criterion such as sum-squared error, common in logistic regression and neural networks. However, Breiman et al. (1984) argue against error as a splitting criterion, and our example in Figure 2 demonstrates a shortcoming of error. Figure 2 shows two linear discriminant functions each optimizing some splitting criterion. (In this case they happened to be soft linear discriminants, but the same phenomenon occurs with hard splits.) The entropy-minimizing discriminant does an excellent job of partitioning the data, while the error-minimizing discriminant fails to partition the data at all.

36.3.2 Finding the Best Discriminant

To build trees using the soft entropy criterion, we must combine several ingredients. Perhaps most important is the agenda for constructing the tree itself. We will use the standard recursive partitioning approach, at each node finding the splitting function that minimizes some impurity measure, splitting the data with this splitting function, and then recursively building trees from the resulting subsets.

The splitting criterion we wish to minimize is still $I(S/g_\theta)$, where I is now specified to be the entropy measure. (Note that I is a function on a pair of sets while entropy is a function of a single set—as in Breiman et al. (1984) we take the weighted average of entropy on the two sets.) The splitting function g is defined in Equation 36.2. In this case, calculating $I(S/g_\theta)$ and $\partial I/\partial \theta$ is straightforward, so we used the simplest unconstrained function optimization method using first derivative information: steepest descent.

It is important to note that once the best θ is found, we use a hard split to partition the data into a left and right subset, which are themselves recursively split. All of the work on soft splitting functions was done just to discover a good hard split. In the case of the soft linear discriminant, it is replaced by its hard cousin:

$$g_\theta(\vec{x}) = 1 \text{ if } \theta^T \vec{x} \geq 0, 0 \text{ otherwise .} \tag{36.3}$$

36.4   Pruning a Tree

Our regularization algorithm contains a pruning step and a retraining step. The pruning step replaces a decision node with a leaf whenever the number of positive or negative instances matching that node drops below k. (In our experiments, k was arbitrarily set to 5.) This is applied to all nodes in the tree in a top down fashion.

In what appears to be a novel approach, the regularization algorithm then *retrains* on the reduced training set, building a completely new tree. The reduced training set is defined to the original training set minus the instances which the pruned tree now classifies incorrectly (the "confusing" instances). While retraining may seem odd, it is in fact just an extension of the assumptions underlying pruning. By pruning the tree we essentially assume that these confusing instances are locally not useful. Retraining merely takes this assumption a step further by completely removing these instances from the training set. The regularization algorithm continues pruning and retraining until no further pruning can be done.

In statistics, estimators resistant to the effect of outliers, *robust* estimators, have been studied in some depth. Points with high *leverage*, points with disproportionately high effect on the fitted model, are often removed from training data in order to better fit the remaining points. In the context of decision trees, we may identify a set of points with high leverage by examining the difference in number of nodes between a tree built with and without the set of points. This difference is estimated by starting with a pure tree built using all the points and then pruning. The training instances that it now classifies incorrectly are the high leverage points that were removed. However, the removal of the points by pruning was only approximate—the obvious step is to remove the points from the training set and retrain.

36.5   Experiments with DT-SEPIR

We ran several cross-validation experiments comparing the DT-SE algorithms with CART (Breiman et al. 1984) and a CART-like algorithm, OC1 (Murthy, Kasif & Salzberg 1994, Murthy, Salzberg & Kasif 1993). Experimental methodology is first discussed, then results are presented and analyzed.

36.5.1   Methodology

Five datasets were gathered from the UCI (Murphy & Aha 1994) and StatLog (Michie, Spiegelhalter & Taylor 1994) collections[2]. In the Vote1 dataset, the task is to guess the political party of congress members given their votes on key issues. To make the problem more difficult, the most predictive attribute has been removed. In the Australian database, the task is to decide whether or not to give someone a credit card. The Breast-Wisc is the Wisconsin breast cancer database. The task is to classify a lump as benign or malignant. The task in the Diabetes dataset is to classify a Pima Indian female as having diabetes or

---

[2] All datasets were processed (either by the author, or in the case of the Statlog datasets, by others) so that all unordered categorical attributes were encoded using 0-1 indicator variables and all ordered categorical attributes were encoded as integers. All missing values were either removed or replaced with their mean or mode.

TABLE 36.2. Test set accuracy and tree size results from 10-fold cross-validation.

		Vote1	Australian	Breast-Wisc	Diabetes	Heart
DT-SEPIR	Acc	89.0±4.0	85.0±4.6	95.1±2.2	73.7±5.8	77.8±10.5
	Size	8.8±3.7	8.0±3.0	6.2±2.3	16.6±5.1	7.4±2.3
DT-SEP	Acc	86.9±5.6	81.8±3.6	95.3±2.2	71.9±6.7	75.6±10.8
	Size	11.8±2.3	35.4±6.3	8.2±1.0	58.2±6.7	13.2±2.0
DT-SE	Acc	85.5±4.8	80.2±4.3	94.3±1.9	68.9±6.7	74.5±11.0
	Size	20.4±3.3	85.8±11.0	19.8±4.3	127.6±10.1	25.0±2.7
OC1-pruned	Acc	89.9±4.2	85.1±2.9	95.6±2.7	74.3±3.5	75.5±6.8
	Size	5.8±6.0	10.2±9.6	3.6±1.0	13.6±12.1	9.6±10.2
OC1	Acc	87.1±3.5	81.2±4.6	94.9±1.6	68.4±4.3	72.6±7.5
	Size	36.2±5.3	83.6±11.0	35.2±5.4	149.8±7.7	42.2±6.3
CART-pruned	Acc	90.8±6.4	85.7±4.3	96.8±2.5	74.0±4.1	78.5±10.9
	Size	3.0±0.0	4.8±3.5	3.0±0.0	5.4±5.2	3.2±0.6
CART	Acc	87.4±5.2	79.9±2.4	95.3±2.7	68.3±4.5	79.3±7.2
	Size	67.0±9.5	102.6±10.2	29.4±4.7	167.8±14.4	41.8±7.9

not. In the Heart database, the algorithms must guess whether or not a patient has heart disease. All algorithms were run using their default parameter values, and CART was set to use linear splits and to continue partitioning the training set until pure regions were found. "CART-pruned" results are from using ten-fold cross-validation and the 1-SE rule to select the correct pruned tree. "CART" results are from using the full tree initially built by CART. Similarly "OC1" results refer to the pure tree initially built. "OC1-pruned" randomly holds out 10% of the training set to use for evaluation during pruning.

Ten-fold cross-validation was used to estimate the accuracy of each algorithm on each dataset. (This is a separate step from CART's own internal cross-validation.) In 10-fold CV, the dataset is divided into 10 blocks of roughly equal size, then the learning algorithm is repeatedly trained on nine out of the ten blocks and tested on the block not seen during training. From the ten accuracy results we report the mean and standard deviation.

36.5.2 Results and Discussion

Table 36.2 gives 10-fold cross-validation results for the DT-SE algorithms, the OC1 algorithm, and CART. Each box contains 2 measurements: the top is the accuracy, and the bottom is the tree size for the ten runs. Within each box we report the mean over the 10 cross-validation runs, and the standard deviation over the 10 runs.

Within the DT-SE algorithms, we note that the DT-SEP (pruned) trees are roughly half the size of the DT-SE (unpruned) trees in each domain, while the accuracy is always about 1% higher (or more). Going from DT-SEP to DT-SEPIR (i.e., adding iterative re-filtering), the trees that were small tend to shrink just a bit, while the larger trees shrink dramatically (from 35 to 8 nodes in the Australian domain, and from 58 to 17 in Diabetes) while accuracy is improved by roughly 2% in all domains except Breast-Wisc. Thus it seems wise to add pruning and iterative re-filtering to the basic tree-building algorithm.

Comparing the DT-SE algorithm with OC1 and CART (all of which used no pruning for this experiment), we notice that DT-SE appears to be the winner in terms of producing smaller trees. CART often produced the largest trees by far, perhaps validating the claims of the OC1 authors that its optimization algorithm is susceptible to local minima (*e.g.*, CART's average size on the Vote1 dataset is 67, versus 36 for OC1 and 20 for DT-SE), and DT-SE produced trees that were about 60% the size of OC1's. Looking at accuracy, all 3 unpruned algorithms performed about the same, except for CART's superior accuracy on the Heart dataset. From these results we would like to conclude that our arguments in favor of soft entropy have been supported—the ability of DT-SE to reach the same accuracy level as OC1 and CART while building smaller trees supports the claim that DT-SE is using a superior splitting criterion.

As we mentioned in the introduction, we must also compare the behavior of splitting criteria coupled with some pruning strategy, since this should correspond to the most frequent use of the algorithm in practice. Looking at DT-SEPIR, OC1-pruned and CART-pruned, CART-pruned selected extremely small trees in most cases—a tree with three nodes is simply one test and two leaves, so in the Vote1 and Breast-Wisc domains, CART-pruned consistently found a single linear discriminant. CART-pruned's accuracy was roughly 1% better than OC1-pruned and DT-SEPIR on each domain. None of the differences in accuracy were statistically significant (95%), except that CART-pruned had higher accuracy than OC1-pruned on the Heart data. From these results we conclude that CART's cross-validation is a superior pruning method, giving higher accuracy and smaller trees. DT-SEPIR, although handicapped by a completely ad-hoc pruning method and arbitrary choice of pruning threshold (5), was statistically indistinguishable from CART-pruned. Since the pruning method is clearly not responsible for any good performance by DT-SEPIR, we conclude that using soft entropy in building the tree, and iterative re-filtering in pruning the tree results in good performance, and should be investigated with a pruning method based on cross-validation.

36.6   Related Work

John (1994) presents preliminary experiments with DT-SEPIR, comparing against other neural network and decision tree algorithms. Friedman (1977) gives an excellent discussion of practically every issue related to the induction of decision trees from data, including linear discriminant splits using Fisher's linear discriminant (Duda & Hart 1973). The earliest attempt at building linear discriminant trees seems to be Henrichon, Jr. & Fu (1969), who use the maximum eigenvector of the covariance matrix to determine the split. Lin & Fu (1983) use k-means clustering to find the splitting function. Bennett & Mangasarian (1992) use a parallel pair of hyperplanes as a splitting function and use linear programming to minimize their splitting criterion.Sankar & Mammone (1991) present "Neural Tree Networks," which are simply soft linear discriminant trees using mean squared error as the criterion. Loh & Vanichsetakul (1988) present linear discriminant trees where the optimal discriminating hyperplane may be determined in closed form by making normality and homoscedasticity assumptions about the data. Quing-Yun & Fu (1983) present a sophisticated method for not only finding linear splits at each node (based on regression) but also selecting the best subset of variables to use in the split. Brodley & Utgoff (1995) present a similar algorithm, also based on linear regression with subset selection.

Although the work in this paper was developed independently, Brent (1991) discusses essentially the same technique for softening a related splitting criterion. However, Brent's focus was not on finding a better splitting criterion for multivariate decision trees; rather, he used this impurity measure to train neural networks one node at a time. Sethi (1990) gives the original description of the mapping between neural nets and decision trees, and proposes converting the tree (built with univariate splits) to neural net form for further refinement. Along the same lines, Jordan & Jacobs (1993) present a method that allows simultaneous optimization of all coefficients in the tree by defining the model likelihood and using the EM algorithm (Dempster, Laird & Rubin 1977) for optimization. Soft splits are used not only for tree construction but also during classification. However, their method requires the user to specify the tree structure a priori. Sahami (1993) investigates various optimization algorithms for finding error-minimizing linear discriminants at each node in a decision tree.

Bichsel & Seitz (1989) give an algorithm for training a tree-structured net very similar to the approach by Heath, Kasif & Salzberg (1993), using simulated annealing to search the space of hyperplanes at each node. The procedure for adding nodes is somewhat similar in spirit to that used in TDIDT methods, but their actual construction algorithm assumes the tree will be very small. A related approach in the neural net community was by Koutsougeras & Papachristou (1988). They discuss the benefits of learning entropy-minimizing hyperplanes in a tree-structured neural net, but in the actual algorithm they instead implement Fisher's linear discriminant function.

Regarding robust methods and outlier rejection, Huber (1977) states "I am inclined to ... prefer technical expertise to any 'statistical' criterion for straight outlier rejection." We are guilty of this sin in our work, but Guyon, Boser & Vapnik (1993) have proposed an interesting method for making use of human expertise in outlier removal to "clean" a dataset. John (1995) presents further experiments with iterative re-filtering in the context of the C4.5 (Quinlan 1993) decision tree learning algorithm.

36.7 Future Work

A known problem with decision trees is that as one goes deeper in the tree, less data is available to each node, due to the recursive hard-partitioning. Since DT-SE already deals with partial assignments of instances to the left and right subsets, it would be a simple matter to use the soft partitions in the recursive step of the tree-building algorithm and during classification as in Jordan & Jacobs (1993).

As mentioned in the introduction, the procedure for finding a good splitting function is just one part of a whole decision tree learning algorithm. The most important issue with respect to generalization ability is regularization (pruning). The simple pruning strategy described should be extended to use cross-validation to pick the threshold, rather than arbitrarily setting it at five. As Brodley & Utgoff (1995) point out, pruning in multivariate trees can be carried out at a finer granularity than removing entire nodes: subset selection should be investigated as an alternate means of regularization.

Finally, another benefit of this approach is that it can be easily generalized to learn splitting functions other than linear splits. Neither the function optimizer nor the objective function depend on the actual splitting function used, so this is a completely independent module. We have investigated the use of more complex splitting functions (neural networks with few hidden units) with mixed results but plan to investigate this further.

36.8   Conclusion

We have provided a general framework to address the problem of finding a splitting function in recursive-partitioning classification tree algorithms. Within this framework we have identified a splitting criterion, soft entropy, which appears to combine the good properties of its hard counterpart entropy as well as soft criteria such as mean squared error in the logistic regression setting. Experimental results indicate that soft entropy is a better objective function than its hard counterpart. A rudimentary pruning algorithm was employed to address the overfitting problem, and a novel data filtering method was used. Accuracy results for the resulting DT-SEPIR algorithm as compared with CART and OC1 are inconclusive. More experiments are needed to isolate the effects of the splitting criterion and pruning methods.

36.9   Acknowledgments

This work was supported by a National Science Foundation Graduate Research Fellowship and NSF Grant IRI-9116399. We would like to thank Pat Langley, Wray Buntine, Jerome Friedman, Scott Benson, Ron Kohavi and Nils Nilsson for comments on the ideas presented here. Thanks to Richard Olshen and Charles Stone for providing the CART software.

36.10   REFERENCES

Bennett, K. P. & Mangasarian, O. L. (1992), Neural network training via linear programming, *in* P. M. Pardalos, ed., "Advances in Optimization and Parallel Computing", North Holland, Amsterdam, pp. 56–67.

Bichsel, M. & Seitz, P. (1989), "Minimum class entropy: A maximum information approach to layered networks", *Neural Networks* **2**, 133–141.

Breiman, L., Friedman, J., Olshen, R. & Stone, C. (1984), *Classification and Regression Trees*, Chapman & Hall, New York.

Brent, R. P. (1991), "Fast training algorithms for neural networks", *IEEE Transactions on Neural Networks* **2**(3), 346–354.

Brodley, C. E. & Utgoff, P. E. (1995), "Multivariate decision trees", *Machine Learning* **19**, 45–76.

Dempster, A. P., Laird, N. M. & Rubin, D. B. (1977), "Maximum likelihood from incomplete data via the EM algorithm", *Journal of the Royal Statistical Society B* **39**, 1–38.

Duda, R. & Hart, P. (1973), *Pattern Classification and Scene Analysis*, Wiley.

Fayyad, U. M. & Irani, K. B. (1992), The attribute selection problem in decision tree generation, *in* "AAAI-92: Proceedings of the Tenth National Conference on Artificial Intelligence", AAAI Press/MIT Press, pp. 104–110.

Friedman, J. H. (1977), "A recursive partitioning decision rule for nonparametric classification", *IEEE Transactions on Computers* pp. 404–408.

Guyon, I., Boser, B. & Vapnik, V. (1993), Automatic capacity tuning of very large VC-dimension classifiers, *in* S. J. Hanson, J. Cowan & C. L. Giles, eds, "Advances in Neural Information Processing Systems", Vol. 5, Morgan Kaufmann, pp. 147–154.

Hastie, T. J. & Tibshirani, R. J. (1990), *Generalized Additive Models*, Chapman and Hall.

Heath, D., Kasif, S. & Salzberg, S. (1993), Induction of oblique decision trees, *in* R. Bajcsy, ed., "Proceedings of the Thirteenth International Joint Conference on Artificial Intelligence", Morgan Kaufmann.

Henrichon, Jr., E. G. & Fu, K.-S. (1969), "A nonparametric partitioning procedure for pattern classification", *IEEE Transactions on Computers* **C-18**(7), 614–624.

Huber, P. J. (1977), *Robust Statistical Procedures*, Society for Industrial and Applied Mathematics, Pittsburgh, PA.

John, G. H. (1994), Finding multivariate splits in decision trees using function optimization, *in* "AAAI-94: Proceedings of the Twelfth National Conference on Artificial Intelligence", AAAI Press/MIT Press, p. 1463.

John, G. H. (1995), Robust decision trees: Removing outliers in databases, *in* "First International Conference on Knowledge Discovery and Data Mining (KDD-95)", AAAI Press, Menlo Park, CA, pp. 174–179.

Jordan, M. I. & Jacobs, R. A. (1993), Supervised learning and divide-and-conquer: A statistical approach, *in* P. Utgoff, ed., "Proceedings of the Tenth International Conference on Machine Learning", Morgan Kaufmann.

Koutsougeras, C. & Papachristou, C. A. (1988), Training of a neural network for pattern classification based on an entropy measure, *in* "IEEE International Conference on Neural Networks", IEEE Press, pp. 247–254.

Lin, Y. K. & Fu, K. S. (1983), "Automatic classification of cervical cells using a binary tree classifier", *Pattern Recognition* **16**(1), 69–80.

Loh, W.-Y. & Vanichsetakul, N. (1988), "Tree-structured classification via generalized discriminant analysis", *Journal of the American Statistical Association* **83**(403), 715–725.

Michie, D., Spiegelhalter, D. J. & Taylor, C. C. (1994), *Machine Learning, Neural and Statistical Classification*, Prentice Hall.

Morgan, J. N. & Messenger, R. C. (1973), *THAID: a sequential analysis program for the analysis of nominal scale dependent variables*, University of Michigan.

Murphy, P. M. & Aha, D. W. (1994), "UCI repository of machine learning databases", Available by anonymous ftp to `ics.uci.edu` in the `pub/machine-learning-databases` directory.

Murthy, S. K., Salzberg, S. & Kasif, S. (1993), "OC1", Available by anonymous ftp in `blaze.cs.jhu.edu:pub/oc1`.

Murthy, S., Kasif, S. & Salzberg, S. (1994), "A system for induction of oblique decision trees", *Journal of Artificial Intelligence Research* **2**, 1–32.

Qing-Yun, S. & Fu, K. S. (1983), "A method for the design of binary-tree classifiers", *Pattern Recognition* **16**(6), 593–603.

Quinlan, J. R. (1986), "Induction of decision trees", *Machine Learning* **1**, 81–106.

Quinlan, J. R. (1993), *C4.5: Programs for Machine Learning*, Morgan Kaufmann.

Sahami, M. (1993), Learning non-linearly separable boolean functions with linear threshold unit trees and madaline-style networks, *in* "AAAI-93: Proceedings of the Eleventh National Conference on Artificial Intelligence", AAAI Press/MIT Press, pp. 335–341.

Sankar, A. & Mammone, R. J. (1991), Optimal pruning of neural tree networks for improved generalization, *in* "IJCNN-91-SEATTLE: International Joint Conference on Neural Networks", IEEE Press, Seattle, WA, pp. II: 219–224.

Sethi, I. K. (1990), "Entropy nets: from decision trees to neural networks", *Proceedings of the IEEE* **78**(10), 1605–1613.

37
Tree Structured Interpretable Regression

David Lubinsky

Department of Computer Science
University of The Witwatersrand
Johannesburg, South Africa
E-Mail: david@cs.wits.ac.za

ABSTRACT We describe a new method of tree-based regression. The response is estimated by building an adaptive linear model which varies for different paths through the tree. The method has the following potential advantages over traditional methods: the method can naturally be applied to very large datasets in which only a small proportion of the predictors are useful, the resulting regression rules are more easily interpreted and applied, and may be more accurate in application, since the rules are derived by means of a cross-validation technique which maximizes their predictive accuracy. The system is evaluated in an empirical study and compared to traditional regression and CART systems.

37.1 Introduction

Broadly speaking, the problems of inducing rules from data can be divided into two categories. When the response or predicted variable is continuous the problem is called a *regression* problem, while problems where the response is categorical[2] (taking only one of a few possible, usually non-ordered) values are called *classification* problems.

There has been much work on building classifier systems that produce classification rules that are both powerful in their predictive ability and can also be easily understood (for surveys, see [1, 2] and [3]). The rules produced by these systems take the form of simple logical predicates or *if ... then* rules, rather than linear or non-linear combinations of variables.

In regression recent work has attempted to increase the power of available models by taking advantage of cheaper computer horse-power, to fit more flexible models, to decrease the influence of outliers, or to fit models which are insensitive to the distributions of the variables but these methods have concentrated little on improving the interpretability of the resulting models. Interpretability is especially important in two settings:

- when the aim of the analysis is not only to produce rules but also to get insight into the underlying process,

- when the rules are to be applied in situations where it is essential to understand all aspects of their operation. For example, when using rules for patient diagnosis, doctors must have rules that they can understand and agree with.

---

[1] *Learning from Data: AI and Statistics V*. Edited by D. Fisher and H.-J. Lenz. ©1996 Springer-Verlag.
[2] Other terms for categorical are symbolic and discrete.

In this paper we introduce an approach to regression that produces sets of rules which, since they have tree structure, can be easily interpreted and are also flexible in the forms of function that they can fit. The rules are derived by maximizing their predictive ability and hence are robust and give good results in practical use.

We begin with a high level description of the proposed approach. This is followed by a more detailed description of how the system is implemented. The third section contains an experiment in which the system is evaluated by comparing it to traditional regression methods on a set of real-world datasets. The fourth section discusses a set of desiderata for an interpretable regression system and the extent to which these are met by this approach. The appendix discusses some of algorithms used in organizing the computation efficiently.

37.2   Description of the TSIR (Tree Structured Interpretable Regression) system

37.2.1   The tree building process

The TSIR (pronounced teaser) system builds a model which is structured as tree. Unlike traditional tree-growing system, some internal nodes in the tree have only one child. At these nodes the data are not split, but residuals are taken from a single variable regression.

At a high level the algorithm proceeds as follows:

1. Consider all possible split steps and single variable regressions. Pick the best from among these, call it σ, by evaluating the predictive ability of each one. If σ is not significant then create a leaf node and return, otherwise proceed to 2. The exact method by which this is all done is discussed in detail below.

2a. If σ is a split then split the data accordingly and recurse on each subset.

2b. If σ is a regression step then take the residuals from the regression and continue from step 1.

This algorithm results in a tree structure. The method by which we predict the response for a new unseen case is to traverse the tree from its root to a leaf based on the values of the predictors. The algorithm is as follows:

1. Let v be the root node. Let $r = 0$.

2. If v is a split then let v' be the left or right child of v depending on the value of the predictor associated with the split. If v is a regression node then let $r = r + p_x$, where p_x is the predicted value based on the single variable x which was used in the regression step of v, and let $v' = $ the child of v.

3. If v' is a leaf then let $r = r+$(the median of the values of the response of the cases in the training set which reached v') and return r. If v' is not a leaf then let $v = v'$ and continue from step 2.

Figure 1 is a simplified example of the output of the system in a heart-disease domain. At A the median of the response variable is given. B shows an example of a rule with a continuous regressor, C shows and example of a split, D is an example of a discrete regressor and E is a leaf.

37.2.2   Comparison to CART

The system is a similar in concept to recursive partitioning systems such as CART [4] and uses predictive maximization techniques similar to those used in the PVM system[5]. The main differences between the CART system and the TSIR system are that in CART

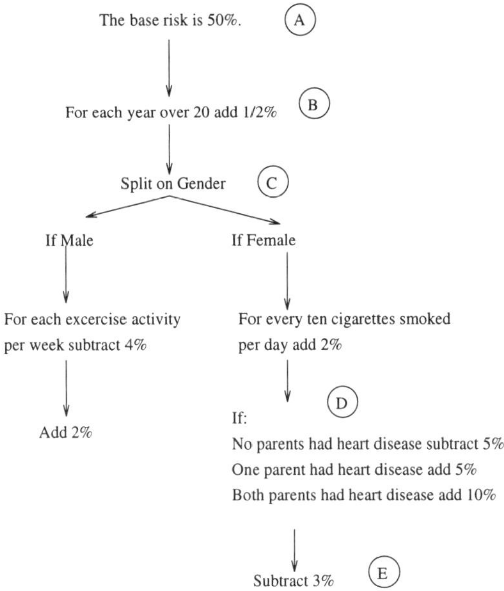

FIGURE 1. Example of a TSIR tree for probability of heart disease

each case which arrives at a leaf of the tree is given the same predicted value. This ignores information about the case. In the TSIR system, all the relevant information is used to make the prediction. The regression steps in TSIR add to the expressive power of the system building an adaptive multivariate regression. Each regression step adds one variable and its coefficient to an incrementally growing model. This means that each leaf of the TSIR tree corresponds to a different multivariate linear regression. The final difference is that split and regression steps are chosen on the basis of their predictive power by cross validation rather than by maximizing a local criterion.

37.2.3   The four types of steps

In step 2 of the algorithm we mention finding the best step of all those considered. We consider four possible types of steps at each stage. These are:

1. **Numerical Split** The data are split into two sets on the basis of a threshold value for a numeric variable. The form of the split is $x <= a$.

2. **Discrete Split** The data are split into two sets using a discrete variable. The form of the split rules is $x \in A$ where A is a subset of the possible values of x.

3. **Numerical Residual** An L_1 line is found and the residuals of the response are taken from the line. The new values of the response are $y_i' = y_i - (a + bx_i)$.

4. **Discrete Residual** The medians of the response are found for each possible value of the predictor and these medians are subtracted from the response: $y_i' = y_i - median(x_i)$.

37.3   The split and regression statistics

Four new statistics are introduced to choose between regressions and splits, each corresponding to one type of step. They are comparable, since all measure the reduction in cross-validated predicted variance.

37.3.1   The predictive correlation for regression steps

The data consists of a n cases, indexed from $1\ldots n$. Each case consists of a set of measured variables and a continuous response variable. The measured variables may be discrete or continuous. We use the letter X to refer to a measured variable and Y to refer to the response. We use capital letters indicate vectors of values, or all values of a particular variable, and small letters indicate scalars. To indicate subsets of vectors, we use the notation $X_{<a}$ to indicate all values in X that are less than a, and similarly for other relational operators. If s is a set of indices, then X_s refers to the values in X corresponding to those indices. In either case, we place a bar over the subscript to indicate all values of the vector except those that are referred to by the subscript. For example, $X_{\le a} \equiv X_{\overline{>a}}$.

Since the system uses an extensive search for good descriptors, it is important to guard against the possibility of spurious fitting of noise. The method used to do this to evaluate all potential descriptors in terms of their predictive ability by cross validation.

We define a cross-validation partition $V = \{v_1 \ldots v_k\}$, where each v_i is a set of indices, as a k partition of a random permutation of the indices $1\ldots n$, such that $|v_i| = \lceil n/k \rceil$ for $1 \le i < k$ and the final partition v_k contains the remaining indices.

The predictive correlation $\pi(X, Y)$ is a measure of the ability of variable X to predict Y. The variable X may be a continuous or discrete variable and Y is the continuous response variable. For each set in the cross-validation partition a predictor is calculated, based only on the cases not in the set, and the predictor is then used to predict the values, for the cases in the set. As an analog of traditional correlation, predictive correlation is defined as follows.

$$\pi(X, Y) \;=\; 1 - \frac{\sum_{v \in V} \sum_{i \in v} |R(X_{\overline{v}}, Y_{\overline{v}}, x_i) - y_i|}{\sum_{i=1}^{n} |y_i - median(Y)|} \quad \text{where}$$

$$R(X, Y, x) \;\Leftrightarrow\; \text{The predicted value at } x \text{ based on the predictor of } Y \text{ on } X$$

The value of $\pi(X, Y)$ is bounded above by one, but may be less than zero if R performs worse than the median as a predictor.

The form of the function R depends on whether X is a numeric or discrete variable. If X is numeric then R is the prediction based on the L_1 regression[3] of Y on X. If a and b are the coefficients of this regression then $R(X, Y, x) = a + bx$. If X is discrete then let $X(x')$ be the set of indices of X whose value is x', then we define

$$R(X, Y, x') = median(Y_{X(x')}).$$

37.3.2   The predictive correlation for split steps

We also define split correlation $\sigma(X, Y)$. The definition of $\sigma(X, Y)$ where the response variable is Y again depends on the type of X.

---

[3]L_1 regression minimizes the sum of the absolute deviations of the points from the fitted line.

If X is continuous, then let U_x be the set of values at midpoints between adjacent values of X. We define the *value* of the split on the continuous variable X as

$$V(X,Y) = \min_{z \in U_x} \sum_{v \in V} \left[\sum_{y \in (Y_v)_{\leq z}} |y - median((Y_{\bar{v}})_{\leq z})| + \sum_{y \in (Y_v)_{> z}} |y - median((Y_{\bar{v}})_{> z})| \right]$$

If X is discrete, then let S_X be the set of possible values of X. For a subset $s \subset S_X$ let $X(s)$ be the set of indices of values of X in s. Then the split value is

$$V(X,Y) = \min_{s \subset S_X} \sum_{v \in V} \left[\sum_{y \in (Y_v)_{X(s)}} |y - median((Y_{\bar{v}})_{X(s)})| + \sum_{y \in (Y_v)_{\overline{X(s)}}} |y - median((Y_{\bar{v}})_{(\overline{X(s)})})| \right] .$$

Using the appropriate version of V(X,Y) we define

$$\sigma(X,Y) = 1 - \frac{V(X,Y)}{\sum_{i=1}^{n} |y_i - median(Y)|}$$

At each step, we pick the best of the four possible steps by picking the step that maximizes the predictive correlation. The procedure terminates when no step yields a *significant* correlation. The various forms of the predictive correlation are not amenable to analytic derivation of their distributions so a large simulation study was run to determine the thresholds of the predictive correlations under various null scenarios. This experiment is described below.

37.3.3 Description of the experiment

Four sub-experiments were run to evaluate the distribution of the various forms of the predictive correlation. In each experiment the predictive correlation was evaluated under a wide range of parameters where there was in fact no relationship between the predictor and the response. These values were then used to get a null distribution of the predictive correlation. The 95'th percentile of this distribution was calculated and compared for each scenario.

To evaluate the predictive correlation for numeric splits and numeric regression steps, random datasets were generated with sizes ranging from 100 to 10000. At each size 1000 datasets were generated. Also, three different distributions (normal, uniform and exponential) were used at each size. There significant no differences between the null distributions of the predictive correlation for the three distributions at each dataset size. As would be expected, the value of the 95'th percentile does decrease as the size of the dataset increases.

For discrete splits and regression steps, the number of categories is introduced as a further parameter. Datasets were generated with two, five and ten categories, varying the proportion of cases in the dominant category from 0.1 to 0.9, and again taking sizes between 100 and 10000. The number of datasets at each size was also 1000. The distributions of the 95'th percentile were again remarkably similar and also similar to the distributions for numeric splits and regression.

To simplify the system, it was decided to use a single distribution function which was similar to all the distributions derived in the experiments. The 95'th percentiles of this function are shown on a log-log scale in Figure 2.

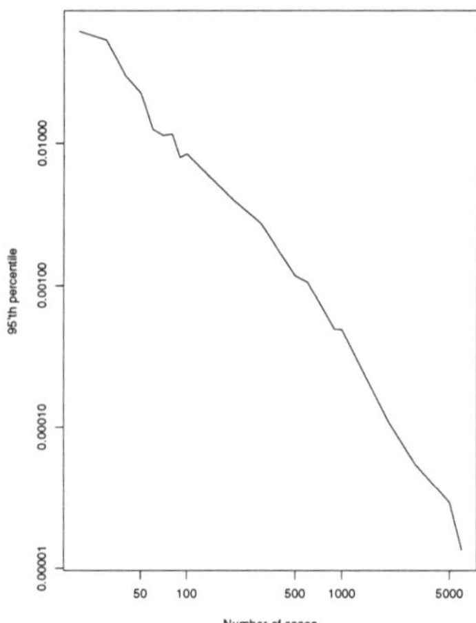

FIGURE 2. The function used to check the significance of steps.

37.4   Empirical Evaluation

37.4.1   An example TSIR tree

Figure 3 shows an actual TSIR tree built from the Boston housing dataset [6]. This display format allows for more interpretation and diagnosis than the simpler format used in Figure 1. The response variable is the median cost of homes in areas around Boston. There are twelve numeric and one categorical predictor.

The tree contains nine nodes including four leaves. The root node is a linear regression step in which a clear downward trend based on LSTAT, the percentage of lower status of the population, is evident. The next node is a split on the mean number of rooms in the house. Cases where the mean is fewer than seven rooms go left. The large majority of areas fit into this category and no further steps were found to be significant. Of those that went right another significant split is found at about 7.5 rooms. The areas with the larger houses have no further steps. Those in the intermediate category have another numeric regression step, on property tax, followed by a final split on pupil-teacher ratio.

37.4.2   Empirical comparison

Table 37.1 shows the results of a empirical study comparing the TSIR methodology to two variations of linear regression as well as CART regression on a number of synthetic and real datasets. Each dataset was divided into a training set and a test set. The size of the training set is given in the column marked 'Train'. The size of the test set is in the column marked 'Test'. The two methods of linear regression were a full linear model including all two-variable interactions and a stepwise-regression method using forward selection. The results for these two methods show the average deviation from the true response for the cases in test set in the columns marked 'Lin' and 'Stepwise'. For CART and TSIR, in addition to the average deviation, the number of nodes in the tree is also shown.

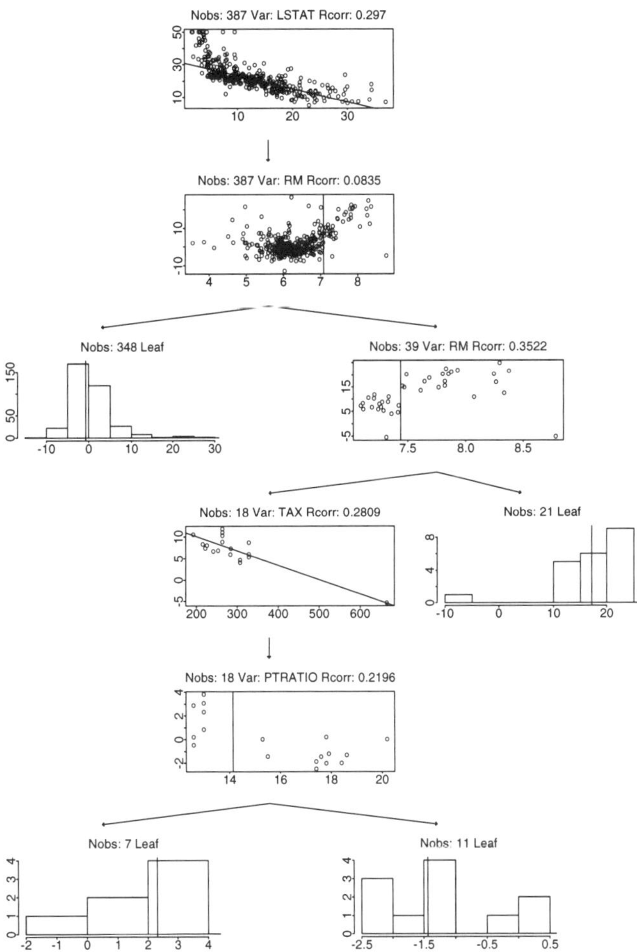

FIGURE 3. Output for the Boston dataset

Name	Num. Obs.		Ave. Deviation		Num. Nodes		Ave. Deviation	
	Train	Test	Lin.	Stepwise	CART	TSIR	CART	TSIR
Interact	1000	2000	0.086	0.6205	37	15	0.495	0.6510
Linear	1000	2000	0.088	0.0645	35	2	0.494	0.0655
Split	1000	2000	0.532	0.5215	15	3	0.328	0.0835
Iris	108	42	0.257	0.2738	10	6	0.298	0.2857
Soil	1745	6896	20.398	21.1753	19	11	13.550	16.9128
Ethanol	64	24	1.113	1.0833	3	12	0.346	0.2708
Environ	79	32	13.938	13.1875	4	7	13.344	13.0938
Boston	415	91	3.066	3.0110	43	9	3.044	3.0330

Each of the synthetic datasets was created by randomly selecting the value of the predictor variables and then adding a Gaussian noise component. The formulas used for each dataset are shown below.

Interact $y = x1 + x2 - x1 * x2 + 0 * x3 + 0 * x4 + \epsilon$

Linear $y = x1 + x2 + 0 * x3 + 0 * x4 + \epsilon$

Split $y = x1 + sign(x2) + 0 * x3 + 0 * x4 + \epsilon$

TABLE 37.1. Results of empirical comparison. The Iris,Soil, Ethanol, Environ and Boston datasets are available from *lib.stat.cmu.edu*.

37.4.3   Analysis of the results of the empirical study

The three synthetic datasets each test a different aspect of predictive ability. The 'Interact' dataset contains two linear terms and a two-variable interaction term. The 'Lin' methods performs best on this set as would be expected. It is interesting that CART outperforms TSIR though it uses more than twice the number of nodes. This suggests that the stopping criterion we use in TSIR may be too strict.

The 'Linear' dataset is simply a linear combination with some noise variables. On this set TSIR outperforms all other approaches, even the linear regression based methods. With only two nodes it exactly captures the true model. The poor performance of CART on this dataset highlights the difference between the two systems.

The 'Split' dataset contains a linear term, a step term and some noise variables. Again, TSIR outperforms the other systems, in this case by far. The combination of linear terms and terms needing splits defeats the other approaches but is exactly captured by TSIR in a three node tree.

The results on the real datasets also show the strengths and weaknesses of the TSIR system. For the 'Iris' data TSIR narrowly beats CART, but a simple linear regression proves the best approach. For the soil data the two tree-based methods perform the best but CART outperforms TSIR, again with a bigger tree pointing to the possibility that a less restrictive stopping criterion should be used. For both the 'Ethanol' and 'Environ' datasets TSIR performs best of all methods, but with a larger trees than CART. In these case it appears that TSIR found some structure in the data that was missed by CART. Finally, for the 'Boston' data all systems have very similar performance with the stepwise approach being marginally the best. TSIR is slightly better than CART with a dramatically smaller tree.

These results indicate that the TSIR approach would be a useful new tool in the analysts tool-box, and is essential in some cases.

37.5   Desiderata for an interpretable regression system

In this section, we give a set of desirable attributes for an interpretable regression system and discuss to what extent each one is met by TSIR.

An interpretable regression system should:

- maximize performance in terms of predictive ability. We attempt to do this by choosing each step to maximize performance in terms of prediction rather than explanation.

- give all rules on untransformed variables. This aids users of systems to interpret the results.

- give indication of variable importance. The TSIR system can output the best predictive correlation for each variable at each step, giving an indication of the variables importance.

- give rules that can be interpreted piecewise (or locally). The power of tree models, like all hierarchies, is that they allow the mind to concentrate on small parts of the model without too much though for what happens in other parts of the tree. This allows them to be interpreted more easily.

- cope with large datasets. Since each step takes $O(mn \log n)$ time, where n is the number of cases and m the number of variables, the TSIR system is quite efficient and can be used on large datasets.

- cope with missing values. The fact that each variable is dealt with individually means that only those cases which are missing on a the variable need be omitted.

- cope with redundant variables. Since each variable is dealt with individually redundant variables are no problem except for in the increased computational cost.

- cope with variables that are linearly related. The numeric regression steps add power to the original CART formulation.

37.6   Further Directions

There is a need for more empirical evaluation. The current method of using a stopping rule rather than building a large tree and then pruning back as is done in CART needs to be re-evaluated. The stopping threshold also needs to be examined.

37.7   Acknowledgements

The reimplimentation of the regression part of CART used in the empirical analysis is by Terry Therneau and is available from *lib.stat.cmu.edu*.

A. Algorithmic speedups

A.1 Updating medians

For each categorical variable X, we need to maintain the set of medians of the response variable for each element of S_X, the set of possible values for X. If the values associated with each element of S_X are stored in a balanced binary tree then the median can be obtained in $O(\log n)$ time and each insert and delete can also be achieved in $O(\log n)$ time. This means that if we consider a k-way cross-validation, the initial trees can be set up in $O(n \log n)$ time and each subsequent partition cost $O((n \log n)/k)$ giving a total cost of $O(n \log n)$, compared with $O(kn \log n)$ for a naive implementation.

A.2 Saving sorting information

When evaluating potential splits on numeric variables, it is necessary that the variable be sorted first. This need only be done once in the beginning, and the sorting information can be saved across splits as follows. Assume the sort order for variable X is O_X, i.e. the i'th element in sorted order of X is $X[O_X[i]]$. Now assume we are splitting on another variable Z, and the set of indeces which *go left* is I_L and the indeces which go right is I_R, then we have two new subsets of the X's $X_L = X[I_L]$ and $X_R = X[I_R]$. To get sort order of X_L, we need a function that for each element x of X_L returns its position in X_L. This function can be computed in linear time. Let this function which returns the index of x in X_L be $I_L(x)$, then we apply the following algorithm: After this algorithm has completed,

---

```
seq = 0
for i = 1 to n
  if O_X[i] ∈ L
    seq = seq + 1
    O_L[seq] = I_L(x_i)
```

FIGURE 4. Finding the sort order after a split

O_L will contain the correct sort order for X_L. The algorithm clearly runs in $O(n)$ time so the sort orders can be maintained with minimal cost.

A.3 Calculating the optimal LAD splits

To calculate the optimal LAD split we must find the index such the following is minimized:

$$\min_s \sum_{i=1}^{s} |y_i - median(y_1 \ldots y_s)| + \sum_{i=s+1}^{n} |y_i - median(y_{s+1} \ldots y_n)|$$

We would like to avoid taking the n median explicitly, since the naive method of doing this would cost $O(n^2 \log n)$ or, even using the linear median algorithm, $O(n^2)$.

We need to be able to update the running median as each new point is added. The way we do this is to maintain a the running median as well as two sets of points *above* and *below*. The *above* set contains all points above the current median and the *below* set contains those below it. The important thing is that the sizes of the two sets should never differ by more than one, and that the median is therefore in the interval $[max(below), min(above)]$. When a new point is added it is either added to the *above* or *below* set depending on its relationship to the current median. This might cause an imbalance in the sizes of the two sets and then a point must migrate across. This point will either be $min(above)$ or $max(below)$. These operations can be achieved in $O(\log n)$ time per operation by storing the two sets in balanced binary trees. The inserts and deletions of smallest or largest can all be done in $O(\log n)$ time leading to a total time of $O(n \log n)$. The algorithm in Figure 5 implements this idea and makes sure that all book-keeping is done correctly. The values of ra and rb maintain the sum of residuals above and below the current median. These are also updated as each point is inserted in constant time, so that the total cost of finding the running medians as well as the sum of absolute residuals is still $O(n \log n)$.

This function gives the running medians in the forward direction. The medians in the backward direction can be obtained simply by reversing the data.

The function $current\_med(above, below)$ returns the current median as either the smallest of the above group, the largest of the below group, or the midpoint between them depending on the relative sizes of the two groups.

37.8   REFERENCES

[1] R.O. Duda and P.E. Hart. *Pattern classification and scene analysis*. Wiley, 1973.

[2] Sholom Weiss and Casimir Kulikowski. *Computer Systems that Learn: Classification and Prediction Methods from Statistics, Neural Nets, Machine Learning, and Expert Systems*. Morgan Kaufmann, 1990.

[3] S.C. Choi, editor. *Statistical Methods of Discrimination and Classification*. Pergamon Press, 1986.

[4] Leo Breiman, Jerome H. Friedman, Richard A. Olshen, and Charles J. Stone. *Classification and Regression Trees*. Wadsworth, Belmont, California, 1984.

[5] Sholom M. Weiss, Robert S. Galen, and Prassad V. Tadepalli. Maximizing the predictive value of production rules. *Artificial Intelligence*, 45:47–71, 1990.

[6] D. Harrison and D.L Rubinfeld. The boston house-price data, 1988. From lib.stat.cmu.edu.

```
below.insert(y[1]) assuming y[0] < y[1]
above.insert(y[2])
ra = (y[2]-y[1])/2
rb = ra
for i = 3 to n
  old = current
  switched = 0
  n_above = above.size()
  n_below = below.size()
  v = y[i]
  min_above = minval(above)
  max_below = maxval(below)
  if v ≥ min_above
    above.insert(v)
    if n_above > n_below
      above.del_min()
      below.insert(min_above)
      switched = 1

    current = current_med(above,below)
    ra = ra + (n_above-switched) * (old - current)
    rb = rb + n_below * (current - old)
    ra = ra + v - current
    if switched
      ra = ra - (min_above - old)
      rb = rb + (current - min_above)

  else if v ≤ max_below
    Do the same as above reversing all and inserting in below
  else new value is between.
    if n_above < n_below insert in above
      above.insert(v)
      current = current_med(above,below)
      ra = ra + n_above * (old - current)
      rb = rb + n_below * (current - old)
      ra = ra + v - current
    else
      Do the same as above reversing all and inserting in below
```

FIGURE 5. Finding the running median and sums of absolute deviations

38
An Exact Probability Metric for Decision Tree Splitting

J. Kent Martin

Department of Information and Computer Science
University of California, Irvine
Irvine, CA
jmartin@ics.uci.edu

ABSTRACT ID3's information gain heuristic is well-known to be biased towards multi-valued attributes. This bias is only partially compensated by the gain ratio used in C4.5. Several alternatives have been proposed, notably orthogonality and Beta. Gain ratio and orthogonality are strongly correlated, and all of the metrics share a common bias towards splits with one or more small expected values, under circumstances where the split likely ocurred by chance. Both classical and Bayesian statistics lead to the multiple hypergeometric distribution as the posterior probability of the null hypothesis. Both gain and the chi-squared significance test are shown to arise in asymptotic approximations to the hypergeometric, revealing similar criteria for admissibility and showing the nature of their biases. Previous failures to find admissible stopping rules are traced to coupling these biased approximations with one another or with arbitrary thresholds; problems which are overcome by the hypergeometric. Empirical results show that pre-pruning should be done, as trees pruned in this way are simpler, more efficient, and no less accurate than unpruned trees. Average training time is reduced by up to 30%, and expensive post-pruning avoided.

38.1 Introduction and Background

Top-Down Induction of Decision Trees (TDIDT [13]) is a method for inferring classification rules (in the form of a decision tree) from a set of examples. The goals of inductive learning are varied, and criteria for judging the success of a learning algorithm are equally varied. There is no one-size-fits-all learning algorithm, nor is there a single universal criterion for evaluating the effectiveness of learning algorithms. The TDIDT family of algorithms makes a greedy heuristic choice of one of a set of candidate splits for a data set and then recursively partitions each of the subsets produced by that split. The recursive splitting process terminates when all members of a subset are in the same class or when the set of candidate splits is empty.

Some TDIDT algorithms have included additional criteria to stop splitting when the incremental improvement in utility is deemed not to be significant. These stopping criteria are collectively referred to as pre-pruning. Other algorithms have added recursive procedures for post-pruning, *i.e.*, replacing one or more of the splits with a terminal node, or leaf. Often, procedures described as post-pruning may go beyond mere pruning, replacing a split with some other split (typically a child of the replaced decision node).

At each decision point in top-down tree building or bottom-up post-pruning, the split selection problem is addressed as two separate but interdependent problems: (1) choosing

---

[1]*Learning from Data: AI and Statistics V.* Edited by D. Fisher and H.-J. Lenz. ©1996 Springer-Verlag.

the set of candidate splits, and (2) selecting one of the candidates (or none of them, if pre-pruning is used). Note that, for example, there are more than 10^{13} distinct decision trees for a set containing only 20 items. Practical algorithms can explore only a small subset of these vast solution spaces. The TDIDT strategy is to build up more complex trees by recursive, greedy refinement of simpler trees, and to explore only relatively simple splits at each decision point.

The process of choosing a split candidate takes place in the context of the choice of the candidate set. At each decision point in the tree, both of these processes take place in the context of all of the split selection choices made at higher levels in the tree. The interactions of the two phases of split selection creates a very complex environment, one in which it is very difficult to determine what the impact would be of changing some aspect of split selection.

Early algorithms restricted candidates to splits on the values of a single attribute and only binary splits for continuous attributes. More recent algorithms extend the candidate space in various ways, while still restricting candidates to relatively simple splits.

The strong interaction of the choice of the set of candidates and the selection among candidates is exemplified by pre-pruning the exclusive-or of two boolean attributes. Neither attribute split appears to have any utility in separating the classes. Yet, the combination of the two attributes will completely separate the classes. If only single-attribute candidate splits are allowed, and any form of pre-pruning is used, the resulting decision tree will consist of a single leaf and will have only 50% accuracy. This example is often cited as an argument against pre-pruning. The difficulty actually is the result of the interaction of pre-pruning and allowing only single-attribute splits, and might equally well be blamed on an overly-restricted choice of candidates as on pre-pruning. Since the point of restricting the set of candidates is to reduce the search space to a practicable level, it seems hard to justify an argument against pre-pruning solely because it precludes a subset of potential trees.

The main focus of this paper is on the second phase of split selection, the use of heuristic evaluation functions to guide selection of the best split from among a set of candidates. A secondary objective is to explore possible causes (other than the exclusive-or difficulty) of the poor performance of pre-pruning in early empirical studies [3, 14].

Evaluation criteria involve issues of accuracy, complexity, and accuracy/complexity tradeoffs, and there is no single measure which combines these in an appropriate manner for every application. See [9] for a review of these issues. Measures of complexity include the number of nodes or leaves and average depth, and the run-time and storage requirements of the learning algorithm.

Within rather broad bounds of complexity, the dominant goal of the learning algorithms is to infer decision trees where the population instances covered by each leaf are, as nearly as possible, members of the same class. In most cases, the accuracy of a decision tree can only be estimated, and it is important to report the variance or confidence interval as well as the point estimate of the accuracy. Either cross-validation or bootstrapping techniques should be used to estimate the accuracy and confidence interval. See [11] for a review of these techniques and issues relating to specifying confidence intervals for classifier accuracy.

38.2   Impact of Different Choices of Candidate Sets

We have seen in the exclusive-or example that the choice of a candidate set can interact strongly with other factors, especially pre-pruning, to preclude or strongly bias against discovering correct decision trees for some problems. A different aspect of the choice of candidate sets occurs when the boundary between the classes is a linear function of two continuous attributes. If the candidate splits are restricted to splitting on a single attribute, the boundary between the classes can only be approximated as a step function, and the accuracy of the tree is directly related to the complexity of the tree and to the sample size (the more leaves and the smaller the area covered by the leaves, the better). If the candidates are further restricted to only binary splits, a deeper tree will be required in order to achieve the same accuracy.

If splits based on linear combinations of continuous attributes are allowed then, for the same sample size, both better accuracy and a much simpler tree can be obtained, and trees with accuracy equivalent to that of single attribute splits can be obtained from smaller samples. Expanding the set of candidate splits is a very powerful tool and can permit discovering decision trees that are both more accurate and less complex. In terms of increasing the number of problems for which reasonably accurate and simple trees can be learned, expanding the set of candidates is likely to be more effective than is using post-pruning rather than pre-pruning.

38.3   Impact of Different Choices Among Candidates

Examination of various scenarios shows that, for unpruned trees, the order in which various splits are made is largely a matter of complexity and efficiency, rather than of accuracy. However, accuracy may be significantly impacted when redundant attributes are noisy and strongly correlated. This is one of the factors underlying the frequent observations [3, 7] that various heuristic functions are largely interchangeable.

It is important not to overemphasize differences in complexity and efficiency, in the sense that if statistically significant differences in accuracy occur, the difference in accuracy would typically be considered to be of overriding importance. When the accuracies of various trees are equivalent, however, there is certainly a preference for simpler and more efficient trees. The differences in complexity and efficiency in most of the example applications in textbooks and in the UCI data repository are relatively minor. For more complex applications involving scores of attributes and thousands of instances, these effects will be compounded, and may have a much greater impact. It should also be noted that all of these differences in accuracy and complexity are being explored in the context of having severely restricted the set of candidate splits for the sole purpose of reducing the complexity to manageable proportions. Differences in complexity and efficiency may be greatly magnified as the set of candidate splits is expanded.

White and Liu [18] discuss the importance of discriminating between attributes which are truly informative and those which are not. Examination of the redundant noisy attributes case shows that accuracy decreases when both, or only the weaker, of two noisy versions of the same feature are used, and that the negative impact is greater if the weaker split occurs nearer to the root of the tree. When this split is made last, it does not improve accuracy (it may even be harmful), and it adds to the complexity of the tree. There is strong evidence that the split overfits the sample data and should be pruned.

When this split is not made last, simple pruning (converting a split to a leaf) would not correct the overfitting (it would, in fact, be very harmful to the accuracy of the output tree). The pruning strategy used in C4.5 [14], replacing the split with one of its children, would be necessary here. This kind of tree surgery is by far more complex and less efficient than simple pruning, and could be avoided if the split were deferred. The presence of this kind of complex tree surgery in an algorithm may be an indication that the algorithm's heuristic does not choose splits in the best order for efficient pruning.

Thus, the following three criteria should be considered in choosing a heuristic:

1. it should prefer splits which most improve the final tree's accuracy and avoid or minimize the impact of those which are harmful to accuracy

2. for splits leading to equivalent accuracies, it should prefer splits which lead to simpler and more efficient trees

3. it should order the splits so as to permit efficient pruning

38.4   Heuristic Functions for Selection

A most natural approach is to label each of the split subsets according to the largest class and choose the split which has the fewest errors. There are several problems with this approach, the most telling being that it simply has not worked out well empirically (see CART [3, pp. 93-98]).

Keeping our three criteria in mind, some measure of split relevance (what CART [3, pp. 32-33] terms impurity functions) would seem to be in order. A convenient representation for splits is a contingency or cross-classification table:

	val-1	val-2	$\cdots$	val-V	Total
cat-1	f_{11}	f_{12}	$\cdots$	f_{1V}	n_1
$\vdots$		$\ddots$			$\vdots$
cat-C	f_{C1}	f_{C2}	$\cdots$	f_{CV}	n_C
Total	m_1	m_2	$\cdots$	m_V	N

C is the number of categories
V is the number of subsets in the split
m_v is the no. of instances in subset v
f_{cv} is the no. of those which are in class c
N is the total no. in the sample
n_c is the total no. in class c

Variants of the information gain heuristic used for the ID3 algorithm [13] have become the *de facto* standard metrics for attribute selection in top-down decision tree learning. Information gain is the difference between the entropy of the population and the weighted average entropy of the subpopulations.

$$\text{gain} = \left(\sum_{c=1}^{C} \left[-\left(\frac{n_c}{N}\right) \log_2 \left(\frac{n_c}{N}\right) \right] \right) - \left(\sum_{v=1}^{V} \left(\frac{m_v}{N}\right) \sum_{c=1}^{C} \left[-\left(\frac{f_{cv}}{m_v}\right) \log_2 \left(\frac{f_{cv}}{m_v}\right) \right] \right) \tag{38.1}$$

The gain ratio function used in C4.5 [14] partially compensates for the bias of gain towards splits having more branches (larger V):

$$\text{gain ratio} = \text{gain} \left/ \sum_{v=1}^{V} \left[-\left(\frac{m_v}{N}\right) \log_2 \left(\frac{m_v}{N}\right) \right] \right. \tag{38.2}$$

Fayyad, *et al* [7] give an orthogonality metric which seeks binary splits that separate as many classes as possible, while keeping members of the same class together.

$$ORT = 1 - \left(\sum_{c=1}^{C} f_{c1} \cdot f_{c2} \right) \bigg/ \left[\left(\sum_{c=1}^{C} f_{c1}^2 \right) \left(\sum_{c=1}^{C} f_{c2}^2 \right) \right]^{1/2} \tag{38.3}$$

Buntine [4] derives a Beta-function splitting rule, and shows that information gain appears as part of an asymptotic approximation to this function:

$$eW(t,\alpha) = \frac{\Gamma(C\alpha)^V}{\Gamma(\alpha)^{CV}} \prod_{v=1}^{V} \frac{\prod_{c=1}^{C} \Gamma(f_{cv} + \alpha)}{\Gamma(m_v + C\alpha)} \tag{38.4}$$

Pearson's Chi-squared statistic [2, pp. 452-462]

$$X^2 = \sum_{c=1}^{C} \sum_{v=1}^{V} \frac{(f_{cv} - e_{cv})^2}{e_{cv}}, \qquad \text{where } e_{cv} = (n_c \, m_v / N) \tag{38.5}$$

is distributed approximately as χ^2 with $(C-1) \times (V-1)$ degrees of freedom. The quantities e_{cv} are the expected values of the frequencies f_{cv} under the null hypothesis that the classification is independent of the split subset. χ^2 is a good approximation when all of the e_{cv} are greater than 1 and no more than 20% of the e_{cv} are less than 5.

Wilks' Likelihood-Ratio Chi-squared statistic [1, pp. 48-49]

$$G^2 = 2 \sum_{c=1}^{C} \sum_{v=1}^{V} f_{cv} \ln \left(\frac{f_{cv}}{e_{cv}} \right) = -2\ln(\Lambda) \tag{38.6}$$

$$\text{where} \qquad \Lambda \equiv \left(\prod_{c=1}^{C} \prod_{v=1}^{V} (n_c m_v)^{f_{cv}} \right) \bigg/ \left(N^N \prod_{c=1}^{C} \prod_{v=1}^{V} f_{cv}^{f_{cv}} \right) \qquad \text{is the likelihood ratio}$$

is also distributed approximately as χ^2 with $(C - 1) \times (V - 1)$ degrees of freedom. The asymptotic convergence of the G^2 statistic is slower than that of X^2, and the χ^2 approximation to G^2 is usually poor when $N < 5 \, CV$.

Replacing e_{cv} by $(n_c \, m_v / N)$ in Equation 38.6 and rearranging gives $G^2 = 2\ln(2)N$ gain. In the arguments supporting adoption of information gain, minimum description length, and general entropy-based heuristics, the product $(N \times \text{gain})$ is approximately the number of bits by which the split would compress a description of the data. Information gain is closely related to conventional log-likelihood analysis; and message compression has a limiting χ^2 distribution, one that converges less quickly than Pearson's X^2. Mingers [12] discusses the G^2 metric (denoted there as G), and White and Liu [18] recommend that the χ^2 approximation to either G^2 or X^2 be used in preference to gain, gain ratio, *etc.*

Quinlan [13] originally proposed using gain to select the split attribute and Pearson's X^2 as a stopping rule. Besides the unfortunate interaction of pre-pruning and split candidates typified by the exclusive-or problem, there are two reasons that this strategy does not work well:

1. for splits with small e_{cv} components the χ^2 approximation for X^2 is not valid, and should not be used (there are similar difficulties with G^2 and with gain) — the

divide-and-conquer strategy of TDIDT creates ever smaller subsets, so that this difficulty is certain to arise after at most $\log_2(N_0/5)$ splits have been made (N_0 is the size of the entire data set)

2. X^2, G^2, and gain converge at different rates (may rank splits in different orders) — gain probably does not order the splits correctly for efficient pruning by X^2

38.5   An Exact Test

Fisher's Exact Test [1, pp. 59-62], [15, pp. 586-592], [8, pp. 332-337] for 2×2 contingency tables is based on the hypergeometric distribution, which gives the exact probability of obtaining the observed data under the null hypothesis, conditioned on the marginal totals (n_c and m_v).

$$P_0 \equiv \left(\begin{array}{c} n_1 \\ f_{11} \end{array} \right) \left(\begin{array}{c} n_2 \\ f_{21} \end{array} \right) \Big/ \left(\begin{array}{c} N \\ m_1 \end{array} \right) \tag{38.7}$$

The level of significance of the test is the sum of the hypergeometric probabilities for the observed data and for all hypothetical data having the same marginal totals which would have given a lower P_0 value. Tocher [16] shows that Fisher's test is uniformly the most powerful unbiased, *i.e.*, that in the significance level approach to hypothesis testing, no other test will out-perform Fisher's exact test (the power of a test is the probability that the null hypothesis will be rejected when some alternative hypothesis is really true).

White and Liu [18] note that, for small e_{cv}, Fisher's exact test should be used in place of the χ^2 approximation, and suggest that a similar test for larger tables could be developed. The extension of Fisher's exact test for contingency tables larger than 2×2 is given by the multiple hypergeometric distribution [1, pp. 62-64]

$$P_0 = \left(\frac{\prod_{c=1}^{C} n_c!}{N!} \right) \prod_{v=1}^{V} \left(\frac{m_v!}{\prod_{c=1}^{C} f_{cv}!} \right) \tag{38.8}$$

This exact probability expression can be derived either from classical statistics, as the probability of obtaining the observed data given that the null hypothesis is true, or from Bayesian statistics using appropriate non-informative priors, as the posterior probability that the null hypothesis is true given the observed data. The Bayesian derivation of P_0 differs from Buntine's Beta analysis [4] primarily by eliminating Buntine's α parameter and by conditioning on both the row and column totals of the contingency table.

For choosing among candidate splits of the same set of data, P_0 is the appropriate metric, rather than the significance level. We are seeking the most relevant feature (the split for which it is least likely that the null hypothesis is true), and that is measured directly by P_0.

The following approximate relationships can be derived:

$$2\ln(2)\, N\, \text{gain} \approx -2\ln(P_0) - (C-1)(V-1)\ln(2\pi N)$$
$$+ \text{(terms increasing as the interaction weighted sum of squares)} \tag{38.9}$$
$$X^2 \approx -2\ln(P_0) - (C-1)(V-1)\ln(2\pi N)$$
$$+ \text{(terms increasing as the main-effects sum of squares)} \tag{38.10}$$

Thus, both X^2, gain, and G^2 arise as terms in alternative approximations to the relevance of a split. In neither case should it be assumed that all the remaining terms vanish, even as $N \to \infty$. Both factors are positive, indicating a tendency for both measures to overestimate the significance of very non-uniform splits.

Values of each of the measures (gain, gain ratio, orthogonality, X^2, Beta, and P_0) have been calculated for 1,067 2×2 tables ($N = 2, 4, 8 \ldots 64$; $n_1 = 1 \ldots N/2$; $m_1 = 1 \ldots n_1$; $f_{11} = 0 \ldots m_1$). Inspection of these data (see [10] for details) confirmed the analytical observations above.

Gain, gain ratio, orthogonality, Beta, G^2, and X^2 are all asymptotic approximations to P_0. For splits with small e_{cv}, they all tend to overestimate the significance of the split, though differing in detail. The null hypothesis probability function P_0 appears to be a measure which properly weighs all the factors (N, C, V, n_c, m_v), and may be a more suitable attribute selection metric than gain, gain ratio, X^2, Beta, or orthogonality.

38.6   Stopping Criteria

Quinlan [13] originally proposed that the X^2 test be used to prevent overfitting in ID3 by pre-pruning, and Breiman, *et al* [3] initially searched for a stopping rule in the form of a minimum gain threshold.[2] Both of these approaches were abandoned in favor of some form of post-pruning ([3, pp 65-81] [13]). There have been a number of studies in this area, among the notable findings are: (1) in general, it seems better to post-prune than to pre-prune, (2) k-fold cross-validation seems to work better than point estimates such as X^2, (3) the decision to prune is a form of bias — whether pruning will improve or degrade accuracy depends on how appropriate the bias is to the problem at hand, and (4) pruning may have a negative effect on accuracy when the data are sparse.

The P_0 function proposed above for split selection might be used for pre-pruning, and is a valid statistic even in cases when the χ^2 approximation is not. The previous negative results concerning pre-pruning appear to be due to use of different inadmissable approximate statistics for attribute selection and stopping, and to the interaction with the restricted split candidates set, rather than to any inherent fault of pre-pruning. Use of the P_0 function for both selection and stopping might permit more efficient construction of decision trees without loss of predictive accuracy.

38.7   Empirical Comparisons of the Measures

Sixteen data sets having widely varying attribute characteristics were chosen (see [10] for a full description of the data sets). None of these data sets has any missing values. Two issues arise with respect to handling the attributes: (1) numeric attributes must be made discrete; procedures for doing this are given in [14, 6], and (2) each attribute must be converted to binary splits for the orthogonality metric — the simplest way to do this is to create a binary feature for each value; other binarization procedures are given in [3, 14].

---

[2]Since ($N \times$ gain) has approximately a χ^2 distribution (which has very complex thresholds), attempting to find a simple threshold for gain was foreordained to fail.

TABLE 38.1. Unpruned Trees, Binary Splits

	Accuracy			No. Leaves			Wtd Avg Depth			Train/Val Time		
	Gain	Ort	P_0	Gain	Ort	P_0	Gain	Ort	P_0	Gain	Ort	P_0
Overall	75.3	74.0	75.0	1186	1233	1213	9.7	15.5	7.3	5658	7269	3028
Natural	72.8	70.9	73.0	371	369	298	13.6	17.1	6.7	1236	1242	746
Quartiles	73.8	72.7	72.6	472	444	502	5.8	6.6	6.2	585	531	428

The particular methods used for handling the attributes have important consequences for both efficiency and predictive accuracy, and can interact with selection and stopping criteria in unpredictable ways. Full consideration of the efficacy of various strategies for handling numeric and multivalued attributes are planned topics for a future paper.

In order to control the splitting context and to avoid bias in comparing the selection metrics, two *a priori*, once-and-for-all, multi-valued strategies were used here for every numeric attribute in every data set: (1) 'natural' cut-points determined by visual inspection of smoothed histograms and (2) arbitrary cut-points at approximately the quartiles. For binarization, all attributes having $V > 2$ values were replaced with V binary attributes. The resulting candidate splits are not intended to be optimum (and may not even be "good"), merely *a priori*, consistent, and unbiased. Results obtained here should be compared only to one another, and not to published results using other strategies. A total of 26 different experimental data sets were generated from the 16 original sets (2 each, natural and quartile, for the 10 sets with continuous attributes, 1 each for the 6 sets having no continuous attributes).

Only the three most different split metrics (gain, orthogonality, and P_0) were evaluated. Separate comparisons of the measures were made using binary and multi-valued splits. For gain and orthogonality the X^2 stopping rule was used when X^2 was valid. This rule never stopped splitting for any of the data sets, even at the 99.9% level, confirming the intuition that Cochran's criteria are rarely satisfied. In each experiment, a tree was grown using all of the instances, and the complexity of this tree was determined. Accuracy was estimated by 10-fold cross-validation.

The results for the unpruned binary trees are summarized in Table 38.1. Details for all 26 data sets can be found in [10]. None of the small differences in accuracy between the split metrics on a given data set is significant. The differences between the arbitrary and 'natural' nominalizations is generally very small[3] and sometimes positive, sometimes negative. The weighted average accuracy for orthogonality using 'natural' subsets is slightly, but significantly, lower than the gain and P_0 accuracies.

The trees have about the same number of leaves on the average. P_0 is more efficient in every case, reducing average depth by more than 25% and training time by about 45% on the average. These data support the conjecture that in every case trees grown using the null hypothesis probability P_0 are more efficient, and no less accurate than the gain and orthogonality trees.

---

[3]With two notable exceptions, the Glass and WAIS data, where the natural cuts resulted in significantly better accuracy.

TABLE 38.2. Stopping Effects on Accuracy

data set	Nominalize	Unpruned	conf lims	Pruning Threshold Level 0.5	0.1	0.05	0.01	0.005	0.001
Finance 1	quartiles	75	57-90	79	79	79	71	64	¶ 44
Obesity	'natural'	58	37-77	47	44	49	40	¶ 33	¶ 13
Obesity	quartiles	51	31-71	42	49	49	40	¶ 29	¶ 36
Pima	quartiles	65	60-70	68	§ 73	§ 73	§ 74	§ 74	§ 75
Servo		95	89-98	93	91	89	89	90	¶ 81
	overall	75.0		75.3	76.4	76.4	75.9	75.5	74.1
	'natural'	73.0		72.5	74.0	73.4	71.7	71.7	70.7
	quartiles	72.6		74.0	74.9	75.1	75.0	74.8	73.1
Number of Leaves									
	overall	1213		895	406	295	192	164	125
	'natural'	298		231	122	82	58	48	39
	quartiles	502		394	151	109	68	60	44
Weighted Average Depth									
	overall	7.30		6.71	5.21	4.78	3.83	3.57	2.96
	'natural'	6.67		5.78	4.29	4.12	3.16	2.95	2.40
	quartiles	6.22		6.06	4.77	4.19	3.30	2.99	2.42
Training & Validation Time (sec)									
	overall	3028		2799	2387	2242	1960	1889	1733
	'natural'	746		684	569	520	461	437	386
	quartiles	428		409	335	307	259	252	221

¶ below the 95% confidence limits   § above the 95% confidence limits

38.7.1 *Effects of Stopping*

The effects of stopping based on P_0 are summarized in Table 38.2. Of the 26 total experiments, only those in which pruning had a statistically significant effect on accuracy are shown in detail (complete results can be found in [10]). Though accuracy for the Servo and Obesity problems is reduced by pre-pruning at the 0.05 level, the differences are not statistically significant. Overall, the accuracy data are mildly concave, peaking at around the $P_0 = 0.05$ level. The results for individual data sets are entirely consistent with the overall results.

These results strongly support the conjecture that growing and stopping decision trees using P_0 at the 0.05 level does no harm and may, in fact, be mildly beneficial to accuracy. Training and validation time is reduced by 25-30% from the unpruned trees, and by 60% from the unpruned trees built using information gain (not including the time that would be required to post-prune those trees). A more detailed study of stopping *vs.* post-pruning by various criteria in controlled contexts is a planned topic for a future paper.

38.7.2 *Binary vs. Multi-way Splits*

Experiments to determine the effects of using binary as opposed to multi-way splits are summarized in Table 38.3. The most striking features of these data are that the multi-way trees have 2 or 3 times as many leaves as the binary split trees, are only one-half to one-third as deep, and reduce training/validation time by 80-85%.

A very substantial time penalty is incurred when V-ary attributes are forced into V binary splits, as was done here to accommodate the orthogonality metric. The time savings for multi-way splits is a straightforward consequence of the increased branching factor reducing the height of the tree and of roughly halving the number of attribute-value

TABLE 38.3. Binary *vs.* Multi-way Splits

	Binary			Multi-way		
	Gain	P_0		Gain	P_0	
		unpruned	0.01		unpruned	0.01
Cross-Validation Accuracy %						
overall	77.1	76.7	77.4	75.5	76.0	77.5
'natural'	72.6	72.8	71.6	71.0	71.1	71.9
quartiles	73.5	72.4	75.0	71.4	72.4	74.8
Number of Leaves						
overall	948	959	220	2161	2390	723
'natural'	366	292	80	880	911	311
quartiles	417	496	107	970	1130	325
Weighted Average Depth						
overall	8.8	6.2	3.9	3.5	3.6	2.6
'natural'	13.9	6.8	4.2	4.3	4.2	3.0
quartiles	5.9	6.3	4.3	3.1	3.3	2.7
Training & Validation Time (sec)						
overall	2086	1361	950	327	204	173
'natural'	1234	745	519	127	83	72
quartiles	583	427	306	126	71	63

pairs (the total number of potential subdivisions that must be assessed in choosing the split feature). Overall, learning time increases quadratically in the dimensionality of the data set. Approaches such as those suggested by Weiss and Indurkhya [17] to reduce dimensionality in a principled way without sacrificing accuracy should be pursued.

There is a slight decrease in accuracy for the multi-way splits, perhaps due to the irrelevant values problem [6], which becomes less significant as the stopping threshold level decreases (and, in fact, is reversed below the 0.01 level). The effect is more pronounced for data sets with lower accuracy.

38.8   Conclusions

1. Whenever one or more of the expected values in a split is small, the information gain, gain ratio, orthogonality, Beta, X^2, and G^2 measures are prone to overestimate the significance of the split. The divide-and-conquer strategy of building classification trees almost inevitably leads to very small subtrees where these measures are inadmissible.

2. The P_0 null hypothesis probability measure proposed here overcomes the difficulties encountered when the classes and attribute values are unevenly distributed or the number of partitions large. The unpruned trees it builds are much more efficient, and no less accurate, than those built by the other measures.

3. The P_0 measure can be used to stop splitting subtrees. The resulting trees are simpler and no less accurate than the unpruned trees. A stopping threshold level of 0.05 or 0.01 is recommended. Training times are reduced by about 30%.

4. The arguments against stopping are, in fact, just arguments against use of overly restricted candidate sets, very sparse (or otherwise ill-conditioned) data, and inadmissible heuristics. There is no point in continuing the inductive process when

$P_0 > 0.5$, and in most domains little point in continuing past the 0.05 level.

5. Dimensionality has a very strong influence on classifier complexity and learning time. Great care should be taken when nominalizing numeric attributes. Approaches such as those suggested by Weiss and Indurkhya [17] to reduce the overall dimensionality of the analysis and optimum binarization techniques such as those used in C4.5 [14] and ASSISTANT [5] shoud be pursued. With the *caveat* that the method of handling numeric attributes and steps to reduce dimensionality can influence accuracy and interact with stopping in unpredictable ways.

38.9 REFERENCES

[1] A. Agresti. *Categorical Data Analysis*. John Wiley & Sons, New York, 1990.

[2] T. W. Anderson and S. L. Sclove. *The Statistical Analysis of Data*. The Scientific Press, Palo Alto, 2nd edition, 1986.

[3] L. Breiman, J. H. Friedman, R. A. Olshen, and C. J. Stone. *Classification and Regression Trees*. Wadsworth & Brooks/Cole Advanced Books & Software, Pacific Grove, CA, 1984.

[4] W. L. Buntine. *A Theory of Learning Classification Rules*. PhD thesis, University of Technology, Sydney, 1990.

[5] B. Cestnik, I. Kononenko, and I. Bratko. ASSISTANT 86: A knowledge-elicitation tool for sophisticated users. In *Progress in Machine Learning — Proceedings of EWSL 87: 2nd European Working Session on Learning*, pages 31–45, Wilmslow, 1987. Sigma Press.

[6] U. M. Fayyad. *On the Induction of Induction Trees for Multiple Concept Learning*. PhD thesis, University of Michigan, 1991.

[7] U. M. Fayyad and K. B. Irani. The attribute selection problem in decision tree generation. In *Proceedings of the 10th National Conference on Artificial Intelligence (AAAI-92)*, pages 104–110, Cambridge, MA, 1992. MIT Press.

[8] J. J. L. Hodges and E. L. Lehman. *Basic Concepts of Probability and Statistics*. Holden-Day, Inc., Oakland, CA, 1970.

[9] J. K. Martin. Evaluating and comparing classifiers: Complexity measures. In *Proceedings Fifth International Workshop on Artificial Intelligence and Statistics*, pages 372–378, Fort Lauderdale, FL, 1995.

[10] J. K. Martin. An exact probability metric for decision tree splitting and stopping. In *Proceedings Fifth International Workshop on Artificial Intelligence and Statistics*, pages 379–385, Fort Lauderdale, FL, 1995.

[11] J. K. Martin and D. S. Hirschberg. Small sample statistics for classification error rates. Technical Report 95-1, University of California, Irvine, Irvine, CA, 1995.

[12] J. Mingers. An empirical comparison of pruning measures for decision tree induction. *Machine Learning*, 4:227–243, 1989.

[13] J. R. Quinlan. Induction of decision trees. *Machine Learning*, 1:81–106, 1986.

[14] J. R. Quinlan. *C4.5: Programs for Machine Learning*. Morgan Kaufmann, San Mateo, CA, 1993.

[15] S. Rasmussen. *An Introduction to Statistics with Data Analysis*. Brooks/Cole Publishing Co., Pacific Grove, CA, 1992.

[16] K. D. Tocher. Extension of the Neyman-Pearson theory of tests to discontinuous variates. *Biometrika*, 37:130–144, 1950.

[17] S. M. Weiss and N. Indurkhya. Reduced complexity rule induction. In *Proceedings of the 12th International Joint Conference on Artificial Intelligence (IJCAI-91)*, pages 678–684, San Mateo, CA, 1991. Morgan Kaufmann.

[18] A. P. White and W. Z. Liu. Bias in information-based measures in decision tree induction. *Machine Learning*, 15:321–329, 1994.

Part VIII
Natural Language Processing

39
Two Applications of Statistical Modelling to Natural Language Processing

William DuMouchel[†], Carol Friedman[††], George Hripcsak[†]
Stephen B. Johnson[†], and Paul D. Clayton[†]

Department of Medical Informatics[†]
Columbia University
161 Fort Washington Avenue
New York, NY 10032, USA

Department of Computer Science[††]
Queens College, CUNY
Flushing, NY 11367, USA

ABSTRACT   Each week the Columbia-Presbyterian Medical Center collects several megabytes of English text transcribed from radiologists' dictation and notes of their interpretations of medical diagnostic x-rays. It is desired to automate the extraction of diagnoses from these natural language reports. This paper reports on two aspects of this project requiring advanced statistical methods. First, the identification of pairs of words and phrases that tend to appear together (collocate) uses a hierarchical Bayesian model that adjusts to different word and word pair distributions in different bodies of text. Second, we present an analysis of data from experiments to compare the performance of the computer diagnostic program to that of a panel of physician and lay readers of randomly sampled texts. A measure of inter-subject distance with respect to the diagnoses is defined for which estimated variances and covariances are easily computed. This allows statistical conclusions about the similarities and dissimilarities among diagnoses by the various programs and experts.

39.1   Empirical Bayes Estimation of Word Collocations

Friedman et al. [Friedman95] describe a natural language processing (NLP) text extraction system, called MEDEXTRA, that was developed with the goal of becoming an integral component of the basic information needs of health care providers at Columbia Presbyterian Medical Center, a large health care facility. The general function of MEDEXTRA is the extraction, structuring and encoding of clinical information in textual patient reports, and the subsequent mapping of the information into a structured patient database which is used by other automated processes within the Clinical Information System (CIS), such as the decision support system or a research database. The first application of MEDEXTRA is to radiological reports, which are typically dictated by a radiologist and typed into the CIS by a clerk as unedited paragraphs of text. These reports typically deviate from the standard format because they are entered by many different typists, and all types of unpredictable variations eventually occur. Friedman et al [Friedman95] describe the many components of MEDEXTRA, but this section focuses on a single aspect of the program, the development of a lexicon for multi-word phrases. The phrasal lexicon is critical to MEDEXTRA because the sublanguage is full of specialized expressions that should be treated as atomic units in order to obtain accurate

---

[1] *Learning from Data: AI and Statistics V.* Edited by D. Fisher and H.-J.. Lenz. ©1996 Springer-Verlag.

interpretations. For example, the phrase *cannot be excluded,* as in *infiltrate cannot be excluded,* sho=uld convey the concept of *low possibility.*

The phrasal lexicon is constructed by a computer search for words that seem to occur together frequently, or *collocate,* after which a physician reviews the candidate phrases and recommends whether or not to include them in the lexicon. This section presents and compares three algorithms for the initial computer screening of word pairs. During analysis of a corpus of text, suppose that n separate word forms have been identified, and thus a potential of n^2 unique word pairs are to be examined for evidence of significant collocation. Let the proportion of times word form i is the first member of a pair be denoted by $p_{i.}$, $i = 1, ..., n$, and also let the proportion of times word form j is the second member of a pair be denoted by $p_{.j}$, $j = 1, ..., n$, where $\Sigma p_{i.} = \Sigma p_{.j} = 1$. Suppose that a sample of N word pairs have been gathered, and let N_{ij} denote the number of times word pair (i, j) has been observed, where $\Sigma_{i,j} N_{ij} = N$. Typically, N is large compared to n, but small compared to n^2, so that most of the $N_{ij} = 0$. For example, in a collection of mammogram reports, we find (see Table 1) $n = 2,043$, $N = 130,737$, and only $10,748$ of the 4 million-plus $N_{ij} > 0$. We desire a measure of how much more frequently than chance a given pair of words arises, with error bars or some measure of statistical significance. Define $E_{ij} = N p_{i.} p_{.j} = N_i N_j / N$ as the expected number of occurrences of word pair (i, j) if the two words occur independently.

The first measure of collocation considered is called the likelihood ratio or mutual information (MI) statistic. Dunning [Dunning93] advocates this measure for assessing word collocation and points out its advantages over the simple Pearson chi-squared statistic when many of the observed counts are small. For pair (i, j) this measure is computed by comparing the observed and expected counts in the 2 by 2 table formed by classifying every observed word pair according to whether or not word i was first and/or word j was second. The likelihood-ratio test statistic for independence in the four-fold table is

$$MI_{ij} = 2[N_{ij} \log \frac{N_{ij}}{E_{ij}} + (N_{i.}-N_{ij})\log \frac{N_{i.} - N_{ij}}{N_{i.} - E_{ij}} + (N_{.j}-N_{ij})\log \frac{N_{.j} - N_{ij}}{N_{.j} - E_{ij}}$$
$$+ (N-N_{i.}-N_{.j}+N_{ij})\log \frac{N - N_{i.} - N_{.j} + N_{ij}}{N - N_{i.} - N_{.j} + E_{ij}}] \tag{1}$$

The second measure is based on the simple observed proportion of times that word j follows word i, given that word i has come first, or vice-versa, whichever is larger. It is the maximum of two conditional probabilities (CP) and is computed as

$$CP_{ij} = N_{ij}/\min(N_{i.}, N_{.j})$$
$$\text{if } N_{ij} > 2, \text{ otherwise set } CP_{ij} = 0 \tag{2}$$

Neither MI_{ij} nor CP_{ij} works well as a measure of word collocation. Since it is a test statistic and not an estimate of a population quantity, the mutual information statistic is quite dependent on sample size, and tends to be very large when there is only a moderate degree of collocation between words i and j but both $N_{i.}$ and $N_{.j}$ are large. The conditional proportion CP_{ij} tends to be large whenever one of $N_{i.}$ or $N_{.j}$ is much larger than the other. The requirement that N_{ij} be at least three is intended to reduce the effect of small frequencies on the conditional probabilities. The CP_{ij} measure is the one used in the version of MEDEXTRA described in [Friedman95]. Next we describe a new measure of collocation that seems to work better than either MI or CP. It focuses on identifying pairs in which the ratio $\lambda = E(N_{ij})/E_{ij}$ is large. To do this, we assume that each pair count N_{ij} has an independent Poisson distribution with mean $\lambda_{ij} E_{ij}$, written as

$$N_{ij} \sim \text{Poisson}(\lambda_{ij}\, E_{ij}) = \text{Poisson}(\lambda_{ij}\, N p_{i.}\, p_{.j})$$

$$P(N_{ij}{=}k \mid \lambda_{ij}{=}\lambda) = (\lambda E_{ij})^k\, e^{-\lambda E_{ij}} / k!$$

The value $\lambda_{ij} = 10$, for example, means that pair (i, j) occurs 10 times as frequently as would be expected by chance collocation. The goal is to estimate the larger values of λ_{ij} and produce confidence intervals for them. The main problem is that there are so many of the λ_{ij} to estimate that further model assumptions are necessary. We take a Bayesian approach to avoid the problem of *multiple comparisons*, whereby taking the largest of thousands of observed values of N_{ij}/E_{ij} at face value ignores the tendency of extreme values to regress toward the mean in future data.

The set of n^2 different λ_{ij} are assumed themselves to vary according to a probability distribution over the interval $0 < \lambda < \infty$. The exact form of this distribution is not known; however we assume that it is approximately a gamma distribution with parameters α and β, where these hyperparameters are estimated from the data. Under this assumption, the mean of all the λs is α/β, and the variance of λ is α/β^2. Thus, whatever the exact distribution of λ, proper choice of α and β can approximate its first two moments. Using Bayes rule, if the prior distribution is $\lambda_{ij} \sim \text{Gamma}(\alpha, \beta)$ and if the likelihood of the data is $N_{ij}|\lambda_{ij} \sim \text{Poisson}(\lambda_{ij}E_{ij})$ then the posterior distribution of λ_{ij} is also Gamma with revised parameters

$$\lambda_{ij}|N_{ij},\, E_{ij} \sim \text{Gamma}(\alpha{+}N_{ij},\, \beta{+}E_{ij})$$

The posterior mean and variance of λ_{ij} are $(\alpha{+}N_{ij})/(\beta{+}E_{ij})$ and $(\alpha{+}N_{ij})/(\beta{+}E_{ij})^2$, respectively. Arbitrary percentiles of the posterior distribution of λ_{ij} can be computed easily, using percentiles of the chi-squared distribution, which is a scaling of gamma distributions. For example, the first percentile of the posterior distribution is interpreted as the value of λ which one is 99% sure that $\lambda_{ij} = E(N_{ij})/E_{ij}$ exceeds, and could be considered a conservative Empirical Bayesian (EB) estimate (a lower confidence bound) of that quantity. It is

$$\text{EB}_{ij} = \chi^2_{.01}(2\alpha{+}2N_{ij})/(2\beta{+}2E_{ij}) \qquad (3)$$

where $\chi^2_{.01}(\text{df})$ is the first percentile of the chi-squared distribution with df degrees of freedom. The values of α and β can be estimated from the marginal distribution of the N_{ij}. Although each N_{ij} has a Poisson distribution conditional on the corresponding λ_{ij}, unconditionally every N_{ij} has a negative binomial distribution that depends only on α, β and E_{ij}. The log-likelihood function is

$$\log L(\alpha, \beta) = \Sigma_{i,j}\, \{[\Sigma_{k=1,\, N_{ij}} \log(\alpha{+}k{-}1)] - N_{ij}\log(1 + \beta/E_{ij}) - \alpha\log(1 + E_{ij}/\beta)\} \qquad (4)$$

The computation and maximization of $\log L(\alpha, \beta)$ is a delicate task when n is very large. Note that the first two terms of the summand in braces above vanish for all $N_{ij} = 0$, but that the last term must be evaluated for all n^2 pairs. The third measure of collocation used here is the quantity (3), where α and β maximize (4).

The resulting estimates are more statistically reliable than using cutoffs based just on a chi-squared or mutual information criterion, or a simple conditional probability, since the estimation of α and β allows the method to adapt to the particular corpus being analyzed. Another advantage is that the interpretation of λ_{ij} is more straightforward than that of a test statistic — for example, its meaning is not dependent on the sample size. The Bayesian estimates are often called *shrinkage estimates*, because all the values N_{ij}/E_{ij} are "shrunk" towards α/β, which, because of the maximum likelihood estimation, will fall in the middle of the distribution of N_{ij}/E_{ij}. Extremely high and low values of N_{ij}/E_{ij} are thus automatically moderated.

Table 1: Statistics for each of the bodies of text.

Domain	n	$N_{ij} > 0$	N	$\alpha \pm st.err.$	$\beta \pm st.err.$	$EB_{ij} > 10$
Abdomen	3,542	21,058	110,415	.0174 ± .0001	.0170 ± .0002	2205
Chest	3,274	19,185	128,924	.0179 ± .0001	.0168 ± .0003	2029
Mammogram	2,043	10,748	130,737	.0133 ± .0001	.0096 ± .0002	1766
Modified Mamm.	2,043	10,748	69,167	.0601 ± .0006	.0533 ± .0013	253

The three methods are compared on three subdomains of the radiology text: abdominal x-rays, chest x-rays, and mammograms. The first three rows of Table 1 show statistics describing these three samples. The values of α and β are all in the range .01 to .02, indicating very high dispersion of λ in the superpopulation model. The fourth row of Table 1 describes the results using an artificial modification of the mammogram data in which all of the $N_{ij} > 2$ that were also greater than $10E_{ij}$ were reduced to $\max(2, 10E_{ij})$.

Table 2 (left) shows the correlation coefficients of the three measures across the 19,185 unique wordpairs occuring in the chest x-ray text. The correlations are small (all < .5) partly because the vast majority of the word pairs are not especially frequent. To give more emphasis to frequent collocations, and to balance the scales of the three measures (1) - (3), the three measures were replaced by their ranks (highest = 1, lowest = 19,185), and the logs of these ranks were correlated. The log scale is intended to ensure that the correlations are mostly determined by the pairs scoring high on the three measures. Table 2 (right) shows that the log-rank correlations range from .68 to .86 with the highest correlation between the Bayesian and mutual information measures. The same pattern held true for the other two subdomains studied.

Table 2: Correlations among the measures (left) and among their log ranks (right) for the chest x-ray text.

	MI	*CP*	*EB*		*lrMI*	*lrCP*	*lrEB*
MI	1.000	0.287	0.269	*lrMI*	1.000	0.677	0.861
CP	0.287	1.000	0.477	*lrCP*	0.677	1.000	0.757
EB	0.269	0.477	1.000	*lrEB*	0.861	0.757	1.000

Table 3 displays examples of the word pairs that each method is best and poorest at detecting. Within each of the three domains, a cutoff ranking of the three measures was based on the number of pairs for which $EB_{ij} > 10$. That is, according to their posterior distribution, we are 99% sure that such pairs are at least 10 times more likely to occur together than if they occurred independently. The last column of Table 1 shows how many word pairs met this criterion in each body of text. Then a cutoff score for the other two collocation measures was set so that the same number of pairs are chosen by each method. Table 3 shows examples of discordant word pairs -- the three measures do not all fall above or below their respective cutoffs.

Table 3: Samples of phrases exclusively chosen and exclusively omitted by each of the three methods.

Mutual Information	*Maximum Conditional Probability*	*Bayesian Posterior Distribution*
Abdomen X-ray Domain -- Exclusively Chosen:		
treatment planning	or artifacts	chronic infection
floor relaxation	impacted in	represent chronic
portable abdominal	is perhaps	multiple mobile
easily palpable	left paravertebral	irregular calcification
pattern appears	of ureteropelvic	some minor
films are	in lumbosacral	loculated effusion
emergency room	or benign	suggest gallstone
normal abdominal	site is	simple appearing
to right	to abscence	markedly limited
and supine	within ovary	patent splenic
Abdomen X-ray Domain -- Exclusively Omitted:		
coarse which	lateral segment	are too
remain clear	nodular densities	of prostate
poor visualization	at 12	diameter of
abnormal endocrinologic	it measures	dimension and
subdiaphragmatic region	radiographs were	from <*NUMBER*>
large hemorrhagic	previous films	pa and
ngt tip	nondilated air	is not
obstructing lesion	pelvic mass	is unremarkable
which correlates	masses can	abdomen and
prior dictation	prostate carcinoma	is normal
Chest X-ray Domain -- Exclusively Chosen:		
night sweats	the pedicles	suggested when
the pneumothorax	is midline	exclude bibasilar
lungs appear	is much	only minimal
obtained in	however the	correlation recommended
and this	be partially	nd through
mild to	origin of	mm metallic
studies are	and ectatic	bibasilar haziness
and are	layering of	important bony
which probably	left humerus	similar examination
house staff	of copd	aortic tortuosity
Chest X-ray Domain -- Exclusively Omitted:		
from september	lower neck	dome of
description :	subclavian vein	the same
somewhat unusual	enlarged but	is centrally
multiple lucencies	defined density	the aortic
operatively demonstrates	cardiac pathology	is identified
continues to	exam :	the trachea
out aspiration	probably due	of both
do not	pericardial pathology	the thoracic
substantial change	rounded opacity	the endotracheal
+) .	pericardial clips	to the
Mammography Domain -- Exclusively Chosen:		
the amount	the eight	dr manson
breasts the	breast ranging	also demonstrated
breast of	showed no	discussed extensively
the mammography	a unilateral	irregular suggestive
or of	minimal to	patient complained
the mammographic	is once	clearly visualized
the microcalcifications	involution of	number when
or parenchyma	a pacemaker	s injury
a symmetric	had a	postradiation changes
scattered microcalcifications	of glandular	lower central
Mammography Domain -- Exclusively Omitted:		
while this	metropolitan hospital	of residual
office was	was indeterminate	which time
be done	changed when	the periareolar
calcifications many	reduction mammoplasty	the retroareolar
was introduced	eosinophilic pneumonia	is suggested
fluctuating cyst	years ago	are two
patient return	suggested particularly	a rounded
left axial	post surgical	the site
completely unchanged	appears mammographically	is an
was notified	cm from	is moderately

For example, at the top of Table 3, the phrase" "treatment planning" ranked high on the MI measure but low on the CP and EB measures, while "or artifacts" ranked high on CP but not on MI or EB, and "chronic infection" ranked high on EB but not on the other two measures. Conversely, the first row of the second block in Table 3 shows that the phrase "coarse which" was not selected by MI, but was by both CP and EB, while "are too" was not selected by EB, although both MI and CP did select it. The phrases in Table 3 are samples of only 10 from each of their respective categories—there were usually between 200 and 500 in each category. But even these small samples are enough to show the pattern: the Bayesian choices tend to be more "interesting" medically, and the Bayesian omissions tend to be less interesting, than those in the other two columns of Table 3. The CP exclusive choices are cluttered by pairs containing common prepositions and conjunctions, while the MI exclusive choices are also biased toward choosing common words having slight tendencies to occur together. The Bayesian choices are better at focusing on reliably higher ratios of $E(N_{ij})/E_{ij}$ (>10 with 99% confidence) because the Bayesian setup allows such word pairs to be explicitly described.

Finally, the last row of Table 1 shows results with the artificial modification of the mammogram text that reduced the frequency of pairs with a high ratio of N_{ij}/E_{ij}. The values of α and β increased by about a factor of 5, and the number of pairs having $EB_{ij} > 10$ dropped by a factor of 7. Although not shown for reasons of space, the same patterns as observed in Table 3 occur for this modified mammogram text. Another advantage of the Bayesian method is that the same cutoff score ($EB_{ij} > 10$) is reasonable for different populations and sample sizes. In order to restrict the choice to just 253 pairs in the modified mammogram text, the cutoff for CP must be increased from .18 in the original mammogram text to .57 and that for MI must be increased from 35 to 132. Cutoff levels for such measures are difficult to choose and somewhat arbitrary.

39.2 Estimating Distances Between Raters

As computers attempt to perform tasks, like those involving natural language, that attempt to mimic human expert performance, our evaluation of the computer's performance resembles a Turing test: we set up an experiment in which both the computer and human experts solve the same problems, and then see if a statistical analysis of the results can pick out the computer from among the humans. The analysis of such data is similar to an interrater reliability analysis, but the focus in not on estimating the overall consistency among raters or judges performing a task, but rather on how the computer's results look in comparison to both the average and the dispersion of the humans' results. One way to attack this problem is to estimate not just an overall interrater reliability score, but to estimate a distance measure between every pair of raters. If we can also estimate standard errors and a covariance matrix for these distance measures, then we will have the statistical tools to answer a variety of questions comparing the performances of the computer and of the human experts.

Hripcsak et al [Hripcsak95] reports on such an experiment using the MEDEXTRA system. There were $n = 200$ independent test items to be assessed or rated, and each is rated by the NLP system (denoted as rater 0) and by some subset of J human experts (denoted as raters $j = 1, 2, ..., J$). Let X_{ij} be the rating score assigned to item i by judge j. Each X_{ij} may itself be a vector of scores, if each item is being rated on more than one characteristic. The experiment may not call on every judge to rate every item, so many of the X_{ij} may be missing at the time of the analysis. We will use subscripted n to denote how many items each judge or combinations of judges have rated. For example n_j denotes the number of items scored by judge j. Usually $n_0 = n$, if the computer rates all items, but this is not necessary to the analysis. Similarly, n_{jk} denotes the number of items rated by both judge j and judge k, and n_{jklm} denotes the number of items rated by all of the judges j, k, l and m. There might be duplicates in the subscripts, in which case, for example, n_{jkjm} is interpreted as n_{jkm}.

Denote d_{ijk} to be some measure of distance between the rating scores, X_{ij} and X_{ik}, that judges j and k assign to item i. By convention, we set $d_{ijk} = 0$ if either of judges j or k did not rate item i. Let

$$\bar{d}_{jk} = \Sigma_i \, d_{ijk} \, / \, n_{jk}$$

$$V(\bar{d}_{jk}) = \text{Cov}(\bar{d}_{jk}, \bar{d}_{jk})$$

$$\text{Cov}(\bar{d}_{jk}, \bar{d}_{lm}) = [\Sigma_i \, d_{ijk}d_{ilm} - n_{jklm} \, \bar{d}_{jk}\bar{d}_{lm}] \, / \, n_{jk}n_{lm}$$

For matrix calculations, we can define the column vector $\bar{d} = (\bar{d}_{01}, \, ..., \, \bar{d}_{J-1,J})^t$ of $J(J+1)/2$ mean distance scores, and the $J(J+1)/2$ by $J(J+1)/2$ covariance matrix C, whose elements are defined above. The estimated variance of any linear combination $b^t\bar{d}$, where $b = (b_{01}, \, ..., \, b_{J-1,J})$ is a vector of coefficients, is $V(b^t\bar{d}) = b^tCb$.

The previous theory enables us to compute estimates and standard errors (and, assuming approximate normality, confidence intervals) for other measures of interrater difference. Some examples:

$$\delta_j = \Sigma_{k \neq j} \, \bar{d}_{jk} \, / \, J \, ; \qquad\qquad j = 0, 1, ..., J$$
[Average distance of judge j from all other judges]

$$\Delta = 2\Sigma_{k<j} \, \bar{d}_{jk} \, / \, J(J+1)$$
[Overall average interjudge distance]

$$\theta_j = \delta_0 - \delta_j \, ; \qquad\qquad j = 1, 2, ..., J$$
[Comparison of computer's mean distance with that of human judge j]

$$\Sigma_j \, \theta_j \, / \, J$$
[Comparison of computer's mean distance to average of all others' mean distances]

In the experiment reported in [Hripcsak95], 18 human subjects and 3 automated algorithms are compared for their agreement in detecting a "reasonably likely" diagnosis of six different conditions from 200 randomly chosen chest x-ray reports. The six conditions are: congestive heart failure (CHF), chronic obstructive pulmonary disease, acute bacterial pneumonia, neoplasm, pleural effusion without CHF, and pneumothorax. The referenced paper provides details of the diagnoses and other aspects of the experimental design. The 18 human subjects included six radiologists, six internists, and six lay persons with no special medical training. Each human subject read 100 of the 200 reports, with each pair of subjects scoring at least 40 reports in common. To read a single report and choose among the six conditions took an average of 70 seconds for human subjects and 2 seconds for the natural language processor (running on a 42 Mhz IBM RS/6000 workstation). The primary computer algorithm of interest will be denoted NLP, and consists the combination of the MEDEXTRA program that reads reports and feeds data to an automated decision support system, and the decision support system which in turn draws conclusions and makes clinical recommendations. Two less sophisticated computer algorithms, based merely on keyword searches, were also included in the study, as well as the null algorithm that merely declares all six medical conditions absent from all reports. Each report was read by six of the twelve medical experts, and, assuming that a condition is present if a majority of experts (four or more out of six) voted for it, the prevalence of conditions ranged from 3% (chronic obstructive pulmonary disease) to 14% (acute bacterial pneumonia).

The distance measure used is the average fraction of diagnoses raters disagree on, and the primary outcome measure is the average distance of each subject to the (other) experts. The average distance of experts from each other was 0.24 (95% CI 0.19 - 0.29) Pairs of experts differed on the interpretation of reports for *at least* one diagnosis about 20% of the time. The average distance of NLP from the experts was 0.26 (95% CI 0.21 - 0.32). No two human experts were significantly more distant from the other experts than the average. The average distance of an expert to another expert of the same specialty (radiology or internal medicine) was almost exactly the same as the average distance of experts across specialties (0.24 vs 0.25, respectively). On the other hand, all of the six lay persons were much more distant from the experts. Their average distance to the 12 experts ranged from 0.51 to 0.73, with standard errors of about 0.03. The distance of the other automated algorithms from the experts was also significantly greater than the experts were from each other.

A multidimensional scaling analysis [Dillon84] helps visualize the interrater distances. Twenty-one raters (the poorest performing keyword search algorithm was excluded) are represented by 21 points in the plane that attempt to preserve the interrater distances $\bar{d}_{jk}$. The Figure displays the results. The 12 experts and the NLP cluster in the lower left of the Figure, while the lay persons are placed far to the right along with the null algorithm, and the complex keyword search stands alone at the upper left of the Figure. The report [Hripcsak95] shows other graphical representations of these data, including a sensitivity-specificity plot of all raters. This shows that dimension 1 of the present multidimensional scaling analysis is primarily variation in sensitivity (the null algorithm has zero sensitivity, of course), and dimension 2 is primarily variation in specificity (the keyword search algorithm and one of the lay persons have many false positives).

39.3 Conclusion

These two examples show that statistical modelling can help develop a more focused approach to developing and evaluating natural language processing algorithms. The Bayesian word collocation analysis provides a more relevent measure of collocation and thus more discrimination among potential word pairs. In the evaluation experiment, the ability to get standard errors and confidence intervals for every interrater distance, and, more importantly, for all linear combinations of these distances, provides convincing evidence of the power of the NLP algorithm to work well in practice.

39.4 References

[Dillon84] Dillon W, Goldstein M (1984) *Multivariate Analysis*, New York: Wiley, 587pp.

[Dunning93] Dunning, Ted (1993) Accurate methods for the statistics of surprise and coincidence, *Computational Linguistics*, 19: 61-74.

[Friedman95] Friedman C, Hripcsak G, DuMouchel W, Johnson S, Clayton P (1995) Natural language processing in an operational clinical information system, *Natural Language Engineering* 1 (1): 1-28.

[Hripcsak95] Hripcsak G, Friedman C, Alderson P, DuMouchel W, Johnson S, Clayton P (1995) Unlocking clinical data from narrative reports: a study of natural language processing. *Annals of Internal Medicine*, 122: 681-688.

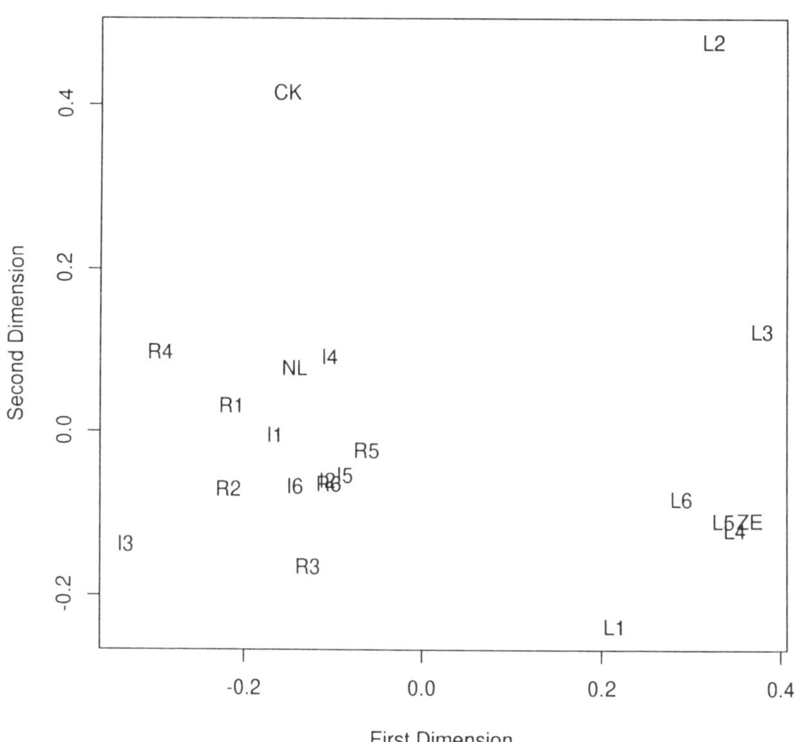

Figure: Location of each "subject" in a two-dimensional MDS fit to the between-rater distances. (Internists: I1...I6, Radiologists: R1...R6, Lay persons: L1...L6, Natural Language Processor: NL, Complex Keyword Search: CK, Constant 0 Rule: ZE.)

40
A Model for Part-of-Speech Prediction

Alexander Franz

Center for Machine Translation
Carnegie Mellon University
Pittsburgh, PA 15213
USA
amf@cs.cmu.edu

ABSTRACT Robust natural language analysis systems must be able to handle words that are not in the lexicon. This paper describes a statistical model that predicts the most likely Parts-of-Speech for previously unseen words. The method uses a loglinear model to combine a number of orthographic and morphological features, and returns a probability distribution over the open word classes. The model is combined with a stochastic Part-of-Speech tagger to provide a model of context. Empirical evaluation shows that this results in significant gains in Part-of-Speech prediction accuracy over simpler methods.

40.1   Introduction

Current natural language analysis systems typically assume a closed vocabulary. For example, the most successful system at a recent data extraction workshop would "simply halt processing when a new word was encountered" [Weischedel et al., 1993]. As natural language analysis systems move out of the realm of small, experimental domains, and toward applications with open-ended vocabulary, robust methods to handle new words become necessary.

This paper describes a statistical model that predicts the most likely Parts-of-Speech (POSs) for unknown words. The model is used to supply the lexical probabilities for unknown words for a stochastic POS tagger. In this way, the most likely POS given the unknown word and the context is found, and the new word becomes amenable to further analysis. Empirical evaluation shows that this method obtains higher accuracy than a simpler method that assumes independence between the features.

40.2   Constructing the Model

The model was constructed in the following way. First, features that could be used to guess the POS of a word were determined by examining the training portion of a text corpus. The initial set of features consisted of the following:

- **INCLUDES-NUMBER.** Does the word include a number? Positive example: *836-901*. Negative example: *Absent-minded*.

- **CAPITALIZED.** Is the word in sentence-initial position and capitalized, in any other position and capitalized, or in lower case?

- **INCLUDES-PERIOD.** Does the word include a period? Positive example: *B.C., 4.2, U.N.* Negative example: *Union.*

- **INCLUDES-COMMA.** Does the word include a comma? Positive example: *500,000.* Negative example: *amazement.*

- **FINAL-PERIOD.** Is the last character of the word a period? Positive example: *B.C., Co.* Negative examples: *U.N, Command.*

- **INCLUDES-HYPHEN.** Does the word include a hyphen? Positive examples: *Poynting-Robertson, anti-party.* Negative example: *answer.*

- **ALL-UPPER-CASE.** Is the word in all upper case? Positive example: *CTCA.* Negative examples: *Fred, accomplish.*

- **SHORT.** Is the length of the word three characters or less? Positive example: *W., Yes, bar, Eta.* Negative example: *Heaven, 100,000.*

- **INFLECTION.** Does the word carry one of the English inflectional suffixes *-ed, -er, -est, -ing, -ly, -s*?

- **PREFIX.** Does the word carry one of a list of frequently occurring prefixes, such as *anti-, dis-, inter-, mal-, micro-, mis-, non-, out-, over-, pre-, pro-, self-, trans-, under-,* etc.?

- **SUFFIX.** Does the word carry one of a list of frequently occurring suffixes, such as *-able, -age, -ation, -ative, -ful, -ish, -land, -ment, -ness, -ster,* etc.?

Exploratory data analysis was performed in order to determine relevant features and their values, and to approximate which features interact. Each word of the training data was then turned into a feature vector, $\vec{v}$, and the feature vectors were cross-classified in a contingency table. The contingency table was smoothed using a loglinear model.

40.2.1 The Contingency Table

A contingency table was used to count the observed values of features and combinations of features. A contingency table is an array with one dimension for each feature; the size of the table is the product of the number of possible values for the features. Each cell in the contingency table records the frequency of data with the appropriate feature values.

40.2.2 Smoothing the Conditional Probabilities

A loglinear model is a statistical model of the effect of a set of categorial features and their combinations on the cell counts in a contingency table. It can act as a "smoothing device, used to obtain cell estimates for every cell in a sparse array, even if the observed count is zero" [Bishop et al., 1975].

Marginal totals of the observed counts are used to estimate the parameters of the loglinear model; the model in turn delivers estimated expected cell counts, which are smoother than the original cell counts. Let m_{ijkl} be the expected cell count for cell (i, j, k, l). The values for the expected cell counts that are estimated by the model are represented by the symbol $\hat{m}_{ijkl}$. The general form of a loglinear model is as follows:

$$\log m_{ijk...} = u + u_{1(i)} + u_{2(j)} + u_{3(k)} + u_{12(ij)} + \ldots$$

In this formula, u denotes the mean of the logarithms of all the expected counts, $u + u_{1(i)}$ denotes the mean of the logarithms of the expected counts with value i of the first feature, $u + u_{2(j)}$ denotes the mean of the logarithms of the expected counts with value j of the second feature, $u + u_{12(ij)}$ denotes the mean of the logarithms of the expected counts with value i of the first feature and value j of the second feature, and so on.

Thus, the term $u_{1(i)}$, for example, denotes the deviation of the expected cell counts with value i of the first feature from the grand mean u. Similarly, the term $u_{12(ij)}$ denotes the deviation of the expected cell counts with value i of the first feature and value j of the second feature from the grand mean u. In other words, $u_{12(ij)}$ represents the *combined effect* of the values i and j for the first and second features on the logarithms of the expected cell counts.

A loglinear model provides a way to estimate expected cell counts that depend not only on the main effects of the features, but also on the interactions between features. This is achieved by adding "interaction terms" such as $u_{12(ij)}$ to the model. For further details, see [Agresti, 1990].

40.2.3   The Iterative Estimation Procedure

For some loglinear models, it is possible to obtain closed forms for the expected cell counts. For more complicated models, the *iterative proportional fitting* algorithm for hierarchical loglinear models [Deming and Stephan, 1940] can be used. Briefly, this procedure works as follows.

The interaction terms in the loglinear models represent constraints on the estimated expected marginal totals. Each of these marginal constraints translates into an adjustment scaling factor for the cell entries. The iterative procedure has the following steps:

1. Start with initial estimates for the estimated expected cell counts. For example, set all $\hat{m}_{ijkl} = 1.0$.

2. Adjust each cell entry by multiplying it by the scaling factors. This moves the cell entries towards satisfaction of the marginal constraints specified by the model.

3. Iterate through the adjustment steps until the maximum difference ϵ between the marginal totals observed in the sample and the estimated marginal totals reaches a certain minimum threshold, e.g. $\epsilon = 0.1$.

After each cycle, the estimates satisfy the constraints specified in the model, and the estimated expected marginal totals come closer to matching the observed totals. Thus, the process converges. This results in Maximum Likelihood estimates for both multinomial and independent Poisson sampling schemes [Agresti, 1990].

Tag	Part of Speech	Example
CD	Cardinal (number)	500,000
FW	Foreign word	Fahrvergnügen
JJ	Adjective	yellow, large
JJR	Comparative Adjective	larger, nicer
JJS	Superlative adjective	largest, nicest
LS	List item marker	1. ..., a) ...
NN	Singular or mass noun	water, rock
NNS	Plural noun	rocks, cars
NP	Singular proper noun	English, March
NPS	Plural proper noun	The English
RB	Adverb	quickly, quite
RBR	Comparative Adverb	wiser, deeper
RBS	Superlative adverb	nearest, best
SYM	Symbol	%, *
UH	Interjection	uh, hmpf
VB	Base form verb	do, go
VBD	Past tense verb	did, went
VBG	Present participle verb	doing, going
VBN	Past participle verb	gone, flown
VBP	Non-3sg present verb	do, go
VBZ	3sg present verb	does, goes

TABLE 40.1. Open class Parts-of-Speech. It can be assumed that all grammatical function words are known and listed in the dictionary, so an unknown word must fall into one of the open classes.

40.2.4 Bayesian Part-of-Speech Prediction

Let $P(t)$ be the prior probability distribution over the possible Part-of-Speech (POS) tags, and $P(\vec{v})$ the prior distribution for the feature vector representations of words. The training data are annotated with the correct POS. From this, Maximum Likelihood estimates for $P(\vec{v}|t)$, the conditional probability distribution for $\vec{v}$, given the POS tag t, are derived. Applying Bayes rule and maximizing the conditional probability $P(t|\vec{v})$ leads to *minimum error rate* POS prediction [Duda and Hart, 1973]. The conditional distribution $P(t|\vec{v})$ can be derived directly from the smoothed contingency table.

40.3   Empirical Evaluation

This section describes the results of an empirical evaluation of the model. During the evaluation the model was used to predict the Parts-of-Speech (POSs) for rare words from an online natural language corpus. The open class POSs are summarized in Table 40.1.

40.3.1 Data

Sets of training and evaluation data were obtained from the Penn Treebank Brown corpus [Marcus et al., 1993]. The characteristics of unknown words differ from the characteristics of words in general, so a two-step procedure was employed a first time to obtain a set of training data, and a second time to obtain a separate set of training data. The two-step procedure uses one randomly selected set of words to form a "dictionary" of known words, and a second randomly selected set of words to represent new text. Then, the words from the new text that are not in the dictionary form a sample of rare words, whose characteristics and distribution are similar to unknown words. The details of this procedure are as follows:

1. A set S_1 containing 400,000 words was selected to form the first dictionary of known words. This set is used to derive the set S_2 in the next step, and then discarded.

2. A set S_2 containing 300,000 words was selected to represent some new text. The words in S_2 but not in S_1 form the *training data*.

3. To obtain the evaluation data, the set S_2 is used as the second dictionary of known words. A third set S_3 containing 300,000 words is selected to represent the second batch of new text. The words in S_3 but not in S_2 form the *evaluation data*.

There were 17,000 words in the training data, and 21,000 words in the evaluation data.

40.3.2   Determining the Feature Set

A set of data does not come with instructions for its analysis. Which aspects of the data should be modeled? Which features should be chosen? What should be chosen as the possible values for the features? For the loglinear model, which features appear independent, and which features interact?

We explored the training data to look for variables that are good discriminators, tried out different combinations of variables, and compared the performance of different models. Based on this exploration, some features were added and deleted, and the number of possible values was changed for others.

40.3.3   Measuring Performance

The performance of the model was evaluated on the evaluation data using a number of different accuracy measures:

- **Overall Accuracy.** Percentage that the most likely POS tag is correct.

- **Cutoff Factor Accuracy.** Accuracy of the *answer set*, which consists of all POS tags whose probability lies within a factor F of the most likely POS [de Marcken, 1990].

- **Residual Ambiguity.** Mean residual ambiguity for the POS tags in the answer set, measured by the perplexity of the answer set [Jelinek et al., 1977]. (The perplexity corresponds to the number of equi-probable members in the answer set.)

40.3.4   Accuracy Results

[Weischedel et al., 1993] describe a model for unknown words that uses four features, which were treated as independent. We reimplemented this model by using four features: **POS, INFLECTION, CAPITALIZED**, and **HYPHENATED**. In Figures 1-2, the results for this model are labeled **Independent Features**. For comparison, we also created a model with the same four features, but using a contingency table that was smoothed with a loglinear model including all two-way interactions. The results for this model are labeled **Loglinear Features**.

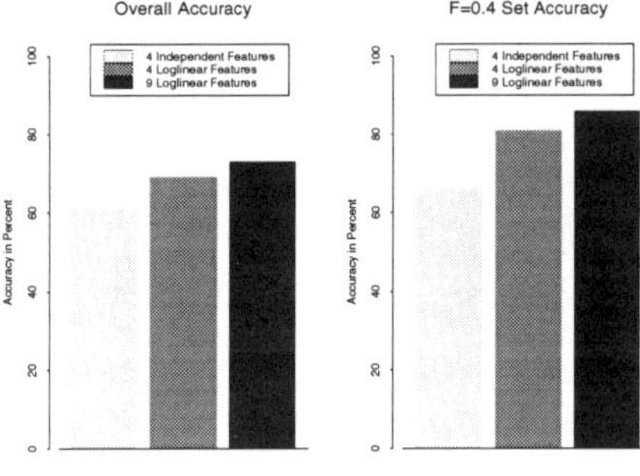

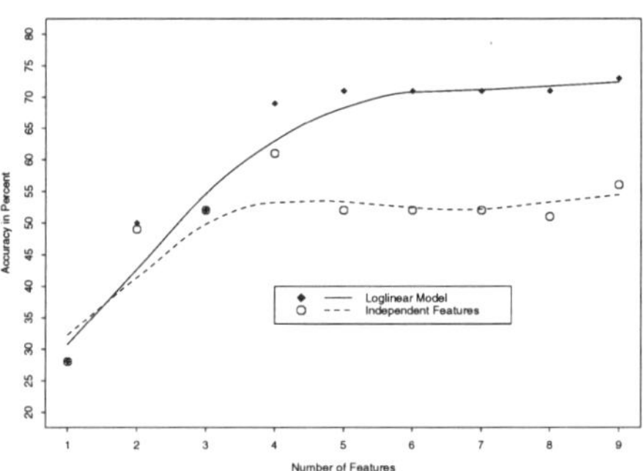

FIGURE 1. Accuracy of the different models. The left diagram compares the accuracy of different models for unknown words. The two groups of bars corresponds to a different accuracy measure. In order of increasing accuracy, the three different models are: A simple probabilistic model using four independent features, a loglinear model using four features, and a loglinear model using nine features. The graph on the right shows how overall accuracy varies with increasing number of features for the loglinear model (upper line), and for the model assuming independent features (lower line).

The highest accuracy was obtained by the loglinear model that includes all two-way interactions and consists of two contingency tables with the following features: **POS, ALL-UPPER-CASE, HYPHENATED, INCLUDES-NUMBER, CAPITALIZED, INFLECTION, SHORT, PREFIX, SUFFIX**. The results for this model are labeled **9 Loglinear Features**. The parameters for all three unknown word models were estimated from the training data, and the models were evaluated on the evaluation data (see Section 40.3.1 for details.)

The accuracy of the different models in assigning the most likely POSs to words is summarized in Figure 1. In the left diagram, the two barcharts show two different accuracy measures: Percent correct (**Overall Accuracy**), and percent correct within the F=0.4 cutoff factor answer set (**F=0.4 Set Accuracy**). (This answer set consists of all POS tags whose probability lies within a factor of 0.4 of the most likely POS.)

In both cases, the loglinear model with four features obtains higher accuracy than the method that assumes independence between the same four features. The loglinear model with nine features further improves this score.

40.3.5 Effect of Number of Features on Accuracy

The performance of the loglinear model can be improved by adding more features, but this is not possible with the method that assumes independence between features. The right diagram in Figure 1 shows the performance of the two types of models with feature sets that ranged from a single feature to nine features.

This diagram shows that the two methods of combining features have very different effects. The accuracies for both methods rise with the first few features, but then the two methods show a clear divergence. The accuracy of the method assuming independent features levels off around at around 50–55%, while the loglinear model reaches an accuracy of 70–75%.

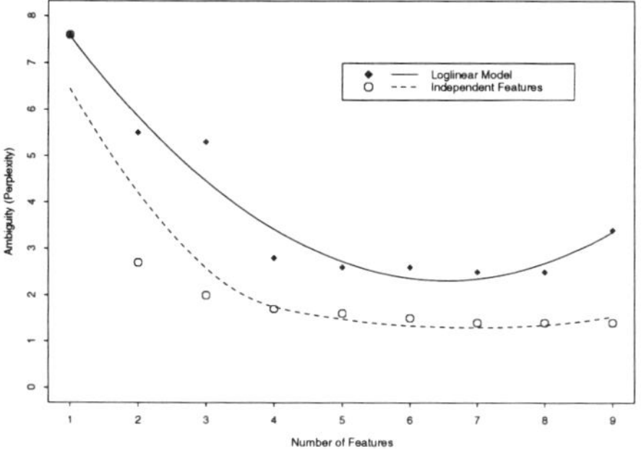 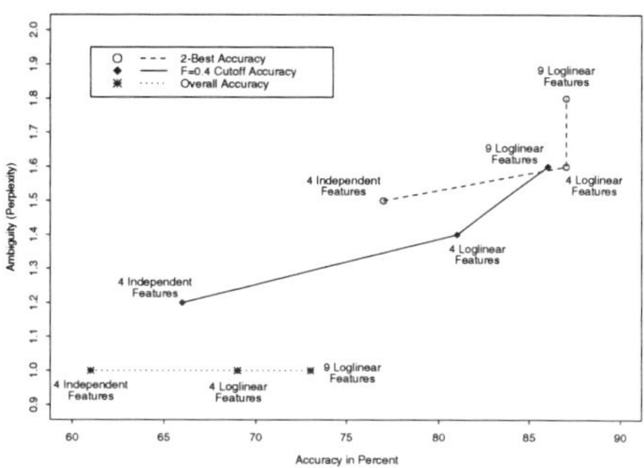

FIGURE 2. Residual Ambiguity of different models. The left diagram shows the effect of the number of features on the residual ambiguity of the answer set of the loglinear model (upper line), and of a model that assumes independence between features (lower line). Residual ambiguity is measured as the perplexity of the answer set. The right diagram shows how residual ambiguity varies with respect to accuracy for the different models.

40.3.6 Tradeoff between Accuracy and Ambiguity

The predictions of a model may reflect higher accuracy, but at the expense of some residual ambiguity. The effect of the number of features on F=0.4 cutoff factor residual ambiguity is shown on the left in Figure 2. Residual ambiguity decreases as more features are added, and the loglinear model shows somewhat higher residual ambiguity than the simple method assuming independence.

The scatter plot on the right in Figure 2 depicts the relation between accuracy and residual ambiguity. The connected series of points correspond to a single accuracy measure with a series of different models. For example, the point in the upper right-hand corner labeled **9 loglinear features** corresponds to the loglinear model with nine features where the answer set is derived using the F=0.4 cutoff factor. The ideal model would be located in the bottom right-hand corner, with high accuracy and low residual ambiguity. The best accuracy/residual ambiguity tradeoff depends on the utility function for each specific natural language processing application.

40.4 Adding Context to the Model

The previous sections discussed a model for unknown words in isolation. This section describes the addition of a stochastic POS tagger [Charniak et al., 1993] to provide a model of context.

40.4.1 Part of Speech Tagging

A POS tagger assigns POS labels to words in a sentence. It uses two types of parameters:

- **Lexical Probabilities:** $P(t|w)$ — the probability of observing tag t given that the word w occurred. If a word w is not known, and it is therefore not possible to estimate a lexical distribution for the individual word, then the more general distribution $P(t|\vec{v})$ from the unknown word model is used. (See Section 40.2.4 for

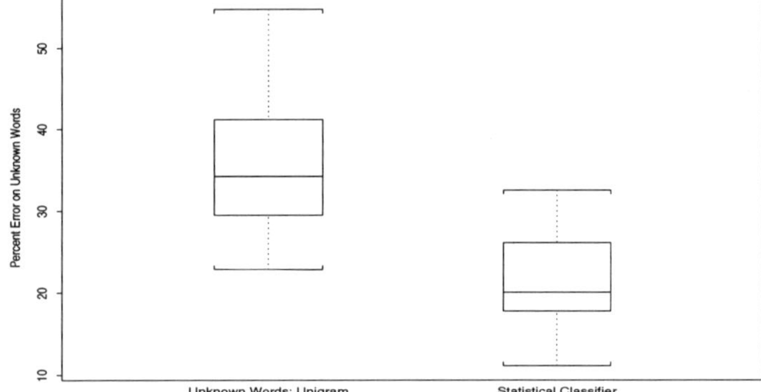

FIGURE 3. Error Rate on Unknown Words. This diagram compares the Part-of-Speech prediction error rate on unknown words using independent features (left) with the error rate obtained with the loglinear model for unknown words (right). There were approximately 35 samples of 4,000 words drawn from Wall Street Journal articles. The boxes are drawn between the quartiles, and the bars indicate the medians.

---

further details.)

- **Contextual Probabilities:** $P(t_i|t_{i-1}, t_{i-2})$ — the probability of observing tag t_i given that the two previous tags t_{i-1}, t_{i-2} occurred.

The tagger maximizes the probability of the tag sequence $T = t_1, t_2, \ldots, t_n$ given the word sequence $W = w_1, w_2, \ldots, w_n$, which is approximated as follows:

$$P(T|W) \approx \prod_{i=1}^{n} P(w_i|t_i)P(t_i|t_{i-1}, t_{i-2})$$

Note that the path T_{MAX} that maximizes $P(T|W)$ also maximizes the right-hand-side approximation.

40.4.2 Evaluating Combined Accuracy

The accuracy of the combination of the loglinear model for local features and the stochastic POS tagger for contextual features was evaluated empirically by comparing three methods of handling unknown words:

- **Unigram:** Using the prior probability distribution $P(t)$ of the POS tags for rare words.

- **Probabilistic UWM:** Using the probabilistic model that assumes independence between the features.

- **Classifier UWM:** Using the loglinear model for unknown words.

Separate sets of training and evaluation data for the tagger were obtained from from the Penn Treebank Wall Street corpus. Evaluation of the combined system was performed

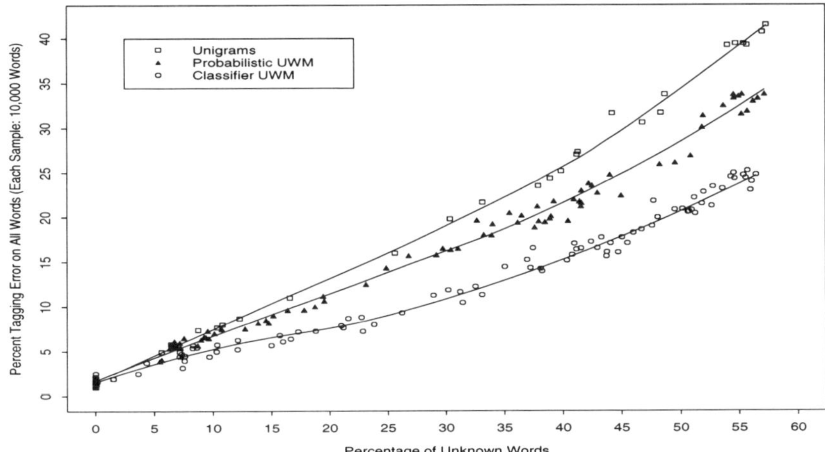

FIGURE 4. Percentage of unknown words versus Part-of-Speech prediction accuracy. This diagram shows how the overall error rate varies with the proportion of unknown words for three different models: Using the prior distribution for infrequent words (upper line), using independent features (middle line), and using the loglinear model for unknown words (lower line). Each plotted symbol denotes one 4,000-word sample of Wall Street Journal text.

on different configurations of the POS tagger on 30–40 different samples containing 4,000 words each. The tagger displays considerable variance in its accuracy in assigning POS to unknown words in context. Figure 3 compares the tagging error rate on unknown words for the unigram method (left) and the loglinear method with nine features (labeled **statistical classifier**) at right.

This shows that the loglinear model significantly improves the Part-of-Speech tagging accuracy of a stochastic tagger on unknown words. The median error rate is lowered considerably, and samples with error rates over 32% are eliminated entirely.

40.4.3 Effect of Proportion of Unknown Words

Finally, we investigated the following questions: How does the overall tagging accuracy depend on unknown words that are not in the system's dictionary? How does the accuracy vary when there are different proportions of unknown words in the text to be tagged?

Samples of text that contained different proportions of unknown words were tagged using the three different methods for handling unknown words. The overall tagging error rate increases significantly as the proportion of new words increases. Figure 4 shows a graph of overall tagging accuracy versus percentage of unknown words in the text. The graph compares the three different methods of handling unknown words. The diagram shows that the loglinear model leads to better overall tagging performance than the simpler methods, with a clear separation of all samples whose proportion of new words is above approximately 10%.

40.5   Conclusions

A simple statistical modeling technique for handling words that have never been encountered before has been described. The results show that the loglinear method is better than a simpler method that assumes independence between the orthographic and morphological features. The loglinear model achieves higher accuracy, and it handles larger feature sets, which may contain nuisance features.

This work has shown that a loglinear model is able to combine a number of linguistic categorial features. Thus, the modeling technique described in this paper is also applicable to a number of other problems in natural language that require the integration of morphological, lexical, syntactic, and semantic clues. In particular, this method holds promise for further advances in automatic natural language ambiguity resolution.

40.6   REFERENCES

[Agresti, 1990] Agresti, A. (1990). *Categorical Data Analysis.* John Wiley & Sons, New York.

[Bishop et al., 1975] Bishop, Y. M., Fienberg, S. E., and Holland, P. W. (1975). *Discrete Multivariate Analysis: Theory and Practice.* MIT Press, Cambridge, MA.

[Charniak et al., 1993] Charniak, E., Hendrickson, C., Jacobson, N., and Perkowitz, M. (1993). Equations for part-of-speech tagging. In *AAAI-93*, pages 784–789.

[de Marcken, 1990] de Marcken, C. G. (1990). Parsing the LOB corpus. In *Proceedings of ACL-90*, pages 243–251.

[Deming and Stephan, 1940] Deming, W. E. and Stephan, F. F. (1940). On a least squares adjustment of a sampled frequency table when the expected marginal totals are known. *Ann. Math. Statis,* (11):427–444.

[Duda and Hart, 1973] Duda, R. O. and Hart, P. E. (1973). *Pattern Classification and Scene Analysis.* John Wiley & Sons, New York.

[Fienberg, 1980] Fienberg, S. E. (1980). *The Analysis of Cross-Classified Categorical Data.* The MIT Press, Cambridge, MA, second edition edition.

[Jelinek et al., 1977] Jelinek, F., Mercer, R. L., Bahl, L. R., and J, K. B. (1977). Perplexity — a measure of difficulty of speech recognition tasks. In *94th Meeting of the Acoustical Society of America*, Miami Beach, FL.

[Marcus et al., 1993] Marcus, M. P., Santorini, B., and Marcinkiewicz, M. A. (1993). Building a large annotated corpus of English: The Penn Treebank. *Computational Linguistics*, 19(2):313–330.

[Weischedel et al., 1993] Weischedel, R., Meteer, M., Schwartz, R., Ramshaw, L., and Palmucci, J. (1993). Coping with ambiguity and unknown words through probabilistic models. *Computational Linguistics*, 19(2):359–382.

41
Viewpoint-Based Measurement of Semantic Similarity between Words

Kaname Kasahara, Kazumitsu Matsuzawa, Tsutomu Ishikawa and
Tsukasa Kawaoka

NTT Communication Science Laboratories
1-2356 Take, Yokosuka-shi
Kanagawa, 238-03
Japan. (e-mail: kaname@nttkb.ntt.jp)

ABSTRACT A method of measuring semantic similarity between words using a knowledge-base
constructed automatically from machine-readable dictionaries is proposed. The method takes into
consideration the fact that similarity changes depending on situation or context, which we call 'viewpoint'. Evaluation shows the proposed method, although based on a simply structured knowledge-
base, is superior to other currently available methods.

41.1   Introduction

Measuring semantic similarity between words is important for natural language processing
when searching text, performing analogical and cases-based reasoning, designing flexible
human interfaces to databases, and other tasks. We are exploring methods for measuring
the similarity between large numbers of daily-use words with an aim toward general applications. In measuring similarity, however, we consider that similarity changes depending
on situation or context, which we call 'viewpoint'. For example, 'horse' is more similar to
'pig' than 'car' from the viewpoint of 'animal'. On the other hand, 'horse' is more similar
to 'car' from the viewpoint of 'vehicle'.

There have been many studies on measuring the semantic similarity of words (for example [Tversky77, Suzuki92]), but methods for acquiring knowledge of words were not
considered in these studies. Babaguchi et al., proposed finding causes of similarity, which
is defined as the 'viewpoint' of similarity, from similar cases and executing similar case
retrieval by using the obtained viewpoint of similarity [Babaguchi94]. However, they assumed that the attributes of the cases were already known, since they focused on applications to an existing database. It is difficult to use their method for measuring similarity
between daily-use words, for which the defining attributes may be unclear.

In this paper, we propose a method that considers viewpoint when measuring the similarity between words. The method uses a knowledge base that is constructed automatically
from machine-readable dictionaries. When computing the similarity between two words,
conditioned on view, aspects of the words' definitions that are important in the view are
emphasized when calculating similarity.

We constructed an experimental knowledge base and evaluated our method by its *pre-*

---

cision during similar-word retrieval. Experimental results indicate that our method is superior, along the precision dimension, to a conventional method that exploits a tree-structured thesaurus.

41.2   Method for measuring similarity between words

41.2.1   The knowledge-base of words

Knowledge of a word is typically represented by a list of attribute-value pairs. For example, the word 'apple' may be represented as

$$'apple' = \{('shape', 'sphere'),\ ('color', 'red'), ('taste', 'sour'), \cdots \}.$$

During the past several years, the Cyc [Guha90] and EDR [EDR93] projects have attempted to acquire an enormous amount of such semantic knowledge. However, it is difficult to manually record all the attributes and values for a large number of daily-use words [Ide93a]. On the other hand, machine-readable dictionaries are a source of semantic knowledge and considerable research on ways of extracting knowledge from them is underway [Ide93b, Nakamura88]. However, no adequate method for acquiring the attributes and values completely and fully has been found yet. Moreover, viewpoint is not considered in any of the research projects referenced above. Therefore, we automatically constructed a simply-structured knowledge base from machine-readable dictionaries only, and tried to compute judgements of viewpoint-based similarity between words.

In our method, each word Word_i in the knowledge base is defined by a list of weighted keywords: a keyword p_{ij} of Word_i and a weight q_{ij} of p_{ij}, where

$$\text{Word}_i \ = \ \{(p_{i1}, q_{i1}), (p_{i2}, q_{i2}), \cdots, (p_{in}, q_{in})\}. \tag{41.1}$$

We choose the morphemes contained in the definition or explanation of a word in the dictionary as a keyword of the word, and choose the number of times that the morpheme appears as the weight of the keyword. Note that each keyword is also a word that appears in the dictionary.

41.2.2   An Overview of the Similarity measurement procedure

The degree of similarity S is calculated using only lists of the keywords and weights of the words included in the knowledge base (Figure 1). Here, Word_1 and Word_2 are words between which the degree of similarity is calculated, and View is also a word, which represents the viewpoint (we call it the 'viewpoint word').

$S(\text{Word}_1, \text{Word}_2, \text{View})$ is calculated as follows.

- **STEP 1: Standardization of keywords**
  Keywords that are semantically similar to each other should be identified, because similarity between words is computed based on their defining keywords, We call this process 'standardization'. Here, each keyword (of Word_1, Word_2, and View) is standardized by placing it into a pred-defined category of the thesaurus.

- **STEP 2: Normalization of weights**
  The number of a word's defining keywords, and the weight of each keyword, depends

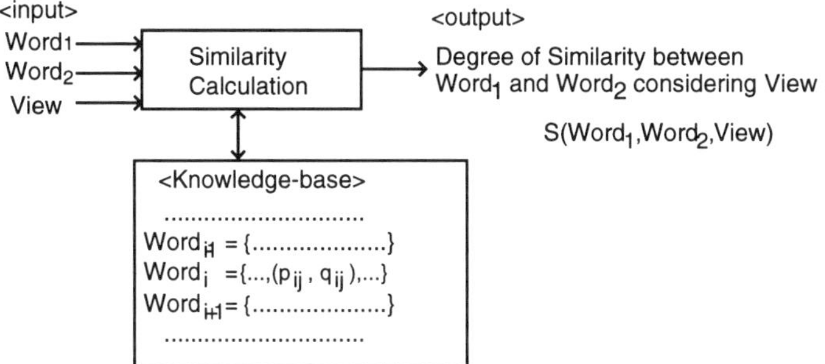

FIGURE 1. A similarity measurement scheme

on the length of the definition or explanation about the word in the dictionary. That is, words in the knowledge base have different numbers of positively-weighted keywords. Since the similarity between two words is based on their positively-weighted keywords, certain undesirable biases in the similarity computation are mitigated by 'normalizing' each word's definition.

- **STEP 3: Modulation based on viewpoint**
  To compute the degree of similarity based on viewpoint, we assume that keyword categories of the viewpoint word (see Step 1) that have a positive weight are important. Therefore, certain keyword categories of each word, Word_1 and Word_2, are emphasized if they are positive-weight categories shared by the viewpoint word. We call this emphasis 'modulation'.

- **STEP 4: Calculation of similarity (S)**
  The degree of similarity, S, between two modulated words is defined based on the angle between two vectors representing the modulated words.

41.2.3 Keyword Standardization

Similarity is calculated by comparing keywords. Therefore, keywords that are nearly equal in meaning should be regarded as being the same keywords. For example, the keyword 'shape' should equal 'form'. We use a thesaurus, T, consisting of k categories for this standardization. c_m denotes the category containing words, p_{ml}, which are considered similar to each other.

$$T = \{c_1, c_2, \cdots, c_m, \cdots, c_k\} \qquad (c_m = \{p_{m1}, p_{m2}, \cdots p_{ml}, \cdots\}). \tag{41.2}$$

We regard the categories in T as being independent of each other from the semantic point of view. In this case, the categories of T span vector space V of k dimensions.

Each keyword of Word_i is standardized by placing it into the category in which it is included. Using these standardized keywords, vectorized-word $\mathbf{Word_i}$ is generated as follows:

$$\mathbf{Word_i} = (Q_{i1}, Q_{i2}, \cdots, Q_{ij}, \cdots, Q_{ik}) \qquad (Q_{ij} = \sum_{\substack{l=1 \\ p_l \in c_j}}^{n} q_{il}). \tag{41.3}$$

Here, the second equation shows that the weights, q_{ij}, of the keywords are summed up, if there are some keywords included in the same category. Note that each word is now represented in terms of the same k categories, though the weight, Q_{ij}, of many of these categories may be 0 for a particular word.

41.2.4 Normalization of weights

The number of positively-weighted keyword categories and the particular weight of each category that now define a word depends on the length of the definition or explanation of a word in the dictionary; thus, the value and number of positive weights vary with word. When calculating the similarity based on the distance between **Word₁** and **Word₂** in vector space V, the value and number of positive weights may undesirably bias the similarity computation. For example, words that have a small number of small positive weights are disposed *a priori* to be nearer and more similar to each other than to a word that has a large number of large positive weights. Therefore, the weights of the word should be normalized so that the length of **Wordᵢ** is constant. We normalize the weight of **Wordᵢ** by dividing it by the the vector norm $\|\mathbf{Word_i}\|$ of the **Wordᵢ**:

$$\mathbf{Word_i} = (\hat{Q}_{i1}, \hat{Q}_{i2}, \cdots, \hat{Q}_{ij}, \cdots, \hat{Q}_{ik})$$

$$\hat{Q}_{ij} = \frac{Q_{ij}}{\|\mathbf{Word_i}\|} = \frac{Q_{ij}}{\sqrt{\sum_{m=1}^{k} Q_{im}^2}}. \tag{41.4}$$

41.2.5 Modulation based on viewpoint

To take the viewpoint into consideration in calculating the degree of similarity, words compared with each other should be emphasized using the viewpoint word. The viewpoint word (**View**) is also standardized and normalized as:

$$\mathbf{View} = (\hat{Q}_{v1}, \hat{Q}_{v2}, \cdots, \hat{Q}_{vj}, \cdots, \hat{Q}_{vk}) \qquad (\|\mathbf{View}\| = \mathbf{1}). \tag{41.5}$$

We assume that categories which have positive weight in **View** should be important when the degree of the similarity is calculated. Therefore, we define 'modulation' as the procedure in which the weights of such categories in the **Wordᵢ** are emphasized when calculating the degree of similarity. A modulated word **Wordᵢᵛ** is generated by the following formula:

$$\mathbf{Word_i^v} = (Q_{i1}^v, Q_{i2}^v, \cdots, Q_{ij}^v, \cdots, Q_{ik}^v) \qquad (Q_{ij}^v = \hat{Q}_{ij} \cdot M(\hat{Q}_{vj})). \tag{41.6}$$

Because all categories in T are regarded as being independent of each other, the weights of **Wordᵢᵛ** Q_{ij}^v can be calculated only from weights of the same category in **Wordᵢ** and **View**. M is a function of modulation, which decides the degree of emphasis for each category. We assume that small weights in the viewpoint are noisy, and have no relation with the viewpoint. Therefore, modulation is based on the step function of Figure 2, where a is a threshold value that filters noise. Next the modulated word **Wordᵢᵛ** is normalized again by Equation 41.4, because the length of the vector norm is changed by modulation.

41.2.6 Calculating the degree of similarity (S)

We calculate $S(\text{Word}_1, \text{Word}_2, \text{View})$ as the similarity between two modulated words.

$$S(\text{Word}_1, \text{Word}_2, \text{View}) = R(\mathbf{Word_1^v}, \mathbf{Word_2^v}). \tag{41.7}$$

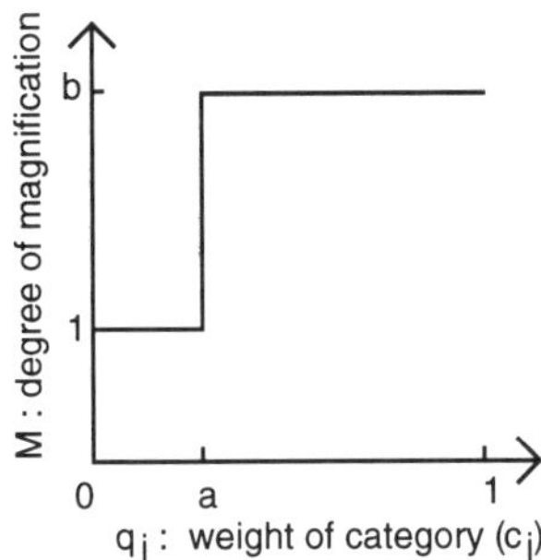

FIGURE 2. Modulation function example

R is the function which indicates the nearness between two modulated words in the vector space V. R is constrained as follows:

- $0 \leq R(\mathbf{Word_a}, \mathbf{Word_b}) \leq 1$
- $R(\mathbf{Word_a}, \mathbf{Word_b}) \equiv R(\mathbf{Word_b}, \mathbf{Word_a})$
- $R(\mathbf{Word_a}, \mathbf{Word_a}) \equiv 1$
- $\mathbf{Word_b}$ is more similar to $\mathbf{Word_c}$ than $\mathbf{Word_a}$,
  if $R(\mathbf{Word_a}, \mathbf{Word_b}) \leq R(\mathbf{Word_c}, \mathbf{Word_b})$.

We select the cosine of the angle θ between $\mathbf{Word_1^v}$ and $\mathbf{Word_2^v}$ as R, because the length of the modulated vector is constant. This satisfies the conditions above.

$$S \;=\; cos\theta = \mathbf{Word_1^v} \cdot \mathbf{Word_2^v}$$
$$= \sum_{j=1}^{k} Q_{1j}^v \, Q_{2j}^v. \qquad (41.8)$$

41.3   Evaluation

41.3.1   Experimental knowledge-base

We selected 40,000 words as Japanese daily-used words by using entries from four dictionaries ([Hamanishi90, Kenbou92, Takabe77, Sanseido92]). We made an experimental knowledge base of the 40,000 Japanese daily-use words using four Japanese dictionaries [Izuru91, Matsumura92, Hamanishi90, Kenbou92] to evaluate the proposed method. To achieve the standardization described in Equation 41.2 we use a thesaurus [Ikehara93] containing 370,000 words grouped into 3,000 categories. Each word in the constructed knowledge base is defined by fifty weighted keywords on average.

41.3.2   Similarity-judgement Examples

In the following example, we calculate the degree of viewpoint-based similarity between 'water' and 'oil'. Table 41.1 shows that the degree of similarity between the two words based on an inadequate viewpoint word ('molecular') is even smaller than the degree

TABLE 41.1. A degree of similarity between 'water' and 'oil'

viewpoint-word $View$	Degree of similarity S('water', 'oil', $View$)
(no word)	0.30
'molecular'	0.24
'liquid'	0.48

TABLE 41.2. An example of selecting a word similar to 'silver'

viewpoint-word $View$	Degree of similarity	
	S('silver', 'gold', $View$)	S('silver', 'lead', $View$)
metal	0.82	0.81
cash	0.89	0.24

calculated when no viewpoint word is given at all. On the other hand, if an adequate viewpoint word ('liquid') is selected, the similarity is considerably larger than that calculated without a viewpoint word. Given the viewpoint word 'molecular', a person might not answer that 'oil' is similar to 'water'. Given the viewpoint word 'liquid', however, the person very likely will answer that 'oil' is similar to 'water'. In this case, the judgement of Table 41.1 matches human intuition.

Another example of similarity judgement is in Table 41.2. The degree of similarity between 'gold' and 'silver' is as large as that between 'lead' and 'silver' when 'metal' is selected as a viewpoint word, which means both 'lead' and 'gold' are similar to 'silver'. On the other hand, 'silver' is much more similar to 'gold' than 'lead' is when 'cash' is selected as the viewpoint word. This result, too, appears to be very "human-like".

When the view-conditioned degrees of similarity between an indicated word and each of the 40,000 words in the knowledge-base are calculated, words can be ordered by their similarity to the indicated word. Table 41.3 is a list of the highest-ranking words which are similar to the Japanese word "banshū"(=late fall) from the viewpoint of "owari"(=the end of something; an ending), as obtained by using our method as a retrieval system. Most of the words retrieved in Table 41.3 are similar to the given word ('banshū'). However, some words are not similar ('shijū' and 'yagate'). This motivates a more systematic study described in the next section.

41.3.3 Evaluation of the proposed method

We evaluated measurement of similarity with the parameter F ($0 < F \le 1$), which is calculated using the precision P_i used for evaluation in information retrieval. These values are given as follows:

$$F = \frac{1}{n} \sum_{i=1}^{n} P_i$$

$$P_i = \frac{A \cap B_i}{B_i}. \tag{41.9}$$

TABLE 41.3. An example of related word retrieval

Rank	Related words	Related words (translated into English)	Degree of Similarity to "ban-shū" from the viewpoint of "owari")
1	banshū	late fall	1.000
2	oshimai	conclusion	0.784
3	banshun	late spring	0.772
4	bantō	late winter	0.771
5	banka	late summer	0.765
6	yūkoku	evening	0.744
7	boshun	late spring	0.739
8	higure	sunset	0.730
9	yūmagure	evening	0.729
10	kure	end of the year	0.727
11	shijū	always	0.723
12	hitomoshigoro	evening	0.721
13	yagate	soon	0.715
14	tettōtetsubi	thoroughly	0.714
14	shoshū	early fall	0.714

When a word and the viewpoint-word are given, A is the (unordered) set of n similar words selected by humans , and B_i is the minimum set of similar words which are selected by the retrieval system and which contains i words of A as in Figure 3. The system is regarded as good when F is near to 1.

Next, let us consider a way of comparing the proposed method with another method. As a subject for comparison, we selected a conventional method which calculates the degree of similarity based on the distance between words in the thesaurus, because a thesaurus is generally thought to be a tool for judging similarity. We use the same tree-structured thesaurus which is used for standardization, and define the distance L as the number of categories which exist on the path between two categories, each of which includes words compared with each other. We define the degree of similarity S' in a conventional manner using L as follows:

$$S' = 1 - \frac{L}{L_{max}}. \tag{41.10}$$

Here L_{max} is the maximum distance between every two categories in the thesaurus, and S' satisfies the degree of similarity function conditions mentioned in Section 2.2.4.

We show an evaluation example in Fig 4. In this figure, F for the proposed method and the conventional method are compared for fifty-eight samples. Each point in Figure 4 means that evaluations of retrievals based on our method and the conventional method are done with a given word and a viewpoint word. The value of the horizontal axis of the point is F calculated based on our method , and the value of the vertical axis on it is F calculated based on the conventional method . From this figure, the value of F based on our method is larger than that of the conventional method for about 80% of samples. Morever, the ratio of F based on the proposed method to F based on the conventional

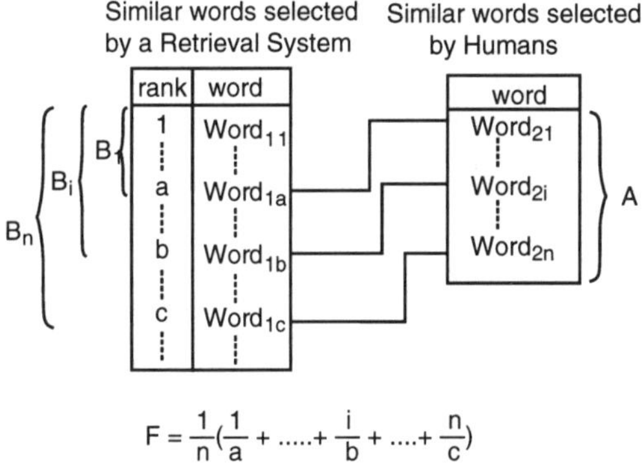

$$F = \frac{1}{n}\left(\frac{1}{a} + \ldots + \frac{i}{b} + \ldots + \frac{n}{c}\right)$$

FIGURE 3. A scheme of calculating F

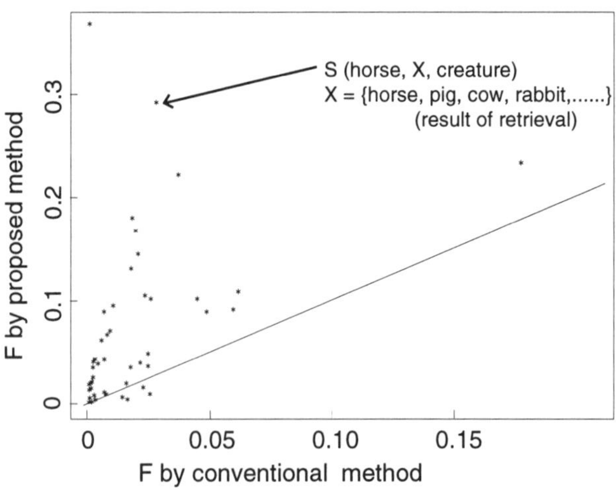

FIGURE 4. Result of evaluation

one at each point is about fifteen to one on average. These facts mean that our proposed method is superior to the conventional one which adopted the thesaurus directly.

41.4   Conclusion

We have proposed a method for viewpoint-based measurement of the semantic similarity between words based on a knowledge-base constructed automatically from machine-readable dictionaries. We constructed the experimental knowledge-base of 40,000 Japanese words and evaluated our method in terms of the efficiency of retrieval of similar words. Test results confirmed that the proposed method, although based on the simply structured knowledge-base, is superior to the conventional one.

41.5   REFERENCES

[Tversky77] Tversky, A. (1977). Features of Similarity. *Psychological Review, 84,* 327 352.

[Suzuki92] Suzuki, H., Ohnishi, H., & Shigemasu, K. (1992). Goal-directed processes in similarity judgements. *Proceedings of the Fourteenth Annual Conference of the Cognitive Science Society* (pp. 327–352).

[Babaguchi94] Babaguchi, N., Sawada, Y., & Ohkawa, T. (1994). Association mechanism based on viewpoints. *Transactions of Information Processing Society of Japan,* 35(5), pp.714–724.

[Guha90] Guha, R. V., & Lenat, D. B. (1990) Cyc: A Midterm Report. *AI Magazine, 11(3),* 32–59.

[EDR93] Japan Electronic Dictionary Research Institute, Ltd. (1993). *EDR Electronic Dictionary Technical Guide,* TR-042.

[Ide93a] Ide, N., & Veronis, J. (1993). Extracting Knowledge Base from Machine-readable Dictionaries: Have we wasted our time? *Proceedings of the First International Conference on Building and Sharing of Very Large-Scale Knowledge Bases* (pp. 257–266).

[Ide93b] Ide, N., & Veronis, J. (1993). Combining Dictionary-Based and Example-Based Methods for Natural Language Analysis. *Proceedings of the Fifth International Conference on Theoretical and Methodological Issues in Machine Translation* (pp. 69–78).

[Nakamura88] Nakamura, J., & Nagao, M. (1988). Extraction of Semantic Information from an Ordinary English Dictionary and its Evaluation. *Proceedings of the 13th International Conference on Computational Lingusitics* (pp. 459–464).

[Ikehara93] Ikehara, S., Ikehara, M., & Yokoo, A. (1993). Classification of language knowledge for meaning analysis in machine translations. *Transactions of Information Processing Society of Japan, 34(8)* 1692–1704.

[Izuru91] Shinmura, I.(Ed.) (1991). *Koujien,* Iwanami Shoten.

[Matsumura92] Matsumura, A. (Ed.) (1992). *Daijirin,* Sanseido Co., Ltd.

[Hamanishi90] Hamanishi, M., & Ono, S. (1990). *Ruigo Kokugo Jiten*, Kadokawa Shoten.

[Kenbou92] Kenbou, H. (Ed.) (1992). *Sanseidou Genndai Kokugo Jiten*, Sanseido Co., Ltd.

[Takabe77] Takebe, T. (Ed.) (1977) *Hikkei Ruigo Jitsuyou Jiten*, Sanseido Co., Ltd.

[Sanseido92] Sanseido (Ed.) (1977) *Hikkei Yougo Youji Jiten*, Sanseido Co., Ltd.

42
Part-of-Speech Tagging from "Small" Data Sets

Eric Neufeld and Greg Adams

Department of Computer Science
University of Saskatchewan
Saskatoon, SK, Canada, S7N 0W0

ABSTRACT   Probabilistic approaches to part-of-speech (POS) tagging compile statistics from massive corpora such as the Lancaster-Oslo-Bergen (LOB) corpus. Training on a 900,000 token training corpus, the hidden Markov model (HMM) method easily achieves a 95 per cent success rate on a 100,000 token test corpus. However, even such large corpora contain relatively few words and new words are subsequently encountered in test corpora. For example, the million-token LOB contains only about 45,000 different words, most of which occur only once or twice. We find that 3–4 per cent of tokens in a disjoint test corpus are unseen, that is, unknown to the tagger after training, and cause a significant proportion of errors. A corpus representative of all possible tag sequences seems implausible enough, let alone a corpus that also represents, even in small numbers, enough of English to make the problem of unseen words insignificant. Experimental results confirm that this extreme course is not necessary. Variations on the HMM approach, including ending-based approaches, incremental learning strategies, and the use of approximate distributions, result in a tagger that tags unseen works nearly as accurately as seen words.

42.1   Introduction

Although probabilistic approaches to linguistic problems were attempted early this century [Zipf,1932], early work was hampered by real difficulties of collecting and managing statistics, not to mention challenges to probabilistic methods in principle. Fast inexpensive computer technology and the availability of tagged electronic corpora such as the million-word Lancaster-Oslo-Bergen (LOB) Corpus [Johansson,1980, Johansson *et al.*,1986] changed this situation, and probabilistic approaches to a variety of natural language processing problems have been popular for some time.

A striking success of the probabilistic approach has been the use of hidden Markov models (HMMs) to attach part-of-speech (POS) tags to unrestricted text, for example, [Kupiec,1992]. Given an actual stream of text, a sequence of tokens (instances of words) $w_1 \ldots w_n$, the HMM method computes the word-tag sequence (or simply tag sequence) $t_1 \ldots t_n$ that most probably generated the sequence. That is, the HMM method finds the tag sequence that maximizes

$$P(t_1 \ldots t_n | w_1 \ldots w_n). \tag{42.1}$$

Such probabilities are impossible to collect in practice; furthermore their number is exponential in n. However, assuming that 1) the probability of a tag t_i directly depends only on the tag immediately preceding it, and that 2) the probability of any word w_i depends only upon the tag t_i that produced it, the preceding, after an application Bayes' rule and

---

factoring, reduces to maximizing

$$\prod_{i=1}^{n+1} P(w_i|t_i)P(t_i|t_{i-1}), \qquad (42.2)$$

where t_0, t_{n-1} are dummy word-tags marking the beginning and the end of a sequence. $P(w_i|t_i)$ is a *lexical probability*; $P(t_i|t_{i-1})$ is a *contextual probability*. Equation 42.2 defines a *bi-tagger*; in a $k+1$-tagger, the probability of a tag depends on the k preceding tags, but this only improves performance marginally[Foster,1991]. See [Charniak *et al.*,1993] for a good overview of this approach to part-of-speech tagging.

Typically, a POS tagger trains by collecting lexical and contextual probabilities from a large subset of a tagged electronic corpus such as the LOB corpus [Johansson,1980] and is tested on a smaller disjoint subset. In earlier work [Adams & Neufeld, 1993], a training corpus of 900,000 tokens and a test corpus of 100,000 tokens was used.

A difficulty arises when attempting to attach part-of-speech tags to *unseen* tag sequences and words [Adams & Neufeld, 1993, Church,1989, Foster,1991, Merialdo,1990, Meteer *et al.*,1991, Kupiec,1992], that is, tag sequences or tokens not occuring in the training corpus, since no lexical probabilities are known for them. A variety of solutions to the problem of unseen tag sequences have been studied by other authors; our work has addressed the problem of unseen words. About half of the words in the LOB only appear once, so many words (about 3–4 per cent [Adams & Neufeld, 1993] are encountered for the first time in the test corpus. Most of these (for example, *outjumped* and *galaxy*) are neither exotic nor highly specialized but simply reflect the vastness of human experience.

Managing the unseen word problem seems to create the greatest overhead when constructing such taggers. The ideal might be to collect probabilities from a tagged corpus sufficiently large to represent all tag sequences and all word/tag pairs. As implied above, however, this would seem to required a tagged training corpus of astronomical size. Even if a corpus of such size was readily obtainable, the value of tagging it must be addressed, since even the success rate of taggers with unseen words artificially eliminated (for example, by testing the tagger on subsets of the training corpus) is about 97 per cent.

Experimental results show that good performance can result from compiling statistics on fixed length *word endings*, instead of whole words. In subsequent experiments, statistics for the n most frequently occurring words in the training set were added to the tagger, where, given a choice, the tagger prefers the whole word statistic to the ending statistic. The experiments suggest that the marginal improvement resulting from increasing the number of whole word statistics is low; this suggests that ending-based approaches are sufficient when it is not possible to store many parameters. These results are reported below. It is also shown that performance can be maintained in the presence of infrequent and unseen words, without necessarily increasing the size of the training corpus, and possibly even reducing it.

42.2   Ending-based strategies

POST [Meteer *et al.*,1991] reports a success rate for tagging unseen words that is significantly improved by compiling word/tag statistics for about 35 preselected word-endings, such as *-ology* and *-tion*. Of course, selecting such a set of word-endings requires expert knowledge of the language and it is reasonable to ask whether dependence on domain knowledge could be avoided by simply compiling statistics on all word endings of some fixed length.

TABLE 42.1. 12 ending-based methods for tagging unseen words

Technique of Word-Feature Identification	Use of External Lexicon	
	Use no external lexical lookup	Use Moby POS database as external lexicon
None (assign any open class)	Method 1 (for control experiment)	Method 2
Poss./non-poss. endings; 2-valued capitalization	Method 3	Method 4
2-character word endings; 3-valued capitalization	Method 5	Method 6
3-character word endings; 3-valued capitalization	Method 7	Method 8
4-character word endings; 3-valued capitalization	Method 9	Method 10
80 predefined word endings; 3-valued capitalization	Method 11	Method 12

This idea was tested exhaustively. Table 42.1 describes the matrix of experiments. *Any open class* means that unseen words were tagged with any a randomly selected open class, assuming equal lexical probabilities for all unseen-word/tag combinations. (The open classes are those classes whose membership is indefinitely extendable, for example, adjectives and verbs, as opposed to the closed classes, such as pronouns and conjunctions.) *Poss. endings* means a possessive form, *80 endings* means the 80 predefined endings similar to those used in [Weischedel *et al.*,1993]. The capitalization values indicate whether a word is capitalized, contains capital letters or contains no capital letters. An external lexicon (*Moby*, copyright by Grady Ward) was used to prune away "impossible" (as defined by the external lexicon) tag assignments.

Whole-word statistics were used to tag seen words; word-endings, capitalization statistics and the external lexicon were used to tag unseen words.

Each unseen word was assigned to an equivalence class used to generate lexical probabilities. Each equivalence class is characterized by a unique combination of capitalization and word ending. Experiments also compared fixed-length and predefined word endings. Where possible, lexical probabilities are estimated by the product of the estimated *feature probability*, the probability of a given pair of word features occuring with a given word class, and the estimated *held-out probability*, the probability of any unseen word occuring with a given word class. A second disjoint corpus of about 100,000 tokens, the *held-out corpus*, was used to estimate held-out probabilities.

Table 42.2 gives the results. The table shows the overall performance, the performance on seen and unseen words as well as performance on words from the external lexicon. The significant effect of unseen words is clear from this table. Generally, the table shows that more information results in better performance. In particular, word-ending statistics significantly improve performance on the tagging of unseen words. Interestingly, taggers based on 3-letter endings do better than taggers based on 2 or 4 letters. On the one hand, shorter endings result in larger and hence more accurate samples, but on the other, longer endings are more specific. In this case, a compromise between sample size and specificity yielded the best result.

TABLE 42.2. For the 12 ending-based experiments: percentage of tokens correctly tagged (SR)

Exp. No.	Tagging Method	For Whole Test Corpus	Tokens of Seen Words	Tokens of Unseen Words	Tokens of Unseen Words Identifiable by External Lexicon	
1	Any open class	95.4%	97.4%	43.9%	47.4%	
2	Any open class; ext. lexicon	96.0%	97.5%	59.6%	73.1%	25.7% rise
3	Poss. endings	96.2%	97.5%	61.9%	63.5%	
4	Poss. endings; ext. lexicon	96.6%	97.5%	74.4%	84.0%	20.5% rise
5	2-char. endings	96.9%	97.5%	81.3%	84.6%	
6	2-char. endings; ext. lexicon	97.0%	97.5%	84.9%	90.5%	5.9% rise
7	3-char. endings	96.9%	97.5%	82.5%	86.4%	
8	3-char. endings; ext. lexicon	97.0%	97.5%	85.2%	90.9%	4.5% rise
9	4-char. endings	96.8%	97.5%	79.8%	86.4%	
10	4-char. endings; ext. lexicon	96.9%	97.5%	82.2%	90.2%	3.8% rise
11	80 endings	96.8%	97.5%	78.8%	84.4%	
12	80 endings; ext. lexicon	96.9%	97.5%	82.2%	89.9%	5.5% rise

It has been noted that many words in the LOB occur infrequently and it is reasonable to ask what the marginal value of extending the lexicon by increasing the size of the data set is. In another set of experiments, this language-independent approach was used. It is possible to look at this problem in another way. That is, initially collect statistics on all L-letter endings, $L = 1, \ldots 4$, but then collect statistics for just the n most frequently occuring words in the corpus, and use the whole-word statistic whenever possible in the tagging process. The first approach gives a measure of value of the ending statistics; the second approach gives a sense of the value of whole-word information, and perhaps more clearly expresses the tradeoff between additional knowledge and improved performance as rare words are added to the model.

	Number of Most Frequent Words Put Back In LOB									
	0	5K	10K	15K	20K	25K	30K	35K	40K	45K
2-letter endings										
AC - Full ETL	83.4	95.0	95.6	95.9	96.1	96.2	96.3	96.3	96.3	96.4
GT - Full ETL	83.4	95.0	95.6	95.9	96.1	96.2	96.3	96.3	96.3	96.4
AC - Unit ETL	88.6	96.0	96.2	96.3	96.3	96.4	96.4	96.4	96.4	96.4
GT - Unit ETL	88.7	96.0	96.3	96.3	96.4	96.4	96.4	96.4	96.4	96.4
3-letter endings										
AC - Full ETL	90.8	95.5	95.9	96.1	96.2	96.3	96.3	96.4	96.4	96.4
GT - Full ETL	90.8	95.5	95.9	96.1	96.2	96.3	96.3	96.4	96.4	96.4
AC - Unit ETL	93.4	96.2	96.3	96.4	96.4	96.4	96.4	96.4	96.4	96.4
GT - Unit ETL	93.4	96.2	96.3	96.4	96.4	96.4	96.4	96.4	96.4	96.4
4-letter endings										
AC - Full ETL	94.3	95.7	95.9	96.0	96.1	96.1	96.1	96.2	96.2	96.2
GT - Full ETL	94.4	95.7	95.9	96.0	96.1	96.1	96.2	96.2	96.2	96.2
AC - Unit ETL	95.2	96.0	96.1	96.2	96.2	96.2	96.2	96.2	96.2	96.2
GT - Unit ETL	95.3	96.1	96.1	96.2	96.2	96.2	96.2	96.2	96.2	96.2

TABLE 42.3. Tokens Correctly Tagged

In Table 42.2 , **AC** means *augmented corpus* and refers to the technique of adding one to all tag-sequence counts and all seen word counts before computing probabilities. **GT** means Good-Turing method [Goo53, CG91] and is a more sophisticated way of adjusting unseen tag sequences and unseen word/tag pairs. *Full ETL* means considering for each token the full set of tag possibilities for the ending; *Unit ETL* means attaching only whole-word possibilities when there only is a single possibility.

The table shows the low marginal value of an increased lexicon. Because about half of the vocabulary occurs only once, probabilities on the last $20,000$ words added back are based on unit samples.

Another variation considered mixed sets of ending statistics. This is another way of examining the tradeoff between *specificity* (ending length) and *accuracy* (sample size) observed in the first set of experiments. The tradeoff can be resolved [Kyburg,1983] by using the narrowest reference class for which there are adequate statistics. Roughly, this means using the narrower statistic unless the difference between it and a more general statistic can be explained by chance error. Statistics were collected for whole words and word-ending lengths from unity to four. (A tagger using just one letter endings can achieve a success rate of 70 per cent.) A $1 - \alpha$ confidence interval was computed for each statistic (although point values were used in the actual tagging process) and the narrowest reference term for which we had adequate statistics was used during testing. The intuition here is to find points at which specificity is more valuable than accuracy, as defined by sample size. Several values of α were tried; the best success rates were obtained by always using the narrowest statistic available.

42.3   Sensitivity analysis

Is the success rate sensitive to the actual numerical values or is it sensitive to their rank orderings? If the latter, it may be possible to construct qualitative tagging algorithms that avoid the calculations.

To test this, lexical probabilities were rounded to the nearest value within some small fixed subset of probabilities. Among many others, the following quantization sets were used:

$$R_1 = \{0.125, 0.25, 0.375, 0.5, 0.625, 0.75, 0.875, 1\}$$
$$R_2 = \{0.0001, 0.01, 0.125, 0.25, 0.375, 0.5, 0.625, 0.75, 0.875, 1\}$$
$$R_7 = \{0.00000001, 0.000001, 0.001, 0.1, 0.2, 0.3, 0.4, 0.5, 0.6, 0.7, 0.8, 0.9, 1\}$$
$$R_{15} = \{0.00001, 0.001, 0.5, 1\}$$

| | Number of Most Frequent Words Put Back In LOB | | | | | | |
	0	3000	5000	10000	20000	30000	45000
R_1	84.0	87.4	88.8	90.1	91.1	91.5	91.8
R_2	90.8	92.9	93.6	94.4	94.9	95.1	95.3
R_7	89.7	92.1	93.0	93.9	94.6	94.8	95.0
R_{15}	88.3	91.1	92.1	93.1	93.8	94.1	94.3

FIGURE 1. Tokens Correctly Tagged

This gives a crude measure of the value of exact probabilistic information. On one hand, in the best cases, it only gives up 1.2 per cent success rate. (On the other hand, it is very difficult to get this much accuracy using other techniques.) The quantization sets were constructed with several issues in mind. One is expressiveness, the number of values in the set. As expressiveness increases, so does accuracy. The other issue is the distribution of values near zero. An earlier set of experiments suggested is it a bad idea to include zero as a possible value, because many of the probabilities are small and simply get rounded down to zero.

Table 42.3 gives results of a final carefully monitored set of experiments. It is surprising how well the tagger performs given such a small set of numbers. It is interesting that adding some very small values to R_1 (giving R_2) gives a dramatic improvement; in fact, R_2 consistently gives the best performance of the four quantization sets. It makes sense that performance will increase with expressiveness, but the role of the distribution is important. It is remarkable that R_{15}, with only four values results in as little as 1 per cent worse performance.

This suggests that the accuracy of the statistics is not as important as the rank ordering, and opens important possibilities. For example, in [Adams & Neufeld, 1993], unseen words were tagged by first consulting an external dictionary that gave legal tag types but contained no statistics. Some performance improvement could be obtained by attaching numbers from one of these fixed subsets to the list of legal tag types thus obtained.

A final conclusion about this work ties in with the work on external lexicons. Performance improves simply from being able to prune away alternatives known by a language expert to be impossible. There is a tradeoff in allowing this if we are tagging imperfect text, since a language expert may rule out choices a language novice might use anyway. However, it seems reasonable to expect an overall performance gain. Here, it is shown that significant performance can be achieved just by knowing rank orderings.

42.4   Conclusions

An unadorned hidden Markov model works well on POS tagging, up to a certain point (say, 95 per cent). It seems difficult to push performance beyond that point. The marginal increase in performance given extra training corpus or vocabulary is very small. The techniques here suggest that it may be possible to achieve success rates of close to 95 per cent without having to train on massive corpora or carry about a massive lexicon. Not only do ending-based approaches work well, they learn quickly. To give an example, it was shown that using ending-based strategies on a very small (100,000 token) training corpus resulted in a decline in the success rate of only 3 per cent.

Of practical interest are the implications the results have as regards the tradeoff between the number of parameters in the HMM and the success rate, if one is interested in constructing a minimal lexicon tagger with reasonable performance. The marginal value of, say, doubling the lexicon is not great. This work also supports the kind of observations in [Meteer *et al.*,1991] that taggers perhaps don't require the huge training sets originally conjectured. In [Meteer *et al.*,1991], it is noted that a tagger trained on 64,000 words rather than 1,000,000 suffers a relatively small decline in performance when tested on known words only. In a similar vein, experiments showed that an ending-based tagger trained on a corpus of 100,000 2-letter endings tagged a 300,000 token corpus with a success rate of 83.3 per cent, as compared to 83.9 per cent on 900,000 words. To put it another way, the gains using the inexpensive techniques described here compare well with the gains achieved by increasing the training data tenfold.

This work shows that there are many other techniques to exploit. For example, maximizing specificity helps if many kinds of ending statistics are available and furthermore, the range of numerical values is relatively unimportant. Other results on incremental learning and tag similarities hint at a hierarchical approach that may improve overall performance.

Acknowledgements

This research was supported by the Natural Science and Engineering Research Council of Canada, the Institute for Robotics and Intelligent Systems, the University of Saskatchewan, and the City of Saskatoon. George Foster made software available.

42.5 REFERENCES

[Adams & Neufeld, 1993] Adams, G. and Neufeld, E. (1993) Automated word-class tagging of unseen words in text. In *Proceedings of the Sixth International Symposium on Artificial Intelligence*, pages 390–397.

[Charniak *et al.*,1993] Charniak, E., Henrickson, C., Jacobson, N., and Perkowitz, M. (1993) Equations for part-of-speech tagging. In *Proceedings of the Eleventh National Conference on Artificial Intelligence*, pages 784–789.

[Church,1989] Church, K. W. (1989) A stochastic parts program and noun phrase parser for unrestricted text. In *Proceedings of IEEE International Conference on Acoustics, Speech, and Signal Processing*, Glasgow, U.K.

[CG91] Kenneth W. Church and William A. Gale. A comparison of the enhanced Good-Turing and deleted estimation methods for estimating probabilities of English bigrams. *Computer Speech and Language*, 5:19–54, 1991.

[Foster,1991] Foster, G. F. (1991) Statistical lexical disambiguation. Master's thesis, McGill University, Montreal.

[Goo53] I. J. Good. The population frequencies of species and the estimation of population parameters. *Biometrika*, 40:237–264, 1953.

[Johansson,1980] Johansson, S. (1980) The LOB Corpus of British English texts: Presentation and comments. *ALLC Journal*, 1(1):25–36.

[Johansson *et al.*,1986] Johansson, S., Atwell, E., Garside, R., and Leech, G. (1986) *The Tagged LOB Corpus: Users' Manual.* Norwegian Computing Centre for the Humanities, Bergen, Norway.

[Kupiec,1992] Kupiec, J. (1992) Robust part-of-speech tagging using a hidden Markov model. *Computer Speech and Language*, 6:225–242.

[Kyburg,1983] . Kyburg, Jr., Henry E. 1983. The reference class. Philosophy of Science, **50**:374–397.

[Merialdo,1990] Merialdo, B. (1990) Tagging text with a probabilistic model. In *Proceedings of the IBM Natural Language ITL*, pages 161–172, Paris.

[Meteer *et al.*,1991] Meteer, M., Schwartz, R., and Weischedel, R. (1991) POST: Using probabilities in language processing. In *IJCAI 91: Proceedings of the 13th International Joint Conference on Artificial Intelligence*, pages 960–965, Sydney, Australia.

[Weischedel *et al.*,1993] Weischedel, R., Meteer, M., Schwartz R., Ramshaw, L. and Palmucci, J. (1993) Coping with Ambiguity and Unknown Words through Probabilistic Models, Computational Linguistics **50**:359–382.

[Zipf,1932] Zipf, G. K. (1932) *Selected Studies of the Principle of Relative Frequency in Language.* Harvard University Press, Cambridge, Massachusetts.

Lecture Notes in Statistics

For information about Volumes 1 to 25
please contact Springer-Verlag

Vol. 69: N.J.D. Nagelkerke, Maximum Likelihood Estimation of Functional Relationships. V, 110 pages, 1992.

Vol. 70: K. Iida, Studies on the Optimal Search Plan. viii, 130 pages, 1992.

Vol. 71: E.M.R.A. Engel, A Road to Randomness in Physical Systems. ix, 155 pages, 1992.

Vol. 72: J.K. Lindsey, The Analysis of Stochastic Processes using GLIM. vi, 294 pages, 1992.

Vol. 73: B.C. Arnold, E. Castillo, J.-M. Sarabia, Conditionally Specified Distributions. xiii, 151 pages, 1992.

Vol. 74: P. Barone, A. Frigessi, M. Piccioni, Stochastic Models, Statistical Methods, and Algorithms in Image Analysis. vi, 258 pages, 1992.

Vol. 75: P.K. Goel, N.S. Iyengar (Eds.), Bayesian Analysis in Statistics and Econometrics. xi, 410 pages, 1992.

Vol. 76: L. Bondesson, Generalized Gamma Convolutions and Related Classes of Distributions and Densities. viii, 173 pages, 1992.

Vol. 77: E. Mammen, When Does Bootstrap Work? Asymptotic Results and Simulations. vi, 196 pages, 1992.

Vol. 78: L. Fahrmeir, B. Francis, R. Gilchrist, G. Tutz (Eds.), Advances in GLIM and Statistical Modelling: Proceedings of the GLIM92 Conference and the 7th International Workshop on Statistical Modelling, Munich, 13-17 July 1992. ix, 225 pages, 1992.

Vol. 79: N. Schmitz, Optimal Sequentially Planned Decision Procedures. xii, 209 pages, 1992.

Vol. 80: M. Fligner, J. Verducci (Eds.), Probability Models and Statistical Analyses for Ranking Data. xxii, 306 pages, 1992.

Vol. 81: P. Spirtes, C. Glymour, R. Scheines, Causation, Prediction, and Search. xxiii, 526 pages, 1993.

Vol. 82: A. Korostelev and A. Tsybakov, Minimax Theory of Image Reconstruction. xii, 268 pages, 1993.

Vol. 83: C. Gatsonis, J. Hodges, R. Kass, N. Singpurwalla (Editors), Case Studies in Bayesian Statistics. xii, 437 pages, 1993.

Vol. 84: S. Yamada, Pivotal Measures in Statistical Experiments and Sufficiency. vii, 129 pages, 1994.

Vol. 85: P. Doukhan, Mixing: Properties and Examples. xi, 142 pages, 1994.

Vol. 86: W. Vach, Logistic Regression with Missing Values in the Covariates. xi, 139 pages, 1994.

Vol. 87: J. Møller, Lectures on Random Voronoi Tessellations. vii, 134 pages, 1994.

Vol. 88: J. E. Kolassa, Series Approximation Methods in Statistics. viii, 150 pages, 1994.

Vol. 89: P. Cheeseman, R.W. Oldford (Editors), Selecting Models From Data: AI and Statistics IV. xii, 487 pages, 1994.

Vol. 90: A. Csenki, Dependability for Systems with a Partitioned State Space: Markov and Semi-Markov Theory and Computational Implementation. x, 241 pages, 1994.

Vol. 91: J.D. Malley, Statistical Applications of Jordan Algebras. viii, 101 pages, 1994.

Vol. 92: M. Eerola, Probabilistic Causality in Longitudinal Studies. vii, 133 pages, 1994.

Vol. 93: Bernard Van Cutsem (Editor), Classification and Dissimilarity Analysis. xiv, 238 pages, 1994.

Vol. 94: Jane F. Gentleman and G.A. Whitmore (Editors), Case Studies in Data Analysis. viii, 262 pages, 1994.

Vol. 95: Shelemyahu Zacks, Stochastic Visibility in Random Fields. x, 175 pages, 1994.

Vol. 96: Ibrahim Rahimov, Random Sums and Branching Stochastic Processes. viii, 195 pages, 1995.

Vol. 97: R. Szekli, Stochastic Ordering and Dependence in Applied Probablility. viii, 194 pages, 1995.

Vol. 98: Philippe Barbe and Patrice Bertail, The Weighted Bootstrap. viii, 230 pages, 1995.

Vol. 99: C.C. Heyde (Editor), Branching Processes: Proceedings of the First World Congress. viii, 185 pages, 1995.

Vol. 100: Wlodzimierz Bryc, The Normal Distribution: Characterizations with Applications. viii, 139 pages, 1995.

Vol. 101: H.H. Andersen, M.Højbjerre, D. Sørensen, P.S.Eriksen, Linear and Graphical Models: for the Multivariate Complex Normal Distribution. x, 184 pages, 1995.

Vol. 102: A.M. Mathai, Serge B. Provost, Takesi Hayakawa, Bilinear Forms and Zonal Polynomials. x, 378 pages, 1995.

Vol. 103: Anestis Antoniadis and Georges Oppenheim (Editors), Wavelets and Statistics. vi, 411 pages, 1995.

Vol. 104: Gilg U.H. Seeber, Brian J. Francis, Reinhold Hatzinger, Gabriele Steckel-Berger (Editors), Statistical Modelling: 10th International Workshop, Innsbruck, July 10-14th 1995. x, 327 pages, 1995.

Vol. 105: Constantine Gatsonis, James S. Hodges, Robert E. Kass, Nozer D. Singpurwalla(Editors), Case Studies in Bayesian Statistics, Volume II. x, 354 pages, 1995.

Vol. 106: Harald Niederreiter, Peter Jau-Shyong Shiue (Editors), Monte Carlo and Quasi-Monte Carlo Methods in Scientific Computing. xiv, 372 pages, 1995.

Vol. 107: Masafumi Akahira, Kei Takeuchi, Non-Regular Statistical Estimation. vii, 183 pages, 1995.

Vol. 108: Wesley L. Schaible (Editor), Indirect Estimators in U.S. Federal Programs. iix, 195 pages, 1995.

Vol. 109: Helmut Rieder (Editor), Robust Statistics, Data Analysis, and Computer Intensive Methods. xiv, 427 pages, 1996.

Vol. 110: D. Bosq, Nonparametric Statistics for Stochastic Processes. xii, 169 pages, 1996.

Vol. 111: Leon Willenborg, Ton de Waal, Statistical Disclosure Control in Practice. xiv, 152 pages, 1996.

Vol. 112: Doug Fischer, Hans-J. Lenz (Editors), Learning from Data. xii, 450 pages, 1996.